Nicole Wellbrock • Andreas Bolte
Editors

Status and Dynamics of Forests in Germany

Results of the National Forest Monitoring

Editors
Nicole Wellbrock
Thünen Institute of Forest Ecosystems
Eberswalde, Germany

Andreas Bolte
Thünen Institute of Forest Ecosystems
Eberswalde, Germany

https://doi.org/10.1007/978-3-030-15734-0

This book is an open access publication.

Preface

Temperate forest ecosystems in Central Europe are changing. This is due to anthropogenic induced site modification through the deposition of air pollutants as well as to the effects of a changing climate. In Germany, the National Forest Monitoring Program consists of a forest condition survey and a soil inventory as well as the intensive forest monitoring studying ecosystem and their responses to changing site and climate conditions. The results allow the assessment of the forest ecosystem status and its dynamics.

This volume compiles and integrates scientific findings of the German Forest Monitoring Program covering major challenges of future preservation and management of forest ecosystems under environmental change. What are the main impact factors affecting forest stand and soil integrity? How, and how dynamic, are forest ecosystems changing due to site and climate variation? How diverse are the changing patterns in Germany? What are the main risks for a sustainable forest ecosystem management in the future? How can policy support the maintenance and development of adaptive and resilient forest ecosystems providing the required ecosystem services, today and in the future?

In this comprehensive analysis, 49 scientists from different research institutes and universities joined hands.

Mainly based on the National Forest Soil Inventory (NFSI), the analysis is supported by case studies from the Intensive Forest Monitoring (Level II) Program in Germany and the results of the nation-wide Crown Condition Survey (CCS), all three parts of the National Forest Monitoring Program. Combining a system of representative plots ($n = 1900$) within a regular grid sample of 8 km width (NFSI) with process and flux studies on 69 exemplary plots of the Intensive Forest Monitoring Program (Level II), this program is part of the International Co-operative Programme on Assessment and Monitoring of Air Pollution Effects on Forests (ICP Forests).

The history of the convention can be traced back to the 1960s, when scientists demonstrated the interrelationship between sulfur emissions in continental Europe and the acidification of Scandinavian lakes. In the 1980s, several studies confirmed the link and public awareness rose linked to the phenomenon of *Waldsterben*, forest die-off. Those discussions resulted in the signature of the International Convention on Long-range Transboundary Air Pollution (UNECE CLRTAP). Already prior to this, the Federal States of Germany and further countries in Europe installed a system to monitor forest condition. Later this method was applied and since then further defined and harmonized to serve as a proxy at an international level. In the 1990s, the ICP Forests Programme was substituted with the Intensive Forest Monitoring (Level II) and a soil survey. Since 2010, the Forest Monitoring Program is part of the German Forest Act (*Bundeswaldgesetz* § 41 a) and under the responsibility of the Federal Ministry of Food and Agriculture (BMEL). Forest research institutes of the Federal states conduct and finance the measurements and assessments on the monitoring plots and prepare regional reporting, whereas the Thünen Institute of Forest Ecosystems manages the data and coordinates the method harmonization, analyses, and reporting at national and international level.

Under the name "National Forest Soil Inventory (NFSI)," a large set of information is combined such as information on forest stand structure, element content of needles or leaves, ground vegetation composition, and crown condition information (from CCS on the same plots). The comparison of the results of the first (1989–1992) and the second NFSI (2006–2008) enabled the assessment of dynamics of forests and forest soils.

Additional process-related parameters are measured within the Intensive Forest Monitoring Program (Level II), e.g., soil solution chemistry, atmospheric deposition, or meteorology. The inclusion of process studies provided a deeper understanding of cause–effect relationship of ecosystem processes triggered by air pollution and climate variation.

May this volume help to understand the importance of soils and related ecosystem processes for the further discussion of sustainable forestry. The functions of forest soils are closely related to forest ecosystem integrity and productivity. Thus, both the knowledge of environmental change and the related change of forest status and dynamics represent a basic principle for successful forest management and forest policy.

Eberswalde, Germany
November 2018

Nicole Wellbrock
Andreas Bolte

Acknowledgments

Our special thanks go to the numerous technicians and researchers of the forest research institutes who have run the National Forest Monitoring Program for decades by tending the plots and their equipment, gathering data, and managing the data at a regional and a national level. We are also very grateful to the responsible ministries of the Federal states and the Federal Ministry of Food and Agriculture (BMEL) for the continuous financing of the program. This also includes various authorities of the European Union (EU) who co-funded parts of the monitoring until 2011 under various conventions and programs (e.g., latest project "FutMon" under EU LIFE+ or BIOSOIL). The network of ICP Forests facilitates discussions above the borders of countries within the expert groups leading to a continuous improvement of the research conducted.

We would also like to thank all involved research organizations, the Federal Environmental Agency (UBA), and the Federal Institute for Geosciences and Natural Resources (BGR) for the productive collaboration. Without all their efforts, we would not have been able to perform the analyses and evaluations presented in this volume.

Contents

Chapter 1
Concept and Methodology of the National Forest Soil Inventory

Nicole Wellbrock, Bernd Ahrends, Rebekka Bögelein, Andreas Bolte, Nadine Eickenscheidt, Erik Grüneberg, Nils König, Andreas Schmitz, Stefan Fleck, and Daniel Ziche

1.1 Introduction

What is the state of our forests and forest soils today? How have they changed over the past 20 years? What actions have had an effect on their status and how? What are the risks that continue to play a role or become relevant in the future? The present report based on the National Forest Soils Inventory (NFSI) as well as on case studies from Intensive Forest Monitoring sites in Germany (see below) addresses these questions and provides a nationwide summary on the current state and development of forests and forest soils. The results offer scientific evidence supporting the evaluation of forest management approaches and policy options at different spatial scales.

Forest soils are the basis for productive and resilient forests that in turn create opportunities for sustainable and successful forest management. Soils provide water

N. Wellbrock (✉) · R. Bögelein · A. Bolte · E. Grüneberg · A. Schmitz · D. Ziche
Thünen Institute of Forest Ecosystems, Eberswalde, Germany
e-mail: nicole.wellbrock@thuenen.de; rebekka.boegelein@gmx.de; andreas.bolte@thuenen.de; erik.grueneberg@thuenen.de; andreas.schmitz@thuenen.de; daniel.ziche@thuenen.de

B. Ahrends · N. König
Northwest German Forest Research Institute, Göttingen, Germany
e-mail: bernd.ahrends@nw-fva.de; nils.koenig@nw-fva.de

N. Eickenscheidt
State Agency for Nature, Environment and Consumer Protection of North Rhine-Westphalia, Recklinghausen, Germany
e-mail: nadine.eickenscheidt@lanuv.nrw.de

S. Fleck
Thünen Institute of Forest Ecosystems, Eberswalde, Germany

Northwest German Forest Research Institute, Göttingen, Germany
e-mail: stefan.fleck@nw-fva.de

N. Wellbrock, A. Bolte (eds.), *Status and Dynamics of Forests in Germany*, Ecological Studies 237, https://doi.org/10.1007/978-3-030-15734-0_1

and nutrients required for forest growth, buffer loading of toxins and acidification and compensate for water shortages during droughts. Forests and their soils represent one of Germany's most natural ecosystems and contribute an important share to its biodiversity. As carbon (C) sinks (Höhle et al. 2018), forests and their soils play a key role in climate protection and compensating for greenhouse gas emissions (Leitgeb et al. 2013).

The current condition of forest soils is the result of both natural changes over long periods of time and anthropogenic influences. Natural factors that contribute to the formation of soils include the parent material, climate, relief and the flora and fauna (Blume et al. 2010).

Over the past decades, impacts from atmospheric pollution caused by humans have had a significant impact on forests (Ellenberg 1971; Ulrich 1987). At the end of the 1970s and in the early 1980s, air pollution effects on forests became evident based on the condition of the crowns of the trees, followed by the discussion about "forest dieback" and new types of forest damage (Kauppi et al. 1990; Ulrich 1983). In addition, for many German forest stands, an increasing risk of drought stress has been identified over the last 60 years, due to climate change (von Wilpert et al. 2016; Schmidt-Walter et al. 2017). Thus, forests can suffer also from interactive effects of both atmospheric pollution and climate change impacts like increasing drought (Bytnerowicz et al. 2007; Hickler et al. 2012).

1.2 The National Forest Soils Inventory as a Part of the Forest Monitoring in Germany

Following the discussion on forest decline, a national forest monitoring was established, consisting of the periodic assessment of crown condition, the establishment of selected Intensive Forest Monitoring sites as well as the nationwide crown and soil condition inventory (National Forest Soils Inventory; NFSI). In 1984, the crown condition plots were established on a 16 × 16 km grid. The 8 × 8 km grid of NFSI plots was subsequently installed based on the same grid. As a consequence, every fourth NFSI plot is one of the original crown condition plots that became part of the European Level I net of the UN/ECE "International Co-operative Programme on Assessment and Monitoring of Air Pollution Effects on Forests" (ICP Forests) and were also part of the EU BioSoil project. In 2006, a soil inventory took place on these plots. All data had been submitted to the Joint Research Centre (JRC) of the european commission.

The NFSI was launched at the end of the 1980s/beginning of the 1990s with surveys taking place at approximately 1900 sampling sites (Fig. 1.1). While the data gathered in this survey are an important pool of information on its own, integrated evaluations, for example, with data from the Intensive Forest Monitoring programme (Level II), allow for comprehensive interpretation of the NFSI data. The Level II programme was initiated by the UN/ECE under the "International Co-operative

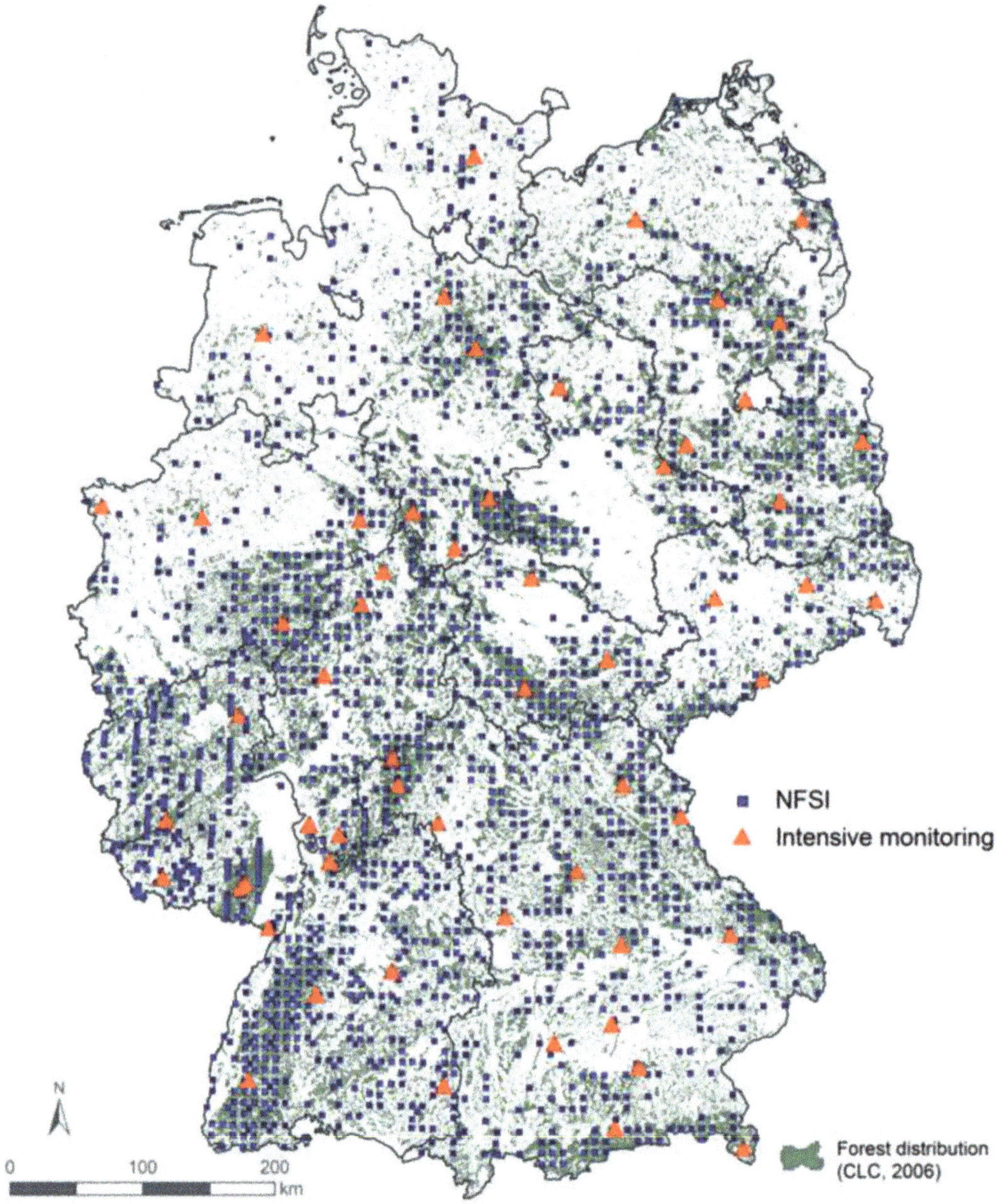

Fig. 1.1 Spatial distribution of Level II and NFSI study sites

Programme on Assessment and Monitoring of Air Pollution Effects on Forests" (ICP Forests), too. Most German sites were established during the mid-1990s by the federal states, and many of them have been active since then, partly co-funded by the EU. In 2014, a German legislation secured the long-term operation of 68 Level II sites (Fig. 1.1). The case studies of the Level II programme aim at understanding cause-effect relationships in forest ecosystems, based on measurements of a comprehensive set of parameters. For example, continuous assessments of deposition, growth and seepage water allow quantifying input-output element budgets.

Table 1.1 Structure of the federal forest monitoring programme

Monitoring programme (frequency)	Grid/plot no.	Assessments
Extensive monitoring (every 10 years)		
Crown condition (annual)	16 × 16 km 420 plots	Crown condition Impacts factors, e.g. insects
National forest soil inventory Level I	8 × 8 km 1859 plots 16 × 16 km 420 plots	Soil chemistry Soil reaction Aqua regia (K, Mg, Na, K, P) C, N, S, P 1:2 extraction nitrogen Cation exchange capacity Soil water Tree growth Ground vegetation Tree nutrition (leave/needle chemistry)
Intensive monitoring (continuously) Level II	Case studies 68 plots	Crown condition Impacts factors, e.g. insects Soil chemistry Soil reaction Aqua regia (K, Mg, Na, K, P) C, N, S, P Cation exchange capacity Soil solution Tree growth Ground vegetation Tree nutrition (leave/needle chemistry) Litterfall Deposition Meteorology Air quality

In line with the federal structure of Germany, data collection for the NFSI plots and Level II sites is organized on the level of federal states. The data are compiled at the Thünen Institute for Forest Ecosystems for evaluations and assessments on a nationwide level. Data analysis took place in a collaborative effort involving the Thünen Institute, representatives of state forestry research institutes or environmental agencies and external experts. Special focus analyses on heavy metals and organic pollutants were conducted by the Federal Institute for Geosciences and Natural Resources (BGR) and the Federal Environmental Agency (UBA), respectively (Table 1.1).

1.3 Legal Framework

The second NFSI was initiated in 2001 in a resolution at the Forest Executive Conference (Forstchefkonferenz; FCK). Data and analyses from the NFSI provide an important basis for the national reporting on greenhouse gases under the

Framework Convention on Climate Change (UNFCCC) and the Kyoto Protocol, as well as succeeding regulations in the "land use, land-use change and forestry" (LULUCF) soil group in the areas of soil and litter. The NFSI provides information useful in implementing the Federal Soil Protection Act (Bundes-Bodenschutzgesetz) in particular §9 which averting the risk of harmful soil contaminations. An important link to international forest monitoring is presented within the framework of the UNECE Convention on Long-range Transboundary Air Pollution (also known as the Air Convention; CLRTAP). The NFSI data represent an essential part of Germany's national contribution to the "International Co-operative Programme on Assessment and Monitoring of Air Pollution Effects on Forests" (ICP Forests) as a CLRTAP programme specific to forests. These commitments to national and international policy advice and monitoring programmes require periodic updates on the condition of forest soils. A step towards a legal basis for this effort has been done by the 2010 amendment to §41a of the Federal Forest Act. It allows the Federal Ministry for Food and Agriculture (BMEL) to collect data on the supply of nutrients and pollution loading in forest soils through legislative decree with the approval of the states. However, corresponding legislation for the third NFSI will be tabled in 2019.

1.4 Objectives and Key Questions

The goal of the NFSI is to generate reliable data on the current state and changes in forest soils and selected features of the forests based on a systematic 8 × 8 km grid and a repetition of the inventory every 15 years. The information collected is representative on regional scale and comparable across the country. Analysis of the status and temporal changes is differentiated by region, allowing for the identification of areas of particular risk. The comparison with previous surveys, especially the NFSI I, is used to identify changes over time. The national grid of the NFSI sites can be also used to regionalize processes and findings that have been identified on the smaller scale of the Intensive Forest Monitoring (Level II programme). The NFSI therefore allows for estimating risks like soil acidification or eutrophication of forest stands. Furthermore, the effects of large-scale soil treatments, such as liming, can be analysed based on NFSI results. As such, the NFSI is able to contribute to planning of sustainable forest management, including balanced soil C and nutrient budgets. Besides providing advice for practical forest management approaches, the NFSI results offer scientific evidence supporting the assessment of higher-level forest and environmental measures and strategies.

The report consists of (I) a text volume presenting the methods and findings as well as (II) a map volume containing point analyses, composite curves and statistical parameters. Specifically, the study addresses questions in the following areas:

- Soil acidification
- Nitrogen status and dynamics in forest soils and soil sensitivity to ongoing inputs of nitrogen

- Current carbon storage and changes of carbon stocks in forest soils (Framework Convention on Climate Change and the Kyoto Protocol)
- Background loading of soils with heavy metals and organic trace elements
- The interplay of soil characteristics and forest nutrition, crown condition and vegetation
- Risks due to negative changes such as nutrient removal or soil acidification in relation to nature conservation and sustainable use of forests
- Effects in terms of soil chemistry and nutritional status on measures to stabilize forest ecosystems (efficiency monitoring, in particular for soil liming and semi-natural silviculture practices)
- The extent of change of soil and forest conditions and the need for a subsequent inventory

1.5 Survey Parameters and Data Harmonization

The guidelines for the Second National Forest Soils Inventory were developed by the NFSI II federal-state working group (Wellbrock et al. 2006). The guidelines were designed to create a comprehensive working document for the field surveys for the NFSI II. Building on the guidelines for the First Soil Inventory in Forests (BMELF 1994), adjustments and enhancements were introduced based on new knowledge and requirements. As much as possible, these changes took into account the conventions of the NFSI I. The compatibility of the methods and their harmonization are presented in a separate volume (Höhle et al. 2018). The guidelines for the survey were supplemented with the "Handbuch Forstliche Analytik" ("Handbook of Forest Analysis"; HFA) prepared by the Forest Analysis Advisory Committee (GAFA) of the Federal Ministry for Food and Agriculture (BMEL), which describes the harmonized methodologies for laboratory analyses (GAFA 2005, 2009, 2014).

As specified by the federal-state working group, the NFSI II comprises the following focal areas (Wellbrock et al. 2006):

- General description of the survey sites: point data, georeferencing, data on the environmental situation, forest inventory and data on factors causing changes to the soil
- Soils: profile description, soil chemistry (including heavy metals and organic compounds) and soil physics separated according to mineral soil and organic layer
- Sampling of needles or leaves
- Collection of forest growth data (additional inventories modified after Hilbrig et al. 2014)
- Crown condition
- Ground vegetation

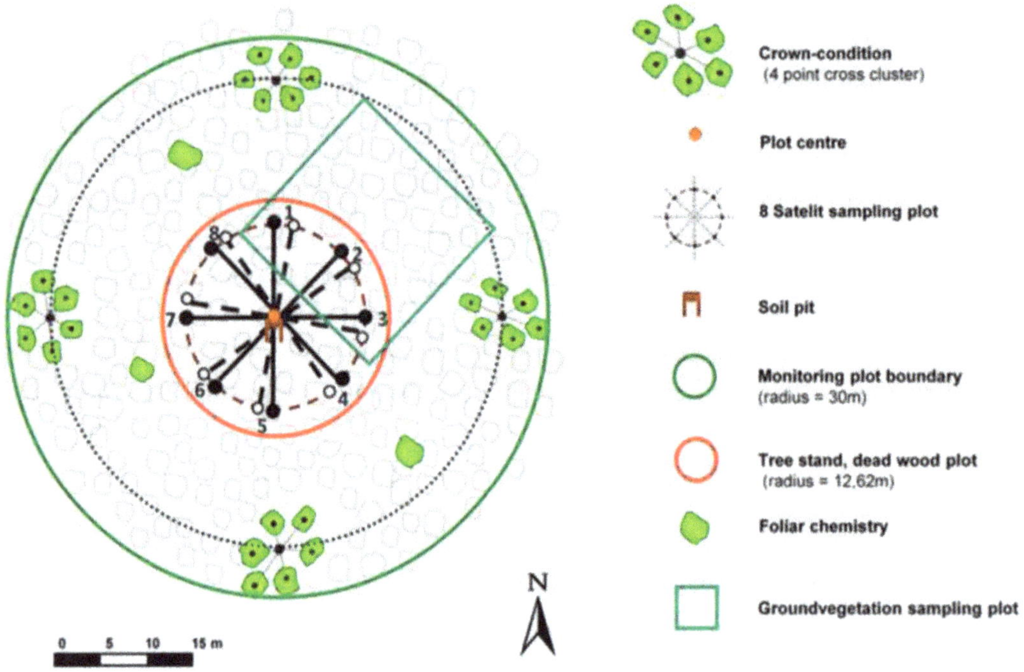

Fig. 1.2 Sampling design of NFSI (Höhle et al. 2018, modified)

1.6 Inventory Design

The NFSI was conducted as a systematic sampling inventory of the condition of forest soils at a nationwide grid of 8 × 8 km, yielding at about 1900 soil sampling sites. The first survey took place from 1987 to 1992 (1991–1992 in the youngest federal states of Germany), and resampling was done from 2006 to 2008. As far as additional findings from the intensive forest monitoring programme are integrated, they refer to the current network of 68 "Level II" sites (Fig. 1.1) and sometimes to additional plots sampled according to the same methodology.

1.7 Soil Sampling

1.7.1 National Forest Soils Inventory

The evaluation of the soil condition of the NFSI I was carried out according to an issued soil survey manual from (BMELF 1994). The standard methods for sampling soils at the NFSI grid plots involved sampling at eight satellites, with a soil profile at the centre of a plot (Fig. 1.2). Volume-based samples taken at the satellites were mixed for every depth level and plot. Satellites were distributed to main and minor geographic directions with a 10 m distance from the profile centre. Soil sampling plots of the NFSI II were shifted about 10 Gon from NFSI I satellites.

Soil profiles at the centre of the plot were used to designate soil horizons and classify soil types according to the manual of soil mapping (GAFA 2005, 2009, 2014). For both inventories, the forest floor litter with the fraction <20 mm was classified as organic layer. In general, the forest floor was divided into a litter (L), a moderately decomposed (Of) and a highly decomposed (Oh) layer if the layer thickness was >1 cm. At about 80% of the NFSI II plots, also the fraction >20 mm was sampled as forest floor material. The sampling of the forest floor was carried out by mixed samples at satellites with metal frames of different sizes. Different organic layers were accounted for as one layer and measured in thickness. Sampling of the mineral soil was obligatory for both inventories at satellites for the depth layers 0–5 cm, 5–10 cm, 10–30 cm, 30–60 cm and 60–90 cm either as mixed samples from the satellites or from the profile. Depending on stone concentration, volume-based mineral soil samples were taken with cylindrical core or cap cutter, with an AMS core sampler or a motor-driven auger.

1.7.2 Level II

The analysis of soil at individual German Intensive Forest Monitoring (Level II) sites is conducted every 10 years. There is no fixed year of sampling that applies to all sites. The sampling site design as well as the sampling procedures, chemical analysis and quality checks are documented in the ICP Forests Manual (http://icp-forests.net/page/icp-forests-manual). Accordingly, soil samples are taken at a minimum of 24 locations within the plot, which are aggregated to at least 3 composite samples per site.

On most plots, the organic layer was divided into a litter (L), a moderately decomposed (Of) and a highly decomposed (Oh) layer and the mineral soil is sampled at fixed depths of 0–10 cm, 10–20 cm, 20–40 cm and 40–80 cm. The top 10 cm of the mineral soil shall be further divided into the layers 0–5 cm and 5–10 cm. Other methodological aspects of the Level II soil sampling are directly comparable to the methods described for the German NFSI.

1.8 Laboratory Analytics Quality Management

For the preparation of NFSI II, the Assessment Committee Forestry Analytical Methods (Gutachterausschuss Forstliche Analytik, GAFA) was formed by the Federal Ministry of Food and Agriculture to standardize, determine and document methods of analysis on the one hand and to develop and determine a quality control programme on the other. All analytical methods used by the NFSI I and NFSI II as well as by the German and international forest monitoring programmes like UN/ECE ICP Forests programme, EU BioSoil), DIN and ISO Standards as well as the methods of the German federal states were subsequently documented and first published in 2005 by GAFA (GAFA 2005) in its manual on forestry analytics, the *Handbuch forstliche Analytik* (HFA). The HFA has been variously supplemented and is now available in the latest edition of 2014 (GAFA 2014).

The quality control programme encompassed 5 concomitant ring tests (3 soil and 2 humus tests of 6 samples each) and the entrainment of standard material for every survey parameter, which had to be analysed by every laboratory on at least 20 samples. Large quantities of six standard materials were produced for this purpose, and it was determined which material was to be analysed for which parameter. The evaluation of the ring tests (Blum and Heinbach 2010) and the examination of the standard materials showed that, barring few exceptions, the analytical data of the federal states and the laboratories can be evaluated comparatively (König et al. 2013). Only in 12 individual cases (laboratories of some federal states—a few parameters) the data for particular parameters of a laboratory or a federal state diverge in a directional bias from the data of the other institutions. The divergent data concerns in each case one laboratory/state of the parameter N (soil), Al, Ca, Fe, Mn and Zn in aqua regia digestion (humus), K in aqua regia digestion (soil), K and Na from the exchange capacity determination (humus) as well as pH (H_20) and pH (KCl) (humus). For details see König et al. (2013). A real problem are the K concentrations in aqua regia digestion (soil and humus), because some laboratories have large directionally biased deviations, in one case so much that the data are not comparable to the other laboratories. This is probably caused by the different degrees to which samples are ground (Houba et al. 1993). Concerning all other parameters and analytical methods, it is safe to assume that the NFSI II data of all laboratories/states can be assessed together.

Based on the results, it must be assumed that the variation within as well as between the laboratories amounts to about 10%. Only few parameters show a slightly smaller variation [e.g. elemental analysis (soil) C, aqua regia digestion (soil) Ca and total digestion (soil) (Ca)]. Table 1.2 gives a general view of the average variations of standard analyses and the ring tests summarized for the particular methods of analysis and which parameters vary more markedly within the particular analytical methods.

1.9 Sample Preparation Methods

The sample preparation, analysis and element determination methods used in NFSI I and NFSI II partly differ due to methodological improvements. Apart from this, some laboratories/federal states have made their own methodological modifications or have even used entirely different methods as recommended/instructed. The report on documentation and harmonization of the National Forest Inventory (Höhle et al. 2018) lists in detail which parameters were used and which methods applied, as well as which differing methods were used in some federal states regarding NFSI I or NFSI II. The report also shows how, using available comparative methodology, it was checked to find out which methods were comparable and with which methods it

Table 1.2 Overview of the individual analytical methods (parameter groups) and degree of variation in the analysis of standards

Analytical method/ parameter groups	Average variation of average value of standard analyses	Average variation of average deviation from average value of ring analyses	Parameter with larger variations	Comments
Elemental analytics (C, N)	±10%	±5%	N low concentration	
Effective EC soil	±10%	±10–15%	Na	Na not comparable
EC humus	±20%	±10–15%	H^+, Na	Na not comparable
Total EC soil	±20%	±10–15%	Mg, Na	Na not comparable
pH soil and humus	±20%	±40–50%	H^+	*Attention*: not pH, but parameter H^+
Aqua regia digestion soil: main elements	±10%	±10%	K, Na	K, Na not comparable
Aqua regia digestion soil: heavy metals	±10%	±10%		
Aqua regia digestion humus: main elements	±10–15%	±15%	Na, Al, K	Na not comparable
Aqua regia digestion humus: heavy metals	±20%	±20%	Cr	
NO_3 in water 1:2 extract	±15%	±20%		
Grain size determination	±20%	±20%	Sand (mean, coarse) silt (coarse)	
Total digestion soil: main elements	±10–20%	±5–15%		
Oxalate extract	±10%	±15%		
EC EU method	±15–20%	±15%	H^+, Na	

was possible to convert the data using correlation analysis. In the following, therefore, only the standard methods used will be described, with reference to the exhaustive description in the manual *Handbuch forstliche Analytik* (HFA).

Prior to analysis all soil, humus or plant samples were, as a rule, stored at 4 °C (HFA, method A1.1.1). Humus or plant samples were dried at 60 °C (HFA A1.2.1 and B1.2.1) and soil samples at 40 °C (HFA A1.2.1). The soil samples were sieved by hand or by machine through a 2 mm mesh (HFA A1.3.1). Humus samples were also sieved or ground manually through a 2 mm mesh. However, in NFSI I the residue was discarded (HFA A1.3.1), whereas in NFSI II only the residue larger than 20 mm was discarded, while the particles between 2 and 20 mm were ground to size with suitable machinery and then added to the 2 mm fraction (HA A1.3.2). The admixture of the 2–20 mm fraction, a fraction with high timber content, to the <2 mm fraction results in lower nutrient or heavy metal content but in higher C levels. This effect is most pronounced in the L horizon of the humus layer, because there the 2–20 mm fraction predominates weight by weight over that in the Of and Oh horizon. A calculation of supply extenuates dilution and accumulation effects (<10%), because humus supply in the L horizon is lower in comparison with that of the Of and Oh horizon (Höhle et al. 2018). For this reason the supply rather than the concentration was compared between NFSI I and NFSI II, and the evaluation extended over the whole organic layer, and not so much with regard to individual layers.

Humus and soil samples were milled to a suitable gradation using ball or disc mills (HFA A1.4.1). For NFSI II humus samples of the 2–20 mm and <2 mm fractions were milled together. Plant samples were milled analytically fine (0.25 mesh) using ball or centrifugal mills (HFA B1.3.1).

1.10 Soil Physical Parameters

In soil layers down to 30 cm, the determination of the volumetric proportion of coarse fragments >2 mm (subscript "cf", HFA A2,8) and of bulk density of fine earth was based on fixed volume samples taken with a cylinder (100 cm^3) or equivalent devices. Coarse fragments were in most cases weighed and their volume calculated based on density of the minerals (D_{cf}, mostly 2.65 g cm^{-3}); alternatively their volume was measured. In soil layers below 30 cm with their usually higher proportion of coarse fragments, it was also accepted to use estimated volumetric proportions of coarse fragments from the soil profile ($V\%_{cf>20\ mm}$) based on visually estimated and subsequently converted area proportions. Estimated values from NFSI I for layers above 30 cm were discarded in the evaluation process and replaced with the measurements from NFSI II.

Bulk density of fine earth (subscript "fe") was measured depending on the volumetric proportion of coarse fragments: If it was below 5%, the fixed volume sample was completely dried at 105 °C and weighed such that bulk density (*m*/*V*, mass/volume) of the whole sample was taken as bulk density of fine earth.

Fixed volume samples with higher proportions of only smaller coarse fragments (2–20 mm) were sieved (2 mm including washing of remainder) in order to separate coarse fragments before determination of weight of fine earth. Bulk density of fine earth (BD_{fe}, g cm^{-3}) and fine earth stock ($stock_{fe}$, t ha^{-1}) were then calculated as:

$$BD_{fe} = \frac{m_{sample} - m_{cf2-20\ mm,\ sample}}{V_{sample} - \frac{m_{cf2-20\ mm,\ sample}}{D_{cf}}} \quad (1.1)$$

$$stock_{fe} = BD_{fe} \times V_{layer} \times \left(1 - \frac{m_{cf2-20\ mm,\ sample}}{D_{cf} \times V_{sample}}\right) \quad (1.2)$$

In case of any proportion of larger coarse fragments (20–63 mm), they were visually estimated at the soil profile or by samples taken with a spade (Höhle et al. 2018). In that case, fine earth stock was calculated as:

$$stock_{fe} = BD_{fe} \times V_{layer} \times \left(1 - \frac{V\%_{cf>20\ mm}}{100} - \frac{m_{cf2-20\ mm,\ sample}}{D_{cf} \times V_{sample}}\right) \quad (1.3)$$

Bulk density estimates from NFSI I based on penetration resistance were considered too rough estimates (Wirth et al. 2004) and not accepted in the evaluation process, but replaced by measurements or estimations from NFSI II. In case of yet available older bulk density measurements according to the described methodology, the sampling was not repeated in NFSI II (see Sect. 1.17.3).

1.11 Chemical Analysis of Soil and Humus

In order to determine the pH values of the NFSI I samples, these were mixed at a ratio of 1:2.5 (weight for mineral soil, volume for humus layer) with H_2O (pH(H_2O); HFA A3.1.1.1) or 1 M KCl solution (pH(KCl); HFA A3.1.1.3), and the pH value was determined with a glass electrode. For NFSI II the samples were mixed with a volume ratio sample: solution of 1:5 with H_2O (pH(H_2O); HFA A3.1.1.2), 1 M KCl solution (pH(KCl); HFA A3.1.1.4) or 0.01 M $CaCl_2$ solution (pH($CaCl_2$); HFA A3.1.1.7), and the pH value was determined with a glass electrode. As these methods differ, a conversion of the data had to be undertaken on the basis of available correlation analyses (Höhle et al. 2018), so that the data are comparable and standardized to NFSI II methodology.

The effective cation exchange capacity (ECEC) and the exchangeable cations were determined by percolation of the dried and sieved samples with 1 M NH_4Cl solution and subsequent measurement of the cations in the extract (HFA A3.2.1.1). The federal states of Bavaria (extraction with 0.5 M NH_4Cl solution (HFA A3.2.1.7)) and Schleswig-Holstein (percolation with $SrCl_2$ solution (HFA A3.2.1.6)) have used diverging methods for NFSI I. Here a conversion of the data could also be achieved on the basis of available correlation analyses (Höhle et al. 2018), so that the data are comparable. The total cation exchange capacity (TCEC) of samples containing carbonate with pH(H_2O) >6.2 was determined by percolation with a $BaCl_2$ triethanolamine solution and $BaCl_2$ solution as well as re-exchange of the barium (Ba) ions with a $MgCl_2$ solution and subsequent measurement of the cations in the extract and the re-exchange extract (Ba) (HFA A3.2.1.2); state-specific modifications are described in the report (Höhle et al. 2018). In the humus layer, the cation exchange capacity (CEC) was only determined for NFSI II. This was done by percolation of sieved samples mixed with quartz sand with a 0.1 M $BaCl_2$ solution and subsequent measurement of the cations in the extract (HFA A3.2.1.9).

Diverse methods had been acceptable for the determination of organic C for NFSI I: (1) elemental analysis with elemental analysers (HFA D31.2.1.1–D31.2.2.4), (2) dry combustion with subsequent conductometric CO_2 determination according to Wösthoff, (3) indirect estimation via loss of ignition at 550 °C and factor correction (factor 1.72 for mineral soil and 2 for organic layers) and (4) wet combustion with K dichromate and sulphuric acid with subsequent photometric chromium (III) determination. For NFSI II elemental analysis, only elemental analysers were used (HFA D31.2.1.1–D31.2.2.4). The four methods used for NFSI I are regarded as comparable. This is shown by the preliminary study of NFSI II (Evers et al. 2002), in which NFSI I samples were re-analysed with elemental analysers, for methods (2) and (3) and also by a study by the *Ökologisches Labor der FHS Eberswalde* (Russ and Riek 2011) for method (4). NFSI I carbonate determination was carried out gas volumetrically according to Scheibler (HFA D31.3.1.1). For NFSI II this was done with elemental analysers (HFA D31.3.1.2, D31.3.1.3, D31.3.1.7) or gas volumetrically according to Scheibler (HFA D31.3.1.1). The results are comparable.

Nitrogen was measured in NFSI I through elemental analysis with elemental analysers (HFA D581–D58.1.2.1) or through Kjeldahl digestion with subsequent photometric or titrimetric determination. For NFSI II only elemental analysers were used (HFA D581–D58.1.2.1). The Kjeldahl method is considered to be comparable to elemental analysis. This is shown by the preliminary study of NFSI II (Evers et al. 2002; HFA D581–D58.1.2.1), in which NFSI I samples were re-analysed with elemental analysers.

For the determination of long-term available nutrients and heavy metals, four methods of analysis were permitted for NFSI I to determine Al, calcium (Ca), Fe, K, magnesium (Mg), Mn and phosphorus (P) concentration, as well as that of the heavy metals cadmium (Cd), copper (Cu), lead (Pb) and Zn in the humus layer and P in the mineral soil: (1) aqua regia extract (HFA A3.3.3), (2) nitric acid pressure digestion (HFA A3.3.4), (3) perchloric acid digestion and (4) total digestion with hydrogen-fluoride addition (HFA A3.3.1, A3.3.2, A3.3.5, A3.3.6). Elemental determination in the digestion solutions were done by ICP, ICP-MS, AAS and spectrophotometric methods. The ring tests that accompanied NFSI I (König and Wolff 1993) showed that the four permitted methods for humus sample digestion for NFSI I provided no nationally comparable data, as there were standard deviations of up to 35% and maximum deviations of up to 160%. Hence the possibility of nationwide evaluation of the data was limited. In order to ensure nationwide comparability of element concentration as well as comparability between first (NFSI I) and subsequent (NFSI II) inventory, the NFSI I retention samples were subjected to aqua regia digestion according to HFA A3.3.3, and the solution then analysed once more. Now all data gained from aqua regia extracts can be compared to another, except the K values, which are heavily dependent on the gradation of the ground samples.

The water 1:2 extract yielded information on soil solution properties. The dried and sieved mineral soil samples were mixed 1:2 in proportion to their weight with demineralized water, allowed to stand for 24 h and then filtered (HFA A3.2.2.1). For the element determination in the extracts, element-specific methods such as ICP, ICP-MS, ion chromatography and photometry were used according to HFA, section D. The concentrations from the 1:2 extract are supposed to mirror the properties of the soil solution below the root area. Schlotter et al. (2009) have, however, shown that this is not so with most cations. For the NO_3 concentrations in the 1:2 extract, there is a good correlation to the values from the soil solution (suction lysimeters) after standardization of the measured values on the water-soil ratios of field-fresh samples (Evers et al. 2002; Kohlpaintner et al. 2012). For more details see Chap. 5.

Oxalate extracts yielded information on occluded Fe oxides. For this, an extraction with 0.2 M ammonium-oxalate solution in a soil solution ratio of 1:50 was carried out with the dried and sieved soil samples and Fe and Al measured with ICP, ICP-MS or AAS (HFA 3.2.3.1). This method was restricted to NFSI II.

1.12 Sampling of Leaves and Needles

It was mandatory to sample needle or leaves of the four main tree species beech, spruce, pine and oak at each plot. Sampling other tree species was optional. At least five trees should be sampled from the main species within a stand. A detailed description of the sampling method is described in Wellbrock et al. (2006). The sampling design is comparable to ICP Forest Manual Part XII.

1.13 Chemical Analysis of Leaves and Needles

For NFSI I total N concentration determination, two methods were admitted: (1) elemental analysis with elemental analysers (HFA D58.1.3.1) and (2) Kjeldahl digestion with subsequent photometric or titrimetric determination. For NFSI II, elemental analysis was carried out with elemental analysers only (HFA D58.1.3.1). The Kjeldahl method is considered to be comparable to element analysis. This is shown by the preliminary study of NFSI II (Evers et al. 2002), in which NFSI I samples were re-analysed with elemental analysis devices.

Chloride (Cl) was determined by the combustion digestion method according to Schöninger (HFA B3.2.2) and Cl measured subsequently in the extract. There is, however, data for only one federal state available.

Five methods were admitted for the NFSI I determination of nutrients Ca, K, Fe, Mg, Mn, P, S, and silicon (Si) and the heavy metals Cd, Cu, Pb and Zn: (1) nitric acid pressure digestion (HFA B3.2.1), (2) total digestion with hydrofluoric acid (HFA B3.2.3), (3) aqua regia digestion and (4) perchloric acid digestion and dry ashing (not for heavy metals). Element determination in the digestion solutions was carried out by ICP, AAS and spectrophotometric methods. For NFSI II nitric acid pressure digestion only was used (HFA B3.2.1); element determination in the digestion solutions was carried out by ICP, AAS and spectrophotometric methods according to HFA, section D. As far as nitric acid digestion was used in NFSI I and NFSI II, the data are comparable. Total digestion is, in part, not comparable for the elements Al and K. All other elements should be comparable, because the digestion methods used cover total concentration.

The determination of 1000 needle dry weight was made by counting from 3 × 50 to 3 × 100 needles, which were then dried at 105 °C and weighed. The weight was then extrapolated to 1000 needles (HA B2.2). The weight of 100 leaves was determined by counting from 50 to 100 leaves, which were then dried at 105 °C and weighed, whereupon the 100 leaf weight was extrapolated (HA B.3).

1.14 Tree Crown Condition

The Forest Condition Survey represents a basic part of the nationwide and Europe-wide forest monitoring besides the NFSI and the Intensive Forest Monitoring (Level II). In Germany, the condition of forest trees was recorded first in 1984 and has been conducted annually throughout Germany since 1990. The methods are widely harmonized and standardized throughout Germany (Wellbrock et al. 2018) and also throughout Europe (Eichhorn et al. 2016). The survey is performed on the extensive monitoring plots (Level I), which were established wherever the nodes of a 16 × 16 km grid over Europe coincided with forest land cover. In Germany, some federal states performed the Level I sampling on a denser grid in addition to the 16 × 16 km grid (approx. 430 Level I plots). Between 2006 and 2008, the forest condition survey took place nationwide on the denser grid of the NFSI II (mainly 8 × 8 km). In addition, changes of the grid over time occurred, such as spatial shifts of the grid in Bavaria in 2006 and Brandenburg in 2009. The most common sampling design in Germany is a cross cluster with four satellites each of which comprising six trees. Thus, usually 24 trees are assessed at 1 sample plot. The trees are permanently marked. They have to belong to Kraft class 1 (dominant) to 3 (subdominant), and suppressed trees are not sampled. The Forest Condition Survey is primarily based on defoliation, which describes the loss of needles or leaves in the crown of a tree compared to a local or absolute reference tree with full foliage. Defoliation represents the most widely used indicator for the assessment of tree condition and vitality. The estimation of defoliation takes place visually using binoculars, and defoliation is recorded in 5% classes from 0% (no defoliation) to 100% (dead tree). A quality assurance programme including national training courses (Eickenscheidt and Wellbrock 2014; Eichhorn et al. 2016) has been initiated in order to monitor consistency and reproducibility of defoliation data. In addition to defoliation, several other parameters (e.g. discolouration, insect infestation and fructification) are recorded. A detailed description of the survey can be found in Wellbrock et al. (2018) for Germany and in the ICP Forests Manual Part IV (Eichhorn et al. 2016) for Europe.

1.15 Critical Loads

Critical loads of pollutants are defined as the annual deposition below which no long-term harmful effects on ecosystem structure and function are expected according to present knowledge (Nilsson and Grennfelt 1988). Site-specific critical loads were calculated for acidity and for nutrient N after the simple mass balance method (SMB; Sverdrup and de Vries 1994) according to the ICP manual for modelling and mapping (CLRTAP 2016). For a detailed description of critical loads calculation on NFSI plots, see Höhle et al. (2018).

1.15.1 Critical Loads of Acidity

Critical loads of acidifying depositions (CL_{aci}) are calculated from the most important proton sources and sinks (CLRTAP 2016):

$$\mathrm{CL_{aci}(S+N) = BC^*_{dep} - Cl^*_{dep} + BC_w - Bc_u + N_i + N_u + N_{de} - ANC_{le,crit}} \tag{1.4}$$

where BC are base cations (Ca + Mg + K + Na) while Bc are nutrient cations (Ca + Mg + K), asterisks mark sea salt corrected (Posch et al. 2003, 2015) deposition (subscript dep), and the subscripts w, u, i and de stand for weathering, net growth uptake by harvested trees, immobilization and denitrification, respectively. $ANC_{le,crit}$ is the critical leaching of acid neutralizing capacity. Because sulphur (S) and N depositions both contribute to the acid load, no unique critical loads for either element can be determined. As N sinks cannot compensate S deposition, the maximum critical load for S $CL_{max}(S)$ is given by:

$$\mathrm{CL_{max}(S) = BC^*_{dep} - Cl^*_{dep} + BC_w - Bc_u - ANC_{le,crit}} \tag{1.5}$$

Equation (1.5) applies only if N depositions are lower than the N sinks ($CL_{min}(N)$), i.e. if $N_{dep} \leq CL_{min}(N) = N_i + N_u$. Together with the maximum critical load for acidifying N (at $S_{dep} = 0$) $CL_{max}(N) = CL_{min}(N) + \frac{CL_{max}(S)}{1-f_{de}}$, where f_{de} is the denitrification factor, these quantities define the critical loads function for acidity (CLF_{aci}) for pairs of N_{dep} and S_{dep} (CLRTAP 2016).

The leaching (subscript le) of acid neutralizing capacity (ANC) is defined as:

$$\begin{aligned} ANC_{le} &= HCO^{-}_{3,le} + RCOO^{-}_{le} - H^{+}_{le} - Al^{3+}_{le} \\ &= Q \times \left(\left[HCO_3^- \right] + \left[RCOO^- \right] - \left[H^+ \right] - \left[Al^{3+} \right] \right) \end{aligned} \tag{1.6}$$

where Q is the precipitation surplus. The critical leaching of acid neutralizing capacity $ANC_{le,crit}$ is calculated from either a critical proton or Al concentration. Both are related through the gibbsite equilibrium constant $K_{gibb} = [Al]/[H]^3$. The critical limit for Al or hydrogen (H) is chosen from the most vulnerable receptor out of four criteria. (1) To reduce phytotoxic effects, a critical Bc/Al ratio is calculated as weighted mean of tree species-specific tolerated ratios (with growth reductions of approx. 20%; Sverdrup and Warfvinge 1993) for plots on mineral soil and a respective critical Bc/H ratio for plots on peat and peaty mineral soil. To calculate $ANC_{le,crit}$ from these ratios, site-specific nutrient cation leaching is given by $Bc_{le} = Bc_{dep} + Bc_w - Bc_u$ (CLRTAP 2016). (2) For soils with base saturation >30%, these criteria are not sufficient to conserve base-rich ecosystems. Therefore, a critical pH is set at the lower boundary of the site-specific buffer range (Ulrich 1981; Balla et al. 2013). Buffer range was determined from weighted actual pH (H_2O) of the rooted zone. (3) To maintain soil structural stability, Al leaching should not exceed Al weathering, which can be related to base cation weathering by $Al_w = 2 \times BC_w$ (CLRTAP 2016). (4) Vegetation-specific critical limits for base saturation have been derived from the BERN model by Balla et al. (2013). These were converted to critical H concentrations with a GAPON exchange equilibrium (de Vries and Posch 2003; CLRTAP 2016).

1.15.2 Critical Loads of Nutrient Nitrogen for Soils

Critical loads of nutrient N ($CL_{nut}(N)$) are calculated as (CLRTAP 2016):

$$CL_{nut}(N) = N_i + N_u + \frac{N_{le(acc)}}{1 - f_{de}} \tag{1.7}$$

The acceptable N leaching $N_{le,acc}$ is calculated from acceptable N concentrations in soil solution (N_{crit}):

$$N_{le(acc)} = Q \times N_{crit} \tag{1.8}$$

The critical limit of N concentration in soil solution (N_{crit}) is set for each plot as the minimum of two criteria (Balla et al. 2013; ARGE_StickstoffBW 2014): (1) To prevent vegetation changes, $N_{crit(plant)}$ was taken from the BERN model (ARGE_StickstoffBW 2014; Balla et al. 2013) based on CLRTAP (2016) for near-natural forest ecosystems. To this end, NFSI plots were characterized according to their plant sociology based on vegetation relevés (Ziche et al. 2016) and assigned to the reference plant communities of the BERN model (Andreae et al. 2016). $N_{crit(plant)}$ was set as 3 mg l^{-1} for forest plantations. (2) To prevent nutrient imbalances, $N_{crit(nut)}$ was derived from the vegetation-specific critical Bc/N ratio after the BERN model (Balla et al. 2013; ARGE_StickstoffBW 2014). For this purpose, nutrient cation concentrations in the rooting zone were calculated from Bc_{le}, Q, available field capacity, soil bulk density and cation exchange capacity from the NFSI II soil samples. Additionally, we compared the results from the described ("modified") approach to the conservative and save approach where N_{crit} is set to 0.2 mg l^{-1} in broadleaf and 0.4 mg l^{-1} in coniferous forests to prevent nutrient imbalances (CLRTAP 2016).

Exceedance of Critical Loads
Exceedance of critical loads for pollutant X by atmospheric deposition X_{dep} is defined as:

$$\mathrm{Ex}(X) = X_{\mathrm{dep}} - \mathrm{CL}(X) \tag{1.9}$$

Exceedance of CL_{aci} depends on two pollutants (S and N) and is defined by the critical loads function CLF_{aci} (CLRTAP 2015, 2016). A multiple-criteria CLF regarding acidifying S and N together with eutrophying N is derived by substituting $CL_{max}(N)$ through $\min(CL_{max}(N), CL_{nut}(N))$. Because critical loads are determined to evaluate anthropogenic emissions, the deposition fluxes were sea salt (Posch et al. 2015) corrected.

1.15.3 Derivation of Input Data

Time series of atmospheric deposition of S, N, Ca, Mg, K, Na and Cl as well as sea salt corrected deposition rates (Alveteg et al. 1998) were modelled for each NFSI plot (see Sect. 1.13). To estimate Bc_{le}, we used the long-term mean (1995–2015) total deposition except for the derivation of $N_{crit(nut)}$, where only the sea salt born deposition is considered (Balla et al. 2013).

As gibbsite equilibrium constant K_{gibb}, the widely used default value of 300 m^6 eq^{-2} was applied on mineral soils and 100 m^6 eq^{-2} on peat and peaty mineral soils (CLRTAP 2016).

Precipitation surplus (Q), rooting depth (86% cumulated root mass) and available field capacity was estimated for every NFSI plot by von Wilpert et al. (2016).

Bicarbonate concentration ($[HCO_3]$) was calculated from the first dissociation constant, Henry's constant and a temperature-dependent estimation of CO_2 partial pressure according to CLRTAP (2016).

Concentration of dissociated organic acids ([RCOO-]) was estimated according to CLRTAP (2016) from reference concentrations of dissolved organic C (de Vries et al.

2005), an exemplary amount of functional groups per molC (Santore et al. 1995) and a pH-dependent estimate of the first dissociation constant (Oliver et al. 1983).

Base cation weathering rate (BC_w) was derived according to Posch et al. (2015) as a function of texture and parent material classes with a temperature correction. The weathering rate of nutrient cations (Bc_w) was approximated by multiplying BC_w with a soil richness factor (0.7–0.85) dependent on cation exchange capacity and available field capacity to distinguish between poor sandy (0.7) and rich soils (0.85).

Net growth uptake (N_u, Bc_u) was calculated based on the average (100 years) increment in wood and bark as estimated from Harmonized Stand Inventory and yield tables (Schober 1975). Recent growth rates are higher due to increased N input (Laubhann et al. 2009) and other environmental factors (Pretzsch et al. 2014); therefore it is recommended to use "old" yield tables for critical load calculations (Posch et al. 2015). Volume increments were transformed in wood and bark mass using wood and bark densities (Ahner et al. 2013) and bark/wood ratios (Rademacher et al. 1999). Element concentrations in biomass compartments were taken from Jacobsen et al. (2002).

Denitrification (N_{de}) is formulated as deposition-dependent quantity: $N_{de} = N_{le,acc} \times \frac{f_{de}}{1-f_{de}}$. The denitrification factor f_{de} was derived from soil texture and drainage status according to CLRTAP (2016) as described in Andreae et al. (2016).

Long-term N immobilization rate (N_i) was derived from actual N stocks in organic and mineral soil layers (0–90 cm) and estimated soil age (10,000 a for glacial and 24,000 a for periglacial soils according to the extension of the last glacial maximum; Höhle and Wellbrock 2017).

1.16 Atmospheric Deposition

Following Thiele et al. (2017), the long-term trends for the deposition of N, S and base cations were calculated with the model MAKEDEP (Alveteg et al. 1998). For details on methodology, see Schaap et al. (2017). To reconstruct the deposition before 2009, we used the regional trend from the EMEP database (Tarrasón and Nyiri 2008) and standard time series from Alveteg et al. (1998). These time series provide only trend information for S and N. For the long-term development of emissions and deposition of Ca, Mg, K and Cl, very little information is available. In addition, there are a number of different local sources of emission that can contribute to the deposition of these elements. These sources can vary widely depending on local and regional factors. This is especially the case for K (Dämmgen et al. 2013a). However, long-term studies of the deposition process (Dämmgen et al. 2013a, b; Hedin et al. 1994; Meesenburg et al. 1995) show that Ca and Mg at least partially follow the trend of S deposition. Therefore, it is assumed that the non-marine portion of the deposition of these elements can at least be partly associated with human activities and have the same trend as the stand S curve

(Johansson et al. 1996). For K and Cl, the influence of emission-reducing measures was also considered (Dämmgen et al. 2013a), but this is not quite as pronounced as for S. More details on the assumptions and evaluation are given in Höhle et al. (2016). Examples of this approach to generate time series of N and S are described in Ahrends et al. (2010), Fleck et al. (2017) and Hauck et al. (2012). In Eastern Germany high depositions rates (fly ashes from brown coal-fired power plants) occurred in the 1970s and 1980s and are accordingly no longer relevant for the deposition rates between the two NFSIs but of course for the soil chemical state and its dynamics (Riek et al. 2012).

1.17 Statistics

1.17.1 Weighting

Weighting of the NFSI plots according to the proportion of forest area they represent was carried out in order to receive an area-representative result. Weighting was necessary because grid densities partly deviated from the 8 × 8 km grid. The weighting factors were calculated as follows (see Eq. 1.10):

$$W_l = \frac{A_l}{\sum_{l=1}^{16} A_l} \times \frac{1}{n_l} \tag{1.10}$$

with:

W_l = Weighting factor of the NFSI plots in the federal state l
A_l = Forest area of federal state l
n_l = Number of NFSI plots in the federal state l

This weighting approach assumes that the plot density within the forested area of a federal state is homogeneous. The proportion of the forest area of a federal state on the forest area of entire Germany was derived from the Corine land cover data 1990 (EEA 2010a) and 2006 (EEA 2010b).

1.17.2 Basic Evaluations

Statistical parameters (e.g. arithmetic means, standard deviations, 10., 25., 50., 75. and 90. percentiles) were estimated for each soil depth layer (organic layer, 0–5 cm, 5–10 cm, 10–30 cm, 30–60 cm, 60–90 cm). An aggregation of soil depth layers, e.g. the entire profile (organic layer and mineral soil up to a maximum of 90 cm soil depth), was conducted for some parameters. For concentration, a weighted aggregation was carried out according to stocks (organic layer stock and fine soil stock). In case of pH values, the values were delogarithmized, and the aggregation was done on basis of the H^+ concentration.

Both the soil parameters for the NFSI II and the temporal change of these parameters between the two NFSIs were analysed. The annual rate of change was calculated as difference between the NFSI I and NFSI II divided by the corresponding time difference. Peats and highly organic soils (63 plots) were not considered for the analyses of the temporal changes but for the description of the state at the NFSI II. The majority of the plots have been sampled at both surveys. These plots represent paired samples. Calculation of changes was done for paired samples as well as for the entire collective (paired and unpaired samples). For paired samples, change rates were calculated as the difference of values between NFSI II and NFSI I divided by the number of years between both surveys. For the entire collective, the calculation of the change rates was based on federal state means. Hence, at most 16 values were available for statistical evaluations.

Differences were tested for significance using a weighted differences *t*-test. The forest-area weighted values of the state at the NFSI II and the temporal changes were presented as boxplots according to soil depths. A solid line indicates the median and a diamond the arithmetic mean. The length of the box reflects the interquartile range. The whiskers represent the lowest or highest value within the 1.5-fold interquartile range starting from the lower or upper quartile. Outliers are not shown but are included in the calculations for the boxplots. Within the boxplots significant differences were indicated by a star. The state at the NFSI II and temporal change rates were furthermore visualized using maps and cumulative frequency distributions. The classification is again based on weighted percentiles. Therefore, in general data of both surveys were used for estimating weighted percentiles; however, in cases where only data of the NFSI II were available, percentiles were calculated from these data.

Furthermore, values were stratified using a number of potential strata alternatively: grouping variables, e.g. forest stand type, types of acidification and liming. The stratification aims at identifying cause-effect relations, and therefore, unweighted values were used. The comparison of means was performed using ANOVA. The group means were compared using Bonferroni corrected *t*-tests.

Further details on the statistical methods are explained in the respective chapters. Statistical significance is defined at $p < 0.05$. Results are presented as arithmetic means $\pm$ standard error, unless indicated differently. The statistical basis evaluations were conducted using R 3.3.3 (R Core Team 2017) and SAS 9.4. The maps were made using ArcGis 10.3.1 for desktop.

1.17.3 Challenges and Solutions

The derivation of trends in element stocks from element concentration and fine earth content measurements at two points in time is challenging due to the numerous possibilities for measurement errors. High spatial and temporal variability of element concentrations on the same plot, the representativeness of subsamples of sieved soil taken for measurements, the relevance of extremely low concentrations in the deeper soil layers, the difficulty of accurate volumetric estimations of fine earth contents and bulk density of soil samples, the variable accuracy of laboratory measurements at both points in time and the inaccuracy of depth delineations are the main error sources (Vanguelova et al. 2016).

Several measures were taken from the beginning to counteract these error sources in NFSI II:

Spatial and Temporal Variability
Spatial variability on the plot scale was counteracted by sampling repetition on eight satellite points, at least up to a depth of 30 cm and in nearly all of the federal states. Below 30 cm it was allowed to reduce the number of sampling locations to four, and this is in accordance with the expected depth gradient in spatial variability of element concentrations from organic layers to the deeper layers of the mineral soil (Bekele et al. 2013; Wang et al. 2017). In certain cases, however, spatial variability may require even more sampling locations to reach low standard errors (Kirwan et al. 2005; Schrumpf et al. 2011), thereby potentially introducing an unavoidable uncertainty. Also the seasonality of soil processes may introduce uncertainties, since samples needed to be taken throughout all seasons in usually more than 1 year in order to cover all sampling locations. However, these uncertainties are not expected to introduce directed errors, and should thus decrease when means over several plots are reported. Sampling dates were recorded in order to enable later analysis of such influences.

Representativeness of Subsamples
The smallest-scaled part of spatial variability was counteracted by the mere size of soil samples taken in the field (2 cm to 4 cm in diameter), thereby integrating over earthworm burrows, fine root channels and bulk soil. The soil samples produced were, thus, a mixture of material from these origins and were well mixed due to the combination with material from the other satellite points, sieving, milling and drying. In most federal states, the dry samples were stirred when subsamples were taken in order to improve the representativeness of subsamples; only one federal state (North Rhine-Westphalia) employed a sample divider (HFA A1.3.1) in both NFSIs to partition subsamples from whole samples.

Low Concentrations in Deeper Soil Layers
The laboratory-dependent limit of quantification was recorded together with analysis results in order to be able to judge the reliability of very low concentrations.

Accuracy of Fine Earth Stocks and Bulk Density
Fine earth stocks were derived from bulk density and coarse soil fractions that were estimated or measured with different approaches (see Sect. 1.9). It was assumed that change of these parameters between NFSI I and II is negligible, and, therefore, their

values were held constant in all evaluations of the paired sample, in most cases employing the NFSI II value.

Variable Accuracy of Laboratory Measurements
The variable accuracy of laboratory measurements at both points in time was investigated by frequently repeated inter-laboratory ring tests that forced laboratories to improve their methods (König and Wolff 1993; Blum and Heinbach 2010; König et al. 2013).

Inaccuracy of Depth Delineations
Due to the operator-dependent delineation of organic layer and upper mineral soil, the evaluation of each respective layer (organic layer and 0–5 cm of mineral soil) was for the most critical parameters compared with a combination of both layers.

References

Ahner J, Ahrends B, Engel F, Hansen J, Hentschel S, Hurling R, Meesenburg H, Mestermacher U, Meyer P, Möhring B, Nagel J, Nagel R, Pape B, Rohde M, Rumpf H, Schmidt M, Schmidt M, Spellmann H, Sutmöller J (2013) Waldentwicklungsszenarien für das Hessische Ried. Entscheidungsunterstützung vor dem Hintergrund sich beschleunigt ändernder Wasserhaushalts-und Klimabedingungen und den Anforderungen aus dem europäischen Schutzgebietssystem Natura 2000. Beiträge aus der Nordwestdeutschen Forstlichen Versuchsanstalt, vol 10. Nordwestdeutsche Forstliche Versuchsanstalt, Göttingen

Ahrends B, Meesenburg H, Döring C, Jansen M (2010) A spatio-temporal modelling approach for assessment of management effects in forest catchments. Status and perspectives of hydrology in small basins. IAHS Publ 336:32–37

Alveteg M, Walse C, Warfvinge P (1998) Reconstructing historic atmospheric deposition and nutrient uptake from present day values using MAKEDEP. Water Air Soil Pollut 104 (3–4):269–283. https://doi.org/10.1023/a:1004958027188

Andreae H, Eickenscheidt N, Evers J, Grüneberg E, Ziche D, Ahrends B, Höhle J, Nagel H-D, Wellbrock N (2016) Stickstoffstatus und dessen zeitliche Veränderung in Waldböden. In: Wellbrock N, Bolte A, Flessa H (eds) Dynamik und räumliche Muster forstlicher Standorte in Deutschland: Ergebnisse der Bodenzustandserhebung im Wald 2006 bis 2008, vol Thünen Report 43. Johann Heinrich von Thünen Institute, Federal Research Institute for Rural Areas, Forestry and Fisheries, Braunschweig, pp 135–180

ARGE_StickstoffBW (ed) (2014) Ermittlung standortspezifischer Critical Loads für Stickstoff – Dokumentation der Critical Limits und sonstiger Annahmen zur Berechnung der Critical Loads für bundesdeutsche FFH-Gebiete, vol ID Umweltbeobachtung U26-S7-N12. Arbeitsgemeinschaft Stickstoff, federal state Baden-Württemberg, Karlsruhe

Balla S, Müller-Pfannenstiel K, Uhl R, Kiebel A, Lüttmann J, Lorentz H, Düring I, Schlutow A, Schleuschner T, Förster M, Becker C, Herzog W (2013) Untersuchung und Bewertung von straßenverkehrsbedingten Nährstoffeinträgen in empfindliche Biotope. Forschung Straßenbau und Straßenverkehrstechnik, vol 1099. Federal Highway Research Institute (BASt), Bremen

Bekele A, Kellman L, Beltrami H (2013) Plot level spatial variability of soil organic carbon, nitrogen, and their stable isotopic compositions in temperate managed forest soils of Atlantic Canada. Soil Sci 178(8):400–416. https://doi.org/10.1097/ss.0000000000000003

Blum U, Heinbach R (2010) Gesamtauswertung der Datengrundlage sämtlicher BZE Ringversuche 2005 bis 2009. Bavarian State Institute of Forestry, Freising

Blume H-P, Horn R, Kandeler E, Kögel-Knabner I, Kretzschmar R, Stahr K, Wilke BM (2010) Scheffer/Schachtschabel: Lehrbuch der Bodenkunde. Spektrum Akademischer Verlag, Heidelberg

BMELF (1994) Arbeitsanleitung zur ersten Bodenzustandserhebung im Wald, 2nd edn. Federal Ministry of Food, Agriculture and Forestry, Berlin

Bytnerowicz A, Omasa K, Paoletti E (2007) Integrated effects of air pollution and climate change on forests: a northern hemisphere perspective. Environ Pollut 147(3):438–445. https://doi.org/10.1016/j.envpol.2006.08.028

CLRTAP (2015) Exceedance calculations. In: Manual on methodologies and criteria for modelleding and mapping critical loads and levels and air pollution effects, risks and trends. UNECE Convention on Long-range Transboundary Air Pollution. http://www.icpmapping.org/. Accessed 12 Jan 2018

CLRTAP (2016) Mapping of critical loads for ecosystems. In: Manual on methodologies and criteria for modelling and mapping critical loads and levels and air pollution effects, risks and trends. UNECE Convention on Long-range Transboundary Air Pollution

Dämmgen U, Matschullat J, Zimmermann F, Strogies M, Grünhage L, Scheler B, Conrad J (2013a) Emission reduction effects on bulk and wet-only deposition in Germany – evidence from long-term observations. Part 2: precipitation, potential sea salt, soil and fly ash constituents. Gefahrst Reinhalt Luft 73(1–2):25–36

Dämmgen U, Matschullat J, Zimmermann F, Strogies M, Grünhage L, Scheler B, Conrad J (2013b) Emission reduction effects on bulk and wet-only deposition in Germany – evidence from long-term observations. Part 3: sulphur and nitrogen compounds. Gefahrst Reinhalt Luft 73 (7–8):330–339

de Vries W, Posch M (2003) Derivation of cation exchange constants for sand, loess, clay and peat soils on the basis of field measurements in the Netherlands. Alterra report. Soil Science Centre at Wageningen University & Research, Wageningen

de Vries W, Schütze G, Lofts S, Tipping E, Meili M, Römkens PFAM, Groenenberg J (2005) Calculation of critical loads for cadmium, lead and mercury – background document to a mapping manual on critical loads of cadmium, lead and mercury. Alterra report. Soil Science Centre at Wageningen University & Research, Wageningen

EEA (2010a) Raster data on land cover for the CLC 1990 inventory (Version 13 of Feb 2010). European Environment Agency (EEA), Kopenhagen

EEA (2010b) Raster data on land cover for the CLC 2006 inventory (Version 13 of Feb 2010). European Environment Agency (EEA), Kopenhagen

Eichhorn J, Roskams P, Potočić N, Timmermann V, Ferretti M, Mues V, Szepesi A, Durrant D, Seletković I, Schroeck H-W, Nevalainen S, Bussotti F, Paloma G, Wulff S (2016) Part IV visual assessment of crown condition and damaging agents. In: UNECE ICP Forests Programme Co-ordinating Centre (ed) Manual on methods and criteria for harmonized sampling, assessment, monitoring and analysis of the effects of air pollution on forests. Thünen Institute of Forest Ecosystems, Eberswalde, p 54

Eickenscheidt N, Wellbrock N (2014) Consistency of defoliation data of the national training courses for the forest condition survey in Germany from 1992 to 2012. Environ Monit Assess 186(1):257–275. https://doi.org/10.1007/s10661-013-3372-3

Ellenberg H (1971) Integrated experimental ecology: methods and results of ecosystem research in the German Solling Project. In: Ecological studies, vol 2. Springer, Berlin

Evers J, König N, Wolff B, Meiwes KJ (2002) Vorbereitung der Zweiten Bodenzustandserhebung im Wald (BZE II): Untersuchungen zur Laboranalytik, Stickstoffbestimmung und zeitlichen Variabilität bodenchemischer Parameter. Federal Ministry of Food, Agriculture and Consumer Protection, Northwest German Forest Research Institute, Bonn

Fleck S, Ahrends B, Sutmöller J, Albert M, Evers J, Meesenburg H (2017) Is biomass accumulation in forests an option to prevent climate change induced increases in nitrate concentrations in the North German Lowland? Forests 8(6):219

GAFA (ed) (2005) Handbuch Forstliche Analytik (HFA). Grundwerk des Gutachterausschuss forstliche Analytik (GAFA). Federal Ministry of Food, Agriculture and Consumer Protection, Northwest German Forest Research Institute, Bonn

GAFA (ed) (2009) Handbuch Forstliche Analytik (HFA). Grundwerk und 1. - 4. Ergänzung des Gutachterausschuss Forstliche Analytik (GAFA). Bundesministerium für Verbraucherschutz, Ernährung und Landwirtschaft, Bonn

GAFA (ed) (2014) Handbuch Forstliche Analytik (HFA). Grundwerk und 1. - 5. Ergänzung des Gutachterausschuss Forstliche Analytik (GAFA). Federal Ministry of Food, Agriculture and Consumer Protection, Northwest German Forest Research Institute, Bonn

Hauck M, Zimmermann J, Mascha J, Dulamsuren C, Bade C, Ahrends B, Leuschner C (2012) Rapid recovery of stem growth at reduced SO_2 levels suggests a major contribution of foliar damage in the pollutant-caused dieback of Norway spruce during the late 20th century. Environ Pollut 164:132–141

Hedin LO, Granat L, Likens GE, Buishand TA, Galloway JN, Butler TJ, Rodhe H (1994) Steep declines in atmospheric base cations in regions of Europe and North-America. Nature 367 (6461):351–354. https://doi.org/10.1038/367351a0

Hickler T, Bolte A, Hartard B, Beierkuhnlein C, Blaschke M, Blick T, Brüggemann W, Dorow WHO, Fritze MM, Gregor T, Ibisch P, Kölling C, Kühn I, Musche M, Pompe S, Petercord R, Schweiger O, Seidling W, Trautmann S, Waldenspuhl T, Walentowski H, Wellbrock N (2012) Folgen des Klimawandels für die Biodiversität in Wald und Forst. In: Mosbrugger V, Brasseur GP, Schaller M, Stribrny B (eds) Klimawandel und Biodiversität – Folgen für Deutschland. Wissenschaftliche Buchgesellschaft, Darmstadt, pp 164–221

Hilbrig L, Wellbrock N, Bielefeldt J (2014) Harmonisierte Bestandesinventur: Zweite Bundesweite Bodenzustandserhebung (BZE II) – Methode. Thünen Working Paper. Johann Heinrich von Thünen Institute, Braunschweig. https://doi.org/10.3220/WP_26_2014

Höhle J, Wellbrock N (2017) Immobilisation of nitrogen in context of critical loads – literature review and analysis of German, French and Swiss soil data. Texte 71/2017. Umweltbundesamt, Dessau-Roßlau

Höhle J, Bielefeldt J, Dühnelt P-E, König N, Ziche D, Eickenscheidt N, Grüneberg E, Hilbrig L, Wellbrock N, Kompa T (2018) Bodenzustandserhebung im Wald – Dokumentation und Harmonisierung der Methoden. Thünen Working Paper. Johann Heinrich von Thünen Institute, Braunschweig. https://doi.org/10.3220/WP1526989795000

Höhle J, König N, Hilbrig L, Bielefeld J, Ziche D, Grüneberg E, Eickenscheidt N, Ahrends B, Wellbrock N (2016) Methodenüberblick und Qualitätssicherung. In: Wellbrock N, Bolte A, Flessa H (eds) Dynamik und räumliche Muster forstlicher Standorte in Deutschland. Ergebnisse der Bodenzustandserhebung im Wald 2006 bis 2008, vol Thünen Report 43. Johann Heinrich von Thünen Institute, Federal Research Institute for Rural Areas, Forestry and Fisheries, Braunschweig, pp 6–43

Houba VJG, Chardon WJ, Roelse K (1993) Influence of grinding of soil on apparent chemical composition. Commun Soil Sci Plant Anal 24(13–14):1591–1602. https://doi.org/10.1080/00103629309368902

Jacobsen C, Rademacher P, Meesenburg H, Meiwes K (2002) Gehalte chemischer Elemente in Baumkompartimenten – Literaturstudie und Datensammlung. Berichte des Forschungszentrums Waldökosysteme. Niedersächsische Forstliche Versuchsanstalt Göttingen, Göttingen

Johansson M, Alveteg M, Walse C, Warfvinge P (1996) Derivation of deposition and uptake scenarios. Paper presented at the international workshop on exceedance of critical loads and levels, Vienna

Kauppi P, Anttila P, Kenttämies K (1990) Acidification in Finland. Finnish Acidification Research Programme HAPRO 1985–1990. Springer, Berlin

Kirwan N, Oliver MA, Moffat AJ, Morgan GW (2005) Sampling the soil in long-term forest plots: the implications of spatial variation. Environ Monit Assess 111(1–3):149–172. https://doi.org/10.1007/s10661-005-8219-0

Kohlpaintner M, Huber C, Göttlein A (2012) Improving the precision of estimating nitrate (NO_3^-) concentration in seepage water of forests by prestratification with soil samples. Eur J For Res 131(5):1399–1409. https://doi.org/10.1007/s10342-012-0606-9

König N, Wolff B (1993) Abschlussbericht über die Ergebnisse und Konsequenzen der im Rahmen der bundesweiten Bodenzustandserhebung im Wald (BZE) durchgeführten Ringanalysen. Reports of the Research Centre for Forest Ecosystems/Forest Decline – Series B. Göttingen University, Göttingen

König N, Schönfelder E, Blum U (2013) Auswertung der Standardmessungen und der Ringversuche im Rahmen der BZE II. Federal Ministry of Food, Agriculture and Consumer Protection, Berlin

Laubhann D, Sterba H, Reinds GJ, de Vries W (2009) The impact of atmospheric deposition and climate on forest growth in European monitoring plots: an individual tree growth model. For Ecol Manag 258(8):1751–1761. https://doi.org/10.1016/j.foreco.2008.09.050

Leitgeb E, Reiter R, Englisch M, Lüscher P, Schad P, Feger K-H (2013) Waldböden: Ein Bildatlas der wichtigsten Bodentypen aus Österreich, Deutschland und der Schweiz. Wiley-VCH, Weinheim

Meesenburg H, Meiwes KJ, Rademacher P (1995) Long term trends in atmospheric deposition and seepage output in northwest German forest ecosystems. Water Air Soil Pollut 85(2):611–616. https://doi.org/10.1007/bf00476896

Nilsson J, Grennfelt P (1988) Critical loads for sulphur and nitrogen. Workshop on "Critical Loads for the effect on soils and groundwater of long term deposition of nitrogen and sulphur compounds and to establish methods to map the geographical areas experiencing higher than critical loads with respect to the sensitivity of different soil types". UN-ECE – Nordic Council of Ministers, Skokloster

Oliver BG, Thurman EM, Malcolm RL (1983) The contribution of humic substances to the acidity of colored natural waters. Geochim Cosmochim Acta 47(11):2031–2035

Posch M, Reinds GJ, Slootweg J (2003) The European background database. In: Posch M, Hettelingh J-P, Slootweg J, Downing RJ (eds) Modelling and mapping of critical thresholds in Europe: CCE status report 2003, vol 259101013. National Institute for Public Health and the Environment (RIVM), Bilthoven, pp 37–44

Posch M, de Vries W, Sverdrup HU (2015) Mass balance models to derive critical loads of nitrogen and acidity for terrestrial and aquatic ecosystems. In: de Vries W, Hettelingh JP, Posch M (eds) Critical loads and dynamic risk assessments: nitrogen, acidity and metals in terrestrial and aquatic ecosystems. Environmental pollution, vol 25. Springer, Dordrecht, pp 171–205. https://doi.org/10.1007/978-94-017-9508-1_6

Pretzsch H, Biber P, Schütze G, Uhl E, Rötzer T (2014) Forest stand growth dynamics in Central Europe have accelerated since 1870. Nat Commun 5. https://doi.org/10.1038/ncomms5967

R Core Team (2017) R: a language and environment for statistical computing. R Foundation for Statistical Computing, Vienna

Rademacher P, Buß B, Müller-Using B (1999) Waldbau und Nährstoffmanagement als integrierte Aufgabe in der Kiefernwirtschaft auf ärmeren pleistozänen Sanden. Forst und Holz 54:330–334

Riek W, Russ A, Martin J (2012) Soil acidification and nutrient sustainability of forest ecosystems in the northeastern German lowlands – results of the national forest soil inventory. Folia For Pol Ser A 54(3):187–195. https://doi.org/10.5281/zenodo.30835

Russ A, Riek W (2011) Pedotransferfunktionen zur Ableitung der nutzbaren Feldkapazität–Validierung für Waldböden des nordostdeutschen Tieflands. Waldökologie, Landschaftsforschung und Naturschutz 11:5–17

Santore RC, Driscoll CT, Aloi M (1995) A model of soil organic matter and its function in temperate forest soil development. In: McFee WW, Kelly JM (eds) Carbon forms and functions in forest soils. Soil Science Society America (SSSA), Madison, WI, pp 275–298

Schaap M, Banzhaf S, Scheuschner T, Geupel M, Hendriks C, Kranenburg R, Nagel H-D, Segers AJ, von Schlutow A, Wichink Kruit R, Builtjes PJH (2017) Atmospheric nitrogen deposition to terrestrial ecosystems across Germany, Biogeosciences Discuss. Dessau-Roßlau. https://doi.org/10.5194/bg-2017-491

Schlotter D, Hildebrand EE, Schack-Kirchner H (2009) Die Bodenlösung – Monitor für den Boden oder für die Methode ihrer Gewinnung? Berichte der DBG. German Soil Science Society (DBG), Bonn

Schmidt-Walter P, Ahrends B, Meesenburg H (2017) Wasserhaushalt und Trockenstress für die BWI quantifiziert. AFZ – Der Wald 72(15):36–39

Schober R (1975) Ertragstafeln wichtiger Baumarten bei verschiedener Durchforstung. Sauerländer, Bad Orb

Schrumpf M, Schulze ED, Kaiser K, Schumacher J (2011) How accurately can soil organic carbon stocks and stock changes be quantified by soil inventories? Biogeosciences 8:1193–1212. https://doi.org/10.5194/bg-8-1193-2011

Sverdrup HU, de Vries W (1994) Calculating critical loads for acidity with the simple mass-balance method. Water Air Soil Pollut 72(1–4):143–162. https://doi.org/10.1007/bf01257121

Sverdrup HU, Warfvinge P (1993) The effect of soil acidification on the growth of trees, grass and herbs as expressed by the (Ca+Mg+K)/Al ratio. Reports in ecology and environmental engineering. Lund University, Lund

Tarrasón L, Nyiri Á (2008) Transboundary acidification, eutrophication and ground level ozone in Europe in 2006. EMEP status report. The Meteorological Institute, Oslo

Thiele JC, Nuske RS, Ahrends B, Panferov O, Albert M, Staupendahl K, Junghans U, Jansen M, Saborowski J (2017) Climate change impact assessment – a simulation experiment with Norway spruce for a forest district in Central Europe. Ecol Model 346(2017):30–47

Ulrich B (1981) Destabilisierung von Waldökosystemen durch Akkumulation von Luftverunreinigungen. Der Forst- und Holzwirt 36(21):525–532

Ulrich B (1983) Soil acidity and its relations to acid deposition. In: Ulrich B, Pankrath J (eds) Effects of accumulation of air pollutants in forest ecosystems. Springer, Dordrecht, pp 127–146

Ulrich B (1987) Stability, elasticity, and resilience of terrestrial ecosystems with respect to matter balance. In: Schulze E-D, Zwölfer H (eds) Potentials and limitations of ecosystem analysis. Ecological studies 61. Springer, Berlin, pp 11–49

Vanguelova EI, Bonifacio E, De Vos B, Hoosbeek MR, Berger TW, Vesterdal L, Armolaitis K, Celi L, Dinca L, Kjonaas OJ, Pavlenda P, Pumpanen J, Puttsepp U, Reidy B, Simoncic P, Tobin B, Zhiyanski M (2016) Sources of errors and uncertainties in the assessment of forest soil carbon stocks at different scales-review and recommendations. Environ Monit Assess 188 (11):11. https://doi.org/10.1007/s10661-016-5608-5

von Wilpert K, Hartmann P, Puhlmann H, Schmidt-Walter P, Meesenburg H, Müller J, Evers J (2016) Bodenwasserhaushalt und Trockenstress. In: Wellbrock N, Bolte A, Flessa H (eds) Dynamik und räumliche Muster forstlicher Standorte in Deutschland: Ergebnisse der Bodenzustandserhebung im Wald 2006 bis 2008, vol Thünen Report 43. Johann Heinrich von Thünen Institute, Federal Research Institute for Rural Areas, Forestry and Fisheries, Braunschweig, pp 343–386

Wang T, Kang FF, Cheng XQ, Han HR, Bai YC, Ma JY (2017) Spatial variability of organic carbon and total nitrogen in the soils of a subalpine forested catchment at Mt. Taiyue, China. Catena 155:41–52. https://doi.org/10.1016/j.catena.2017.03.004

Wellbrock N, Aydın CT, Block J, Bussian B, Deckert M, Diekmann O, Evers J, Fetzer KD, Gauer J, Gehrmann J, Kölling C, König N, Liesebach M, Martin J, Meiwes KJ, Milbert G, Raben G, Riek W, Schäffer W, Wolff B et al (2006) Bodenzustandserhebung im Wald (BZE II) – Arbeitsanleitung für die Außenaufnahmen. Federal Ministry of Food, Agriculture and Consumer Protection, Bonn

Wellbrock N, Eickenscheidt N, Hilbrig L, Dühnelt P-E, Holzhausen M, Bauer A, Dammann I, Strich S, Engels F, Wauer A (2018) Leitfaden und Dokumentation zur Waldzustandserhebung in Deutschland. Thünen Working Paper, vol 84. Johann Heinrich von Thünen Institute, Braunschweig. https://doi.org/10.3220/WP1513589598000

Wirth C, Schulze E-D, Schwalbe G, Tomczyk S, Weber G, Weller E (2004) Dynamik der Kohlenstoffvorräte in den Wäldern Thüringens. Mitteilungen Thüringer Landesanstalt für Wald, Jagd und Fischerei, Gotha

Ziche D, Michler B, Fischer HS, Kompa T, Höhle J, Hilbrig L, Ewald J (2016) Boden als Grundlage biologischer Vielfalt. In: Wellbrock N, Bolte A, Flessa H (eds) Dynamik und räumliche Muster forstlicher Standorte in Deutschland. Ergebnisse der Bodenzustandserhebung im Wald 2006 bis 2008, vol Thünen Report 43. Johann Heinrich von Thünen Institute, Federal Research Institute for Rural Areas, Forestry and Fisheries, Braunschweig, pp 292–342

Chapter 2
Environmental Settings and Their Changes in the Last Decades

Nicole Wellbrock, Nadine Eickenscheidt, Erik Grüneberg, and Rebekka Bögelein

2.1 Introduction

The historical use of European forests has significantly affected forest soil conditions (Wittich 1951; Härdtle 1995; Ludemann 2002; Lüst and Giani 2006; Peters 1990; Rinklebe and Makeschin 2003). As a consequence of centuries of use by humans, until recently forests have been subject to severe losses of nutrients, for example, through harvesting of wood for fuel and lumber, bark utilization and litter raking, forest pasturing, clear cutting and slash-and-burn clearance with subsequent agricultural use and the associated erosion of the soils (Ellenberg and Leuschner 2010; Kreutzer 1972). Today, standard restrictions on forest utilization to timber harvesting have meant that these impacts have been curbed significantly (Bundeswaldgesetz). The majority of modern forests are the result of decisions based on silvicultural management that has left behind stands of multiple generations. The selection and mix of tree species and the structure of the forest stands have had a particularly significant impact (Augusto et al. 2002; Pretzsch et al. 2010). Forest management has exerted a considerable influence on the quantity and distribution of organic substances in the soils, the material cycles, deposition loading and the acid-base ratio, as well as the interior forest climate and water regime.

Over the past decades, inputs from atmospheric pollution caused by humans have had a significant impact on forests (Ellenberg 1971; Ulrich 1987; de Vries et al.

N. Wellbrock (✉) · E. Grüneberg · R. Bögelein
Thünen Institute of Forest Ecosystems, Eberswalde, Germany
e-mail: nicole.wellbrock@thuenen.de; erik.grueneberg@thuenen.de; rebekka.boegelein@gmx.de

N. Eickenscheidt
Thünen Institute of Forest Ecosystems, Eberswalde, Germany

State Agency for Nature, Environment and Consumer Protection of North Rhine-Westphalia, Recklinghausen, Germany
e-mail: nadine.eickenscheidt@lanuv.nrw.de

N. Wellbrock, A. Bolte (eds.), *Status and Dynamics of Forests in Germany*,
Ecological Studies 237, https://doi.org/10.1007/978-3-030-15734-0_2

2014; Waldner et al. 2015). At the end of the 1970s and in the early 1980s, this was first officially identified based on the condition of the crowns of the trees and was initially discussed as "forest dieback" and later as a new type of forest damage in many countries (Kauppi et al. 1990; Ulrich 1983). The three main effects of air pollution from sulphur (S) and nitrogen (N) compounds as well as ozone that were identified and discussed as reasons for this damage were (1) direct harmful effects on the assimilation organs of trees from acidic S emissions (Wentzel 1979, 1982) and ozone (Bucher 1984); (2) root damage and magnesium deficiencies resulting from soil acidification, leaching of bases and release of aluminium and heavy metal ions that are toxic to roots (Ulrich 1986, 1995); and (3) nutrient imbalances accruing due to the deposition of eutrophying N (Nihlgård 1985).

The onset of the industrial revolution has resulted in a continuous increase in anthropogenic depositions of sulphur and nitrogen in Central Europe. These permanent element inputs led to changes in the element budgets of forest ecosystems, which therefore can no longer be considered as constant but as dynamic systems. The acid-base balance and the N budget are particularly affected by anthropogenic depositions. Atmospheric S and N compounds mainly originate from industrial sources such as traffic (NO_y), industrial plants, energy conversion (SO_x, NO_y) as well as intensive agriculture (NH_x). Sulphur dioxide and NO_y form strong acids with the water contained in the atmosphere and are mostly carried to the forest soil by precipitation. The canopy absorbed parts of nitrogen, too. Deposition of NH_x also results in soil acidification due to nitrification to nitrate. Nitrate leaching can further lead to ground water pollution and cation losses. In addition, the greenhouse gas nitrous oxide may be produced during denitrification of nitrate. In natural ecosystems, N is limiting growth. However, eutrophication of forest ecosystems indicated by nutrient imbalances and changes in the species composition may be results of long-term N surplus. Besides S and N depositions, anthropogenic base cation depositions occur and could in particular be observed as alkaline-rich fly ashes in former eastern Germany. Deposition rates of S and N below which no long-term harmful effects on ecosystem structure and function are expected are based on the concept of critical loads (Nilsson and Grennfelt 1988).

The "stress hypothesis" links the effects of the deposition of substances with additional biotic and climatic factors into a complex of interactions among predisposing, triggering and contributing stress factors (Manion 1981; Schütt et al. 1984; Eichhorn et al. 2005). Therefore, the interactive effects of climate change can also be described together with the effects of air pollution and substance depositions; in such a scenario, effects can both be enhanced and mitigated (Bytnerowicz et al. 2007).

Following on the discussion of forest decline in Germany, a nationwide soil condition inventory (National Forest Soils Inventory; NFSI) was launched at the end of the 1980s with surveys taking place at approximately 1900 sampling sites in forests. Systematic inventory sampling in 8 × 8 km grids across the country allowed describing forest soil conditions representative for Germany. Combined with intensive monitoring (Level II), the NFSI represents an integral component of environmental monitoring of forest ecosystems. Results of the first National Forest Soils

Inventory (NFSI I) were published by Wolff and Riek (1996) reporting a substantial and extensive acidification and depletion of basic substances independent of substrate throughout the topsoil layer, as well as a tendency towards levelling of the chemical condition of the topsoil. Accumulation of copper and lead in the organic layer was elevated. The conclusions were supported by critical results for the nutritional status and crown condition of the trees. Elevated depositions of S, N and heavy metals were seen as the cause for the very low contents of magnesium, in particular in spruce trees, which were also evident in higher loss of needles. Transnational action to control air pollution (e.g. Convention of Long-Range Transboundary Air Pollution CLRTAP; UNECE 1979) has meant a decrease of S depositions in particular. Input of fly ash has also dropped significantly since lignite combustion, the principal source of fly ash, has declined considerably following the reunification of Germany and technical efforts (filter systems). In contrast, however, little change has been achieved in inputs of N (Waldner et al. 2014, 2015). Accordingly, atmospheric deposition must continue to be regarded as crucial (Verstraeten et al. 2012). As a consequence raised from the results of the NFSI I, forest stands especially affected by acidification were limed, and several countries began intensive forest conversion efforts.

To describe the changes of forest soil conditions over a period of 15 years, the second National Forest Soils Inventory (NFSI II) was carried out from 2006 to 2008. An integrated evaluation of the forest conditions and their soils was made possible through correlation with other surveys of crown condition, nutritional status of the trees and vegetation. In addition interpolated climate data, information about geology and substrate classes and profile descriptions with soil-type and humus-type classification are available.

2.2 Changes of Atmospheric Deposition on NFSI Plots

The modelled atmospheric depositions on NFSI plots show a considerable decline in element inputs during the last two decades (Fig. 2.1) as consequence of extensive emission reduction measures. The importance of NH_x on the total acid deposition however increased due to the strong decrease of S deposition (Schöpp et al. 2003). Therefore, reduced N inputs from agriculture represent the main source of impact for many forests in Germany in recent years. Specific regional patterns of atmospheric depositions have developed. Spatial differences depend upon the source (e.g. agriculture, traffic infrastructure), the diameter and weight of the particles, the climatic situation and the surface or canopy roughness.

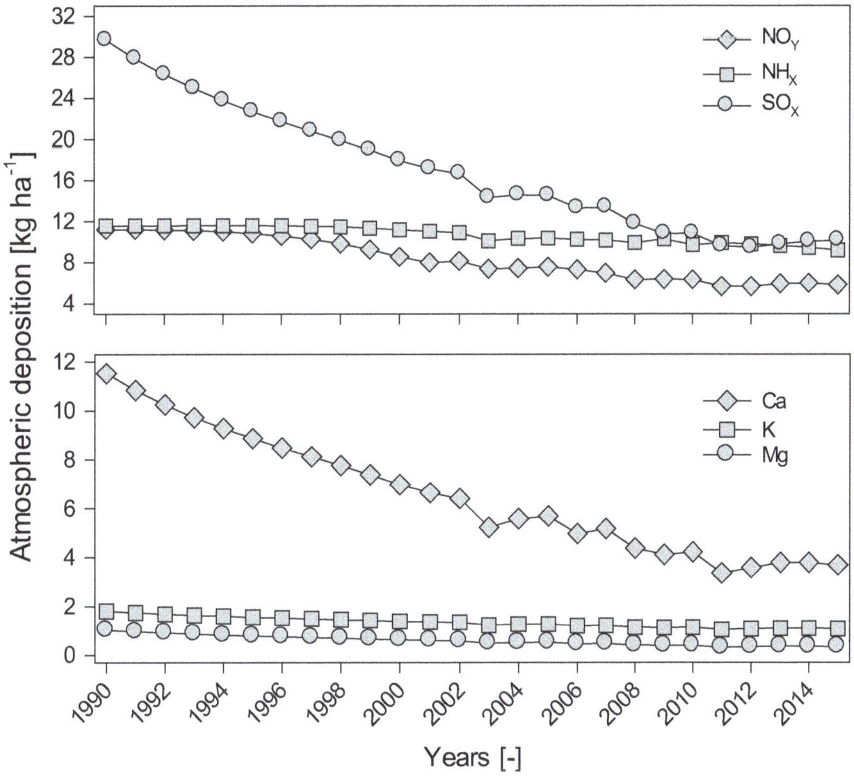

Fig. 2.1 Trends of annual atmospheric depositions of pollutants (SO_x, oxidized S; NH_x, reduced N; NO_y, oxidized N) and base cations (Ca, K, Mg) between 1990 and 2006. Depositions were sea salt-corrected with exception of N depositions

2.3 Climate

Germany is situated in the warm-temperate humid climate zone of middle latitudes in the transition between oceanic climate of Western Europe and continental climate of Eastern Europe. The oceanic climate impact decreases from West to East, while from North to South, the climate is controlled by increasing height above sea level. The lowlands are characterized by relatively warm and dry climate; however, a cool and humid climate is predominant at the mountain ridges. The slopes of the mountain ranges are characterized by locally limited warm climate because those areas are protected against cool winds due to their lower height. Climate date was derived by a method described by Ziche and Seidling (2010). Evaluating climate variables is based on the evaluation criteria by AK Standortskartierung (2003) as well as revaluations by Wolff et al. (2003). The annual mean temperature on sites of the NFSI ranged between 7 and 9 °C. The distribution of annual mean temperature showed with 1% a very small proportion of alpine but with 44% a high proportion of colline sites. Also frequently were submontane sites accounting for 29%, while montane and planare sites were less common sharing 12% and 10%, respectively. The proportion of high-montane and warm-planar sites decreased significantly with a share each of 2% (Fig. 2.2).

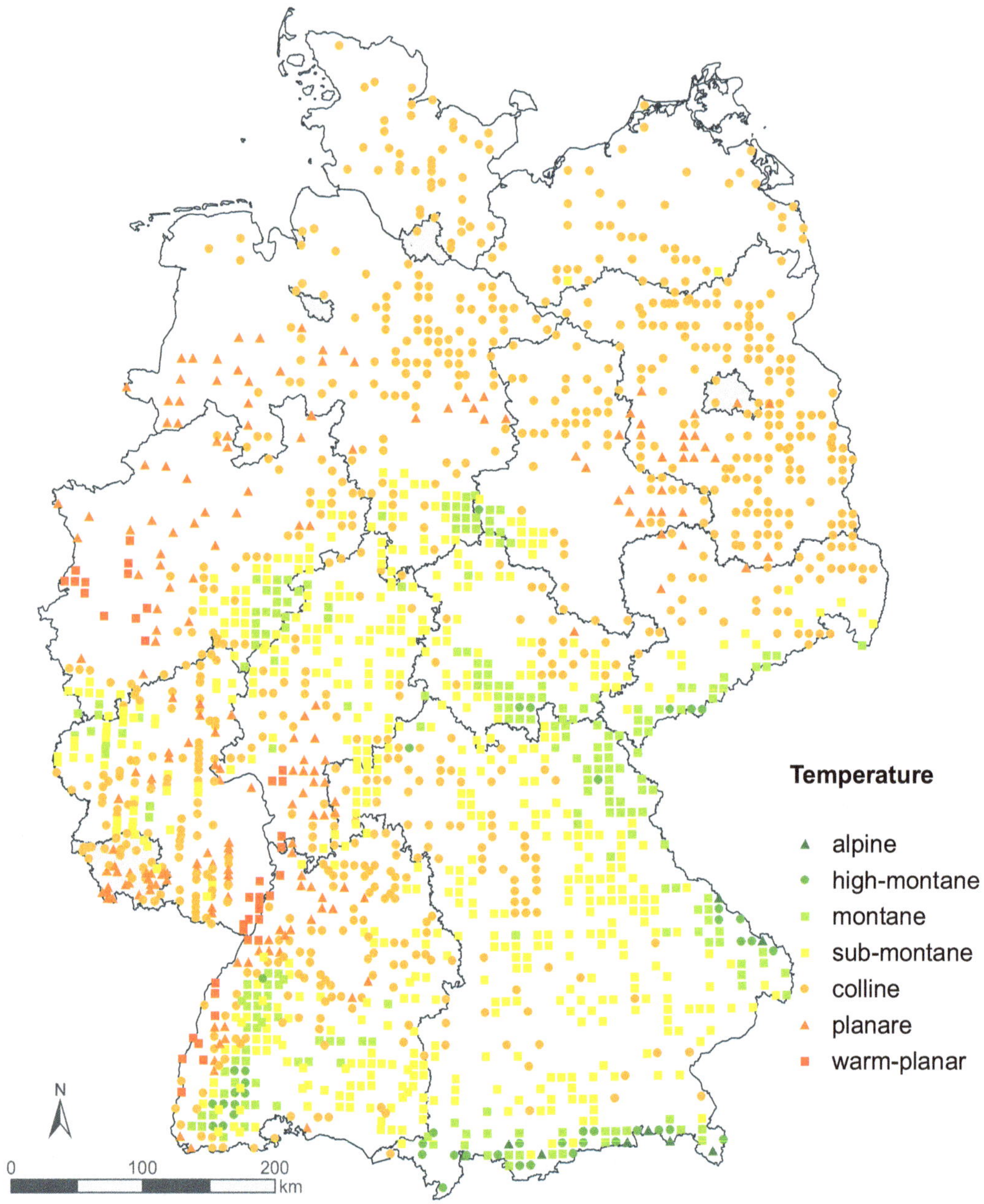

Fig. 2.2 Spatial distribution of the annual mean temperature on NFSI plots (classification according to Wolff et al. 2003)

In case of the annual mean precipitation, there were a proportion of <0.5% with only a few very dry sites; however, much more sites were dry (14%), very weak dry (14%), humid (10%) and very humid (9%). Sites with moderate precipitation were most frequent ranging from weak dry to weak humid accounting for 25% and 28%, respectively (Fig. 2.3).

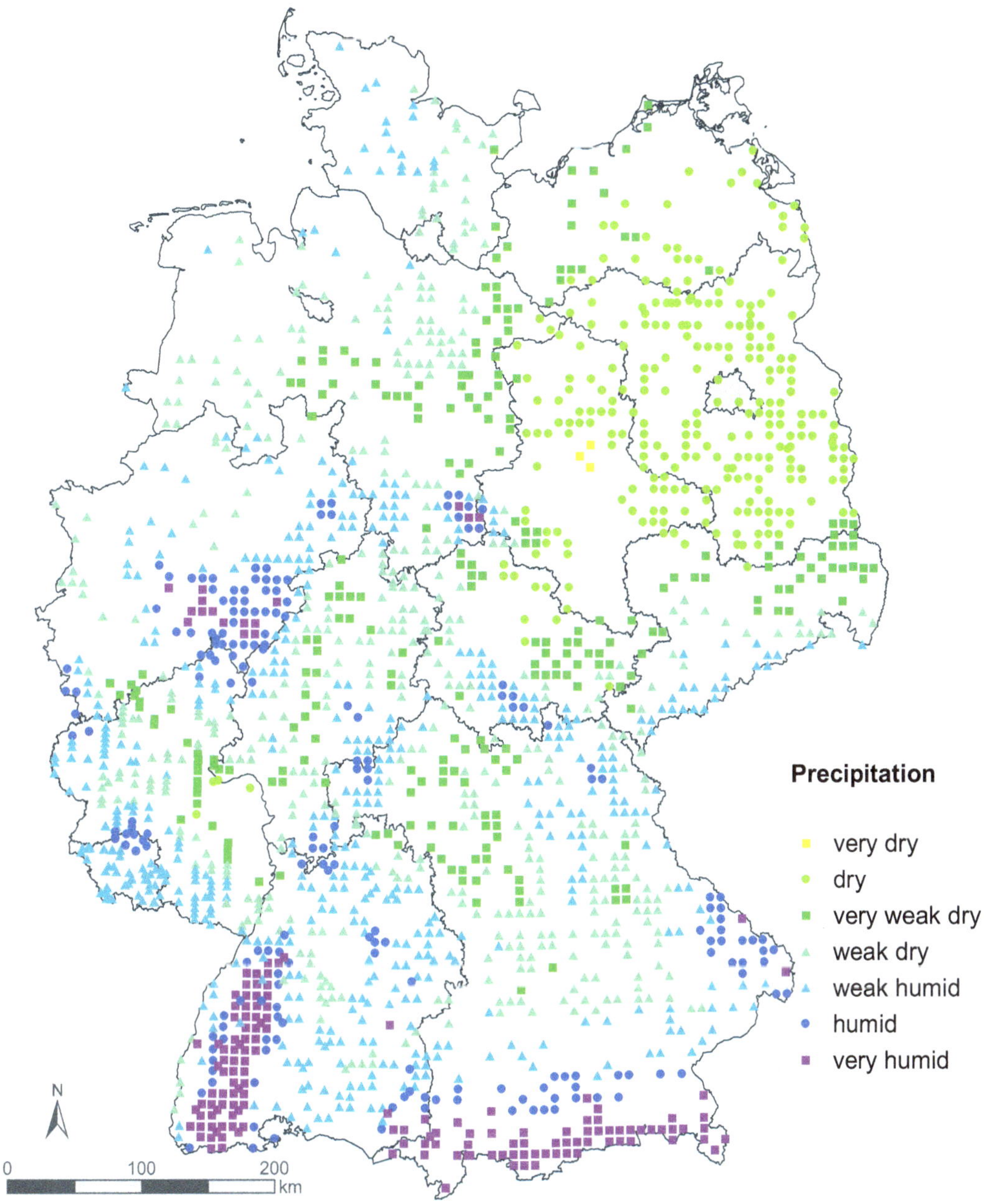

Fig. 2.3 Spatial distribution of the annual precipitation on NFSI plots (classification according to Wolff et al. 2003)

2.4 Soil Parent Material Groups

The growth of forest trees is controlled by water availability and nutrition supply which is especially determined by site properties resulting from parent materials for soil formation. We distinguished among eight groups of various soils parent materials. The largest proportion with 42% was found for soils from base-poor consolidated bedrock, followed by soils from base-poor unconsolidated sediment accounting for 21%. The latter group consists of nutrient-poor, largely sandy river deposits, periglacial sediments or river terrace deposits and is widely distributed in the North German Lowlands, while the former group comprises various acidic, crystalline and sedimentary bedrocks, partly covered or mixed with loess. These sites are most frequent in mountain ranges of middle and south Germany. Loamy soils of the lowland accounted for 13% of the survey plots, covering a broad range of various forest sites. Main distribution areas are the undulating lowlands and hilly areas with loamy-clayey, partly calcareous moraine deposits as well as sites in loess areas. About 9% of the surveyed plots were soils from weathered carbonate bedrock. Main distribution areas are the continuous mountain range built of carbonate and dolomite bedrock with their weathering products and redeposited material. Soils from basic-intermediate bedrock are small-scale distributed in hilly areas and mountain areas accounting for 8% of the surveyed plots. Additionally, soils from marlstone and claystone weathering were integrated into this group. Soils of alluvial plains contain sites in broad river valleys, including terraces and lowlands with a proportion of 3% less frequent. Typical sites comprise of frequently alternating sandy to clayey, partly calcareous fluviatile sediments. Soils from the Alps accounted for 1% of the surveyed plots. The group integrates sites of different altitudes of the Alps from limestone and dolomite as well as of noncalcareous silicate rocks. Organic soils play with a proportion of 2% a minor role in the collective. Moreover, soil development and properties as well as chemical response to environmental conditions are different from soils derived from weathered mineral matter.

2.5 Soil Classes

Soil classes (suborders) differentiate according to soil dynamics and morphological properties caused by specific dominant pedological processes. For each of the surveyed plots soil classes could be designated by describing soil profiles according to Ad-HocAG_Boden (2005). Brown soils (Cambisol/Arenosols) were the most common soil class accounted for 55% of all soils and distributed over a wide range of various soil parent materials. Soils with stagnic properties and Podzols are also characterized by a broad spectrum of site variability. These soil classes are common both in the North German Lowlands and hilly areas and mountain ranges of middle and south Germany, whereas soils with stagnic properties accounted for 11%, while Podzols accounted for 7%. Lessives are particularly suitable for crop-land; however, they accounted for 8% of the surveyed plots. The proportion of Ah/C soils and steppe soils is similar to the lessives. Nevertheless, this soil class comprises soils that are very different in physical and chemical properties. Terra calcis and pelosols amounted to a proportion of 3% and 1%, respectively. Floodplain soils and groundwater-affected soils are soil classes that are influenced by redoximorphic features and have a 4% share of total. This soil class is geographically widespread but often small-scaled distributed under forest due to their soil formation developing independently from bedrock. Terrestrial anthropogenic soils and cultivated peat soils have been altered strongly by human activity and accounted for 2% of the surveyed plots.

2.6 Humus Forms

Humus forms of German forest soils are distinguished according to the environmental conditions in terrestrial and hydromorphic humus forms. Terrestrial humus forms are formed under predominantly aerobic conditions and contain mull, moder and mor as well as their transitional forms. The hydromorphic humus forms hydromorphic mull, hydromorphic moder and hydromorphic mor as well as their transitional forms developed under intermittent anoxic conditions. The latter are hardly relevant in the NFSI collective covering 2% of the surveyed plots. Aerobic humus forms with a continuous fine humus horizon accounted for 52% of the plots, of which 16% is for mor-like moder and typical mor and 36% is for typical moder and rhizo moder. Terrestrial humus forms without a continuous fine humus horizon accounted for 46% of which 8% is for typical mull, 28% is for typical mull with Oe horizon and 10% is for moder-mull. The spatial distribution of humus forms showed a dominance of moder-like humus forms in the North German Lowland, while rhizo moder is less frequent in western parts of this landscape. Mor-like humus forms are present in the north-eastern parts of Germany, in East German mountain ranges and in the Bavarian Forest, while mull-like humus forms are dominating the middle and South of Germany (Fig. 2.4).

2.7 Types of Depth Profiles of Base Saturation

Soil acidification occurs under humid climate naturally very slowly as soil is weathered, but processes like tree uptake and acidic deposition can reduce the base saturation of the cation exchange capacity of the soil (Hartmann and von Wilpert 2016). Depending on soil properties and the supply of alkaline cations by mineral weathering, the soil profile is characterized by a depth-specific distribution of base saturation. The identified types of depth profiles of base saturation correspond with those described by von Wilpert (1996) and Kölling et al. (1996). Admittedly, an additional type which included sites that were limed was designated within the range of poor-base sites. A continuous high base saturation throughout the soil profile is represented by type 1, while type 2 showed in the upper 30 cm of the soil a moderate but below 30 cm a high base saturation. Types 1 and 2 had a share of 14% and 4%, respectively, of the surveyed plots. Moderate to low base saturation above 30 cm but moderate to high values in the soil below 30 cm provide a sufficient base supply of type 3, who accounted for 8% of the surveyed plots. Type 4 is characterized by an increasing base depletion to a depth of 60 cm when the soil below is rich in base cations. In contrast, type 5 showed a base depletion throughout the whole soil profile. The proportions of the types amounted to 21% and 46%,

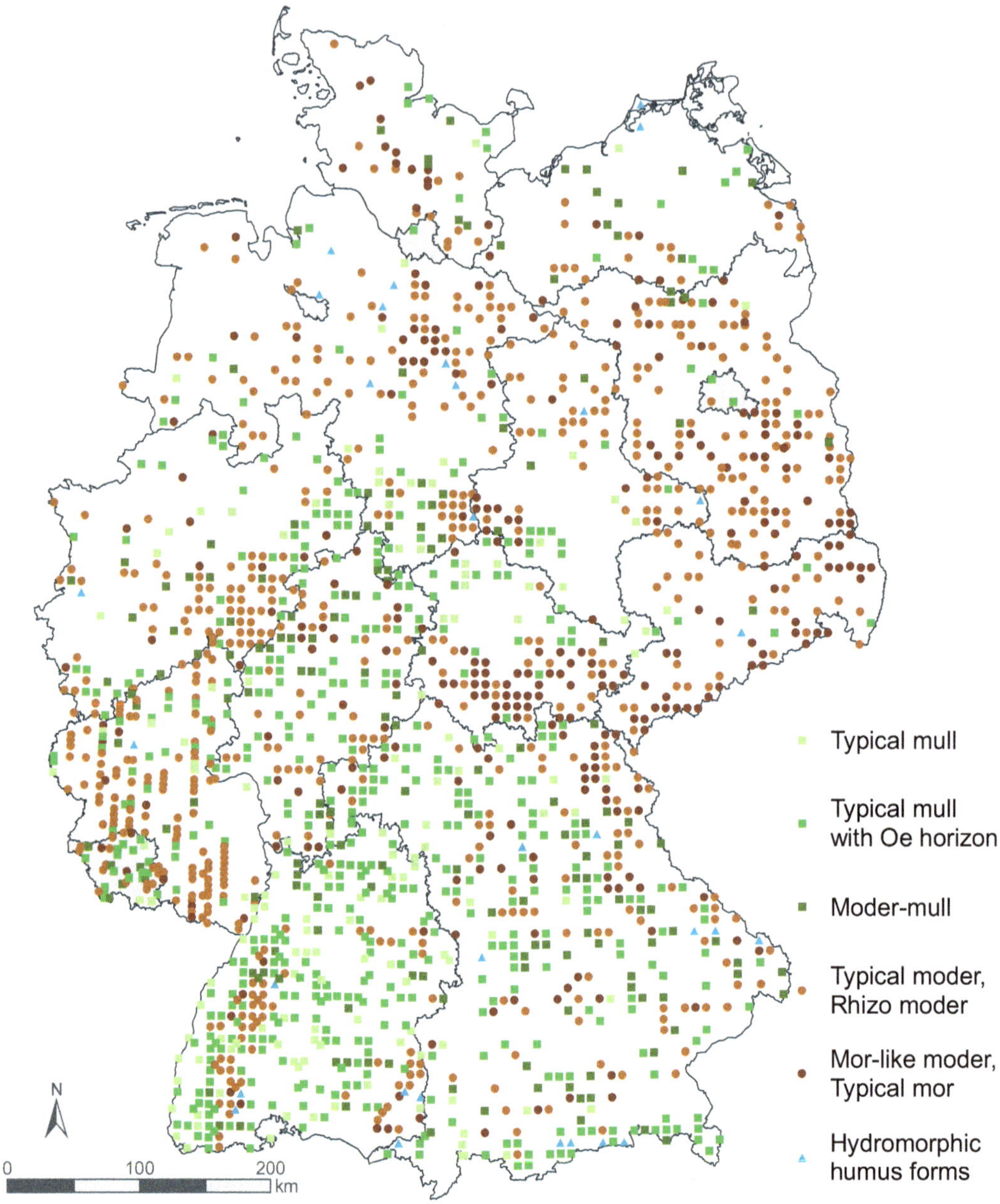

Fig. 2.4 Spatial distribution of humus forms on NFSI plots

respectively. Type 6 accounted for 7% of the surveyed plots. Since this type is very poor in base cations, liming of the forest brought base cations to organic layer and topsoil. Nevertheless, also other sites received lime; the effects, however, showed no clear effects in the mineral soil (Fig. 2.5).

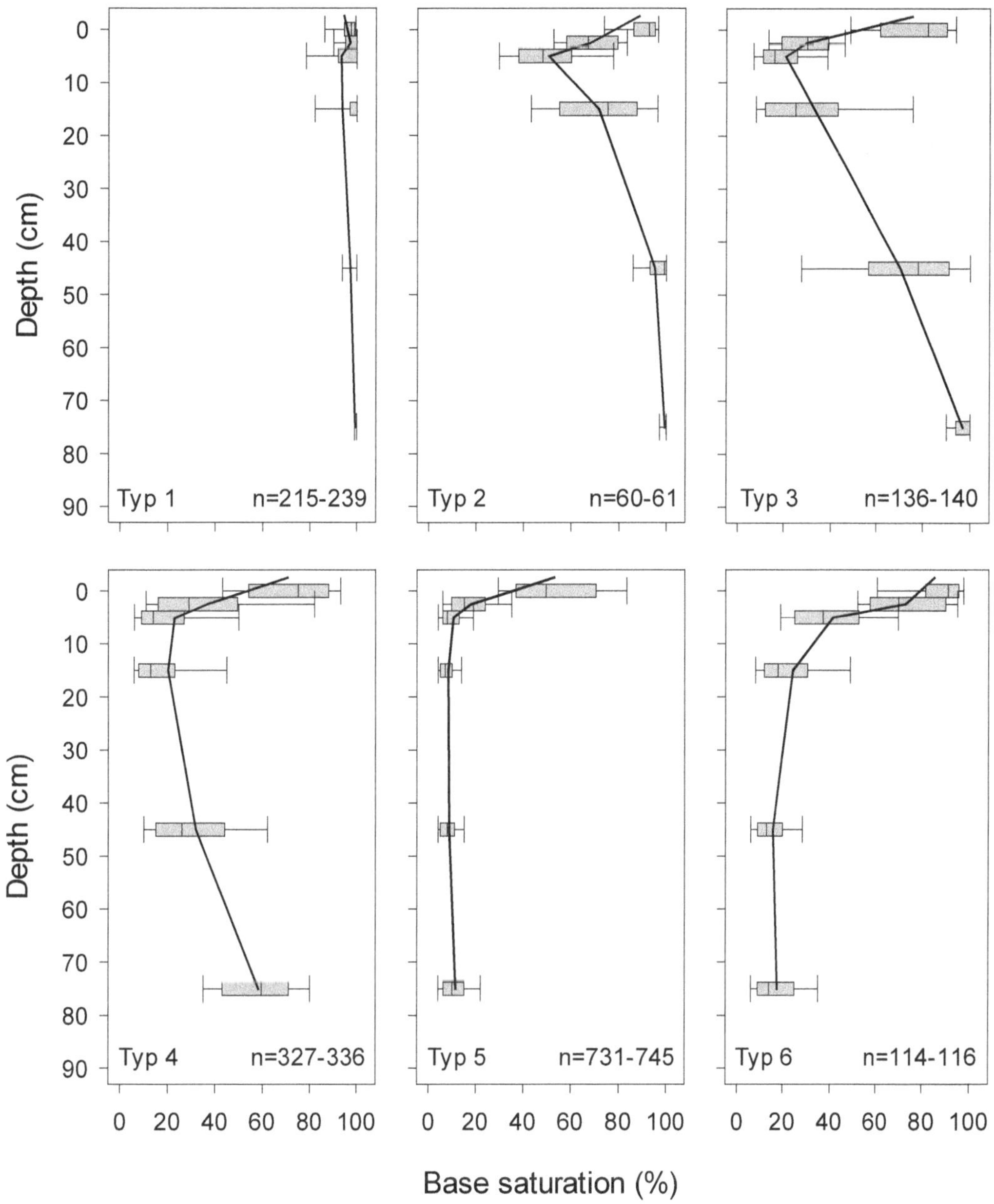

Fig. 2.5 Types of depth profiles of base saturation

2.8 Acid-Sensitive Sites

The purpose of liming forest soils in Germany was primarily the mobilization of N from a biological inactive organic layer to compensate effects of soil acidification. From the mid-1980s, most of the federal states started to lime to (1) compensate soil acid loads and current acid inputs, (2) improve growth conditions for microorganisms and root systems, (3) offset acidification-related loss of nutrient cation and imbalances in forest nutrition due to high N depositions, (4) improve vitality of forest stands and (5) protect spring and groundwater from inputs of heavy metals, toxic aluminium and/or acids. Federal state-specific liming with respect to liming concepts is suitable to lime. We defined all plots which reach the criteria for liming as sensitive to acidification (Höhle et al. 2018, pp. 11–42). Exclusively on such sites, effects of liming can be analysed by comparing limed and non-limed sites of the first and second NFSI. Of the total of 1859 surveyed plots of the NFSI II, 749 sites were acid sensitive. Of those 385 sites were limed at least once since the first inventory. Concepts adapted to site-specific requirements were developed for an efficient and long-lasting regeneration of essential soil functions. The identified acid-sensitive sites comprise soils that in respect to liming concepts are suitable to lime. Exclusively on such sites, effects of liming can be analysed by comparing limed and non-limed sites of the first and second NFSI. Of the total of 1859 surveyed plots of the NFSI II, 749 sites were acid sensitive. Of those 385 sites were limed at least once since the first inventory (Fig. 2.6).

2.9 Forest Stands

Forest stands play an important role in characterizing ecological conditions of a forest site. The forest stand type was designated for each of the NFSI plots according to the dominant principle species and mixing of various tree species, respectively. Specific forest stand information was obtained in more detail on NFSI plots with the harmonized stand inventory (HIS; Hilbrig et al. 2014). Data from the National Forest Inventory was used for surveyed plots of Bavaria (BMEL 2014). The NFSI distinguished pure stands with a share of 70% of principle tree species, while the HSI accounted for 90% of principle tree species to characterize pure stands. Dominant forest stands of the NFSI II were pure spruce “Spruce, pure” and pine “Pine, pure” stands accounting for 25% and 22% of the surveyed plots. The most frequent broadleaf tree species was beech “Beech, pure” with a share of 16%. The stands with other broadleaf species “Broadleaf, others” and mixed coniferous-broadleaf stands “Broadleaf, rich in coniferous” accounted for 10%, with a share of 6% followed oak stands “Oak, pure” and other coniferous stands “Coniferous, others”. The smallest proportion with 5% was found for mixed broadleaf-coniferous stands “Coniferous, rich in broadleaf”. The HSI revealed a proportion of 46% of coniferous stands “Coniferous, pure”. Adding the class “Coniferous, dominated” (13%) to coniferous pure stands, this is demonstrating the dominance of coniferous trees on surveyed plots. The class “Broadleaf, pure” and “Broadleaf, dominated” accounted for 29% and 8% of the surveyed plots. Even more rarely are mixed broadleaf and coniferous stands “Mixed” with a share of 4%. The spatial distribution of forest stands showed that broad areas of Northern German Lowlands are stocked with pine stands. By transgressing the mountain range, the spatial distribution of forest stands is changing towards spruce and beech as well as mixed stands, the former dominates the mountain range and the latter, however, is characterizing the undulating lowlands and hilly areas. Most oak stands are scattered distributed in western parts of Germany (Fig. 2.7).

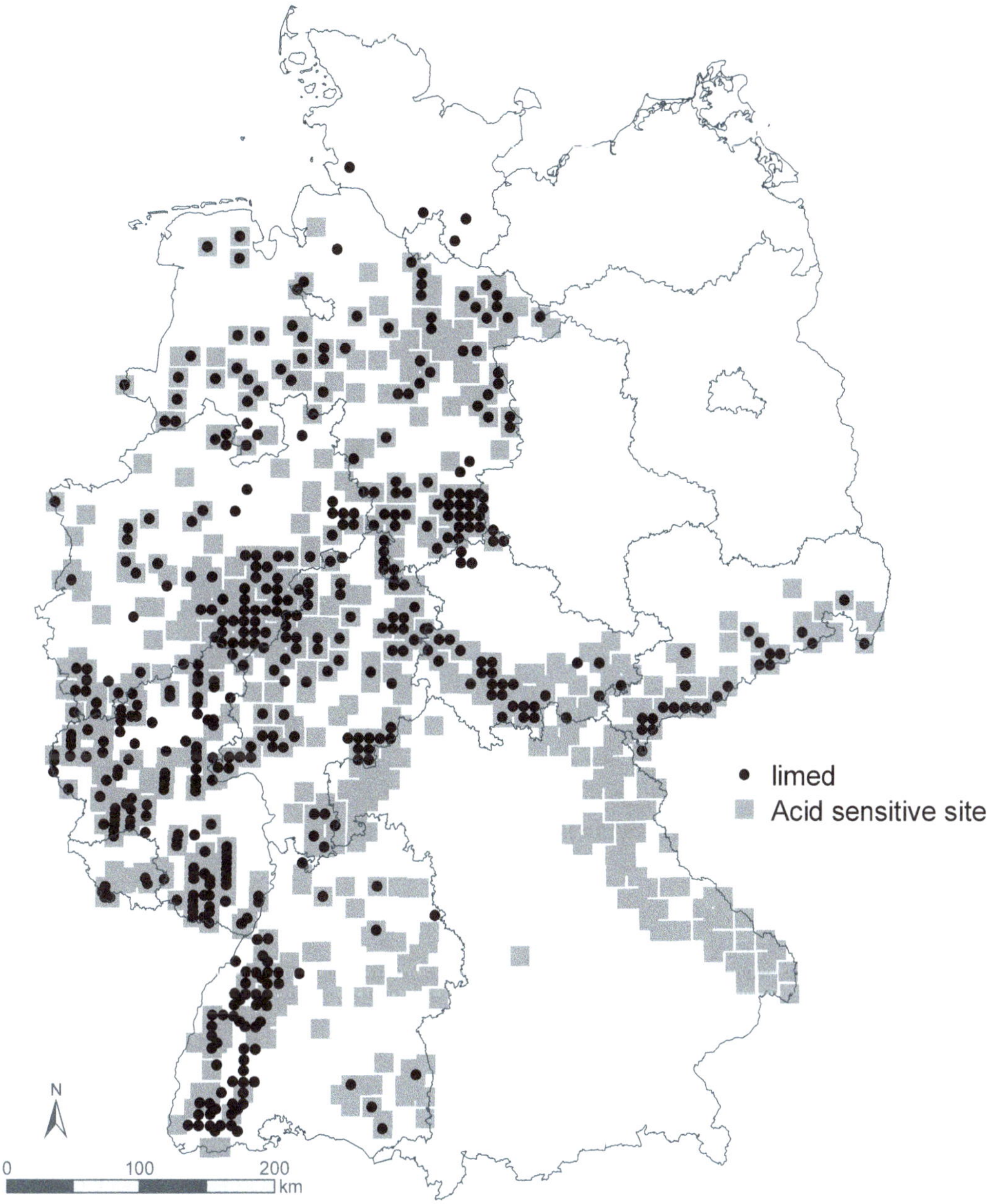

Fig. 2.6 Spatial distribution of acid-sensitive sites and limed NFSI plots

2.10 Classification of Forests Based on the Atmospheric Deposition

In order to determine the characteristic deposition patterns of forests in Germany during NFSI I and NFSI II, a factor analysis linked to a cluster analysis was performed using the modelled deposition data (Fig. 2.1). The mean of the deposition data from 1990 to 1993 were used for the period of NFSI I and the mean of 2004 to 2007 was used for the period of NFSI II. Evaluations were done using the statistic software R 3.4.1 (R Core Team 2017). Explorative maximum-likelihood factor

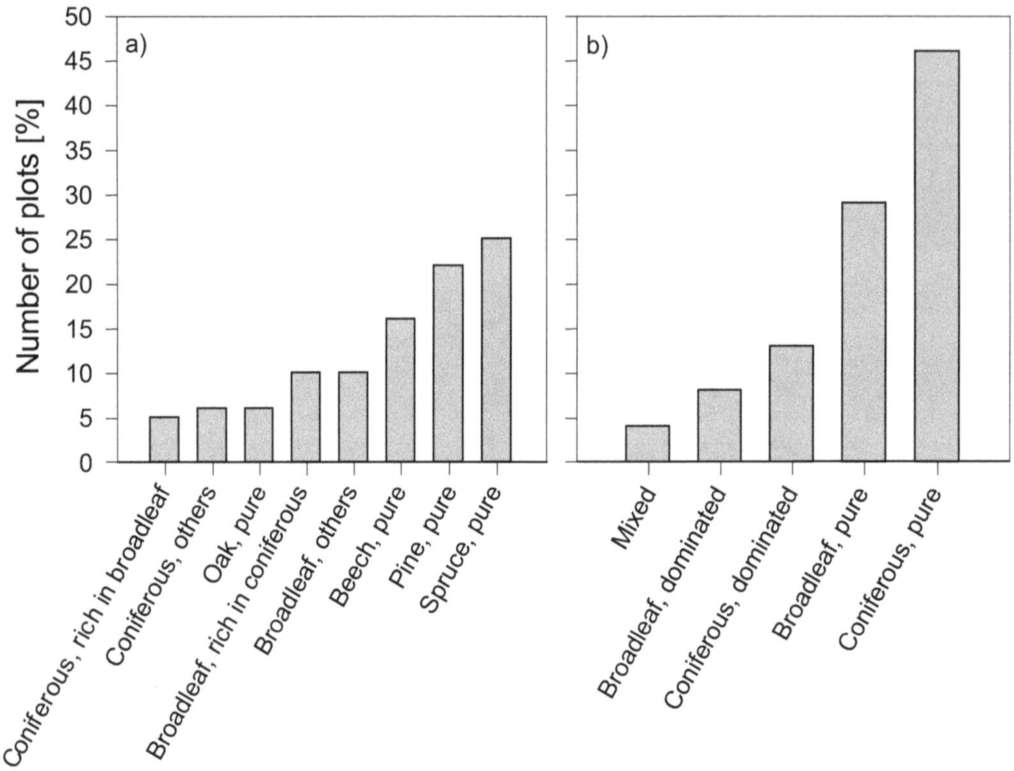

Fig. 2.7 Distribution of forest stand types at NFSI II differentiated according to (**a**) main tree species and (**b**) broadleaves and coniferous stands

Table 2.1 Loading values of the deposition factors obtained from the deposition components by factor analysis with varimax rotation

	Deposition factor 1990–1993			Deposition factor 2004–2007		
Deposition component	1	2	3	1	2	3
N_{tot}	0.63	0.77		0.89	0.43	
NH_x-N	0.21	0.96		0.98		
NO_y-N	0.99				0.98	
SO_x-S	0.61		−0.22		0.64	−0.31
Ca	0.21		0.98		0.22	0.97
Mg			0.72			0.70
K	0.42		0.31	0.32	0.49	0.32
Cl		0.46		0.48		
Explained variance (%)	26	22	21	26	23	22
Σ explained variance (%)	69			71		

analysis with varimax rotation was applied on log-transformed deposition data. The factor analysis identified three independent deposition factors for each inventory which explained 69% (NFSI I) and 71% (NFSI II), respectively, of the total variability within the deposition data (Table 2.1).

During NFSI I the factor which explained most variability was mainly loaded by oxidized N compounds (NO_y-N) but also by S compounds (SO_x-S), followed by a factor which was mainly loaded by reduced N compounds (NH_x-N) and total N

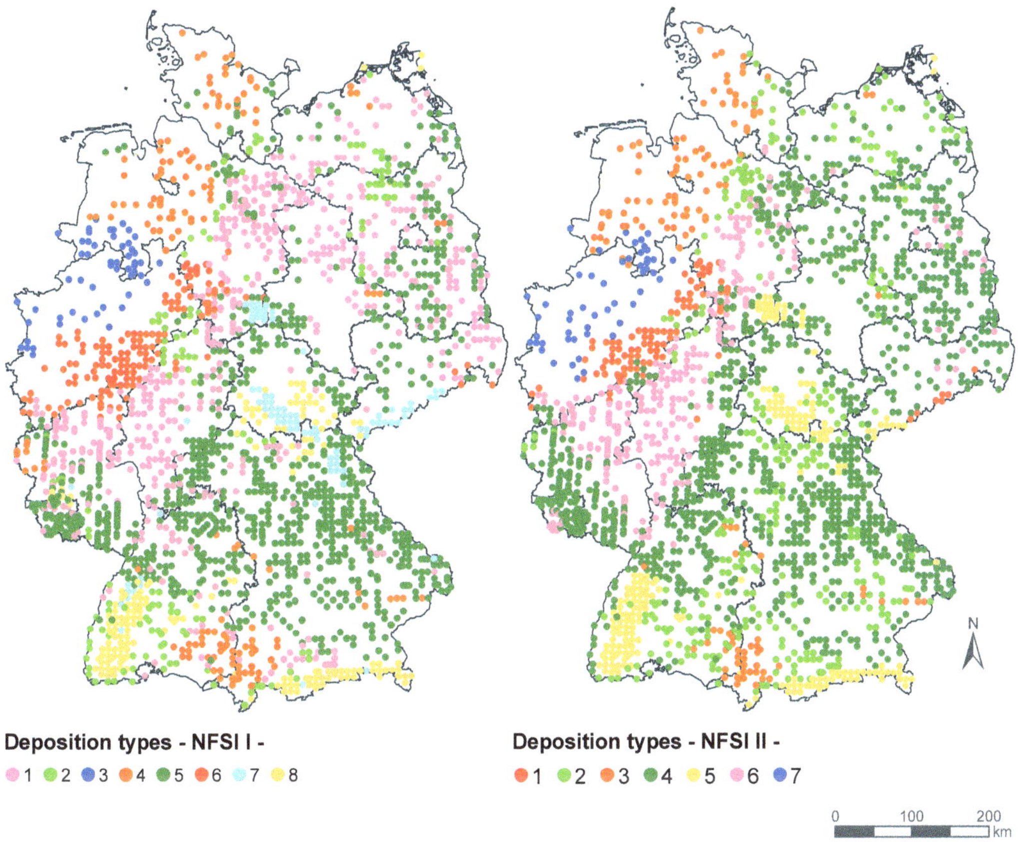

Fig. 2.8 Spatial distribution of atmospheric deposition types obtained by a model-based cluster analysis on basis of the deposition factors for the time period 1990–1993 (NFSI I, left side) and 2004–2007 (NFSI II, right side)

deposition. During NFSI II, the factor which was mainly loaded by reduced N compounds and total N deposition gained importance compared to NFSI I, whereas the factor which was mainly loaded by oxidized N compounds and S compounds was only at second place. A third deposition factor which showed similar importance during both time periods was loaded by base cations particularly by calcium and magnesium.

A model-based cluster analysis (R package mclust, version 5.3; Fraley et al. 2017) was conducted on basis of the three deposition factors for each time period. Eight clusters (atmospheric deposition types) were obtained for the first period (NFSI I) and seven clusters for the second period (NFSI II) (Fig. 2.8).

Cluster 1 (NFSI I) could be found in large areas of former eastern Germany and the adjacent eastern part of Lower Saxony as well as in the majority of Hesse and in north-eastern Rhineland-Palatinate. The cluster was characterized by average depositions from industry and traffic and low depositions from agriculture and base cations. During NFSI II the mentioned part of eastern Germany and north-eastern Lower Saxony fell into Cluster 4. The mentioned parts of south-eastern Lower Saxony, Hesse and north-eastern Rhineland-Palatinate fell into Cluster 6 (still low depositions from agriculture and base cation but higher inputs from industry and traffic compared to Cluster 4 and eastern Germany) during NFSI II.

Cluster 2 (NFSI I) comprised comparably few plots which were spread, e.g. around the Black Forest. The cluster was characterized by average depositions from agriculture and base cation and low inputs from industry and traffic. These plots mainly also fell in Cluster 2 of NFSI II. However, Cluster 2 of NFSI II could be found more widespread than Cluster 2 of NFSI I.

Cluster 3 (NFS II) could be found relative locally in the lowland of North Rhine-Westphalia and the adjacent south-western lowland of Lower Saxony. The cluster showed very high inputs from industry and traffic as well as from agriculture but very low base cation inputs. The cluster still existed during NFSI II where it was Cluster 7.

Cluster 4 (NFSI I) was located in the western part of Lower Saxony, in Schleswig-Holstein and in the border area between Baden-Wuerttemberg and Bavaria. The cluster also existed during NFSI II (Cluster 3) and was characterized by very high depositions from agriculture, average inputs of base cations and low to average inputs from industry and traffic.

Cluster 5 (NFSI I) comprised large areas especially in the southern part of Germany ranging from Bavaria to Saarland and was characterized by average inputs regarding all three input sources. The cluster was the dominating cluster during NFSI II (Cluster 4). Besides the mentioned areas, also large areas of former eastern Germany fell into this cluster during NFSI II.

Cluster 6 (NFSI I) could primarily be found in the mountainous regions of North Rhine-Westphalia but also in the adjacent mountainous regions of Lower Saxony. The cluster was present during NFSI II (Cluster 1), as well. Inputs from industry and traffic were very high, inputs from agriculture were average to high and base cation depositions were low.

Cluster 7 (NFSI I) could be found in the western part of the Harz, in the Thuringian Forest, in the Ore Mountains and in the Fichtel Mountains, which were all characterized by former intensive mining activities and very high inputs from industry and base cations. This cluster did no longer exist during NFSI II. The plots mainly fell into Cluster 5 (NFSI II), which also showed very high base cation inputs but high inputs from industry and traffic and average input from agriculture.

Cluster 8 (NFSI I) comprised the Black Forest, the Alps, parts of the Alpine Uplands and the Thuringian Basin. Base cation inputs were very high due to the calcareous parent material of the mountainous regions themselves or the adjacent areas. This cluster to a great extent still existed during NFSI II (Cluster 5) but was enlarged by the former Cluster 7 (NFSI I). Depositions from agriculture were low, and depositions from industry and traffic were average.

In conclusion, though atmospheric inputs especially from industry and traffic declined in most regions, regional patterns of deposition exhibited little changes between NFSI I and NFSI II in particular regarding former Western Germany. However in Bavaria some plots changed the deposition type (Cluster 5 of NFSI I to Cluster 2 of NFSI II) due to a decrease of depositions from industry and traffic accompanied by an increase of depositions from agriculture. Reductions in inputs from industry and traffic further resulted in a change of deposition type of several plots in the north-eastern lowlands. Stronger alteration in the deposition types however could be observed in the mountainous areas characterized by intensive mining activities and industry in the past (Harz, Thuringian Forest, Ore Mountains, Fichtel Mountains). The western part of the Harz was additionally affected by pollution derived from the Ruhr Area. Cluster 7 became no longer necessary during NFSI II due to a significant decrease in SO_x, NO_y and base cation depositions. A factor which was mainly loaded by reduced N compounds (NH_x-N) and total N deposition became more important. Thus, the importance of emissions from industry and traffic slightly decreased, while the importance of emissions from agriculture increased.

2.11 Critical Loads for Eutrophication and Acidification and Their Exceedance

Continuous immissions of the airborne pollutants sulphate, nitrate and ammonia can cause indirect (soil-mediated) damage to forest ecosystems by acidification, eutrophication and nutrient imbalances. Critical loads of S and N are defined as the annual deposition rate below which no long-term harmful effects on ecosystem structure and function are expected according to present knowledge (Nilsson and Grennfelt 1988). The starting points for critical loads calculations are the receptor systems in need for protection. In forest ecosystems, the considered receptors are trees, ground vegetation and soil. Critical limits are no-effect concentration thresholds in the soil solution used to calculate critical loads with a simple mass balance approach (see Chap. 1). Regarding acidification, critical limits are set to prevent toxic reactions of solved acidogenic cations (Al^{3+}, H^+) on tree roots, the shift to a site-specific atypical buffer range, irreversible damage of soil structure and decline of base saturation under a tolerable threshold for the plant community. Critical limits for nutrient N are set to prevent changes in the plant community and nutrient imbalances in trees. Waldner et al. (2015) showed that exceedance of critical loads for nutrient nitrogen ($CL_{nut}(N)$) leads to nutrient imbalances in the foliage.

2.11.1 Parameters for Critical Loads Calculation Derived from NFSI II Data

Estimated and measured parameters used for critical loads calculation are based on NFSI data (Höhle et al. 2018). Estimated base cation (Ca, Mg, K) weathering rate (Bc_w) varied substantially with soil parent material and rooting depth with a median of 0.8 keq ha^{-1} year^{-1} (Table 2.2). Base cation and N uptake (Bc_u, N_u) amounted to 0.5 ± 0.00 keq ha^{-1} year^{-1} and 5.9 ± 0.04 kg ha^{-1} year^{-1}, respectively.

Table 2.2 Parameters for critical loads calculation and critical limits

Parameter		Min	P_{25}	P_{50}	P_{75}	Max
Bc_w	keq ha^{-1} year^{-1}	0.03	0.44	0.80	1.50	13.44
Bc_u	keq ha^{-1} year^{-1}	0.12	0.37	0.46	0.58	0.85
N_u	kg ha^{-1} year^{-1}	1.54	4.60	5.79	7.05	13.27
N_{de} mod	kg ha^{-1} year^{-1}	0.00	0.11	0.40	1.17	61.70
N_{de} cons	kg ha^{-1} year^{-1}	0.00	0.04	0.11	0.25	17.78
N_i glacial	kg ha^{-1} year^{-1}	0.18	0.32	0.47	0.84	5.41
N_i periglacial	kg ha^{-1} year^{-1}	0.04	0.20	0.27	0.37	1.78
$ANC_{le,crit}$	keq ha^{-1} year^{-1}	−7.77	−1.02	−0.40	−0.03	3.11
$N_{le,acc}$ mod	kg ha^{-1} year^{-1}	0.00	0.81	2.22	5.24	59.99
$N_{le,acc}$ cons	kg ha^{-1} year^{-1}	0.00	0.25	0.62	1.21	7.10

Bc_w base cation weathering rate, Bc_u base cation uptake, N_u nitrogen uptake, N_{de} *mod* denitrification rate in the modified approach, N_{de} *cons* denitrification rate in the conservative approach, N_i *glacial* nitrogen immobilization rate on glacial sites, N_i *periglacial* nitrogen immobilization rate on periglacial sites, $ANC_{le,crit}$ critical leaching of acid-neutralizing capacity, $N_{le,acc}$ *mod* acceptable N leaching in the modified approach, $N_{le,acc}$ *cons* acceptable N leaching in conservative approach

Denitrification rate (N_{de}) was higher in the modified approach according to Balla et al. (2013) and ARGE_StickstoffBW (2014) than with conservative critical limits assumptions (Table 2.2, see Chap. 1). Nitrogen immobilization rate (N_i) was estimated as 0.5 and 0.3 kg ha^{-1} year^{-1} on glacial and periglacial sites referring to the last glacial maximum, respectively.

2.11.2 *Critical Limits and Critical Loads*

The critical leaching of acid-neutralizing capacity ($ANC_{le,crit}$; see Chap. 1) was calculated from the most stringent of five chemical soil criteria. The criterion to prevent aluminium toxicity on tree roots allowed the least leaching rates on most plots in siliceous low mountain ranges and the Northern German Plain. The critical base cation to aluminium ratio $(Bc/Al)_{crit}$ (see Chap. 1) was effective on 62% of the plots. It was most important for pine stands (83%), which are more vulnerable to aluminium toxicity, but only relevant for 23% of other broadleaf stands. The critical base cation to proton ratio $(Bc/H)_{crit}$ (see Chap. 1) was relevant for 71% of the peat and peaty mineral soils (2% of the NFSI plots, mostly covered by other broadleaf stands). A critical pH to preserve the soil typical buffer range was only considered for soils with base saturation >30% and was relevant for 84% of those (33% of the NFSI plots). Accordingly, it was underrepresented on all coniferous and overrepresented on all broadleaf stands. The critical aluminium mobilization rate (Al_w) to maintain soil structure determined $ANC_{le,crit}$ on only 1% of all NFSI plots, mostly other coniferous forests (13% of those). Critical base saturation was also relevant for only 1% of the plots. Median $ANC_{le,crit}$ was at −0.4 keq ha^{-1} year^{-1} (Table 2.2).

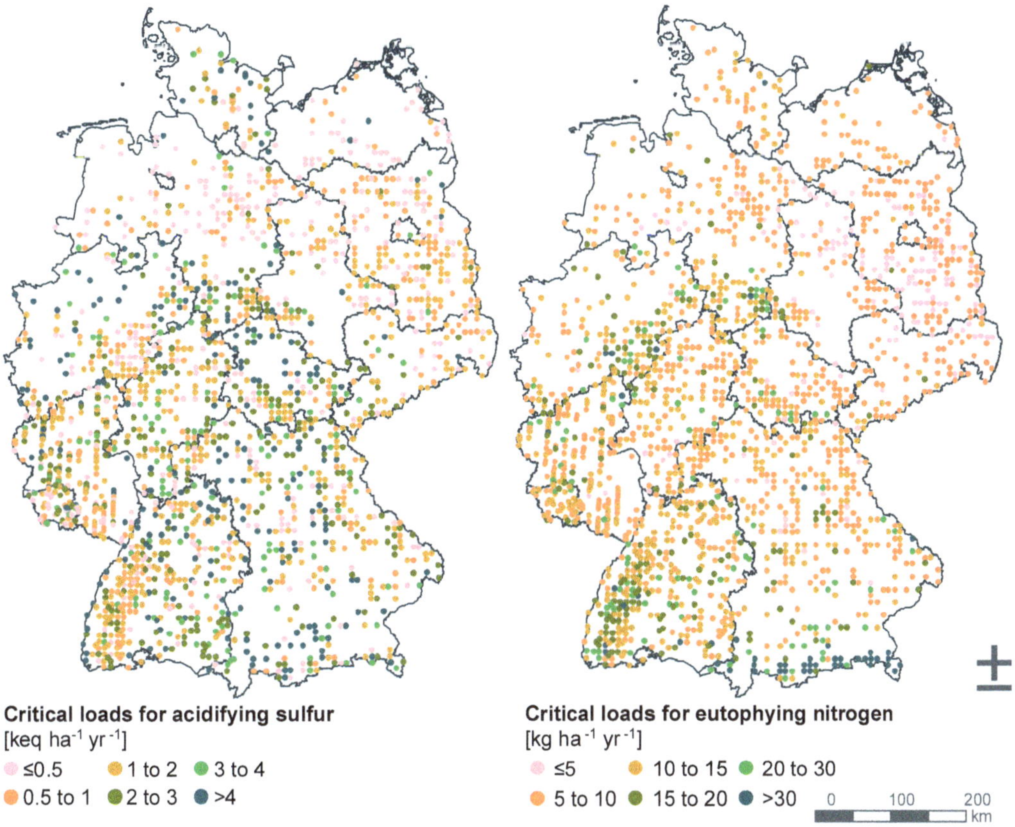

Fig. 2.9 Geographical distribution of critical loads for acidifying sulphur ($CL_{max}(S)$) and eutrophying nitrogen ($CL_{nut}(N)$) on NFSI plots

The magnitude and geographical distribution of critical loads for acidifying sulphur $CL_{max}(S)$ (see Chap. 1, Fig. 2.9) were strongly determined by the weathering rate of base cations ($r = 0.91$). Consequently, $CL_{max}(S)$ was lowest on soils from base-poor unconsolidated sediment and organic soils and peaty mineral soils (1.2 ± 0.1 keq ha^{-1} year^{-1} and 1.1 ± 0.5 keq ha^{-1} year^{-1}, respectively) and highest on loessic loam in the lowlands and calcareous soils (3.3 ± 0.2 keq ha^{-1} year^{-1} and 3.9 ± 0.2 keq ha^{-1} year^{-1}, respectively). Mean $CL_{max}(S)$ ranged from 1.3 ± 0.1 keq ha^{-1} year^{-1} on pine stands up to 3.3 ± 0.4 keq ha^{-1} year^{-1} on other broadleaf forest stands (Fig. 2.10).

Critical limits of the N concentration in the soil solution for eutrophying nitrogen (N_{crit}) were calculated according to the modified approach (see Chap. 1) as 1.3 ± 0.03 mg l^{-1} for near-natural forest ecosystems and 1.5 ± 0.04 mg l^{-1} for planted forests. The resulting acceptable N leaching ($N_{le,acc}$) amounted to 4.0 ± 0.12 kg ha^{-1} year^{-1} as compared to 0.9 ± 0.02 kg ha^{-1} year^{-1} when computed from the conservative approach with N_{crit} = 0.2–0.4 mg l^{-1} N (Table 2.2). The high limits of leaching rates in the modified approach may not protect the ecosystems from nutrient imbalances in all cases. We therefore consider non-exceedance of $CL_{nut}(N)$ but exceedance of $CL_{nut}(N)_{cons}$ from the conservative approach (see Chap. 1) as risk of potential exceedance.

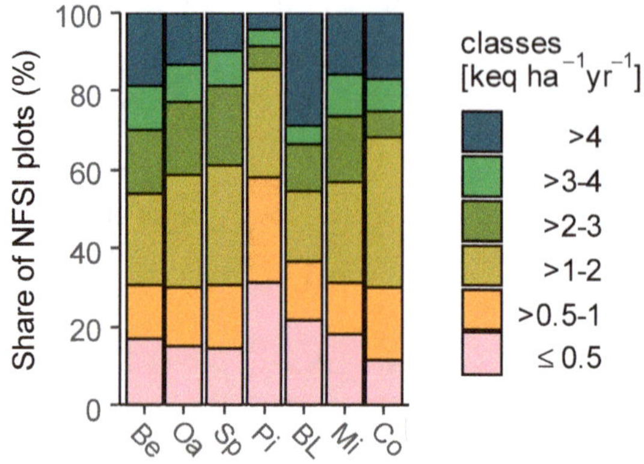

Fig. 2.10 Relative frequency of critical loads for acidifying sulphur ($CL_{max}(S)$) on NFSI plots on different forest stand types. *Be* beech $n = 226$, *Oa* oak $n = 97$, *Sp* spruce $n = 424$, *Pi* pine $n = 353$, *BL* other broadleaf stand $n = 101$, *Mi* mixed stand $n = 503$, *Co* other coniferous stand $n = 60$

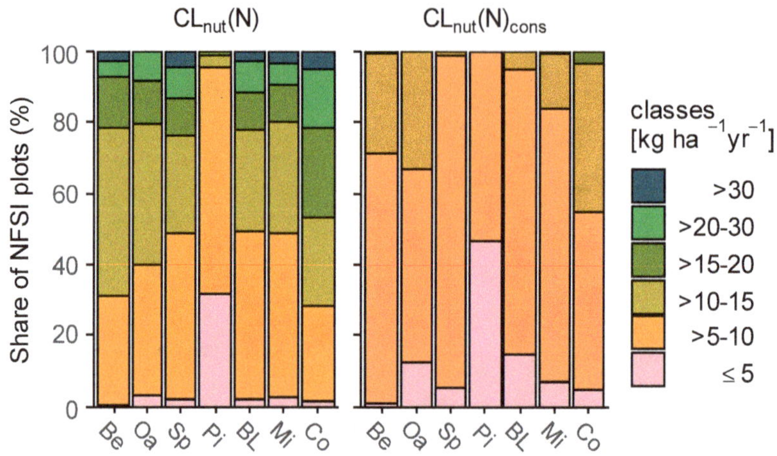

Fig. 2.11 Relative frequency of critical loads for eutrophying nitrogen ($CL_{nut}(N)$) calculated after the modified ($CL_{nut}(N)$) and the conservative approach ($CL_{nut}(N)_{cons}$) on NFSI plots on different forest stand types (Fig. 2.10)

The mean $CL_{nut}(N)$ was determined at 11.3 ± 0.17 kg ha^{-1} year^{-1}, with considerable variation among forest stand types. It ranged from 6.0 ± 0.12 kg ha^{-1} year^{-1} for pine stands, followed by oak, other broadleaf and mixed stands (11.9 ± 0.48, 12.0 ± 0.63 and 12.0 ± 0.34 kg ha^{-1} year^{-1}, respectively), over 13.0 ± 0.40 kg ha^{-1} year^{-1} for beech and spruce stands, up to 15.7 ± 1.19 kg ha^{-1} year^{-1} for other coniferous stands (Fig. 2.11). These modelled values are in good agreement with empirical critical loads (Bobbink and Hettelingh 2011). Values for $CL_{nut}(N)_{cons}$ were generally lower (7.3 ± 0.05 kg ha^{-1} year^{-1}), with similar distribution among the stand types (Fig. 2.11). For the geographical distribution of $CL_{nut}(N)$ (Fig. 2.11), seepage water is the most important influential factor ($r = 0.66$).

Almost all NFSI plots (91%) are more vulnerable to the eutrophying effect of N than to its acidifying effect ($CL_{nut}(N) < CL_{max}(N)$).

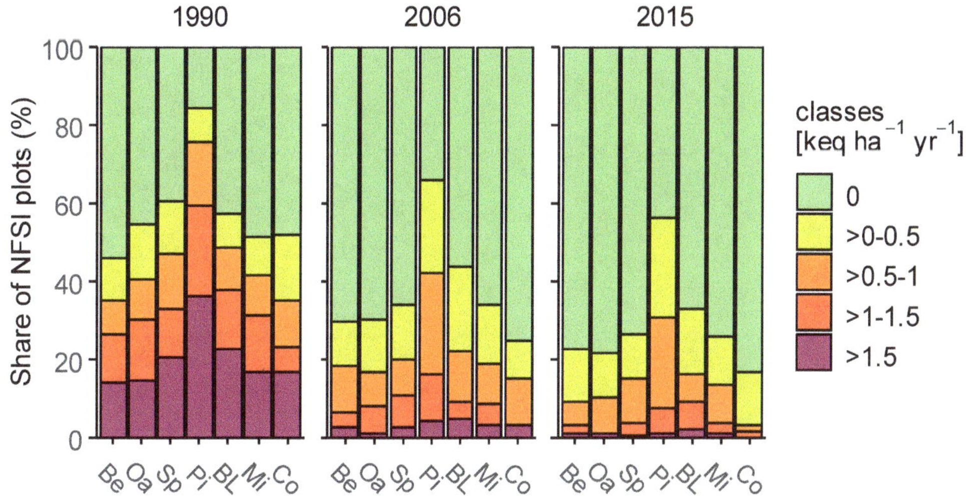

Fig. 2.12 Relative frequency of exceedance of the CL function for acidity (CLF_{aci}) on NFSI plots in 1990, 2006 and 2015 on different forest stand types (Fig. 2.10)

2.11.3 Exceedance of Critical Loads

Since 1990, the measures of air pollution control succeeded in reducing extreme exceedances of the critical load function for acidity (CLF_{aci}, see Chap. 1) (Fig. 2.12). Nevertheless, 40% of the NFSI plots were still exposed to loads of acidifying S and N beyond the long-term buffering capacity in 2006. Most threatened were pine stands with exceedance on 66% of the plots in 2006. Most compromised were the forests in north-western and eastern Germany (Fig. 2.13).

The exceedance of CLF_{aci} was higher for S than for N on 84% of the compromised plots in 1990 and still on 68% in 2006. In the meantime, the share of compromised plots where the exceedance of acidifying N went far beyond S exceedance (ExN-ExS >0.1 keq ha^{-1} year^{-1}) increased from 12 to 25%, while the share of surpassing S exceedance decreased from 21 to 4%. This development reflects the increasing role of N as acidifying pollutant, which is most substantial in the regions of high agricultural N emissions. The further development from 2006 to 2015 is characterized by reduction of extreme exceedances, while a substantial share of plots with moderate exceedance remained (Fig. 2.12).

Since 1990, the extreme exceedances (>10 kg ha^{-1} year^{-1}) of CL_{nut}(N) declined due to the reduction of N emissions (Fig. 2.14). However, in 2006, 85% of the NFSI plots were still exposed to critical N deposition, causing prospective vegetation changes and/or nutrient imbalances, and only 4% were beyond the risk of potential N exceedance. The highest exceedances occurred in the north-western lowlands and in central/eastern Germany on sandy soils and acid bedrocks (Fig. 2.13). The threat was most pronounced for pine stands with N exceedance on 99% of the plots and exceedance >10 kg ha^{-1} year^{-1} on 81% (Fig. 2.13). The share of plots with definite N exceedance of the other forest stand types ranged from 56% for other coniferous stands (2% without risk of exceedance) up to 80% for other broadleaf forests (0%

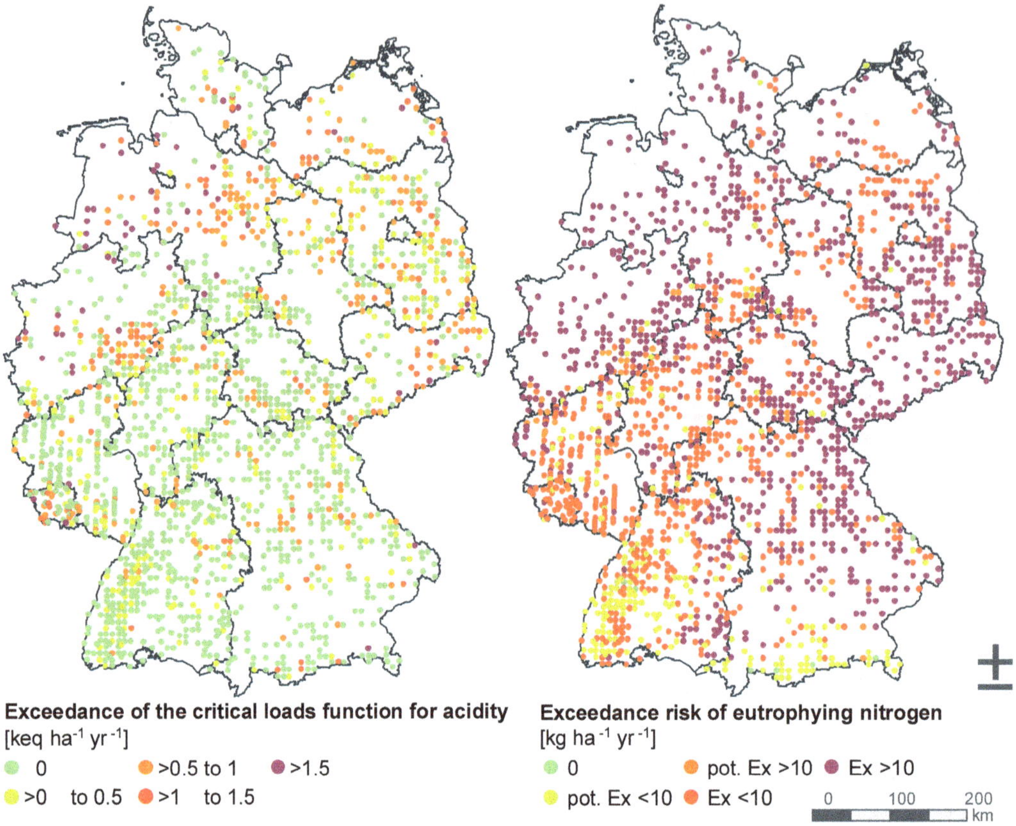

Fig. 2.13 Geographical distribution of the exceedance of the CL function for acidity (CLF_{aci}) and the exceedance risk of eutrophying N (*Ex* exceedance, *pot.Ex* potential exceedance) on NFSI plots in 2006. Potential exceedance was calculated from the conservative approach with $N_{crit} = 0.2–0.4$ mg l^{-1} N

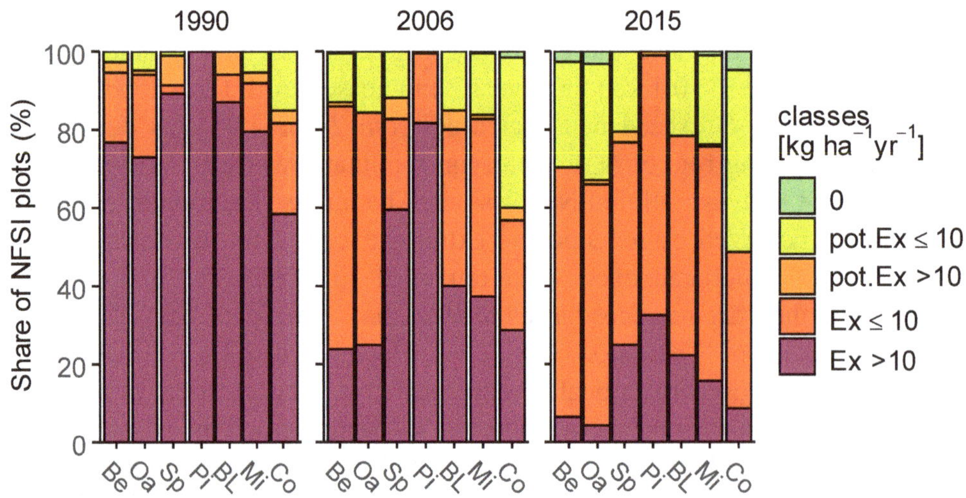

Fig. 2.14 Relative frequency of exceedance risk of eutrophying N (*Ex* exceedance, *pot.Ex* potential exceedance) on NFSI plots in 1990, 2006 and 2015 on different forest stand types (Fig. 2.10). Potential exceedance is calculated from the conservative approach with $N_{crit} = 0.2–0.4$ mg l^{-1} N

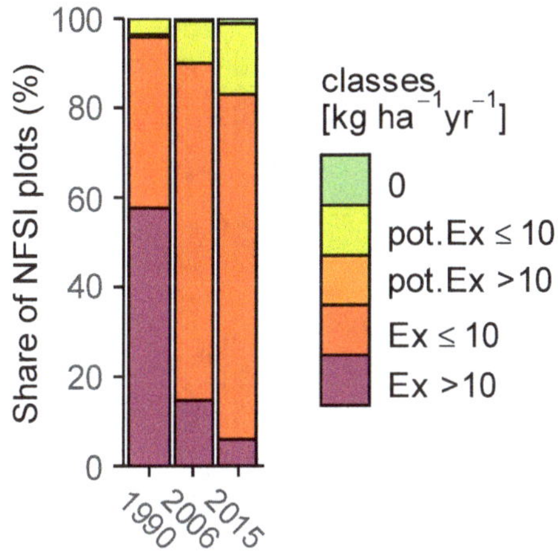

Fig. 2.15 Relative frequency of exceedance risk of the CL function for acidification and eutrophication for two pollutants (Ex, exceedance; pot.Ex, potential exceedance) on NFSI plots in 1990, 2006 and 2015. Potential exceedance is calculated from the conservative approach with $N_{crit} = 0.2–0.4$ mg l^{-1} N. Not visible are negligible shares of the classes "0" (1990: 0.06, 2006: 0.23) and "pot.Ex >10" (1990: 0.51, 2006: 0.17, 2015: 0.06)

without risk of exceedance) in 2006. Exceedances >10 kg ha^{-1} year^{-1} occurred on 24–28% of the beech, oak and other coniferous stands and on 37–59% of the mixed, other broadleaf and spruce stands in 2006 (Fig. 2.14). In 2015, extreme exceedances of CL_{nut}(N) were further reduced, while most of the NFSI plots were still exposed to moderate or at least potential exceedance.

The integrated evaluation of critical loads exceedances of acidifying and eutrophying depositions (CLRTAP 2015) showed that, despite the reduction of S and NO_x emissions, 90% of the NFSI plots were still compromised in 2006 and only 0.2% were beyond the risk of exceedance (Fig. 2.15).

Exposure types (Gauger et al. 2008) classify the plots according to the need for deposition reductions (Fig. 2.16). The sole reduction of S deposition at constant N deposition rates would result in non-exceedance of the CL function on only a minimal share of the NFSI plots (1% in 1990 and 2% in 2006). From 1990 to 2006, only the share of plots with need for reduction of S deposition (at additional need for reduction of N deposition) was reduced substantially (from 35 to 18%), while the share of plots with need for reduction of N input remained at an alarmingly high level (95% in 1990 and 88% in 2006) (Fig. 2.16). This trend continued until 2015.

2.12 Summary and Conclusions

The spatial variability of environmental impact factor like climate or tree species distribution or soil class in Germany is high. That leads to small-scale pattern of characteristics of forest soils. Forest policy like liming is under the responsibility of the federal states; that is why impact factors change at borders of different federal

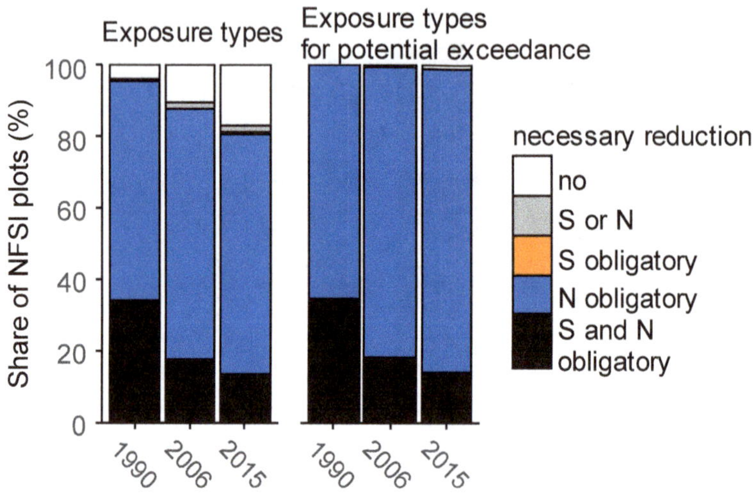

Fig. 2.16 Relative frequency of exposure types for exceedance and for potential exceedance (calculated from the conservative approach) of the CL function for two pollutants on NFSI plots in 1990, 2006 and 2015. Not visible are negligible shares of exposure types of the classes "S or N" (1990: 0.06) and "S obligatory" (1990: 0.34, 2006: 0.28, 2015: 0.40) and exposure types for potential exceedance of the classes "no" (1990: 0.06, 2006: 0.23, 2015: 1.02), "S or N" (2006: 0.11, 2015: 0.06) and "S obligatory" (2015: 0.06)

states. Furthermore many important impact factor changes within the NFSI. The exposure to airborne pollutants decreased in the interval between the first and the second inventory. The situation improved mainly for acidifying S depositions and for extreme exceedances of eutrophying N input. However, more than one third of the investigated plots were still exposed to critical rates of acidifying depositions and almost all plots to critical rates of eutrophying N prior to the second inventory. The contribution of N to acidification increased, and these trends continued since the second NFSI. That leads to small-scale pattern of characteristics of forest soils which were investigated in the following chapters.

References

Ad-HocAG_Boden (ed) (2005) Bodenkundliche Kartieranleitung (KA 5), vol 5. Schweizerbart'sche Verlagsbuchhandlung, Stuttgart

AK Standortskartierung (2003) Forstliche Standortsaufnahme: Begriffe, Definitionen, Einteilungen, Kennzeichnungen, Erläuterungen, 6th edn. IHW-Verlag, Eiching near Munich

ARGE_StickstoffBW (ed) (2014) Ermittlung standortspezifischer Critical Loads für Stickstoff – Dokumentation der Critical Limits und sonstiger Annahmen zur Berechnung der Critical Loads für bundesdeutsche FFH-Gebiete, vol ID Umweltbeobachtung U26-S7-N12. Arbeitsgemeinschaft Stickstoff, federal state Baden-Württemberg, Karlsruhe

Augusto L, Ranger J, Binkley D, Rothe A (2002) Impact of several common tree species of European temperate forests on soil fertility. Ann For Sci 59(3):233–253. https://doi.org/10.1051/forest:2002020

Balla S, Müller-Pfannenstiel K, Uhl R, Kiebel A, Lüttmann J, Lorentz H, Düring I, Schlutow A, Schleuschner T, Förster M, Becker C, Herzog W (2013) Untersuchung und Bewertung von straßenverkehrsbedingten Nährstoffeinträgen in empfindliche Biotope. Forschung Straßenbau und Straßenverkehrstechnik, vol 1099. Federal Highway Research Institute (BASt), Bremen

BMEL (2014) Der Wald in Deutschland – Ausgewählte Ergebnisse der dritten Bundeswaldinventur. Bundesministerium für Ernährung und Landwirtschaft, Berlin, 54 p

Bobbink R, Hettelingh J-P (2011) Review and revision of empirical critical loads and dose-response relationships. Review and revision of empirical critical loads and dose-response relationships. National Institute for Public Health and the Environment (RIVM), Noordwijkerhout

Bucher JB (1984) Bemerkungen zum Waldsterben und Umweltschutz in der Schweiz. Forstwissenschaftliches Centralblatt 103(1):16–27

Bytnerowicz A, Omasa K, Paoletti E (2007) Integrated effects of air pollution and climate change on forests: a northern hemisphere perspective. Environ Pollut 147(3):438–445. https://doi.org/10.1016/j.envpol.2006.08.028

CLRTAP (2015) Exceedance calculations. In: Manual on methodologies and criteria for modelling and mapping critical loads and levels and air pollution effects, risks and trends. UNECE Convention on Long-range Transboundary Air Pollution. http://www.icpmapping.org/. Accessed 1 Dec 2018

de Vries W, Dobbertin MH, Solberg S, Van Dobben HF, Schaub M (2014) Impacts of acid deposition, ozone exposure and weather conditions on forest ecosystems in Europe: an overview. Plant Soil 380(1–2):1–45

Eichhorn J, Icke R, Isenberg A, Paar U, Schönfelder E (2005) Temporal development of crown condition of beech and oak as a response variable for integrated evaluations. Eur J For Res 124 (4):335–347. https://doi.org/10.1007/s10342-005-0097-z

Ellenberg H (1971) Integrated experimental ecology: methods and results of ecosystem research in the German Solling Project. In: Ecological studies, vol 2. Springer, Berlin

Ellenberg H, Leuschner C (2010) Vegetation Mitteleuropas mit den Alpen: In ökologischer, dynamischer und historischer Sicht, 6th edn. UTB, Stuttgart

Fraley C, Raftery AE, Scrucca L, Murphy TB, Fop M (2017) Mclust: gaussian mixture modelling for model-based clustering, classification, and density estimation, r package version 5.3. https://cran.r-project.org/web/packages/mclust/index.html

Gauger T, Haenel HD, Rösemann C, Nagel HD, Becker R, Kraft P, Schlutow A, Schütze G, Weigelt-Kirchner R, Anshelm F (2008) Nationale Umsetzung UNECE-Luftreinhaltekonvention (Wirkungen). Teil 2: Wirkungen und Risiokoabschätzungen Critical Loads, Biodiversität, Dynamische Modellierung, Critical Levels Überschreitungen, Materialkorrosion. Im Auftrag des Umweltbundesamtes, Dessau-Rosslau

Härdtle W (1995) Vegetation und Standort der Laubwaldgesellschaften (*Querco-Fagetea*) im nördlichen Schleswig-Holstein. Mitteilungen der Arbeitsgemeinschaft Geobotanik in Schleswig-Holstein und Hamburg, Kiel

Hartmann P, von Wilpert K (2016) Statistisch definierte Vertikalgradienten der Basensättigung sind geeignete Indikatoren für den Status und die Veränderungen der Bodenversauerung in Waldböden. Allgemeine Forst-und Jagdzeitung 187(3–4):61–69

Hilbrig L, Wellbrock N, Bielefeldt J (2014) Harmonisierte Bestandesinventur: Zweite Bundesweite Bodenzustandserhebung (BZE II) – Methode. Thünen Working Paper. Johann Heinrich von Thünen Institute, Braunschweig. https://doi.org/10.3220/WP_26_2014

Höhle J, Bielefeldt J, Dühnelt P-E, König N, Ziche D, Eickenscheidt N, Grüneberg E, Hilbrig L, Wellbrock N, Kompa T (2018) Bodenzustandserhebung im Wald – Dokumentation und Harmonisierung der Methoden. Thünen Working Paper. Johann Heinrich von Thünen Institute, Braunschweig. https://doi.org/10.3220/WP1526989795000

Kauppi P, Anttila P, Kenttämies K (1990) Acidification in Finland. Finnish Acidification Research Programme HAPRO 1985–1990. Springer, Berlin

Kölling C, Hoffmann M, Gulder H-J (1996) Bodenchemische Vertikalgradienten als charakteristische Zustandsgrößen von Waldökosystemen. Z Pflanzenernähr Bodenkd 159 (1):69–77. https://doi.org/10.1002/jpln.1996.3581590111

Kreutzer K (1972) Über den Einfluß der Streunutzung auf den Stickstoffhaushalt von Kiefernbeständen (*Pinus sylvestris L.*). Forstwissenschaftliches Centralblatt 91(1):263–270

Ludemann T (2002) Historische Holznutzung und Waldstandorte im Südschwarzwald. Freiburger Forstliche Forschung 18:194–207

Lüst C, Giani L (2006) Merkmale von Böden unter rezenten Wäldern, die auf ehemals landwirtschaftlich genutzten Flächen stocken. Drosera 1(2):27–34

Manion PD (1981) Tree disease concepts. Prentice-Hall, Englewood Cliffs, NJ

Nihlgård B (1985) The ammonium hypothesis: an additional explanation to the forest dieback in Europe. Ambio 14(1):2–8

Nilsson J, Grennfelt P (1988) Critical loads for sulphur and nitrogen. Workshop on "Critical Loads for the effect on soils and groundwater of long term deposition of nitrogen and sulphur compounds and to establish methods to map the geographical areas experiencing higher than critical loads with respect to the sensitivity of different soil types". UN-ECE – Nordic Council of Ministers, Skokloster

Peters M (1990) Nutzungseinfluß auf die Stoffdynamik schleswig-holsteinischer Böden: Wasser-, Luft-, Nähr-und Schadstoffdynamik. Kiel University, Kiel

Pretzsch H, Block J, Dieler J, Dong PH, Kohnle U, Nagel J, Spellmann H, Zingg A (2010) Comparison between the productivity of pure and mixed stands of Norway spruce and European beech along an ecological gradient. Ann For Sci 67(7):712. https://doi.org/10.1051/forest/2010037

R Core Team (2017) R: a language and environment for statistical computing. R Foundation for Statistical Computing, Vienna

Rinklebe J, Makeschin F (2003) The impact of arable and forest land use on soil and vegetation – a comparison after 27 years. Forstwissenschaftliches Centralblatt 122(2):81–98. https://doi.org/10.1046/j.1439-0337.2003.00081.x

Schöpp W, Posch M, Mylona S, Johansson M (2003) Long-term development of acid deposition (1880–2030) in sensitive freshwater regions in Europe. Hydrol Earth Syst Sci 7(4):436–446

Schütt P, Blaschke H, Holdenrieder O, Koch W, Lang KJ, Schuck HJ, Stimm B, Summerer H (1984) Der Wald stirbt an Stress. Bertelsmann, Munich

Ulrich B (1983) Soil acidity and its relations to acid deposition. In: Ulrich B, Pankrath J (eds) Effects of accumulation of air pollutants in forest ecosystems. Springer, Dordrecht, pp 127–146

Ulrich B (1986) Die Rolle der Bodenversauerung beim Waldsterben: Langfristige Konsequenzen und forstliche Möglichkeiten. Forstwissenschaftliches Centralblatt 105(1):421–435

Ulrich B (1987) Stability, elasticity, and resilience of terrestrial ecosystems with respect to matter balance. In: Schulze E-D, Zwölfer H (eds) Potentials and limitations of ecosystem analysis. Ecological studies 61. Springer, Berlin, pp 11–49

Ulrich B (1995) The history and possible causes of forest decline in central Europe, with particular attention to the German situation. Environ Rev 3(3–4):262–276

UNECE (1979) Convention on Long-range Transboundary Air Pollution (CLRTAP). United Nations Economic Commission for Europe, Geneva

Verstraeten A, Neirynck J, Genouw G, Cools N, Roskams P, Hens M (2012) Impact of declining atmospheric deposition on forest soil solution chemistry in Flanders, Belgium. Atmos Environ 62:50–63. https://doi.org/10.1016/j.atmosenv.2012.08.017

von Wilpert K (1996) Aus der BZE abgeleitete Indizien einer bodenchemischen Drift in Baden-Württemberg. Mitteilungen der Deutschen Bodenkundlichen Gesellschaft 79:189–192

Waldner P, Marchetto A, Thimonier A, Schmitt M, Rogora M, Granke O, Mues V, Hansen K, Karlsson GP, Zlindra D, Clarke N, Verstraeten A, Lazdins A, Schimming C, Iacoban C, Lindroos AJ, Vanguelova E, Benham S, Meesenburg H, Nicolas M, Kowalska A, Apuhtin V, Napa U, Lachmanova Z, Kristoefel F, Bleeker A, Ingerslev M, Vesterdal L, Molina J, Fischer U, Seidling W, Jonard M, O'Dea P, Johnson J, Fischer R, Lorenz M (2014) Detection of temporal trends in atmospheric deposition of inorganic nitrogen and sulphate to forests in Europe. Atmos Environ 95:363–374. https://doi.org/10.1016/j.atmosenv.2014.06.054

Waldner P, Thimonier A, Pannatier EG, Etzold S, Schmitt M, Marchetto A, Rautio P, Derome K, Nieminen TM, Nevalainen S, Lindroos AJ, Merila P, Kindermann G, Neumann M, Cools N, de Vos B, Roskams P, Verstraeten A, Hansen K, Karlsson GP, Dietrich HP, Raspe S, Fischer R, Lorenz M, Iost S, Granke O, Sanders TGM, Michel A, Nagel HD, Scheuschner T, Simoncic P, von Wilpert K, Meesenburg H, Fleck S, Benham S, Vanguelova E, Clarke N, Ingerslev M, Vesterdal L, Gundersen P, Stupak I, Jonard M, Potocic N, Minaya M (2015) Exceedance of critical loads and of critical limits impacts tree nutrition across Europe. Ann For Sci 72 (7):929–939. https://doi.org/10.1007/s13595-015-0489-2

Wentzel KF (1979) Schwefel-Immissionsbelastung der Koniferenwalder des Raumes Frankfurt/ Main. Forstarchiv 50(6):112–121

Wentzel KF (1982) Immissionen oder Saurer Regen – Wovon sterben Wälder und Seen. Der Forst- und Holzwirt 37:410–413

Wittich W (1951) Der Einfluß der Streunutzung auf den Boden. Forstwissenschaftliches Centralblatt 70(2):65–92

Wolff B, Riek W (1996) Deutscher Waldbodenbericht 1996 – Ergebnisse der bundesweiten Bodenzustandserhebung im Wald von 1987–1993. Bundesministerium für Ernährung, Landwirtschaft und Forsten, Bonn

Wolff B, Erhard M, Holzhausen M, Kuhlow T (2003) Das Klima in den forstlichen Wuchsgebieten und Wuchsbezirken Deutschlands. Mitteilungen der Bundesforschungsanstalt für Forst-und Holzwirtschaft, vol 211. Kommissionsverlag Max Wiedenbusch, Hamburg

Ziche D, Seidling W (2010) Homogenisation of climate time series from ICP Forests Level II monitoring sites in Germany based on interpolated climate data. Ann For Sci 67(8):804. https://doi.org/10.1051/forest/2010051

Chapter 3
Soil Water Budget and Drought Stress

Heike Puhlmann, Paul Schmidt-Walter, Peter Hartmann, Henning Meesenburg, and Klaus von Wilpert

3.1 Introduction

The water budget of forest ecosystems is fed by precipitation. Loss into the atmosphere results from direct evaporation from the soil, transpiration, which is far more significant in forests due to their large crown-surface area, and evaporation from interception. The climatic processes governing these types of evaporation are air vapour pressure deficit, air temperature and convective water vapour transport. Another factor influencing the water budget occurs through soil water seepage, which is controlled mainly by the texture of the soil, its rock content and its bulk density. A certain amount of water is stored in the soil depending on texture, bulk density, carbon content, rock content and thickness of the soil layer. In most soils, except for very clayey and very sandy soils, the predominant part of the stored water is available to plants (available water capacity). The third factor determining the distribution and flow of water in the soil is vegetation. On the one hand, a considerable part of rain water is retained by the tree canopy and may evaporate directly from there (interception). On the other hand, water uptake by roots and transpiration influence the soil water budget of forest sites. This shows that, in addition to climatic processes, water-holding capacity and water conductivity of the soil as well as interactions between soil properties and vegetation properties (e.g. regarding root distribution) substantially determine and vary the water budget.

The water phase of the soil is the space in which the solution, exchange and transport processes take place that dictate most soil functions. Furthermore, water availability is a central property for the growth and productivity of forests. The

H. Puhlmann (✉) · P. Hartmann · K. von Wilpert
Forest Research Institute Baden-Württemberg, Freiburg, Germany
e-mail: heike.puhlmann@forst.bwl.de; peter.hartmann@forst.bwl.de

P. Schmidt-Walter · H. Meesenburg
Northwest German Forest Research Institute, Göttingen, Germany
e-mail: paul.schmidt-walter@nw-fva.de; henning.meesenburg@nw-fva.de

N. Wellbrock, A. Bolte (eds.), *Status and Dynamics of Forests in Germany*,
Ecological Studies 237, https://doi.org/10.1007/978-3-030-15734-0_3

reaction equation of photosynthesis shows that water and carbon dioxide are the central components of plant growth. Seepage water flux is the transportation route for substances between the atmosphere, the soil and the hydrosphere. These are substances added by the rain such as nitrogen compounds, as well as reaction products of soil acidification such as aluminium and manganese ions. This shows that the seepage water flux substantially determines how well a soil acts as a filter. Because nearly all plant nutrients are taken up from the soil solution, there is a close connection between forest nutrition and the water budget. For example, a latent potassium deficiency becomes acute on clayey sites during and after a dry period. The reason for this is the accumulation of nutrients inside soil aggregates where they are difficult to access for plant roots (von Wilpert and Hildebrand 1997). Thus, the water budget plays an important role in the soil's ability to provide nutrients for the forest.

To avoid overlapping with other chapters of this book, the water budget of forests will be treated in this chapter by focusing on the direct effects of water availability. To achieve this, soil properties and soil processes are prioritized over climate and vegetation properties, all of which determine the water budget. The aim is to identify in what way water flow and availability are affected by the variability of soil properties. This analysis was done largely on the basis of soil data in the NFSI data set. Soil hydraulic properties such as water retention and water conductivity were derived from measured and estimated soil properties using pedotransfer functions (PTFs) (Puhlmann and von Wilpert 2011, 2012); see Sect. 3.2.1. Additionally, the depth profiles of fine roots for each of the NFSI profiles were estimated in a correlation analysis between fine root density (FRD) and soil properties (Hartmann and von Wilpert 2014); see Sect. 3.3.

Capacitive and dynamic water budget parameters for all NFSI profiles are derived from water budget modelling using LWF-Brook90. For the predominant part of the analysis, the forest stand properties (age, tree species composition, degree of canopy cover) were kept constant in order to clearly identify the significance of the individual soil properties for the result of the water budget modelling at the soil profiles. The vegetation properties described in the NFSI database are exemplarily included in the model for the purpose of comparison only in a last step of the analysis.

Static and dynamic drought stress characteristics were derived from water budget modelling in the concluding paragraph. They are discussed in reference to their ecological relevance on the basis of empirical data on tree growth (inter- and intra-annual tree-ring characteristics) and on mortality in Chap. 3. With this, the NFSI data are made available for comprehensive analyses in climate impact research. The model results are, of course, also applicable in a number of other applications such as predictions about seepage water output. In several federal states (Baden-Wuerttemberg, Bavaria, North Rhine-Westphalia, Rhineland-Palatinate and Saxony), concrete model developments are currently in progress to advance and objectify traditional, analogue site mapping. Because the identification of the water budget in all previous site mapping operations occurred in relative ordinal scales, it is neither possible to derive quantitative information on the water budget from the site maps nor to dynamically fit the site maps to changing climatic conditions (Gauer and Kroiher

2012). Therefore, quantitative and dynamic modelling of the water budget plays a central role when developing model-based site mapping procedures. The NFSI data is, in principle, suitable for that purpose—especially the measured physical soil parameters. The measuring network of the NFSI with its 2430 points does not represent Germany's forested area with its 11.1 million ha for detailed analyses properly; the mean density of the grid is approximately 6.8 × 6.8 km. If, however, multivariate estimation models with sufficient explanatory power are created on the basis of regression relations between measured soil parameters and terrain predictors, it is possible to create water budget models using estimated soil information (Zirlewagen and von Wilpert 2011) and link them to polygons of the forest site survey. The regression models will deliver more realistic results when incorporating the data set of all the German states, because the spatial density of measured soil parameters as well as corresponding co-variables (e.g. terrain attributes) becomes higher than would be possible on the level of single states. In this way, it is possible to substitute the ordinal, qualitative assessment of the site water budget of the individual forest site survey systems in the different federal states with modelled, time-variant quantitative information about the site water budget. Apart from methodological homogenization between the federal states, this would make the states' forest site survey systems also sensitive to climate.

3.2 Soil Properties as Input for Water Budget Modelling

Soil properties are input parameters for estimating and modelling the water budget of soils. They are included in the parameters of the NFSI II or can be derived from the information given in the NFSI. Especially with regard to physical soil characterization, the scope of the obligatory NFSI II parameters was extended significantly compared to the preceding inventory, and the measuring of parameters such as bulk density (using sampling rings or volume replacement samples), skeletal fraction (by volume replacement), texture (% S, U, T using the Köhn-Pipette or Lasersizer method) and carbon content (by element analysis) was required. For parameters, where measuring is extensive work such as counting fine root density with spatially high-resolving counting frames, most states still used estimated data; more precise counts where only done in Baden-Wuerttemberg. In addition to depth-related measured data, horizon-related estimated data regarding, e.g. texture, rock content and bulk density was collected in each case. Water retention and water conductivity, the two soil hydraulic properties required as input to soil water models, can be derived from these baseline pedological data only by using PTFs.

This chapter gives a description of regional variations in the mentioned measured and estimated physical soil parameters and of measurement and/or estimated accuracy.

The best available quality of the soil input data should be applied for modelling the water budget, i.e. the measured data (measured texture data, bulk density, etc.). However, because this data is not available in all states and only for certain depths, the estimated data from the profile descriptions must be included to have a homogeneous database. It would be theoretically possible and necessary to consolidate the information from both data sources (estimated and measured) and to harmonize it with regard to survey artefacts (e.g. systematic bias between different estimating methods). Related harmonization work in individual states (e.g. Baden-Wuerttemberg) showed that this is a labour-intensive process and due to high heterogeneity among the states not possible to conduct nationwide within the framework of NFSI II. Therefore, for the federal soil report, measured data were used from depths in which they were available, and horizon-related estimated data were used where they were not.

3.2.1 Estimating Soil Hydraulic Functions Using Pedotransfer Functions

3.2.1.1 Introduction

Measured data on soil hydrological properties such as available water capacity (AWC) and the parameters for retention and conductivity functions at the NFSI plots were not all collected due to the high cost of measurement. For the characterization of the water budget, they have to be deduced indirectly using PTFs from the measured soil physical properties (texture, bulk density, carbon and humus content). The AWC shows the soil's ability to store water available for plants. Generally, this is specified by the difference between the volumetric water content at a soil water potential of −60 hPa (field capacity, pF 1.8) and −16,000 hPa (permanent wilting point, pF 4.2) and can be used for the static characterization of the water budget of a forest site. The parameters for the retention (van Genuchten 1980) and conductivity (Mualem 1976) models—abbreviated as MvG parameters below—are input parameters for process-based water budget models such as LWF-Brook90 (Hammel and Kennel 2001), which was used in this report (see Sect. 3.3).

A number of different PTFs for estimating characteristic soil hydraulic functions can be found in the literature. There have already been various projects and publications on the validation of PTFs for soil hydraulic characteristics in the past; especially the thorough works by Hangen and Scherzer (2004), Schramm et al. (2006), Mellert et al. (2009) and Russ and Riek (2011) should be mentioned in this context. However, the mentioned studies could not yet include newer developments such as the one by Puhlmann et al. (2009) and Puhlmann and von Wilpert (2011). In addition to that, many authors point out that the quality of the prediction is only valid for certain regions due to the geographical origin of the measured data used for validation or that some soil textures are underrepresented due to an insufficient number of samples. The results therefore do not allow any conclusions about the choice and application of the examined PTFs for the soils at the NFSI plots. In order to assist in choosing a PTF from the currently available PTFs, an individual study on the basis of an extensive data set was conducted. That data was collected on forest sites in the whole of Europe and covers a wide range of soil types.

3.2.1.2 Materials and Methods

The data set available for PTF validation comprises a total of 2075 retention curves of mineral soil horizons, 130 retention curves of organic horizons as well as 866 samples where the saturated conductivity was measured. In addition to the target variables (retention curves, saturated hydraulic conductivity), the corresponding physical/chemical soil parameters (grain size fractions, bulk density and carbon/humus content) were available. These data are necessary for estimating AWC and MvG parameters using PTFs and for comparing them with the measured values. The data were obtained from the European Level II monitoring network (ICP Forests) and the forest research institutions and universities in Göttingen, Freiburg, Freising, Eberswalde and Graupa. It was ensured that no data sets were used that have already been incorporated in the development of the examined PTFs. In many cases, the sample rings for the retention curves did not come from the same soil profile as the physical/chemical soil parameters, only from the same test site. Because of these uncertainties, the decision was made to also exclude those measurements from the analyses and to only use samples where the input data for the PTF and the target variables originate from the same soil profile and horizon. By applying these restrictions, the data basis for validating the PTFs for the retention curve was reduced to 1641 mineral soil samples. One thousand four hundred twelve samples remained for the validation of the AWC, because not all retention curves contained information on the permanent wilting point and the AWC could therefore not be calculated. Figure 3.1 shows the distribution of the remaining retention curves inside the soil texture triangle; the samples of the NFSI II data set are shown for comparison.

Table 3.1 lists the examined PTFs and their abbreviations used below, as well as remarks and adjustments made during their application in this study. Which examined PTFs were chosen is based on the work of Hangen and Scherzer (2004) who, by means of extensive literature research, identified PTFs that seemed especially suitable for deriving hydraulic properties of forest soils. Two types of PTFs can be distinguished. One type of PTF predicts the water content for certain soil water potentials, typically for pF 1.8 (field capacity) and pF 4.2 (permanent wilting point). Because these PTFs predict fixed points on the retention curve, they are also called point PTFs. The tabular approaches by Wessolek et al. (2009) and Teepe et al. (2003) examined in this analysis fall into this category. Parametric PTFs, on the other hand, predict the parameters of retention models such as the van Genuchten model. With a water retention model, the water content can be predicted continuously as a function of the soil water potential. The regression functions by Wösten et al. (1999), Puhlmann and von Wilpert (2011), Vereecken et al. (1989) and Teepe et al. (2003) are of this type. Tabular PTFs that list mean MvG parameters for different texture and density combinations also fall into this category, e.g. the PTF

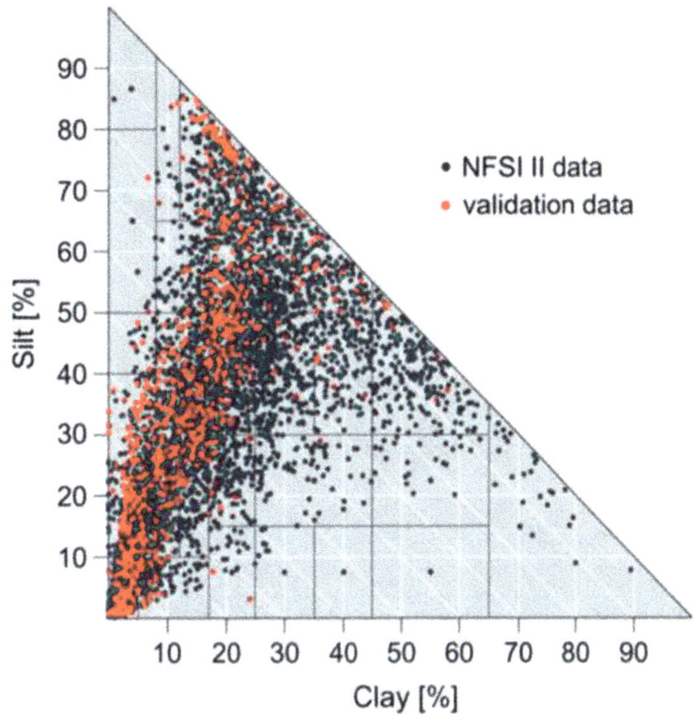

Fig. 3.1 Distribution of retention data in the texture triangle

Teepe.TexTRD or the approach according to DIN 4220 (2008–2011) which was modified for this examination.

In addition to these modifications, adjustments were made to the PTF of Wessolek et al. (2009), as well as to the PTFs *Teepe.TexTRD* and *Teepe.KGA*. It turned out that for the latter, the values for the parameter α with the specified unit (kPa^{-1}) of the MvG curves do not fit to the table values for volumetric water content at field capacity (FC) and permanent wilting point (PWP). It did, however, produce results consistent with the table values when changing the unit for α to hPa^{-1}. Therefore, the α-values were multiplied by 10. The values of the lowest bulk density (TRD) class (TRD <1.2 g cm^{-3}) were not considered in the PTF *Wessolek.TexTRD*, because very high humus contents can often be found in horizons of forest soils in this TRD class. Accounting for humus content led in this TRD class to a significant overestimation of the total pore volume, the FC and the AWC. It turned out that, for soil samples from the TRD class 1.1, the increased water storage is already sufficiently reflected by the large humus content. Accordingly, samples with a TRD lower than 1.2 g cm^{-3} were assigned to the TRD class 1.3 when applying the PTF of *Wessolek.TexTRD*.

To objectively assess the quality of prediction of the individual PTFs, statistical goodness-of-fit measures were calculated based on the difference between estimated

Table 3.1 Overview of the examined pedotransfer functions for available water capacity (AWC)

Description/ literature source	Type/ calculation	Input parameters	Comments/adjustments
Wessolek. TexTRD (Wessolek et al. 2009)	Point PTF, Ks/table	Texture class (KA5), TRD class, humus class	Tables 3 + 5. Density class 1.1 was not considered, and samples with TRD <1.2 g cm^{-3} were assigned level 1.3. For undefined combinations of TRD and texture class, the value of the bordering density level of the respective texture class was assumed
DIN4220. TexTRD (DIN 4220 2008–2011)	Param. PTF/ table	Texture class (KA5), TRD class	MvG parameters from Table 10 in Wessolek et al. (2009) were adjusted according to the water potential/water content values from Table A1 of the DIN 4220, so that MvG parameters were available separately for individual texture and TRD classes
Teepe.KGA (Teepe et al. 2003)	Point PTF + param. PTF/equation	%S/%U/%T, TRD	MvG-α was multiplied by 10
Teepe.TexTRD (Teepe et al. 2003)	Point PTF + param. PTF/table	Texture class (KA5), TRD class, Corg	MvG-α was multiplied by 10; supplementing of the missing values for air capacity, AWC and PWP for sands from DIN 4220. Minimum %S, %U, %T limited to 0.5%
PUH2.KGA (Puhlmann and von Wilpert 2011)	Point PTF + param. PTF, Ks/equation	%S/%U/%T, TRD, Corg	
Vereecken.KGA (Vereecken et al. 1989)	param. PTF/ equation	%S/%U/%T, TRD, Corg	Adjustment of %S and %U to the grain size boundary at 50 μm through log-linear interpolation of the grain size curve
Hypres.KGA (Wösten et al. 1999)	Param. PTF, Ks/equation	%S/%U/%T, TRD, humus content, top-/subsoil	Minimum TRD limited to 0.5 g cm^{-3}, minimum %U, %T to 0.5%. All samples taken from a depth below 30 cm were classified as subsoil. Adjustment of the boundary between %S and %U as for Vereecken.KGA

and measured volumetric water content. Accuracy measures of predictions are the mean error (ME) and the mean absolute error (MAE):

$$\mathrm{ME} = \frac{1}{n}\sum_{i=1}^{n}\left(\widehat{y_i} - y_i\right) \tag{3.1}$$

$$\mathrm{MAE} = \frac{1}{n}\sum_{i=1}^{n}\left(\left|\hat{y_i} - y_i\right|\right) \tag{3.2}$$

The ME as the arithmetic mean of all residuals (difference between estimate $\hat{y}$ and measurement e) indicates systematic over- (positive values) and underestimations (negative values). In the ideal case of an unbiased prediction, ME takes a value of 0.

The precision of predictions was evaluated using the root mean squared error (RMSE):

$$\mathrm{RMSE} = \sqrt{\frac{1}{n}\sum_{i=1}^{n}\left(\hat{y_i} - y_i\right)^2} \tag{3.3}$$

The RMSE quantifies the standard deviation of the residuals and therefore the absolute value of the total error that can be expected on average and ideally also takes a value of 0.

Pearson's correlation coefficient is a statistical measure for comparing individual measured retention curves with estimated ones using parametric PTFs. It is calculated as follows:

$$r = \frac{\sum_{i=1}^{n}\left(\hat{y_i} - \bar{\hat{y}}\right)\left(y_i - \bar{y}\right)}{\sqrt{\sum_{i=1}^{n}\left(\hat{y_i} - \bar{\hat{y}}\right)^2 + \sum_{i=1}^{n}\left(y_i - \bar{y}\right)^2}} \tag{3.4}$$

Thereby, $\bar{y}$ and $\bar{\hat{y}}$ specify the mean values of measurement and estimate. In the case of perfect positive correlation, the different water contents of the individual points of soil water potential lie on a line with a positive slope, and r assumes a value of 1. The correlation coefficient can therefore be seen as a similarity measure of the progression and shape of two retention curves. Another similarity measure is Wilmot's index (w):

$$w = 1 - \frac{\sum_{i=1}^{n}\left(y_i - \hat{y_i}\right)^2}{\sqrt{\sum_{i=1}^{n}\left(\left|\hat{y_i} - \bar{y}\right|\right)^2 + \left(\left|y_i - \bar{y}\right|\right)^2}} \tag{3.5}$$

Multiplied with the correlation coefficient, Wilmot's w produces the confidence index ($\mathrm{CI} = w * r$) that also takes a value of 1 when measurement and estimate are in perfect accordance. Values of >0.85 are seen as excellent predictions, values of >0.75 as good and values of ≤0.75 as less than good (de Camargo and Sentelhas 1997).

The ME and the RMSE are used in the assessment of the prediction quality for for water contents of individual water potential levels (1, 10, 60, 100, 330, 1000, 2500, 5000, 16,000 hPa) and available water capacity. For evaluating parametric PTFs for the retention curve, the CI, the RMSE and the ME are calculated for the whole range of soil water potential. In order to make conclusions about the approximate distribution of the quality criteria of the PTF over a range of soil textures, the quality criteria are calculated separately for the 11 soil texture groups of the German soil texture classification system (Ad-Hoc AG–Boden 2005). After that, the PTFs are ranked for each quality criterion. For the available water capacity, the differences between PTF estimates and measured values are added to the texture triangle as isolines. In this way it is possible to make conclusions about the PTFs' performance for different texture groups. The goal is to identify which PTF is best suited for the NFSI II soil data. To achieve this, the mean rank for each quality criterion is calculated from the ranks of the quality criteria of the individual soil texture groups. This mean rank is weighted using the respective number of observations in the NFSI II data set. Finally, simple not-weighted mean values are produced from these, establishing a final rank order for the PTFs.

3.2.1.3 Results and Discussion

Figure 3.2 shows the deviation between estimated and measured water contents for each PTF as a function of the soil water potential, separated according to soil texture groups. Additionally, the ME is plotted for the various water potential classes, which makes it possible to recognize systematic errors. The graphical display of the results for clayey loams (tl), loamy clays (lt) and sandy silts (su) was refrained from because of the small number of observations. The quality criteria MAE, RMSE and CI which integrate over all pF levels can be found in Table 3.2 for all soil texture groups. The systematic errors of individual soil texture groups are listed in Table 3.2 and were calculated as mean values of the MAE in order to not cancel out opposing deviations between the individual water potential classes.

In general, the variation and therefore the RMSE decrease with increasing water potential due to decreasing water contents. This becomes clear when looking at the narrowing box and whisker ranges (Fig. 3.2). It seems that the characteristic progressions of the retention curves of the individual soil texture groups cannot be reconstructed satisfactorily by each of the investigated PTFs. For example, the PTFs *Hypres.KGA* and *Vereecken.KGA* tend for nearly all soil texture groups to estimate a sharper decrease in water content than the measured data suggest. This becomes apparent from overestimated water contents at low water potentials in combination with underestimated water contents at high water potentials (Fig. 3.2) and can also be seen in the low goodness-of-fit measures (Table 3.2).

The curvature of the van Genuchten retention model depends on the parameters α and n. Since the water content close to saturation and the water content at the permanent wilting point are predicted relatively well, it seems reasonable to hold insufficient estimates of the α and n parameters responsible for the weaknesses of the

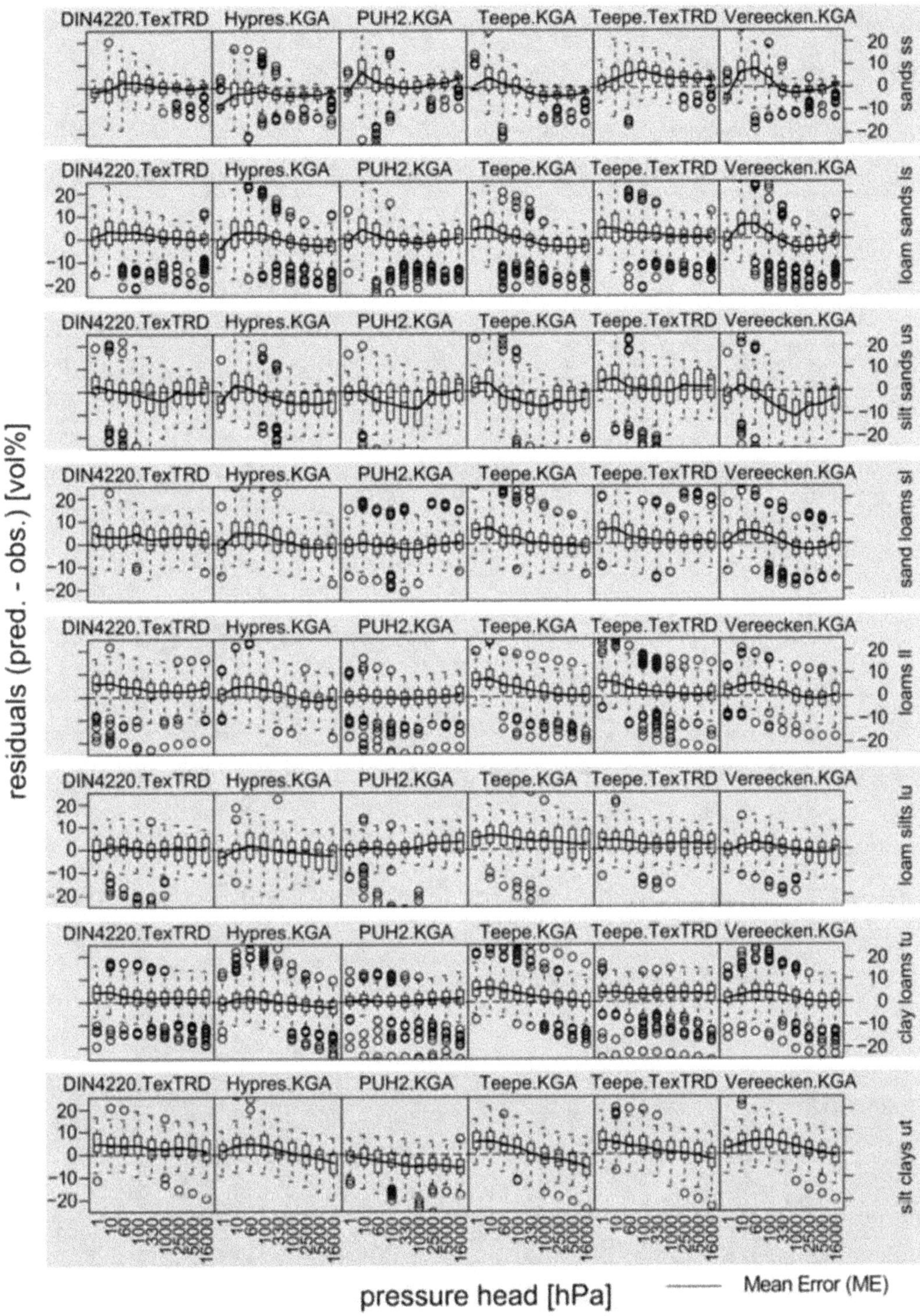

Fig. 3.2 Deviations (boxplots) and mean deviation (solid lines) between PTF estimates and measured water contents (in vol%) as a function of matric potential, divided into soil texture groups

two PTFs. Except for silty-clayey substrates (texture groups lu, tu), the PTFs do not produce satisfactory predictions for the retention curves of the validation data set.

Table 3.2 Mean absolute error (MAE, vol%), root mean squared error (RMSE, vol%) and confidence index (CI, dimensionless) over all pF levels for all soil texture groups

Soil texture class	Sands			Loams			Silts			Clays		Mean
Soil texture group	ss	ls	us	sl	ll	tl	su	lu	tu	ut	lt	
N	132	400	136	272	261	1	4	79	260	82	14	
RMSE (vol%)												
DIN4220.TexTRD	**5.0**	**5.9**	7.9	6.5	6.7	8.8	8.0	6.2	5.7	6.9	**5.3**	6.6
Hypres.KGA	5.6	6.4	**7.6**	7.3	6.4	**3.6**	6.1	7.1	5.8	**6.5**	5.9	6.2
PUH2.KGA	5.9	6.0	10.1	**5.8**	**4.9**	5.4	**6.0**	**5.6**	**4.6**	7.0	13.4	6.8
Teepe.KGA	5.6	6.1	8.8	7.3	7.0	5.5	7.2	8.0	6.7	6.9	8.2	7.0
Teepe.TexTRD	7.0	6.3	7.9	7.5	6.7	5.9	7.8	6.3	5.8	6.8	8.8	7.0
Vereecken.KGA	6.5	6.6	9.9	6.6	5.9	4.5	5.3	5.7	5.7	7.1	5.8	6.3
MAE (vol%)												
DIN4220.TexTRD	**1.1**	1.6	**1.6**	3.1	3.9	8.5	4.8	**0.6**	2.5	3.3	2.9	3.1
Hypres.KGA	3.3	2.5	3.8	2.4	2.3	**3.1**	**3.0**	1.6	1.4	**2.5**	3.9	2.7
PUH2.KGA	2.1	**1.4**	3.1	**1.0**	**0.5**	4.2	3.9	1.2	**0.6**	3.8	12.8	3.1
Teepe.KGA	2.6	3.1	4.3	2.6	3.5	5.2	6.0	4.1	3.1	3.5	7.0	4.1
Teepe.TexTRD	4.2	2.1	2.1	2.7	2.9	5.1	5.9	2.8	3.5	2.7	6.6	3.7
Vereecken.KGA	3.5	3.2	5.0	2.6	2.7	3.9	3.2	1.3	2.0	3.4	**2.4**	3.0
CI (–)												
DIN4220.TexTRD	0.82	0.85	0.74	0.81	0.78	0.81	0.82	0.81	0.80	0.60	0.49	0.76
Hypres.KGA	0.81	0.83	**0.81**	0.78	0.78	**0.95**	0.87	0.77	0.78	**0.66**	**0.56**	0.78
PUH2.KGA	0.78	0.83	0.63	0.82	**0.83**	0.87	0.91	0.85	**0.84**	0.61	0.33	0.76
Teepe.KGA	**0.83**	**0.87**	0.76	0.81	0.80	0.91	**0.92**	0.78	0.79	0.65	0.49	0.78
Teepe.TexTRD	0.76	0.84	0.76	0.79	0.80	0.89	0.87	0.84	0.81	0.65	0.31	0.76
Vereecken.KGA	0.77	0.83	0.70	**0.82**	0.82	0.93	0.91	**0.85**	0.81	0.64	0.53	0.78

The best values for individual quality criteria in the respective soil texture group are highlighted in bold. Below the soil texture group symbols, the number of samples within each group is given

The PTFs *Teepe.TexTRD* and *Teepe.KGA* produce slightly better results. These seem to have their strengths in clayey-silty textures, for which, judging by the CI, the shape of the retention curve can be predicted relatively well. However, the water contents are overestimated for nearly all texture groups, especially in the range of low soil water potentials, and come with a wide uncertainty range. For these reasons and on the basis of the mean RMSE and MAE values, they have to be rated as the least suitable PTFs for predicting the retention properties of forest soils.

Using objective criteria, the best PTF found for predicting the retention function of the examined soil samples, is the PTF *PUH2.KGA* (Table 3.2). It provides nearly unbiased, high-precision estimates for the retention function (Fig. 3.2) over a wide range of textures. With *PUH2.KGA*, only the prediction of retention properties of sands seems problematic. The water contents of the silt sands (us) are significantly underestimated for medium water potentials and come with great uncertainties. Contrary to the loamy-silty substrates, the water contents of the pure sands (ss) and loam sands (ls) at low water potentials as well as the water content at the permanent wilting point (−16,000 hPa) are in part largely overestimated. Additionally, there is a significant underestimate of water retention in clayey soils (ut, lt).

A PTF that, in contrast to all other PTFs, provides satisfactory estimates of the retention function over all soil texture groups, is the PTF *DIN4220.TexTRD*. It is, for example, the only PTF that estimates the retention of pure sands accurately and also provides robust, although slightly too high, estimates for clayey soils. In the loamy-silty range (sl, ll, lu, tu), *DIN4220.TexTRD* is less precise compared to other PTFs and tends to have slightly high, but satisfactory, estimates with a mean RMSE of 6.6 vol% and a MAE of 3.1 vol%. The mean CI over all soil texture groups with a value of 0.76 is also slightly lower than that of other PTFs. This can be explained by a generally larger variation of the deviations of tabular PTFs. On average, however, the shape of the retention curve can be satisfactorily represented. This becomes also clear when looking at the generally uniform distribution of residuals over the entire range of water potentials.

The estimates for the AWC show a picture similar to the estimates for the retention curve. *Hypres.KGA* and *Vereecken.KGA* overestimate the AWC in all soil texture groups on average by approx. 6 vol% (Table 3.3). This was to be expected considering the systematic overestimates of the water content at low, and the underestimates at high water potentials described in the previous paragraphs. Both PTFs show on average the highest uncertainties (*Hypres.KGA*, 9.9 vol%; *Vereecken.KGA*, 8.4 vol%) and will be rated as least suitable for estimating the AWC of forest soils. Slightly better results are found for the PTFs *Teepe.KGA* and *Teepe.TexTRD*. Their estimates also show large uncertainties, but with a mean ME of 3.6 vol% (*Teepe.TexTRD*) and 4.5 vol% (*Teepe.KGA*), they provide less biased estimates. The PTFs *DIN4220.TexTRD* and *Wessolek.TexTRD* show similar results due to originating from the same source, whereby the point PTF *Wessolek.TexTRD* includes the humus contents of the samples, while *DIN4220.TexTRD* does not. This is possibly the reason for the RMSE of the PTF.

Wessolek.TexTRD with 6.5 vol% being slightly lower than the one of *DIN4220.TexTRD* (RMSE 7.2 vol%). Both PTFs slightly overestimate the AWC by 3.4 vol%.

Table 3.3 Mean error (ME, vol%) and root mean squared error (RMSE, vol%) of the estimate of the available water capacity for all soil texture groups

Main soil texture	Sands			Loams			Silts			Clays		Mean
Soil texture group	ss	ls	us	sl	ll	tl	su	lu	tu	ut	lt	
	61	250	67	185	193	0	4	49	210	44	6	
RMSE (vol%)												
DIN4220. TexTRD	7.2	7.2	8.6	7.0	5.0		6.1	7.6	5.6	7.3	10.7	7.2
Hypres. KGA	8.7	10	9.8	9.5	8.0		9.2	10.2	7.8	11.0	14.8	9.9
PUH2. KGA	11.5	7.5	6.9	**6.1**	**4.2**		**3.2**	**7.5**	**5.2**	**5.5**	**7.6**	6.5
Wessolek. TexTRD	**6.3**	**7.1**	8.3	6.4	4.5		6.1	7.7	5.5	7.9	12.4	7.2
Teepe. KGA	7.9	8	7.6	7.0	5.5		5.9	7.8	6.1	9.9	14.3	8.0
Teepe. TexTRD	7.6	7.2	**6.5**	7.3	6.7		3.5	7.9	5.7	10.6	13.1	7.6
Vereecken. KGA	9.7	8.9	8.6	7.9	6.6		9.0	7.8	6.9	9.4	9.4	8.4
ME (vol%)												
DIN4220. TexTRD	3	2.6	4.2	1.4	0.6		5.2	0.7	0.9	5	10.4	3.4
Hypres. KGA	**0.3**	6.1	5.0	4.7	5.4		8.5	4.2	4.5	9.7	14.0	6.2
PUH2. KGA	−9.0	−3.6	−2.5	**−1.0**	**0.0**		2.7	−2.2	**0.0**	**2.4**	**−3.2**	−1.6
Wessolek. TexTRD	−1.0	**2.4**	5.2	2.6	1.3		5.2	0.1	0.6	5.7	12	3.4
Teepe. KGA	4.1	3.9	2.1	2.4	3.1		5.2	**0.0**	1.8	8.5	13.4	4.5
Teepe. TexTRD	3.6	2.9	**1.2**	2.5	2.9		**2.2**	−0.7	**0.0**	8.4	12.6	3.6
Vereecken. KGA	5.9	6.5	5.6	5.1	4.9		8.8	2.8	4.2	8.1	8.2	6.0

Highlighted in bold are the best values of the quality criteria in the respective soil texture group

In contrast to all other PTFs that on average overestimate the AWC of all soil texture groups by varying degrees, *PUH2.KGA* on average slightly underestimates AWC (−1.6 vol%). In the same way as for the retention curve, this PTF provides the most precise and least biased estimates for all soil texture groups.

Figure 3.3 shows the differences (smoothened using a generalized additive model) between the estimated AWC values using the PTFs *PUH2.KGA* (Fig. 3.3a) and *Wessolek.TexTRD* (Fig. 3.3b) and the measured AWC values in the texture triangle. Areas with the same systematic error are framed by isolines (Table 3.3),

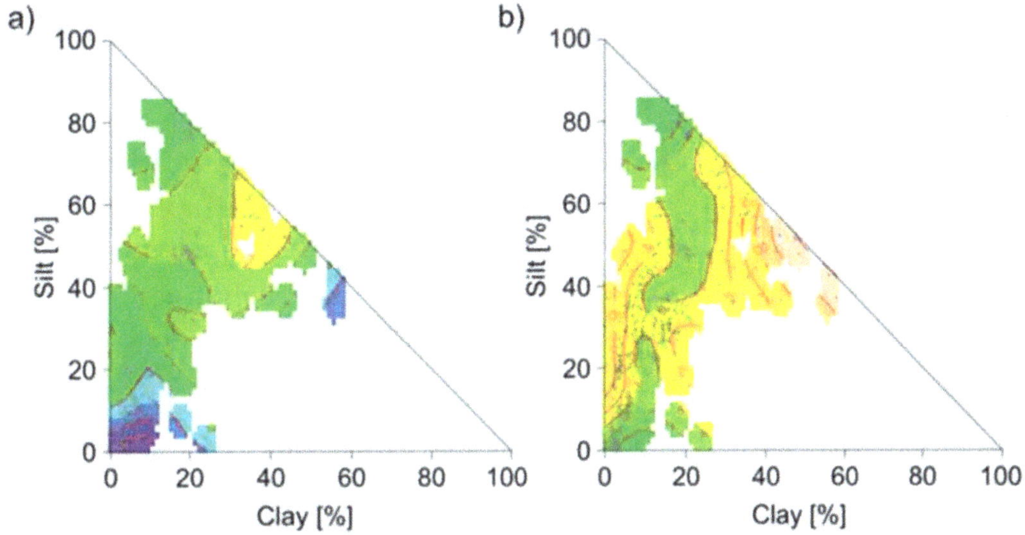

Fig. 3.3 Isoline diagrams generated using a generalized additive model of the smoothened deviations of estimated and measured available water capacity (vol%) as a function of the clay and silt content. Displayed are the pedotransfer functions PUH2.KGA (A) and Wessolek.TexTRD (B). Green and yellow-green areas represent deviations in the range of ±2 vol%, yellow to yellow-brown areas indicate systematic overestimates and blue-green and blue areas indicate an underestimate of the available water capacity

whereby blue areas indicate underestimations of ≤4 vol%, and yellowish-brown areas overestimations of >4 vol%. As shown already in Table 3.3, systematic deviations of the PTF *PUH2.KGA* lie between −2 vol% and +2 vol% in large areas of the texture triangle, the smallest deviations appearing in silty-clayey areas. With decreasing silt and clay content, the PTF increasingly underestimates the AWC, at first only slightly, but then strongly when the sand content becomes larger than 60%. In this way, silt contents of <10% and <5% lead to systematic underestimates of the AWC of more than 8 vol% and more than 10 vol%, respectively. Because the validation data set contains many samples within this texture range, it can be assumed that this trend originates from a systematic error of the PTF *PUH2.KGA*, because the calibration data set used for developing the PTF contained only few pure sand and silty sand samples. The group of loamy clays was also underrepresented in the data set for the development of *PUH2.KGA*, and there are indications for a substantial underestimation of the AWC. However, within the validation data set, the number of samples with clay contents of >50% is too low for making serious conclusions about this response.

Contrary to *PUH2.KGA*, Fig. 3.3b shows a systematic overestimate of 8–12 vol% of the PTF *Wessolek.TexTRD* for the clayey silts and the clayey loams. The least biased estimates lie, for the largest part, within the texture ranges that also have the highest sample density in the NFSI data collection. They extend from the pure and loamy sands to the sandy loams to the silts. Several areas, despite of having a sufficient amount of samples, are affected by significant overestimates in the range of 4–6 vol%, such as the silty sands and the sandy silts as well as regular loams and silty clays. This circumstance can be explained by higher uncertainties of *Wessolek.TexTRD* when compared to *PUH2.KGA* (Table 3.3). The biggest difference between PTFs *PUH2.KGA* and *Wessolek.TexTRD* exists within the pure sands and the silty sands, for which *Wessolek.TexTRD* estimates the AWC with deviations of <2%, while *PUH2.KGA* significantly underestimates the AWC in this range of textures.

3.2.1.4 Conclusions About Choosing the Appropriate Pedotransfer Function for the Water Budget Modelling

The final ranking of the PTFs is based on the quality criteria of Table 3.2 and shows that the most appropriate approach for estimating the retention curve for the depths used in the NFSI data set is the one by Puhlmann and von Wilpert (2011).

For the weighted overall ranking, low prediction quality of the numerous sandy substrates within the NFSI II data set is compensated for by very good prediction properties for loamy and silty substrates, which are very numerous within the data set as well. However, a PTF that is applied to such a diverse data set as the NFSI II is required to satisfactorily predict the retention function for all soil textures. Therefore, the PTF according to DIN 4220 (2008–2011) was finally chosen for the water budget model. This approach is only second choice by objective criteria (Table 3.4). It does, however, provide satisfactory results over a wide range of soil textures. In addition, for estimating the soil hydrological properties FC, AWC and PWP, the best approach according to objective criteria, the *Wessolek.TexTRD*, may be used without the values differing much from the values of the retention curve, because both PTFs are derived from the same source.

Table 3.4 Summary of overall ranking of the pedotransfer functions for the retention curve and the available water capacity, based on the ranking of the quality criteria of Tables 3.2 and 3.3

	Retention curve								AWC					
	Rank RMSE		Rank MAE		Rank CI		Mean rank		Rank RMSE		Rank ME		Mean rank	
DIN4220.TexTRD	2.4	(2)	3.1	(3)	3.7	(5)	3.1	(2)	2.6	(3)	2.6	(2)	2.6	(1)
Hypres.KGA	3.5	(3)	2.9	(2)	4.2	(6)	3.5	(3)	6.5	(7)	5.5	(6)	6.0	(7)
PUH2.KGA	2.4	(1)	2.0	(1)	3.0	(1)	2.5	(1)	2.5	(2)	3.0	(3)	2.8	(3)
Wessolek.TexTRD									2.0	(1)	2.6	(1)	2.3	(2)
Teepe.KGA	4.4	(5)	4.7	(6)	3.0	(2)	4.0	(5)	4.3	(5)	4.3	(5)	4.3	(5)
Teepe.TexTRD	4.5	(6)	4.4	(5)	3.6	(4)	4.2	(6)	4.2	(4)	3.4	(4)	3.8	(4)
Vereecken.KGA	3.7	(4)	3.9	(4)	3.4	(3)	3.7	(4)	5.2	(6)	5.8	(7)	5.5	(6)

The ranks of the quality criteria in the individual soil texture groups were weighted by the number of observations of the NFSI II within the respective soil texture group; the placement is shown in parentheses next to the ranks

3.3 Fine Root Distribution on NFSI Sites

Fine root distributions are essential for the parametrization of soil water budget models of forest sites, because they represent the link between the forest stand and the soil. However, the ideal for tree species-specific rooting patterns is not applicable in practice, because the actual fine root distribution patterns are strongly influenced by soil and site properties (Hartmann and von Wilpert 2014). Therefore, it is necessary to determine the actual depth distribution of fine roots or define reliable transfer functions for sites without any root information.

3.3.1 Fine Root Density Model

Using multivariable statistical analyses, correlation analyses were performed between the fine root density (FRD, measured in classes shown in the profile description according to Ad-Hoc AG–Boden (2005), converted into mean values of each class) as the target variable and soil and site properties as predictor variables. With the help of "boosted regression trees" (BRT), the principle parameters and their influence on FRD were identified. The final BRT model is a linear combination of all calculated regression trees. The quality of the model can be rated by correlation with the training and the validation data, as well as the remaining error. The influence of the individual parameters is interpreted using the percentage of the influence and the absolute effects on the model's result. For the analyses, the "dismo" package of the statistical software "R" was used (Elith et al. 2008).

The final model was limited to the five most important parameters and has a good model quality (correlation training data, 0.701; correlation validation, 0.695). The effect of the individual parameters on the target variable FRD is displayed in Fig. 3.4. With a percentage of 62.2%, the parameter soil depth explains the largest part of the FRD, meaning that the maximum can be found in the topsoil, with a continuous decrease with soil depth. The humus content proved to be another important predictor. On the one hand, it is a result of root growth, especially in the subsoil, and on the other, it facilitates root penetration through loosening the soil structure. Alongside those parameters are also soil physical properties such as bulk density and available water capacity, as well as the relief parameter slope in direct relation with the FRD.

The approximate linear relationships within the model can be summed up in a PTF. To achieve this, a simple linear equation model was created. The resulting PTF for FRD in n dm^{-2} is (adjusted R-squared, 0.3973; $p < 0.001$; Fig. 3.4):

$$\begin{aligned} \mathrm{FRD} = {} & 11.63 - 0.084\ \mathrm{soil_{depth}} + 3.22\ \ \mathrm{humus_{class}} - 3.42\ \mathrm{TRD} \\ & + 0.108\ \mathrm{slope} + 0.095\ \mathrm{AWC} \end{aligned} \tag{3.6}$$

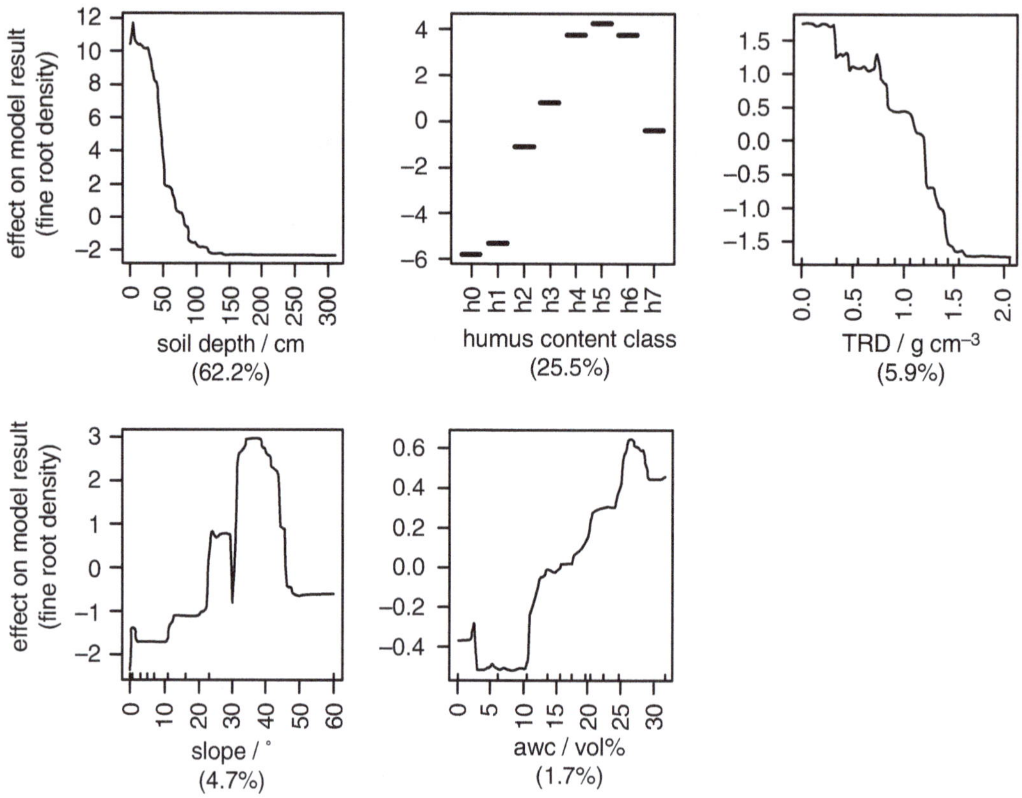

Fig. 3.4 Results of the "boosted regression trees" analysis: The fine root density in the horizon is the models' target variable. The individual diagrams show the influence of the parameters (*x*-axis) on the modelled fine root density (*y*-axis). The percentage determines the weighting of the parameter in the model; soil_depth, depth of the horizons bottom edge (cm); humus = humus class according to Ad-Hoc AG–Boden (2005); TRD, bulk density, measured in different depths and converted to represent the specific soil horizon (g cm^{-3}); slope, slope on site (°); AWC, available water capacity, measured in different depths and converted to represent the specific soil horizon (vol%)

In comparison with the BRT modelling, the PTF performs worse. But although being weaker, the PTF reaches a high model quality as well.

3.3.2 Continuous Fine Root Distribution

For parametrization of the root distribution in water budget models, continuous information on the FRD is required. To achieve this, the exponential function according to Gale and Grigal (1987) may be used. It calculates the depth-dependent, relative root distribution continuously using a shape parameter β (Fig. 3.5). This is especially reasonable for inhomogeneous data sets such as the NFSI II data set, because different data collection methods (Bavaria, estimation at soil auger samples; Baden-Wuerttemberg, counting fine roots in a 5 cm grid at the wall of the profile; rest

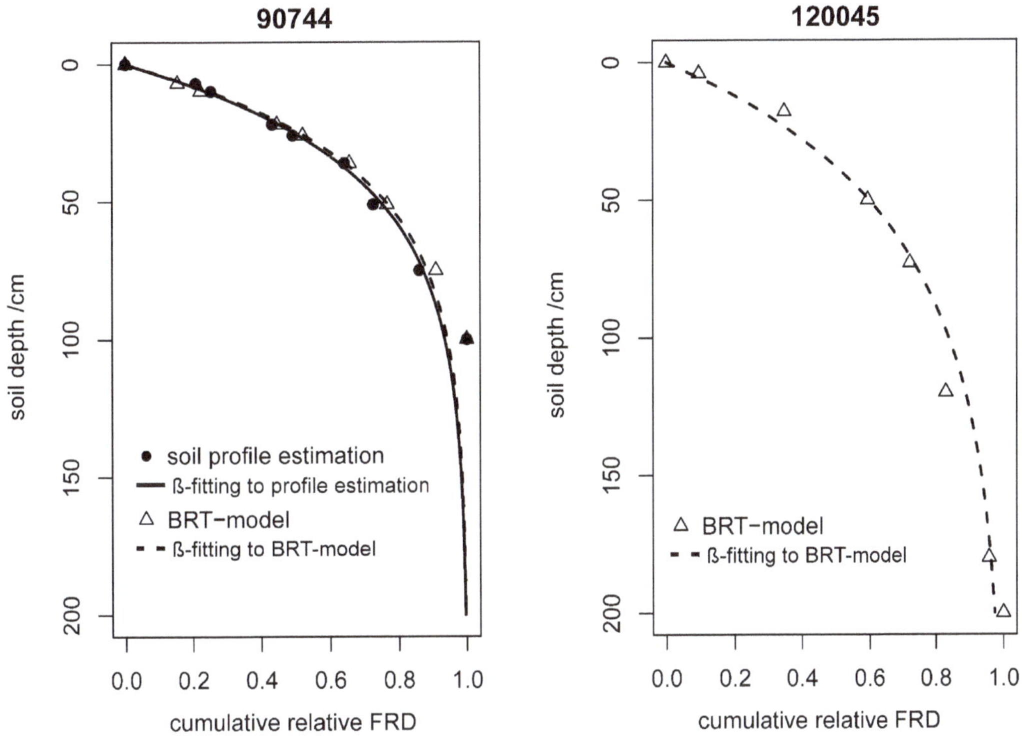

Fig. 3.5 Example of a good adjustment and congruency between estimated values, "boosted regression trees" model values and the respective adjustment to the β-model (left), and example of a site at which the β-model was adjusted to the "boosted regression tree" model values, because soil profile estimates were lacking (right)

of Germany, estimating at the soil profile) lead to different estimates of rooting intensities.

Figure 3.6a shows the number of fine roots (derived from the different estimating procedures) in the recorded root zone for all NFSI profiles. It can be noticed that above-average amounts of fine roots occur in Baden-Wuerttemberg, whereas in Bavaria the amounts are often extremely small. Despite the regional differences, the estimates of the FRD were used for further adjustment, since only the relative depth distribution was taken into account, and therefore, different classification procedures during profile description were "normalized". Where there were no estimated values available, they were filled in with values from the estimates of the BRT. First, the FRDs per horizon were calculated (FRD × thickness) and plotted as the cumulative sum on the lower end of the horizon and then fitted to the β-model according to Gale and Grigal (1987) with a non-linear optimization algorithm (Fig. 3.5):

$$Y = 1 - \beta^z \tag{3.7}$$

where Y is the cumulative relative FRD in depth z and β the parameter to be adjusted.

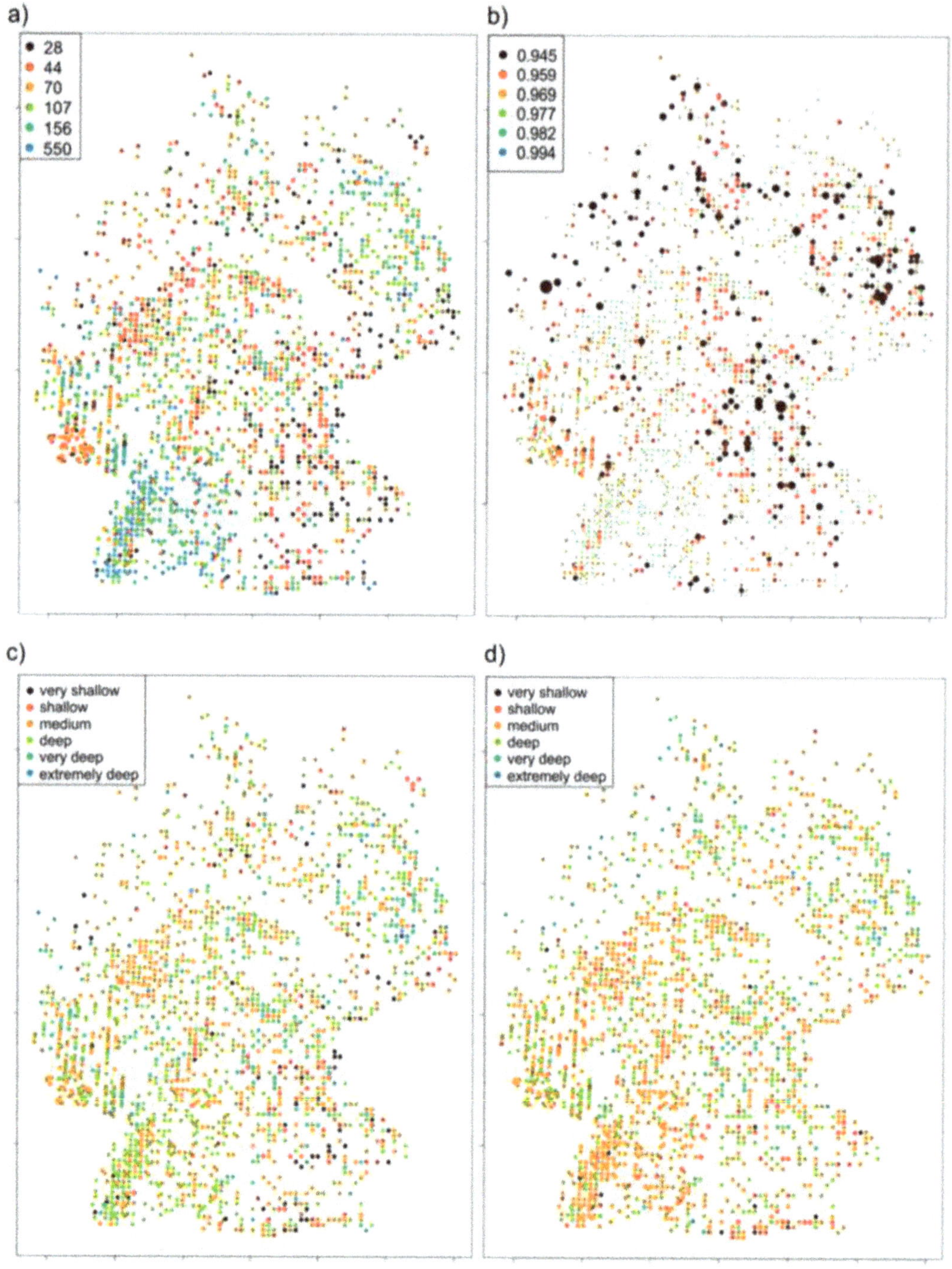

Fig. 3.6 Number of fine roots summed up over the whole soil profile to a width of 1 dm (**a**). The different colours define percentiles (10-25-50-75-90-100); β-values of the NFSI II profiles (percentiles 10-25-50-75-90-100). The size of the dots correlates with the β-value (**b**). Blue dots represent deep distributions, red shallow ones; RDeff according to conventional definition (**c**); RDeff, determined using the depth, in which 86% of the cumulative amount of fine roots are located according to the β-model (**d**)

The relative cumulative amount of fine roots can now be calculated for any depth using this function. The resulting β-values are shown in Fig. 3.6b, where high values represent deeper distributions.

3.3.3 Effective Rooting Depth

The effective rooting depth is a cautious estimation of the main rooting zone used in German forest soil classification, which can be derived from the lower limit of a FRD >2 fine roots per dm^2 according to AK Standortskartierung (2003). By this approach an overestimation of the effective rooting depth is avoided, because the least densely rooted zone in the subsoil is not taken into account. In our data, the effective rooting depth always coincides with the lower boundary of a horizon, because the FRD is defined for each horizon individually. This value is distorted to some extent, as a depth-dependent decrease in FRD can also be assumed within a horizon. Additionally, different data collection methods result in FRDs that are not comparable to each other (Fig. 3.6c). The estimates from the conventional approach reaffirm the differences between Baden-Wuerttemberg and Bavaria caused by different estimation methods.

In order to generate a uniform value for the β-model on the basis of the boundary Wf2/Wf1 (2 fine roots per dm^2) from the conventional model, the cumulative percentage Z was determined, for which the rooting depth from the conventional approach, RDeff(conv), can be reached best. This was done using a non-linear adjustment:

$$\mathrm{RDeff(conv)} \sim \log(1-Z)/\log(\beta) \tag{3.8}$$

This results in a value for Z of 86%, which will be used hereafter as the effective rooting depth according to the β-model, RDeff(β 86%) (Fig. 3.6d).

3.3.4 Effect of Stand Type, Soil Class and Acidification

The forest stand at the NFSI sites are in most cases no pure stands with one tree species only, rather stand types of different degrees of purity and mixed stands are defined. Hence, as the species origin of the fine roots was not determined, the following analyses always refer to the stand type and not a single tree species. The evaluation of the rooting space is based on the RDeff value (Fig. 3.6), which corresponds to the 86% depth in the β-model. No significant differences were found in the rooting depths for the individual stand types (Fig. 3.7). There is a wide variation in all stand types, with median values of approx. 60 cm. The stratification according to soil type also shows only small differences between the groups. Only the sandy soils poor in base cations show increased RDeff values,

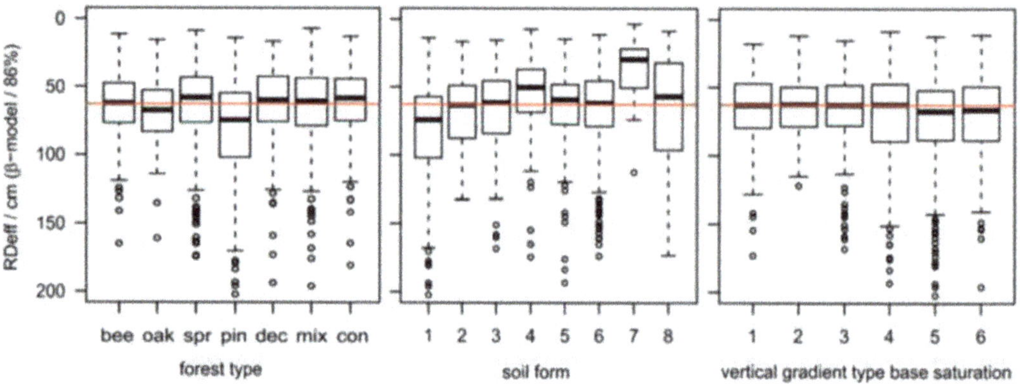

Fig. 3.7 Boxplots showing the effective rooting depth, RDeff (β-model; see Eq. 3.8), stratified according to forest type, soil form (centre) and vertical gradient of base saturation (right); *bee* beech, *oak* oak, *spr* spruce, *pin* pine, *dec* deciduous forest, *mix* mixed forest, *con* coniferous forest; soil forms, 1 = sands poor in base cations, 2 = alluvial soils and gleys of wide river valleys, 3 = lowland soils and loess loams, 4 = limestone-derived soils, 5 = basic to intermediate soils originated from crystalline rock and pelosols, 6 = soils poor in base cations from crystalline rock, 7 = alpine soils, 8 = peat soils; vertical gradients, 1 = complete base saturation (100%) in the root space, 2–4 = base saturation decreasing with depth and qualitatively in the main root space with full base saturation in the subsoil, 5 = base depletion in the whole profile, 6 = increased base saturation in the humus/mineral topsoil by liming, classification according to Hartmann and von Wilpert (2016)

whereas in limestone-derived soils, and especially in alpine soils, several significantly reduced values appear. Surprisingly, the RDeff values do not point towards an influence of an advancing acidification or of forest liming, as the stratification according to vertical gradients of base saturation shows.

3.4 Modelling Dynamic Water Availability in Forests

3.4.1 Model Description, Input Data, Parameterization and Target Variables

The water budget modelling for individual NFSI II points was carried out using the process-based forest hydrological simulation model LWF-Brook90 (Hammel and Kennel 2001), which is a further development of the model Brook90 (Federer et al. 2003). LWF-Brook90 calculates the water budget of a one-dimensional, multilayered soil profile with vegetation cover in daily resolution. Water transport in the soil is described by the Richards equation. Source and sink terms thereby include processes such as flow through macropores and water uptake by roots. The hydraulic properties of individual soil layers are parameterized according to Mualem (1976) and van Genuchten (1980). Evapotranspiration is calculated according to the approach by Shuttleworth and Wallace (1985), which differentiates between the evaporation originating from the soil or a snow cover and the transpiration and interception from a "single big leaf" plant cover, using a conductivity model. The seasonally varying leaf area index (LAI) is an important controlling factor, which plays a key role for interception storage and the distribution of the energy available for evaporation between soil and plant, according to Lambert-Beer's law. It also has a strong influence on transpiration as it is used for upscaling canopy conductance from stomatal conductance, which varies depending on the meteorological conditions. Actual canopy conductance and available energy, in turn, determine the water demand of the vegetation cover (potential transpiration, T_p), which is covered from the soil layers depending on the layers' water availability and root length density (actual transpiration, T_a).

3.4.1.1 Climate Data, Soil and Site

LWF-Brook90 requires meteorological input data in daily resolution (precipitation, temperature, radiation, water vapour pressure, wind speed). For this model application, the data was provided by the chair of physical geography (Prof. Böhner) at the University of Hamburg using a 250 × 250 m grid. For methodical details see Weinzierl et al. (2013). The validation of the regionalized air temperatures shows a deviation of the monthly mean temperatures around ±0.5 °C for 2002 and ± 0.2 °C for 2003. Monthly precipitation sums deviate on average by 5% from the measured values.

For representing the NFSI soil profiles in the model, the measured physical properties of the different depths were used and complemented with information about soil horizons, coarse soil fraction, texture and humus content from the soil profile description. Bulk density of the mineral soil was not available below 90 cm soil depth, and the measured value of the lowest layer was assigned to all subjacent horizons. Very high coarse soil fractions of bedrock horizons were constrained to 92.5% in order to include water retention in cracks and fractures in the rock. From the complete soil profile information and humus layer thickness, the model layer discretization was created, consisting of a varying number of model layers that preserve the original horizon depth boundaries and have an increasing thickness of 1–20 cm. The fine root distribution of the mineral soil model layers was derived from Eq. (3.6) (see Sect. 3.3.2), and the estimated root density of the uppermost mineral soil layer was used for the humus layer. The rooting depth was constrained to the soil depth where on-site field observations reported zero root abundance and all layers below that depth were assumed to be root-free. In case roots were observed down to the bottom of the soil profile, or not reported to the bottom of the profile, the maximum root depth was assumed to be the bottom of the soil profile, but at least 160 cm. Shallow profiles were extended correspondingly. Root penetration at profiles with limited soil depth due to solid bedrock, ground- or stagnant water was further restricted to a soil depth which was derived from soil horizon symbols following a rule-based procedure. The lower boundary of all soil profiles was formed by two standard model layers with a coarse-grained texture and an overall thickness of 40 cm in order to define a uniform lower boundary on the one hand and to facilitate upward water flow to the rooting zone even in shallow soils.

In a last step, the hydraulic properties of the mineral soil layers were derived from soil physical properties using those PTFs, which provided the most reliable predictions when compared with measured hydraulic properties (see Sect. 3.2.1). The parameters concerning the function for soil water retention according to van Genuchten (1980) (θ_s, θ_r, α, n) were derived using the PTF according to DIN 4220 (2008–2011). The parameter m was specified as $1–1/n$. The saturated conductivity (K_s) for the unsaturated conductivity model of Mualem (1976) was derived from tabular values of Wessolek et al. (2009); the parameter τ was set to 0.5. In order to derive hydraulic properties of the humus layer horizons, the PTF of Hammel and Kennel (2001) was used. Peat layers were parameterized using the PTF of Wösten et al. (1999) for organic soils. Apart from water transport in the soil matrix, for which retention and conductivity properties are relevant, the model also allowed macropore-assisted infiltration by distributing net precipitation among the humus layer and the upper 30 cm of the mineral soil with the proportions decreasing exponentially with depths.

Further variables used are latitude, slope and aspect, which determine a site's net radiation. They were taken from the header information of the NFSI II data set. The possible influence of capillary rise of water from shallow groundwater bodies to the rooting zone was not considered explicitly in the simulations.

3.4.1.2 Parameterization of the Vegetation

Two different approaches were chosen for the parameterization of the vegetation. In a first approach, standard parameter sets for theoretical forest stands were defined. They represent the above-ground properties [LAI, stem area index (SAI) and height] of typical stands of full-grown beech, oak, spruce, pine and mixed forests in the model (Table 3.5). The applied theoretical forest stand was determined using the

Table 3.5 Model parameters that differ depending on the forest stands in the model

	Pine	Spruce	Beech	Oak	Mixed forest
Canopy height	25	30	30	30	30
Leaf area index (m^2 m^{-2}) (max/min)	3.5/1.4	5.5/ 4.4	6/0.6	4.5/ 0.45	5/2
Maximum canopy conductance (mm s^{-1}) (max/min)	18/ 10.8	18/ 14.4	25/2.5	20/2	21/8.4
Canopy interception capacity (mm) (max/min)	1.63/ 0.98	1.5/ 1.2	1.2/ 0.12	1.3/ 0.13	1.1/0.44
Interception catch rate (–) (max/min)	0.41/ 0.25	0.8/ 0.64	0.92/ 0.09	0.51/ 0.05	0.8/0.32
Leaf width (cm)	0.04	0.04	05	5	1
Albedo (–)	0.14	0.14	0.18	0.18	0.18
Albedo, ground covered with snow (–)	0.14	0.14	0.23	0.23	0.23

Maximum and minimum values refer to intra-annual variability (summer/winter)

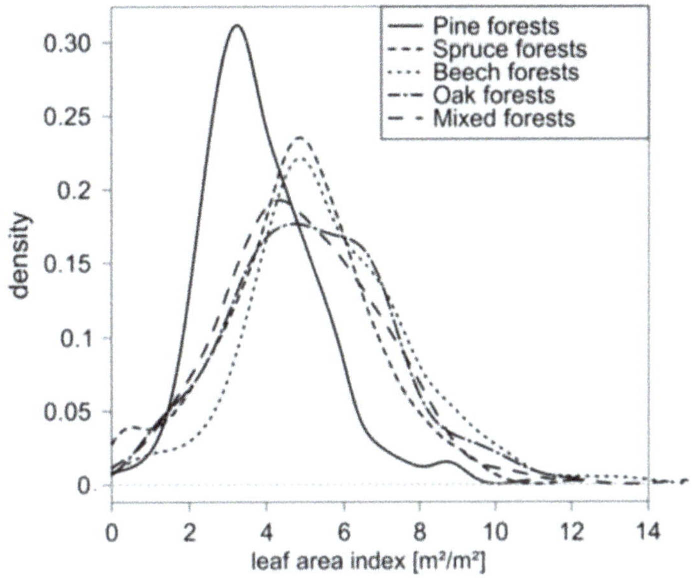

Fig. 3.8 Density plots of the values of leaf area index at the NFSI plots, estimated using individual tree data, shown for the different forest stand types

information about forest stand type contained in the NFSI II header information. Unspecified broadleaf forest stands were assumed to be beech stands, while unspecified coniferous forest stands were assumed to be spruce stands.

A second approach incorporates the effects of the actual forest stand (by estimating the LAI, stem area index and height) on the NFSI plots into the simulation. The results of this second application of the model are not discussed further in the context of this chapter. They are, however, made available as annual values for other evaluation groups and may be used to calculate the nutrient output with seepage water and to estimate nutrient budgets. For deriving the stem area index from the individual tree data in the harmonized forest stand inventory (HBI), the allometric functions by Hammel and Kennel (2001) were applied. The functions distinguish between broadleaf and coniferous trees. For estimating the LAI using the forest stand data, different estimation functions, depending on tree species, were used. For estimating the LAI of pines and larches, the litterfall model of Law et al. (2001) was adopted and parameterized for pine according to Ahrends et al. (2010). The leaf area of beech, oak and other broadleaf trees were estimated using the biomass functions of Weis et al. (2012). The leaf area index for spruce, fir and Douglas fir was estimated using the allometric function by Hammel and Kennel (2001), which is based on reanalyses of data on leaf mass published in Hammel and Kennel (2001). Subsequently, the estimated leaf areas of the individual trees were added together, and the forest stand LAI was calculated. The frequency distribution of the LAI estimates for various forest types is shown in Fig. 3.8 as density plot.

Apart from the forest stand properties above, both model applications (theoretical standardized forest stands and real forest stands) used the same model settings and parameters. The choice of which model parameters and model settings to use is based directly on the suggestions made by Federer et al. (2003) (albedo, light extinction coefficient), as well as the suggestions made by Hammel and Kennel (2001) (length of the foliation and leaf fall phases). Beginning and end of the growing season (i.e. bud burst in spring and start of leaf fall in autumn) were determined dynamically for each NFSI plot using the temperature-based methods described in Menzel (1997) and von Wilpert (1991), which were parameterized for different tree species. Other parameters were adapted to fit interception evaporation measured at Level II intensive monitoring plots in Germany (ICP Forests 2010).

Because no water budget measurements exist for the NFSI plots (e.g. throughfall precipitation, soil water potential or soil water content), a direct validation of the water budget simulation cannot be carried out at the NFSI plots. However, model applications of LWF-Brook90 on intensively monitored sites and the comparison of time series collected there showed that water budget simulations using the described settings and parameters produced satisfactory results (unpublished data). In addition, the comparison between the modelled amounts of seepage water and the values taken from the Hydrological Atlas of Germany showed no systematic variations. Therefore, the assumption can be made that the model settings used provide reasonable estimates about the water budget parameters and drought stress indicators, which reflect physically and physiologically reasonable influences of climate, soil and forest stand properties.

3.4.1.3 Processing the Results

The modelled water flows [e.g. groundwater recharge (GWR), evapotranspiration (ET), actual transpiration (T_a), evaporation from interception (I) and evaporation from the soil (E)] were aggregated to time series of monthly sums and sums over the dynamic growing season for further analysis. Additionally, different drought stress indicators were derived that quantify limitations on water availability. A widely used indicator is the relative water content (RW), which is the ratio of the actual soil water storage, S_t, to the soil water storage at field capacity, FC: RW $= S_t$/FC. The relative extractable water storage, REW, is the ratio of currently plant-available water storage ($S_p = S_t -$ PWP) to the plant-available water capacity, AWC: REW $= S_p$/AWC. From the depth-discrete results of the LWF-Brook90 modelling, values for S_t, S_p, RW and REW were aggregated for the following depth ranges: 0–30 cm for the mineral soil depth ($S_{t,030}$, $S_{p,030}$, RW_{030}, REW_{030}) and 0–90 cm for the effective rooting depth. Mean and minimum values over the dynamic growing season were then calculated using the daily values of the drought stress indicators in order to judge mean water availability in individual years.

Another widely used indicator for water deficiency is the difference or the ratio between actual (T_a) and potential transpiration (T_p). A decrease in T_a below T_p is caused by limited water availability in the root space in LWF-Brook90. The transpiration ratio, Tratio $= T_a/T_p$, with values of <1, and the transpiration difference, Tdiff $= T_p - T_a$, with values of >0, therefore indicate a water shortage in the root space.

From the annual and growing season total water flows and drought stress indicators, distribution statistics (mean, median, quartile, minimum, maximum) for the period 1981–2010 were calculated for each NFSI plot. Also, the relative deviation of the annual values from the mean value in the period 1961–1990 was calculated for each NFSI plot in order to examine time-related changes with the help of distribution statistics (median, quartile, 90%- and 10%-quantile for each year over all NFSI plots).

3.4.2 Results

Figure 3.9 shows distribution statistics of evapotranspiration during the growing season (ET, Fig. 3.9a), annual total groundwater recharge (Fig. 3.9b), transpiration difference (TDIFF, Fig. 3.9c) and mean relative extractable soil water storage in the rooting zone during growing season (REW, Fig. 3.9d) at the NFSI plots, modelled using the standardized forest stand. It shows the cumulative frequency distribution of the medians, as well as the respective ranges and quartiles of the individual annual values of NFSI plots for the period 1981–2010. The spatial distribution of the medians of these dimensions is shown as a map in Fig. 3.10.

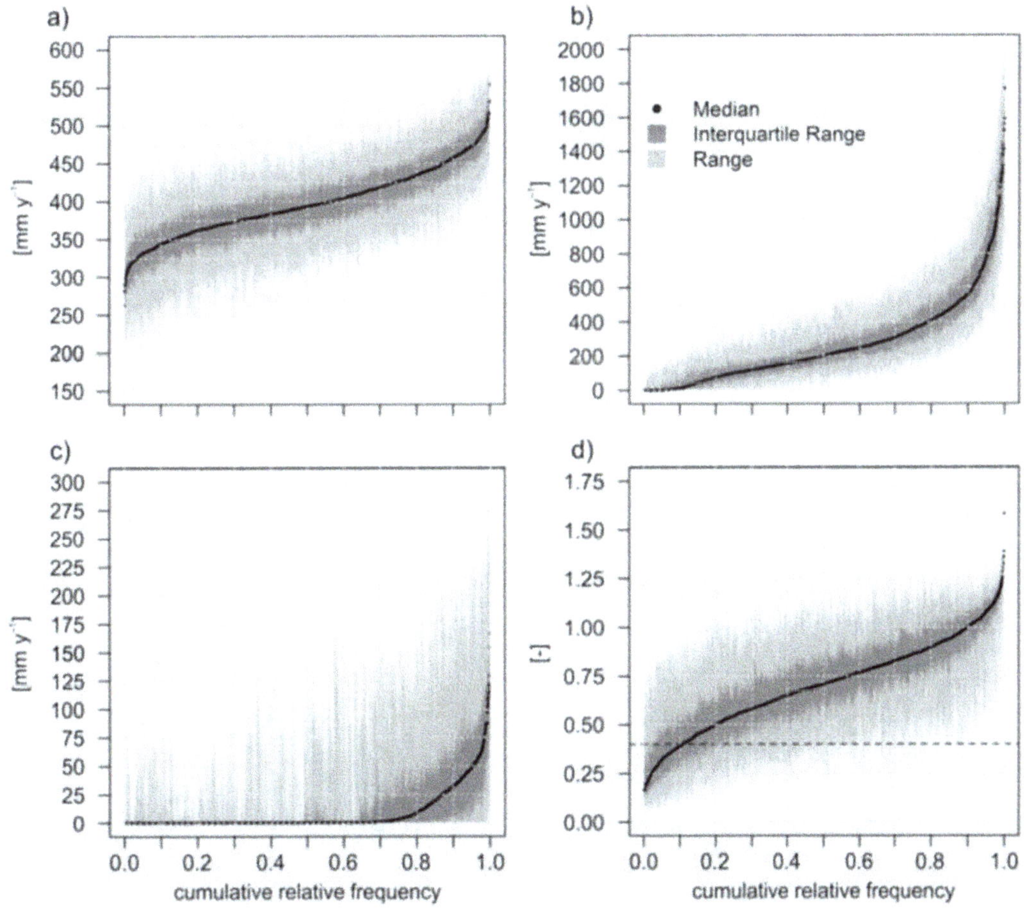

Fig. 3.9 Median values of the actual evapotranspiration during growing season (**a**), the annual groundwater recharge (**b**), the transpiration difference during growing season (**c**) and the mean relative available soil water storage during growing season (**d**) for NFSI plots in the period 1981–2010; modelled with standardized forest stands

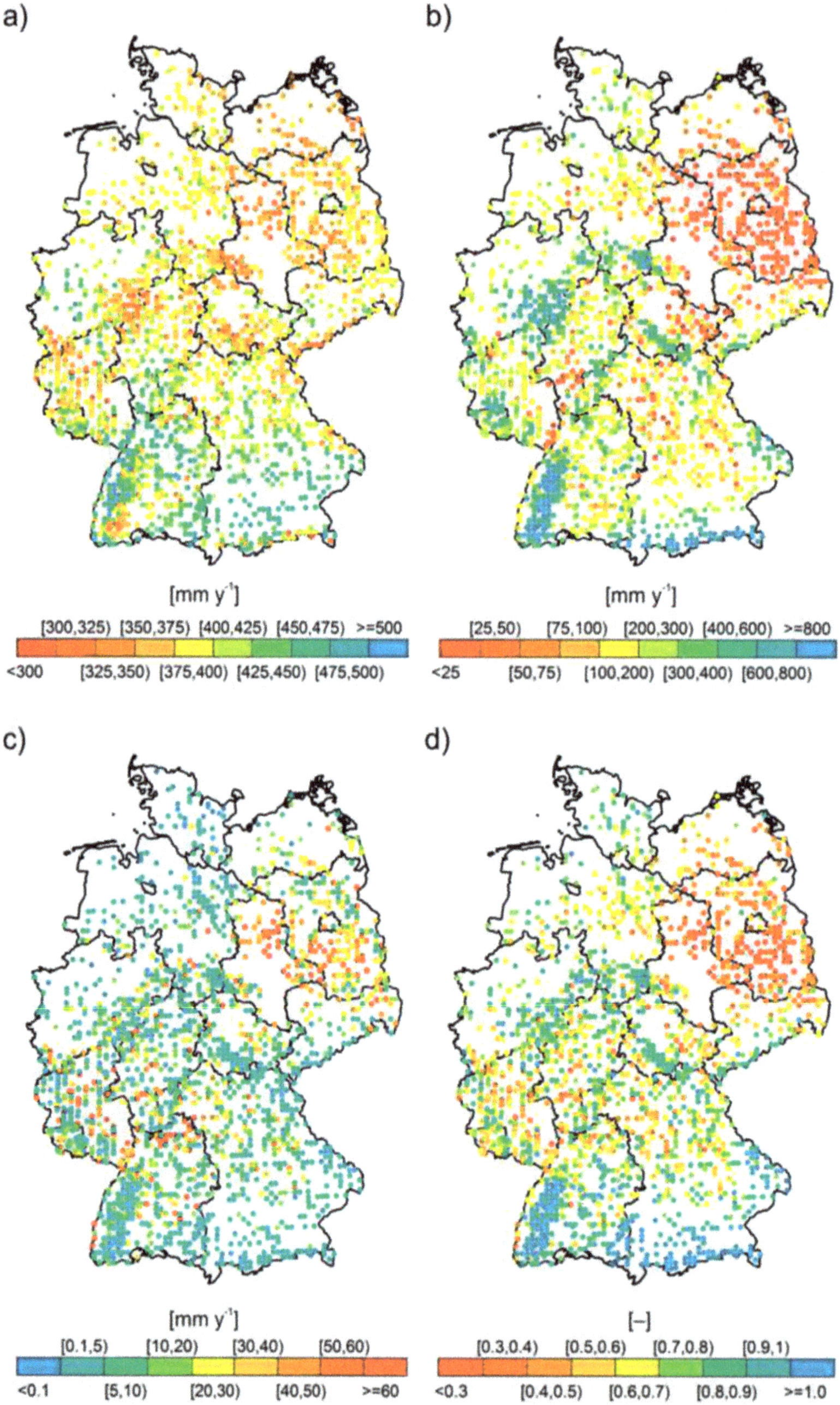

Fig. 3.10 Spatial distribution of the median values of the actual evapotranspiration during growing season (**a**), the annual groundwater recharge (**b**), the transpiration difference during growing season (**c**) and the mean relative available soil water storage during growing season (**d**) for NFSI plots in the period 1981–2010; modelled with standardized forest stands

Figure 3.9a shows that the mean actual evapotranspiration during the growing season of 80% of the NFSI plots lies between 350 and 450 mm. Plots especially with lower mean evaporation have extreme fluctuations, with up to 300 mm between individual years of the represented period (1981–2010). But in half of all years, the fluctuation lies below 60 mm which is the mean difference between the upper and the lower quartile over all plots. About 5% of the NFSI plots show mean evapotranspiration rates of 475 mm or more (Fig. 3.9a). These can be found in the Bavarian late-moraines and molasses and in the foothills of the Alps and the Black Forest (Fig. 3.10a; blue and blue-green dots). High evaporation rates of 450 mm or more can also be found on the western slopes of the Pfälzerwald and Odenwald and in the Bergische Land. On the one hand, the mentioned regions have very high precipitation. On the other hand, they are also warm with high potential evaporation rates that take effect during long growing seasons. The lowest evapotranspiration rates do not show such a distinct geographical pattern. Several areas in the rain shadows of mountain ranges (north-eastern foothills of the Harz, Altmark, Hessisches Schiefergebirge) are distinguishable, and it seems that there is an increase of NFSI plots with less than 350 mm evapotranspiration (yellow, orange and red dots) towards the north. But plots with very low evapotranspiration of less than 300 mm appear more or less scattered everywhere on the map, thereby blurring spatial patterns. On these plots, actual transpiration is substantially limited each year due to a high skeletal fraction and/or shallow tree rooting. Apart from those sites, the mean evapotranspiration follows the interaction between precipitation and temperature. In this way, a site in the cool and moist Hochsauerland (precipitation/temperature during growing season, 420 mm/13 °C) with an average evapotranspiration that is limited by rather low potential transpiration can have the same low evaporation of less than 350 mm as a warm and dry site (250 mm/16.5 °C) in Brandenburg or Saxony-Anhalt during the summer half of the year. There, the low evapotranspiration rate is caused by low precipitation and the low water retention capacity of the typically sandy soils. The transpiration difference (Tdiff, Fig. 3.10c) makes it possible to determine whether or not a site's evaporation is limited by water shortage. Tdiff in the growing season is higher than 30 mm on average in southern Brandenburg and smaller than 5 mm in the Hochsauerland and other top ranges of low mountain ranges and in the proximity of the North Sea. Overall, growing season Tdiff on all NFSI sites is characterized by a large range (Fig. 3.9c), illustrating that water shortage may occur in all plots during dry years in which the vegetation's water demand cannot be entirely met through water reserves in the soil.

About 25% of the NFSI plots show a mean annual groundwater recharge of less than 100 mm (Fig. 3.9b), 10% of the plots have very little groundwater recharge. There, the median is at 0 mm, which means that in half of the years from 1981 to 2010, the groundwater has not been recharged at all for the assumed theoretical forest stand. For most plots, annual groundwater recharge is 50–400 mm, but a substantial decrease in, or even a complete absence of, groundwater recharge in extreme years is possible. Approx. 10% of the plots have a discharge of more than 500 mm per year; it can be as high as 2400 mm in some years, e.g. at sites at high altitude and with high precipitation, where snow masses accumulate in winter and melt in spring. While the actual evapotranspiration during growing season is mainly a product of temperature and precipitation, groundwater recharge clearly follows the precipitation distribution (Fig. 3.10b). Plots with an annual groundwater recharge of more than 400 mm are located—almost without exception—in high altitudes with high precipitation. Large areas with very low (<25 mm) groundwater recharge are located mainly in Brandenburg and Saxony-Anhalt, and smaller areas are situated in eastern Thuringia, southern Hesse and the Rhine-Main area (Fig. 3.10b), all of which experience very low annual rainfall amounts and high potential evaporation.

Figure 3.9d shows the cumulative frequency distribution of the mean available soil water storage in the root space during growing season. It becomes clear that the NFSI plots have a wide range. Approx. 10% of the sites have a mean available soil water storage of more than 100% of the AWC. On the one hand, soil horizons with low hydraulic conductivity cause infiltrating water to accumulate in the root space, so that the FC is regularly exceeded on these sites. On the other hand, large amounts of precipitation provide a positive water balance also during growing season and therefore make sure that loss through transpiration is regularly compensated by abundant precipitation. On approx. 10% of the NFSI plots, however, the mean available soil water storage decreased to below 40% of the maximum value (AWC) in at least half, on 5% of the NFSI plots in three-fourths of the years in the period 1981–2010. Falling below this threshold (drawn in as a reference line in Fig. 3.9d) is interpreted as a water shortage by Bréda and Granier (1996). It occurs mainly on NFSI plots that get low amounts of precipitation during growing season on the one hand, or whose available water capacity is limited by a high skeletal fraction or a shallow rooting on the other hand. Such NFSI plots are mainly found in parts of Brandenburg, Thuringia and Saxony-Anhalt but also scattered in the low mountain ranges on particularly shallow sites. On most NFSI plots, water shortage only occurs in dry years. However, for approx. 90% of the NFSI plots during the period 1981–2010, mean available water retention falls below the Granier threshold of 40% AWC for at least one growing season.

3.5 Deriving the Risk for Drought Stress

The probability of falling below the limit values of soil water availability relevant to tree physiology, such as the threshold value according to Bréda and Granier (1996), is a key indicator for predicting tree damage caused by climate change. In the following, the results of the simulation with LWF-Brook90 are aggregated to get characteristic values for water shortage. They are linked to tree growth observed at the NFSI plots in Chap. 11.

3.5.1 Characteristic Properties of Water Shortage

In addition to the characteristic properties of water shortage defined in Sect. 3.4.1.3, other indicators are derived below, which quantify the intensity and duration of periods with water shortage. The characteristic value d_REW_{CL} indicates the number of days, on which the relative soil water storage (REW) falls below a certain critical limit (CL_{REW}) during growing season; v_REW_{CL} indicates the missing water volume below the threshold value:

$$v\_REW_{CL} = \sum_{i=\mathrm{vegbeg}}^{\mathrm{vegend}} \begin{cases} S_{p,i}/aFC < CL_{REW} : 1 - \left(S_{p,i}/aFC\right)/CL_{REW} \\ S_{p,i}/aFC \geq CL_{REW} : 0 \end{cases} \tag{3.9}$$

$$d\_REW_{CL} = \sum_{i=\mathrm{vegbeg}}^{\mathrm{vegend}} \begin{cases} S_{p,i}/aFC < CL_{REW} : 1 \\ S_{p,i}/aFC \geq CL_{REW} : 0 \end{cases} \tag{3.10}$$

For further evaluation, three different threshold values were tested: $CL_{REW} = 0.2$, 0.4 and 0.6.

Similarly, drought stress indicators based on threshold values were also calculated using transpiration ratios according to the following equations:

$$v\_Tratio_{CL} = \sum_{i=\mathrm{vegbeg}}^{\mathrm{vegend}} \begin{cases} T_a/T_p < CL_{Tratio} : 1 - \left(T_a/T_p\right)/CL_{Tratio} \\ T_a/T_p \geq CL_{Tratio} : 0 \end{cases} \tag{3.11}$$

$$d\_Tratio_{CL} = \sum_{i=\mathrm{vegbeg}}^{\mathrm{vegend}} \begin{cases} T_a/T_p < CL_{Tratio} : 1 \\ T_a/T_p \geq CL_{Tratio} : 0 \end{cases} \tag{3.12}$$

The critical limits used were $CL_{Tratio} = 0.8$ and $CL_{Tratio} = 0.5$.

Further drought stress indicators were derived from the soil water potential in the root space. For this purpose, the mean soil water potential (ψw) (weighted by layer thickness) for the root space was calculated, and the days during growing season were counted on which ψw fell below the threshold value $CL_{\psi} = -1200$ hPa (von Wilpert 1991). Also, a “deficit” (integral of the time series of water potential below CL_{ψ}) was defined similar to v_REW_{CL}:

$$v\_\psi_{CL} = \sum_{i=\mathrm{vegbeg}}^{\mathrm{vegend}} \begin{cases} \psi w_i < CL_{\psi} : \psi_{CL} - \psi w_i \\ \psi w_i \geq CL_{\psi} : 0 \end{cases} \tag{3.13}$$

$$d\_\psi_{CL} = \sum_{i=\mathrm{vegbeg}}^{\mathrm{vegend}} \begin{cases} \psi w_i < CL_{\psi} : 1 \\ \psi w_i \geq CL_{\psi} : 0 \end{cases} \tag{3.14}$$

The amount of deep seepage water below the root space (*vrfln*) was considered as a measure for possible water surplus. In addition to the output parameters of LWF-Brook90, different climatic parameters were examined in regard to their influence on growth in thickness of the trees: air temperature (mean value *tmean*, temperature sum *tsum*, minimum *tmin*), number of days that exceed a temperature threshold of 5, 10 or 20 °C (*gdd5*, *gdd10*, *gdd20*), as well as respective temperature sums (*tsum5*, *tsum10*, *tsum20*), precipitation (*prec*), global radiation (*globrad*), FAO grass reference evapotranspiration (*et0*), climatic water budget (*kwb*), beginning, end and length of the growing season (*vp_start*, *vp_end*, *vp_dauer*). Annual values for all parameters were derived from the modelling. They are either calculated as the mean values/sums of the whole year (ending of the variable is *_y*), of the dynamic growing season of this year which was calculated using LWF-Brook90 (*_vp*), of the months April–September or of the months May–July.

3.5.2 Future Drought Trend

It was shown that soil water availability has substantial influence on the width of year rings of the different trees examined. With the exception of the BRT model for oak, the extent of dry periods with a soil water potential of less than −1200 hPa in the root space $v\_\Psi w_{1200}\_vp$ (see Eq. 3.13) was an important covariate in the BRT models of all tree species. In the following, the results of the LWF-Brook90 modelling are presented in regard to the space-time dynamic of the water shortage index $d\_\Psi w_{1200}\_vp$ which correlates strongly with $v\_\Psi w_{1200}\_vp$, but which is, in contrast, specified by a descriptive unit (number of days during growing season where the value falls below the limit). Figure 3.11 shows the annual values for $d\_\Psi w_{1200}\_vp$ at each NFSI plot for the period 1961–2013, where the red symbols represent intensive water shortage and the blue symbols represent plots where the critical value of −1200 hPa was not exceeded throughout the vegetation period. In extremely dry years (1976, 2003), the modelling shows that nearly all NFSI plots fall—in part substantially and for a very long time—below the critical soil water potential. A distinct spatial pattern of the modelled water shortage can be noticed in medium years. There are two types of areas which have particularly intensive water shortage: the stony soils in the low mountain ranges, where the transpiration of the plants uses up the AWC quickly due to high rock content, and the regions located in rain shadows of mountains (e.g. east of the Harz, Thüringer Becken). Looking at the model results for the individual years, the year 1989 seems to mark a turning point in the water supply of forest stands. Since the beginning of the 1990s, years with increased water stress have been on the rise. Of the 10 years with the best water supply—meaning the highest percentage of plots without water stress ($d\_\Psi w_{1200}\_vp = 0$)—there is only one after 1990. For four of the years within the decade from 1970 to 1979, the median of $d\_\Psi w_{1200}\_vp$ across all plots lies below the long-time median, which means that 4 of the 10 years are dryer than average. This value is particularly low for the decade 1980–1989 (3 out of 10 years) and is increasing ever since (1990–1999, 7 out of 10 years; 2000–2009, 6 out of 10 years; 2010–2013, 3 of the 4 years). In addition to the increasing number of particularly dry years, the results of the model show a tendency towards a decrease in variance between the NFSI plots. This can be attributed to plots that usually have a

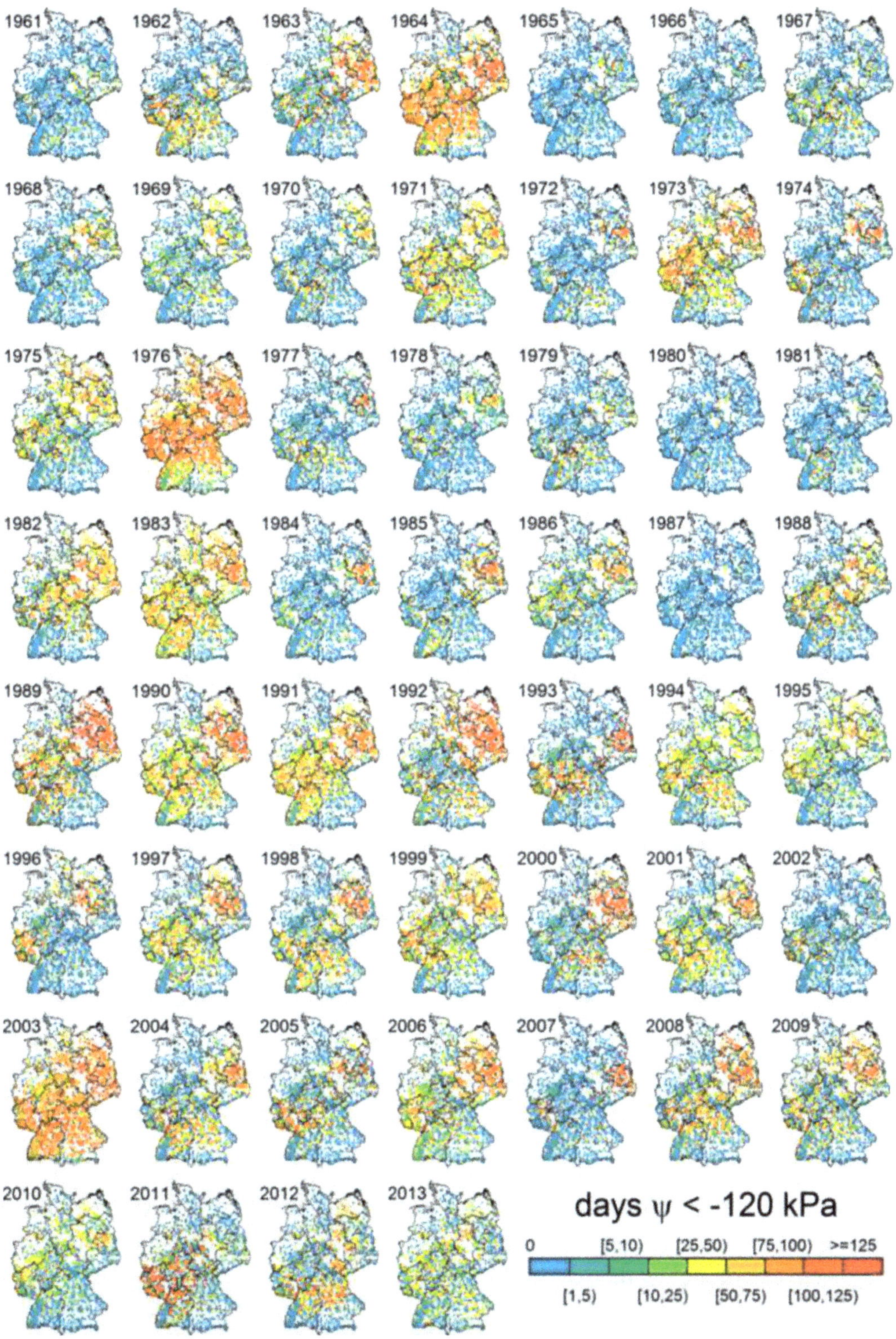

Fig. 3.11 Water shortage index $d\_\Psi w_{1200}\_vp$ (Eq. 3.14), derived from LWF-Brook90 modelling for the years 1961–2013, which quantifies the shortfall below a critical matric potential ($CL_{\Psi} = -1200$ hPa) in the root space

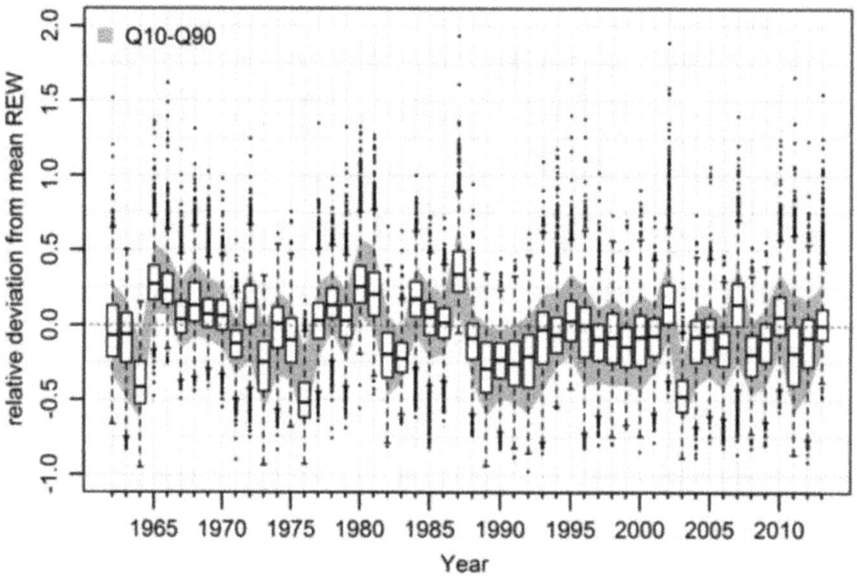

Fig. 3.12 Distribution statistics (medians, quartile areas, 10%- and 90%-quantile) of the available soil water storage during growing season, shown as relative deviation from the long-term mean value of the period 1961–1990

good water supply but experience drought stress in the second half of the simulation period.

The long-term trend of periods of water shortage in the root space also becomes clear in Fig. 3.12, where the chronological trend of the available water storage (*REW_WReff_vp*) is shown as a distribution statistics of the relative deviation from the reference period 1961 to 1990. Years that were exceptionally moist or dry during growing season compared to the reference period can be identified in this way. The general trend of an increasing number of growing seasons with below-average water supply since the end of the 1980s can also be seen here. In most years since 1989, the median was negative, which shows that a below-average soil water storage was registered here for most NFSI plots, compared to 1961–1990. In the time from 1989 to 1992, even the 90%-quantile is negative. This means that 90% of the NFSI plots in four consecutive growing seasons showed a below-average water supply. A fourth of the plots had a soil water storage which was 25% lower compared to the reference period. In the extremely dry years 1976 and 2003, as many as 75% of the NFSI plots had a water storage which was at least 25% less than average. Since 1988, only in the 5 years 1995, 2002, 2007, 2010 and 2013 were the medians of the available water storage clearly above the reference value of the period 1961–1990, while in 20 years, they were clearly below that value.

3.6 Summary and Conclusions

By reprocessing and combining soil, root and forest stand parameters, it was possible to make reliable estimates about the governing factors for the water budget on the individual NFSI plots and to parameterize the one-dimensional water budget model LWF-Brook90 based on this. Special attention was paid to the evaluation of PTFs used for estimating soil hydraulic properties and to the identification of those functions that best represent the complete value range of the very heterogeneous NFSI data set. The decision was made in favour of the PTF according to DIN 4220 (2008–2011) for estimating the van Genuchten parameters, and the PTF by Wessolek et al. (2009) for estimating the FC, AWC and PWP, although the PTF by Puhlmann and von Wilpert (2011) produced the best and most stable estimates for all textures except the sands.

A similarly intensive evaluation was carried out for estimating the relative depth distribution of roots based on data from root estimates and root counting at the NFSI profiles. A multivariate BRT model was created that is able to explain the depth distribution of fine roots (<2 mm) by the determining factors soil depth, humus content, bulk density, slope and AWC. What was unexpected was that there were largely no dependencies on forest stand type and the degree of acidification (depth profile of base saturation) and only a weak dependency on the soil type. Those are the determining factors that are formally attributed to have a significant influence on the rooting depth. A possible explanation for the rooting intensity not being consistent with the hypothesis could be that the chemical site properties are levelled out by acidification of the soil to such an extent that both tree species and trophy of the sites do not have a differentiating effect on the depth profiles of fine roots anymore, and therefore only distinct differences in soil physics and soil structure (TRD, slope inclination, AWC, humus content) are distinguishable.

On the basis of these input parameters, two versions of water budget modelling were carried out using LWF-Brook90. They are different in the treatment of vegetation properties—one version was calculated using regionally adapted standardized forest stand properties (beech, oak, spruce, pine and mixed stands) to focus on the soil properties that diversify the water budget. In a second version, the LAI, the roughness of the bark and the height were taken from the forest stand information at a NFSI point in order to represent the actual drought stress occurring at that point as realistically as possible. All modelling was carried out in daily resolution, so that target variables such as seepage water output, change in soil water storage, evapotranspiration, etc. are available in daily resolution or coarser and so that those variables may be used for applications such as seepage water predictions for contaminant and nutrient output, water availability for the parameterization of climate-sensitive growth models or analysis of the significance of dry years for tree growth and forest health. The emphasis of the examination in this chapter was on the derivation and evaluation of characteristic drought stress variables. The time series of available soil water storage and different drought stress indices conformably show that the intensity of water shortage increased since 1990 and that from then on, years with good water supply occurred only sporadically, while before, years in which the soil water storage was above- or below-average existed in equal parts.

To empirically evaluate the impact of drought on tree growth, the deciding factor is not only the water deficit but also timing, duration and intensity of the drought. For identical weather conditions, not only the tree species but especially also the soil with its retention capacity determine the extent to which droughts can develop. Experiments on young beech and oak trees prove that trees experience acute drought stress when the soil water availability falls below 20%, which can eventually lead to their death.

References

Ad-Hoc AG–Boden (ed) (2005) Bodenkundliche Kartieranleitung (KA 5), vol 5. Schweizerbart'sche Verlagsbuchhandlung, Stuttgart

Ahrends B, Meesenburg H, Döring C, Jansen M (2010) A spatio-temporal modelling approach for assessment of management effects in forest catchments. Status and perspectives of hydrology in small basins. IAHS, Goslar-Hahnenklee

AK Standortskartierung (2003) Forstliche Standortsaufnahme: Begriffe, Definitionen, Einteilungen, Kennzeichnungen, Erläuterungen, 6th edn. IHW-Verlag, Eiching near Munich

Bréda N, Granier A (1996) Intra- and interannual variations of transpiration, leaf area index and radial growth of a sessile oak stand (*Quercus petraea*). Annales Des Sciences Forestieres 53 (2–3):521–536. https://doi.org/10.1051/forest:19960232

de Camargo AP, Sentelhas PC (1997) Performance evaluation of different methods for estimating potential evapotranspiration in the state of Sao Paulo, Brazil – an analytical review of potential evapotranspiration. Revista Brasileira de Agrometeorologia 5(1):89–97

DIN 4220 (2008–2011) Bodenkundliche Standortbeurteilung – Kennzeichnung, Klassifizierung und Ableitung von Bodenkennwerten (normative und nominale Skalierung)

Elith J, Leathwick JR, Hastie T (2008) A working guide to boosted regression trees. J Anim Ecol 77 (4):802–813. https://doi.org/10.1111/j.1365-2656.2008.01390.x

Federer CA, Vörösmarty C, Fekete B (2003) Sensitivity of annual evaporation to soil and root properties in two models of contrasting complexity. J Hydrometeorol 4(6):1276–1290. https://doi.org/10.1175/1525-7541(2003)004<1276:soaets>2.0.co;2

Gale M, Grigal D (1987) Vertical root distributions of northern tree species in relation to successional status. Can J For Res 17(8):829–834

Gauer J, Kroiher F (2012) Waldökologische Naturräume Deutschlands – Forstliche Wuchsgebiete und Wuchsbezirke – Digitale Topographische Grundlagen – Neubearbeitung Stand 2011. Landbauforschung vTI Agriculture and Forestry Research, Braunschweig

Hammel K, Kennel M (2001) Charakterisierung und Analyse der Wasserverfügbarkeit und des Wasserhaushalts von Waldstandorten in Bayern mit dem Simulationsmodell BROOK90. Forstliche Forschungsberichte München, vol 185. Technische Uni München Wissenschaftszentrum Weihenstephan, Munich

Hangen E, Scherzer J (2004) Ermittlung von Pedotransferfunktionen zur rechnerischen Ableitung von Kennwerten des Bodenwasserhaushalts (FK, PWP, nFK, kapillarer Aufstieg). Bundesministerium für Verbraucherschutz, Ernährung und Landwirtschaft (BMVEL), Bonn

Hartmann P, von Wilpert K (2014) Fine-root distributions of central European forest soils and their interaction with site and soil properties. Can J For Res 44(1):71–81. https://doi.org/10.1139/cjfr-2013-0357

Hartmann P, von Wilpert K (2016) Statistisch definierte Vertikalgradienten der Basensättigung sind geeignete Indikatoren für den Status und die Veränderungen der Bodenversaucrung in Waldböden. Allgemeine Forst- und Jagdzeitung 187(3/4):61–69

ICP Forests (2010) Manual on methods and criteria for harmonized sampling, assessment, monitoring and analysis of the effects of air pollution on forests. UNECE, ICP Forests, Hamburg

Law BE, Van Tuyl S, Cescatti A, Baldocchi DD (2001) Estimation of leaf area index in open-canopy ponderosa pine forests at different successional stages and management regimes in Oregon. Agric For Meteorol 108(1):1–14. https://doi.org/10.1016/s0168-1923(01)00226-x

Mellert KH, Rückert G, Weis W, Tiemann J, Brendel J (2009) Validierung von Pedotransferfunktionen – Zebris Projektbericht 2.0. Johann Heinrich von Thünen Institut (TI)

Menzel A (1997) Phänologie von Waldbäumen unter sich ändernden Klimabedingungen – Auswertung der Beobachtungen in den Internationalen Phänologischen Gärten und Möglichkeiten der Modellierung von Phänodaten. Technische Universität München, Wissenschaftszentrum Weihenstephan, Munich

Mualem Y (1976) A new model for predicting the hydraulic conductivity of unsaturated porous media. Water Resour Res 12(3):513–522

Puhlmann H, von Wilpert K (2011) Test und Entwicklung von Pedotransferfunktionen für Wasserretention und hydraulische Leitfähigkeit von Waldböden. Waldökologie, Landschaftsforschung und Naturschutz 12:61–71

Puhlmann H, von Wilpert K (2012) Pedotransfer functions for water retention and unsaturated hydraulic conductivity of forest soils. J Plant Nutr Soil Sci 175(2):221–235. https://doi.org/10.1002/jpln.201100139

Puhlmann H, von Wilpert K, Lukes M, Dröge W (2009) Multistep outflow experiments to derive a soil hydraulic database for forest soils. Eur J Soil Sci 60(5):792–806. https://doi.org/10.1111/j.1365-2389.2009.01169.x

Russ A, Riek W (2011) Pedotransferfunktionen zur Ableitung der nutzbaren Feldkapazität–Validierung für Waldböden des nordostdeutschen. Tieflands Waldökologie, Landschaftsforschung und Naturschutz 11:5–17

Schramm D, Schultze B, Scherzer J (2006) Validierung von Pedotransferfunktionen zur Berechnung von bodenhydrologischen Parametern als Grundlage für die Ermittlung von Kennwerten des Wasserhaushaltes im Rahmen der BZE II. Technical Report, TU Bergakademie Freiberg und UDATA-Umweltschutz und Datenanalyse im Auftrag des Bundesministerium für Ernährung, Landwirtschaft und Verbraucherschutz (BMELV)

Shuttleworth WJ, Wallace JS (1985) Evaporation from sparse crops – an energy combination theory. Q J R Meteorol Soc 111(469):839–855. https://doi.org/10.1256/smsqj.46909, https://doi.org/1002/qj.49711146910

Teepe R, Dilling H, Beese F (2003) Estimating water retention curves of forest soils from soil texture and bulk density. J Plant Nutr Soil Sci 166(1):111–119. https://doi.org/10.1002/jpln.200390001

van Genuchten MT (1980) A closed-form equation for predicting the hydraulic conductivity of unsaturated soils. Soil Sci Soc Am J 44(5):892–898

Vereecken H, Maes J, Feyen J, Darius P (1989) Estimating the soil moisture retention characteristic from texture, bulk density, and carbon content. Soil Sci 148(6):389–403

von Wilpert K (1991) Intraannual variation of radial tracheid diameters as monitor of site specific water stress. Dendrochronologia 9:95–113

von Wilpert K, Hildebrand EE (1997) Kaliummangel in Wäldern durch selektive Kaliumverarmung von Aggregatoberflächen. Mitteilungen der Deutschen Bodenkundlichen Gesellschaft 85 (1):449–452

Weinzierl T, Conrad O, Böhner J, Wehberg J (2013) Regionalization of baseline climatologies and time series for the Okavango catchment. Biodiversity & Ecology 5:235–245

Weis W, Hertel C, Wagner A, Raspe S (2012) Abschlussbericht ST 241 – Verbesserung der Wasserhaushaltsmodellierung mit Daten des forstlichen Umweltmonitorings im Projekt FUTMON (LIFE+). Landesanstalt für Wald und Forstwirtschaft (LWF), Freising

Wessolek G, Kaupenjohann M, Renger M (2009) Bodenphysikalische Kennwerte und Berechnungsverfahren für die Praxis. Rote Reihe, vol 40. TU Berlin, Institut für Ökologie, Fachgebiet Bodenkunde, Standortkunde und Bodenschutz, Berlin

Wösten JHM, Lilly A, Nemes A, Le Bas C (1999) Development and use of a database of hydraulic properties of European soils. Geoderma 90(3–4):169–185. https://doi.org/10.1016/s0016-7061(98)00132-3

Zirlewagen D, von Wilpert K (2011) Regionalisierung bodenphysikalischer Eingangsgrößen für bodenhydraulische Pedotransferfunktionen. Waldökologie, Landschaftsforschung und Naturschutz 12:73–83

Chapter 4
Soil Acidification in German Forest Soils

Henning Meesenburg, Winfried Riek, Bernd Ahrends, Nadine Eickenscheidt, Erik Grüneberg, Jan Evers, Heike Fortmann, Nils König, Amalie Lauer, Karl Josef Meiwes, Hans-Dieter Nagel, Claus-Georg Schimming, and Nicole Wellbrock

4.1 Introduction

The deposition of acidifying pollutants drastically altered the element cycles of forest ecosystems in large regions of the world (Ulrich et al. 1980; de Vries et al. 2014). Sulphur (S) was the major component of acid deposition since the beginning of industrialization until the 1980s, whereas the nitrogen (N) species nitrate (NO_3) and ammonium (NH_4) became more important since the 1990s (Schöpp et al. 2003). Besides its acidifying effect, deposition of N also causes an increasing nitrogen saturation of forest ecosystems (Aber et al. 1989, 1998) (see Chap. 5).

H. Meesenburg (✉) · B. Ahrends · J. Evers · H. Fortmann · N. König · K. J. Meiwes
Northwest German Forest Research Institute, Göttingen, Germany
e-mail: henning.meesenburg@nw-fva.de; bernd.ahrends@nw-fva.de; jan.evers@nw-fva.de; heike.fortmann@nw-fva.de; nils.koenig@nw-fva.de; karl-josef.meiwes@nw-fva.de

W. Riek
Eberswalde Forestry State Center of Excellence, University for Sustainable Development, Eberswalde, Germany
e-mail: winfried.riek@hnee.de

N. Eickenscheidt
State Agency for Nature, Environment and Consumer Protection of North Rhine-Westphalia, Recklinghausen, Germany
e-mail: nadine.eickenscheidt@lanuv.nrw.de

E. Grüneberg · A. Lauer · N. Wellbrock
Thünen Institute of Forest Ecosystems, Eberswalde, Germany
e-mail: erik.grueneberg@thuenen.de; nicole.wellbrock@thuenen.de

H.-D. Nagel
Öko-Data, Ahrensfelde, Germany
e-mail: hans.dieter.nagel@oekodata.com

C.-G. Schimming
Christian-Albrechts University Kiel, Kiel, Germany
e-mail: cschimming@ecology.uni-kiel.de

N. Wellbrock, A. Bolte (eds.), *Status and Dynamics of Forests in Germany*, Ecological Studies 237, https://doi.org/10.1007/978-3-030-15734-0_4

If acids are introduced to soils, a reduction of acid neutralizing capacity (ANC) is caused. However, the decline of ANC isn't accompanied by an equivalent increase of free acidity, since many soil components contain neutralizing properties. Dependent on soil-forming substrate and acid-base status, several buffer ranges differing in buffer capacity and buffer rate become active over time under the influence of acid inputs (Table 4.1). A characteristic change of soil solution composition occurs at the transition between different buffer ranges (Ulrich 1987). If the aluminium (Al) buffer range is approached, the risk of toxic concentrations of Al ions in soil solution is significantly enhanced (Reuss and Johnson 1986). Elevated concentrations of free Al impede the nutrient uptake by tree roots and mycorrhiza (de Wit et al. 2010) and hamper the growth of microorganisms (Piña and Cervantes 1996). Under acid conditions solute Al forms stable complexes with phosphorus (P) and may thus reduce the P availability in soils (Hansen et al. 2007). Whereas in the soil solid phase the reduction of ANC due to acid inputs is common to all soils, adverse impacts of soil acidification such as pH decline and Al mobilization into soil solution are restricted to forest soils with low buffer capacity.

Acids are introduced to soils as strong mineral acids (sulphuric acid H_2SO_4, nitric acid HNO_3) or generated through biochemical processes such as uptake of NH_4 and equivalent release of protons by tree roots or the nitrification of reduced N to HNO_3. Since the anions NO_3^- and chloride (Cl^-) are only marginally adsorbed in soils, they are prone to leaching into deeper soil horizons or to the groundwater. To preserve electroneutrality anions have to be accompanied by an equivalent amount of cations (Seip 1980). The composition of these cations is controlled by the composition of exchangeable cations (Reuss and Johnson 1985). Acidification caused by strong mineral acids may thus affect deep soil horizons and also groundwater or surface waters. On the other hand, natural soil acidification by weak acids such as carbonic or organic acids, which are increasingly protonated if pH declines, is mostly limited to upper soil horizons. Under certain conditions, acidification due to deposition of H_2SO_4 can by delayed by the retention of sulphate $\left(SO_4^{2-}\right)$ in the soil by adsorption to pedogenic hydroxides or by precipitation as Al hydroxy sulphate. The remobilization of previously stored SO_4^{2-} causes an equivalent release of acidity (Ulrich 1994).

Buffering of acids in forest soils in the long term is mainly provided by weathering of primary minerals (Ulrich 1983). In carbonate-free soils, the release of base cations (Ca, Mg, K, Na) from silicate minerals such as feldspars and clay minerals is the major proton consuming process (Fölster 1985; Tarrah et al. 2000). The release of structural Al during the weathering process usually results in the formation of interlayer Al hydroxides in clay minerals, which may reduce the cation exchange capacity drastically (Rich 1968). If acidification continues, proton buffering is also realized by the dissolution of these interlayer Al hydroxides, and cation exchange capacity may increase; however, previous values are usually not approached (Mareschal et al. 2013). For Germany, Ahrends et al. (2018) estimated an average base cation release due to weathering of 1.0 keq ha^{-1} $year^{-1}$, 0.8 keq ha^{-1} $year^{-1}$ of which are nutrient cations (Ca, Mg, K, see Chap. 2).

Table 4.1 Acid buffer ranges in aerated soils (Ulrich 1981, 1983) adjusted according to AK Standortskartierung (2003)

Buffer range	Carbonate ($CaCO_3$)	Silicate	Exchange (exch.)	Aluminium (Al)	Al-Fe	Iron (Fe)
pH value	8 6–6.2	6.2–5.0	5.0–4.2	4.2–3.8	3.8–3.0	<3.0
Buffer rate	High	Low	Very high	High-medium	Medium in presence of soluble organic acids	
Major biogeo-chemical processes	Decalcification	Release of structural cations, regeneration of clay minerals	Leaching of exchange-able cations, decrease of exchange capacity	Resolution of Al silicates, destruction of clay minerals, protolysis of Al hydroxides	Fe mobiliza-tion as organic complexes	Bleaching under oxida-tive conditions

Consequences of soil acidification are the leaching of base cations (i.e. K, Mg, Ca) as counterions of mobile anions, the mobilization of protons, reactive Al and heavy metals and subsequent release into soil solution (Reuss and Johnson 1985). As base cations are important nutrients for forest trees, the loss of base cations may result in nutrient imbalances on base-poor sites (de Vries et al. 2014).

On the other hand, the uptake of nutrient cations may contribute to soil acidification, as an excess of ANC is removed from the soil (Ulrich 1994). The amount of ANC removal by biomass uptake depends on the preference of N uptake either as NH_4 or as NO_3, resulting in an equivalent release of H^+ or OH^- via the root system, respectively (Reuss and Johnson 1986). For German NFSI plots on average, a net uptake of 0.5 keq ha^{-1} $year^{-1}$ of nutrient cations was estimated, if stemwood utilization including bark is assumed (see Chap. 2). If NH_4 and NO_3 net uptake is balanced, this equals the net removal of ANC.

Soil acidification has been early claimed as a possible cause of forest decline, which was observed in central Europe since the late 1970s (Ulrich et al. 1980; Ulrich 1987). A consequence of this debate was the start of clean air policy, beginning with the Large Combustion Plant Directive in 1983 (13. BImSchV). This and other national and international measures (e.g. Convention of Long-range Transboundary Air Pollution CLRTAP; UNECE 1979) resulted in a significant reduction of deposition of acidity in Europe (Waldner et al. 2014).

Besides the CLRTAP, legal instruments for the prevention and compensation of acidification effects are the National Emissions Ceilings (NEC) Directive (2016/2284/EU), the German soil protection law (BBodSchG) and the European Water Framework Directive (WFD 2000/60/EC), which define criteria for the aspired state of soils and freshwater. Measures for the reduction of acid inputs are related to an ecological tolerable load. These critical loads provide "a quantitative estimate of an exposure to one or more pollutants below which significant harmful effects on specified sensitive elements of the environment do not occur according to present knowledge" (Spranger et al. 2015).

From the comparison with actual deposition loads, the compliance with or the exceedance of the critical loads can be derived. Further, the effectiveness of air pollution reduction measures can be controlled through the evaluation of deposition monitoring. The overarching environmental quality target for air pollution is the compliance with the critical loads for acidity (see Chap. 2). Moreover, the National Strategy on Biological Diversity in Germany aims to protect all sensitive ecosystems against acidification and eutrophication (BMU 2007).

Evidence for soil acidification caused by acid deposition proved by repeated soil sampling has been demonstrated in many studies (Falkengren-Grerup et al. 1987; Meesenburg et al. 2016; Blake et al. 1999; Johnson et al. 1994). So far, only few studies are showing recovery of forest soils from acidification (Lawrence et al. 2015), whereas recovery of surface waters due to declining deposition is a widespread phenomenon (Sucker et al. 2011; Wright et al. 2005; Stoddard et al. 1999).

The pH value, either measured in water (pH(H_2O)) or in potassium chloride solution (pH(KCl)), base saturation and the fractions of exchangeable acid cations (H^+, Al, Mn, Fe) are used as indicators for the acid-base status of forest soils in this study. The objective of this chapter is to evaluate if indications of recovery of forest soils from acidification in Germany are visible and to identify main drivers behind possible changes of soil acidity.

4.2 Acid-Base Status of German Forest Soils

4.2.1 Soil Acidity

For the characterization of soil acidity, the pH value in water (pH(H_2O)) and in potassium chloride solution (pH(KCl)) was determined. The pH(H_2O) indicates the effective soil acidity and shows distinct seasonal and episodic fluctuation. Through the introduction of potassium (K) during the pH(KCl) determination, the release of exchangeable aluminium (Al) and protons lowers the pH by 0.4–1.1 units. Thus the pH(KCl) is an indication of the potential soil acidity (Ulrich 1981).

The results of the second NFSI reveal a large spatial variability of pH in forest soils of Germany. On average, pH(H_2O) was 4.6 ± 0.02, and pH(KCl) was 3.9 ± 0.02 in the organic layer. Limed plots showed higher pH values as compared to unlimed plots. The 10th percentile in the organic layer was 3.8 for pH(H_2O) and 2.9 for pH(KCl), and the 90th percentile was 5.8 for pH(H_2O) and 5.3 for pH(KCl). As compared to the organic layer, mean pH was slightly lower in the uppermost mineral soil layer and increased with depth. In 30–60 cm depth a mean pH(H_2O) of 5.2 ± 0.03 and pH(KCl) of 4.4 ± 0.03 was observed. The 10th percentile in this depth interval was 4.3 for pH(H_2O) and 3.6 for pH(KCl), and the 90th percentile was 7.5 for pH(H_2O) and 6.9 for pH(KCl).

The spatial distribution of pH values mainly reflects the different parent material of the soils (Fig. 4.1). High pH values were observed in soils developed from carbonatic bedrock, e.g. at the Swabian-Franconian Alb and in the Alps. Locally, also soils with intermediate basic-intermediate bedrock, e.g. tertiary basalts, show higher pH values in the topsoil (Meesenburg et al. 2009). Plots with acidic soils over the complete depth profile occur mainly in regions with soils developed from base-poor substrates, such as in the Black Forest, the Bavarian Forest, the Ore Mountains, the Harz Mountains and the North German lowlands with early Pleistocene sediments.

With respect to soil parent material groups, NFSI plots with soils from weathered carbonate bedrock revealed highest pH values, whereas lowest pH was observed for soils from base-poor unconsolidated sediments, soils from base-poor consolidated bedrock and loamy soils of the lowland (data not shown). Soils from basic-intermediate bedrock and soils from alluvial plains showed medium pH values, which were significantly different from base-poor and base-rich soil parent material groups in the organic layer.

The pH of the organic layer was lower under coniferous forest stand types than under deciduous and mixed stand types. This becomes more apparent, if unlimed plots for distinct soil parent material groups are compared. In the mineral soil, tree

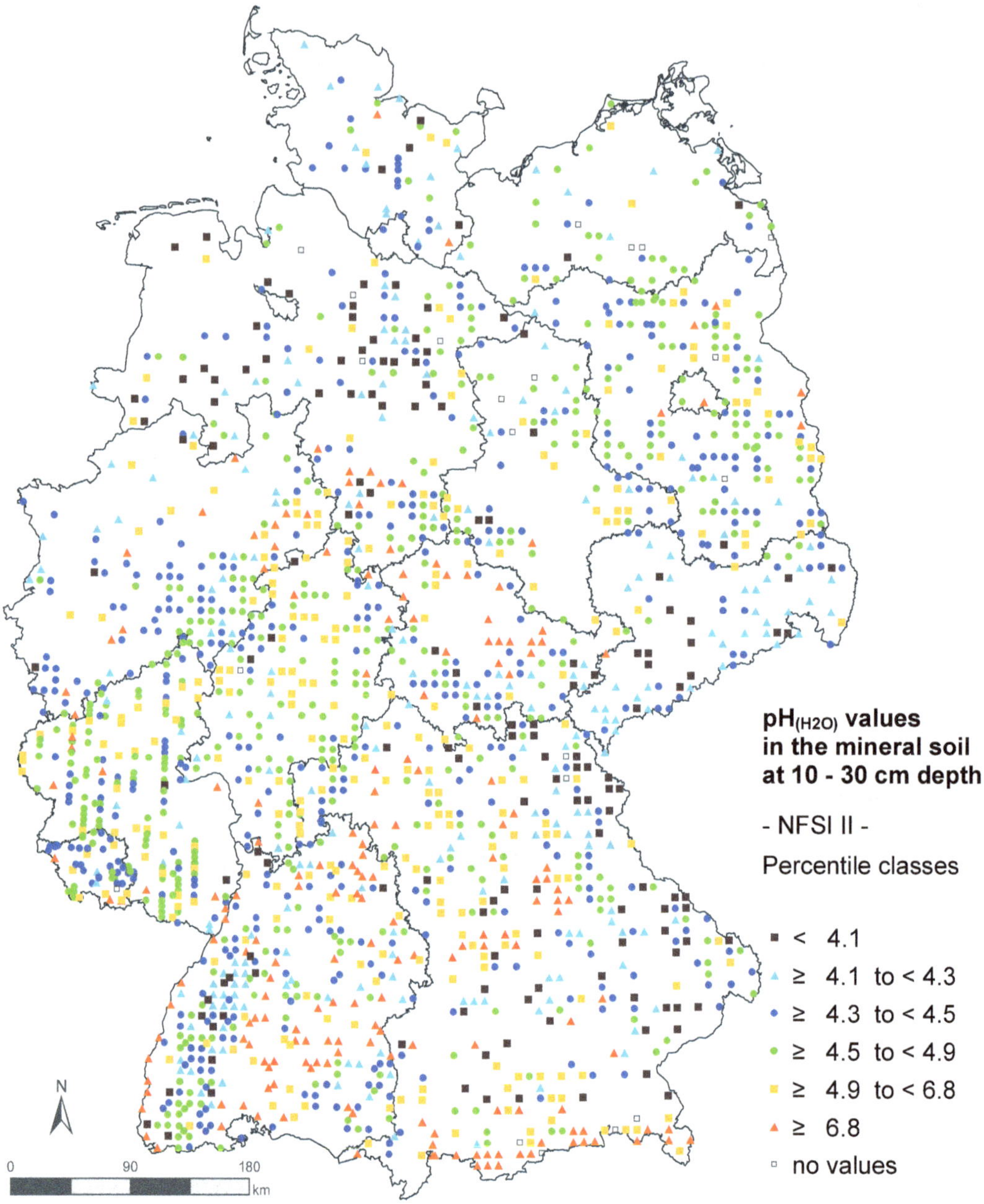

Fig. 4.1 pH(H_2O) values in the mineral soil at 10–30 cm depth of NFSI II

species-specific differences disappear with increasing depth. The observed pattern can be attributed to a higher base cation content of deciduous litter (Augusto et al. 2002; Jacobsen et al. 2002), but also to a preferential cultivation of coniferous tree species on acid organic layers.

The comparison between first and second NFSI reveal on average a significant increase of pH(H_2O) in the organic layer and in all depth intervals of the mineral soil (Fig. 4.2). The pH increase was highest in the organic layer (0.013 ± 0.0014 $year^{-1}$) and in the 0–5 cm interval of the mineral soil (0.011 ± 0.0011 $year^{-1}$). In the deeper mineral soil, the pH change was much smaller. For pH(KCl) a significant increase of

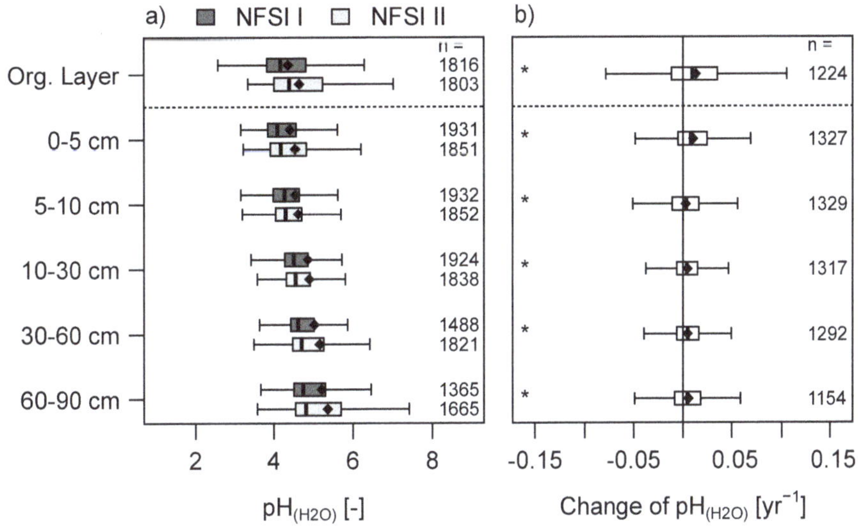

Fig. 4.2 pH(H_2O) values in the organic layer and in the mineral soil layers for NFSI I and NFSI II (**a**, complete sample including bogs and organic soils) and differences of pH(H_2O) values between NFSI I and NFSI II (**b**, paired sample without bogs and organic soils, *denotes significant change of pH(H_2O), $p \leq 0.05$)

0.004 ± 0.0014 year^{-1} pH units could only be observed for the organic layer, whereas the 5–10 cm interval showed a pH decrease of 0.004 ± 0.001 year^{-1} (not shown). The other sampling intervals revealed only insignificant changes. The pH increase in the topsoil is attributed primarily to the reduced deposition of acidity and to liming measures. The reduced leaching of basic cations from the organic layer may have contributed to the pH increase together with an increased contribution of base-rich litter from deciduous tree species, which were increasingly present at the plots since NFSI I.

The higher increase of pH(H_2O) as compared to pH(KCl) in all depth intervals may be attributed to a decrease of ionic strength of soil solution. For a given distribution of exchangeable cations, a decrease of ionic strength results in an increase of pH, which is more prominent for pH(H_2O) than for pH(KCl) as the latter is also determined by the exchanger composition (Reuss and Johnson 1986). A decrease of ionic strength of soil solution during the period from 1990 to 2006/2008 was demonstrated for several forested sites in northwest Germany (Klinck et al. 2012; Meesenburg et al. 2016).

According to the pH(H_2O) value the samples can be attributed to pH buffer ranges (Ulrich 1983). Changes in the frequency of samples in the different buffer ranges between NFSI I and NFSI II illustrate changes in the acid-base status of the NFSI plots. As compared to NFSI I, an increase of samples within the exchange buffer range, silicate buffer range and the $CaCO_3$ buffer range of 9% in the 0–5 cm interval and of 5% in the 5–30 cm interval was observed leading to equivalent reductions in the Fe, the Fe-Al and the Al buffer range (Fig. 4.3). In the 60–90 cm interval, a 5% decrease occurred in the Al and exchange buffer range with a concurrent increase in the silicate and $CaCO_3$ buffer range. The shift between pH

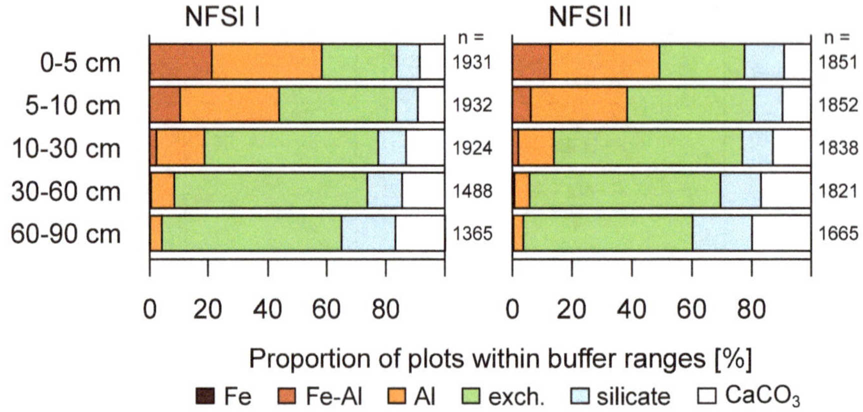

Fig. 4.3 Fractions of plots within different pH buffer ranges (according to Ulrich 1983) in the mineral soil for complete samples of NFSI I and NFSI II (Fe, Fe buffer range; Fe-Al, Fe-Al buffer range; Al, Al buffer range; exch., exchange buffer range; silicate, silicate buffer range; $CaCO_3$, carbonate buffer range)

buffer ranges was higher for limed as compared to unlimed plots. At unlimed plots, distinct changes of the frequency in the buffer ranges cannot be observed. This pattern is most probably a result of liming measures and a general decrease of the ionic strength of soil solution, which was observed for several intensive forest monitoring sites in northwestern Germany (Klinck et al. 2012). For the subsoil of unlimed NFSI plots, on average a balance between processes increasing and decreasing acidity can be assumed.

In the organic layer and in the mineral soil, up to 60 cm depth of acid-sensitive sites pH(H_2O) was significantly higher at limed as compared to unlimed plots (Fig. 4.4). Differences were less obvious for pH(KCl) (not shown). Between NFSI I and NFSI II, pH(H_2O) increased in the organic layer and in the mineral soil up to 10 cm depth both at limed and unlimed plots. Below 10 cm depth, unlimed plots showed no temporal change, whereas limed plots revealed an increase of pH(H_2O). For pH(KCl) no distinct pattern of changes was found. An exception from the observed pH increase in the organic layer is plots with pine as dominating tree species. For these plots pH(KCl) in the organic layer and up to 10 cm of the mineral soil decreased significantly. Pine stands are often cultivated on soil from base-poor unconsolidated sediments. Pine stands located in the northeastern German lowlands are mostly unlimed. In addition, many sites where high loads of basic dust were deposited in the last century are stocked with pine. These sites were acidified between NFSI I and NFSI II due to the leaching of bases (see Chap. 2).

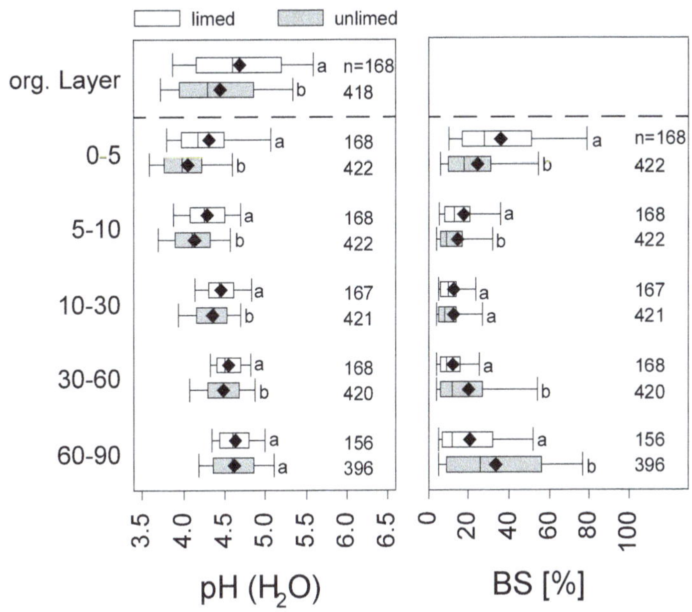

Fig. 4.4 pH(H_2O) values (left) and base saturation (BS [%], right) for limed and unlimed acid-sensitive plots in the organic layer and mineral soil for NFSI II (acid-sensitive plots identified according to exchangeable aluminium pool in the subsoil)

4.2.2 Base Saturation

The acid-base status determines to a large degree the availability of nutrients in the soil, e.g. by the pH-dependent allocation of cations at the exchange sites. Thus, the nutrition of the forest stands is affected by soil acidification (see Chap. 9). The base saturation describes the proportion of effective cation exchange capacity in the soil occupied by Na, K, Mg and Ca. It serves as an indicator for the acid-base status of forest soils as the allocation of the exchange sites is strongly pH dependent. The introduction of acid cations can alter the base saturation significantly especially in the exchange buffer range. The occupation of exchange sites with acid cations such as Al, Mn, Fe and H^+ is reciprocal to base saturation. Among the acid cations, Al usually constitutes the largest proportion.

Some exchange sites may also be occupied by NH_4 (Aber 1992). In the mineral soil, NH_4 can be retained at clay minerals (Davidson et al. 1991). As the ion radius of NH_4 is similar to that of K, in clay-rich soils NH_4 is fixed in the interlayers of 2:1 clay minerals and further not fully available for cation exchange (Nieder et al. 2011). However, as exchangeable cations were extracted with NH_4Cl solution (see Chap. 1), a quantification of the saturation of exchange sites with NH_4 is not possible for NFSI. In the mineral soil, soil solution concentrations of NH_4 are usually very low due to retention in the organic layer and nitrification (Brumme et al. 2009; Schwarz et al. 2016). Corre et al. (2007) found for German forest sites on a N deposition gradient, that microbial NH_4 retention was more important than abiotic NH_4 retention. Thus, it is assumed that exchangeable NH_4 only contributes to a small extent to total exchangeable cations.

Observations of base saturation from NFSI I and NFSI II are only available for the mineral soil. Hence, the medium-term availability of nutrient cations is characterized for the organic layer by their aqua regia extractable stocks as these may be released when the organic layer is decomposed (see Chap. 3).

Mean base saturation in the uppermost 5 cm of the mineral soil was 41.5 ± 0.8% for NFSI II. It approached lowest values in 10–30 cm depth (31.1 ± 0.84%) and increased to deeper soil layers. In all soil layers a large span of base saturation from <4 to 100% was observed. Median values are consistently lower than arithmetic means for the different soil layers, e.g. 28% for the 0–5 cm depth and 12% for 10–30 cm.

Sites with high base saturation in the whole soil profile occur predominantly in regions with carbonate substrates, e.g. in the Alps and at the Swabian-Franconian Alb (Fig. 4.5). Low base saturation can be found in almost every region demonstrating the widespread loss of bases from forest soils in Germany. Forest soils with low base saturation over the whole profile were observed especially at low mountain ranges with base-poor silicate bedrock and at unconsolidated sandy substrates from Saalian sediments in the North German lowlands. Topsoils with low base saturation were observed in many substrates across Germany. Often forest soils with low base saturation occur in short distance to soils with high base saturation, which can be attributed to a high heterogeneity of site condition and to liming measures. Especially loamy soils of the lowlands and soils of alluvial plains showed a great variety of base saturation values and depth profiles (Fig. 4.6). Plots vegetated with deciduous tree species revealed significant higher base saturation in the upper 10 cm of the mineral soil as compared to coniferous tree species (Wellbrock et al. 2016). This pattern may be attributed to a higher base cation content of litter from deciduous tree species (Augusto et al. 2002; Jacobsen et al. 2002).

A significant increase of base saturation between NFSI I and NFSI II of 0.28 ± 0.04% a^{-1} was observed in 0–5 cm depth of the mineral soil for the paired sample (Fig. 4.7). In contrast, a significant decrease of base saturation of 0.16 ± 0.03% a^{-1}, 0.3 ± 0.04% a^{-1} and 0.28 ± 0.04% a^{-1} in 10–30 cm, 30–60 cm and 60–90 cm, respectively, occurred. Accordingly, the fraction of plots within base saturation classes "very low" to "low to medium" (AK Standortskartierung 2003) decreased in the 0–5 cm depth from 65 to 52%, while plots within base saturation classes "medium" to "very high" increased (Fig. 4.8). The observed pattern of improved base saturation in the topsoil but ongoing soil acidification in the subsoil probably points to the effects of liming measures and reduced deposition of acidity, which are most prominent in the topsoil. Nitrification of reduced nitrogen and the remobilization of sulphur may delay the recovery of the subsoil. In addition, the uptake of base cations by the forest stands contributes to the loss of ANC from the subsoil.

Base saturation was higher on limed as compared to unlimed acid-sensitive plots of NFSI II down to 10 cm depth (Fig. 4.4). Acid-sensitive plots were identified according to the exchangeable Al pool in the subsoil. If the pool of exchangeable Al in 60–90 cm depth was higher than an equivalent of 3 to ha^{-1} of dolomitic lime, the respective NFSI plot was classified as acid sensitive (Höhle et al. 2018). For 45 long-term liming trials in northwest Germany, Guckland et al. (2012) found an 11% points increase of base saturation up to 40 cm depth. Unlimed plots of the NFSI II

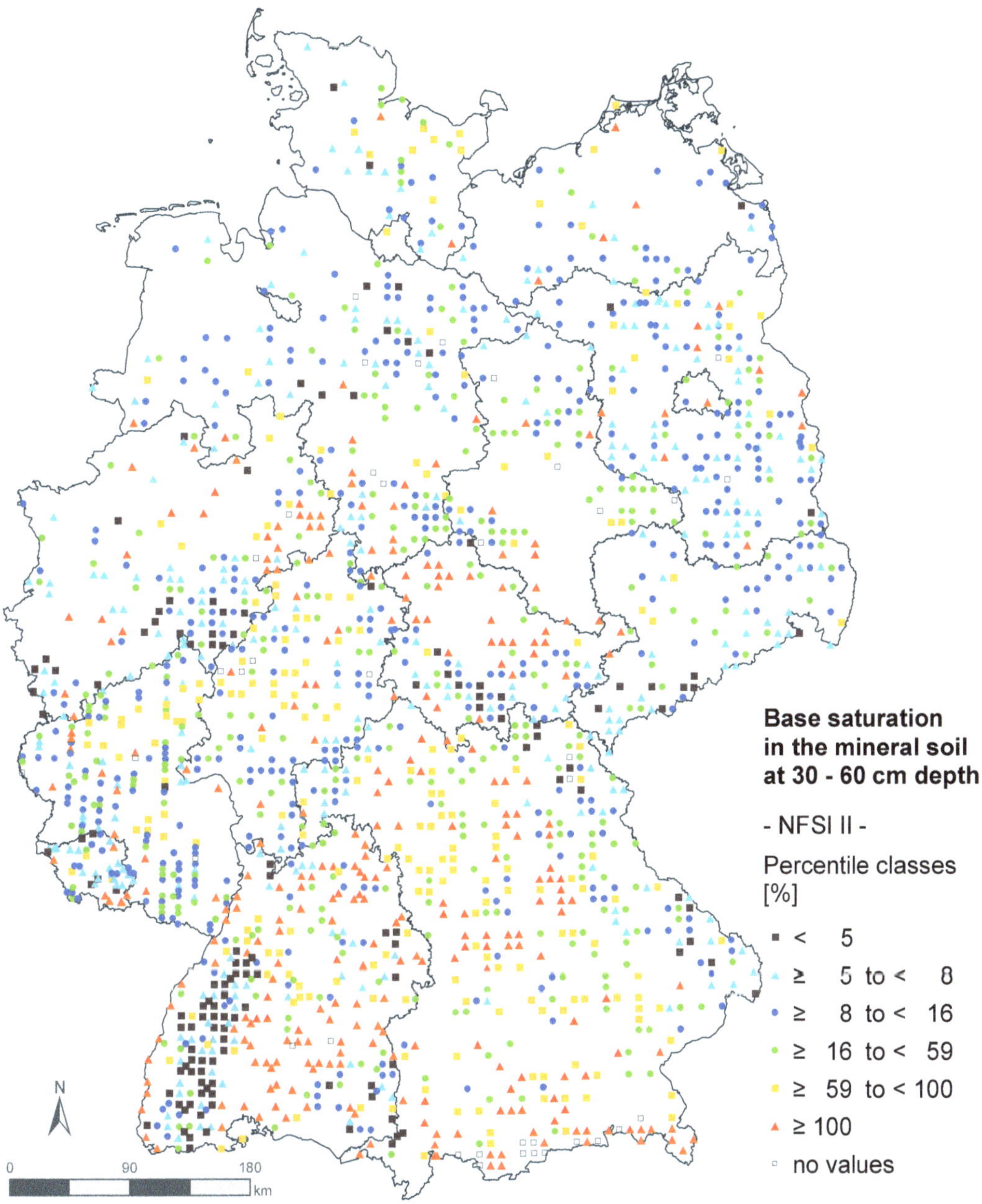

Fig. 4.5 Base saturation in % in the mineral soil at 30–60 cm depth of NFSI II

showed higher base saturation than limed plots in the subsoil (Fig. 4.4). This pattern may be interpreted with the selection of sites for liming, where most acidified sites are preferred. As the liming effect is mainly restricted to the topsoil, the sites selected for liming on average show lower base saturation. Between NFSI I and NFSI II, an increase of base saturation in the upper 30 cm of the limed plots and a decrease in 5–90 cm depth of the unlimed acid-sensitive plots were observed (Wellbrock et al. 2016).

An increase of base saturation between NFSI I and NFSI II occurred predominantly at loamy soils of the lowlands (0–5 cm depth), at soils from base-poor

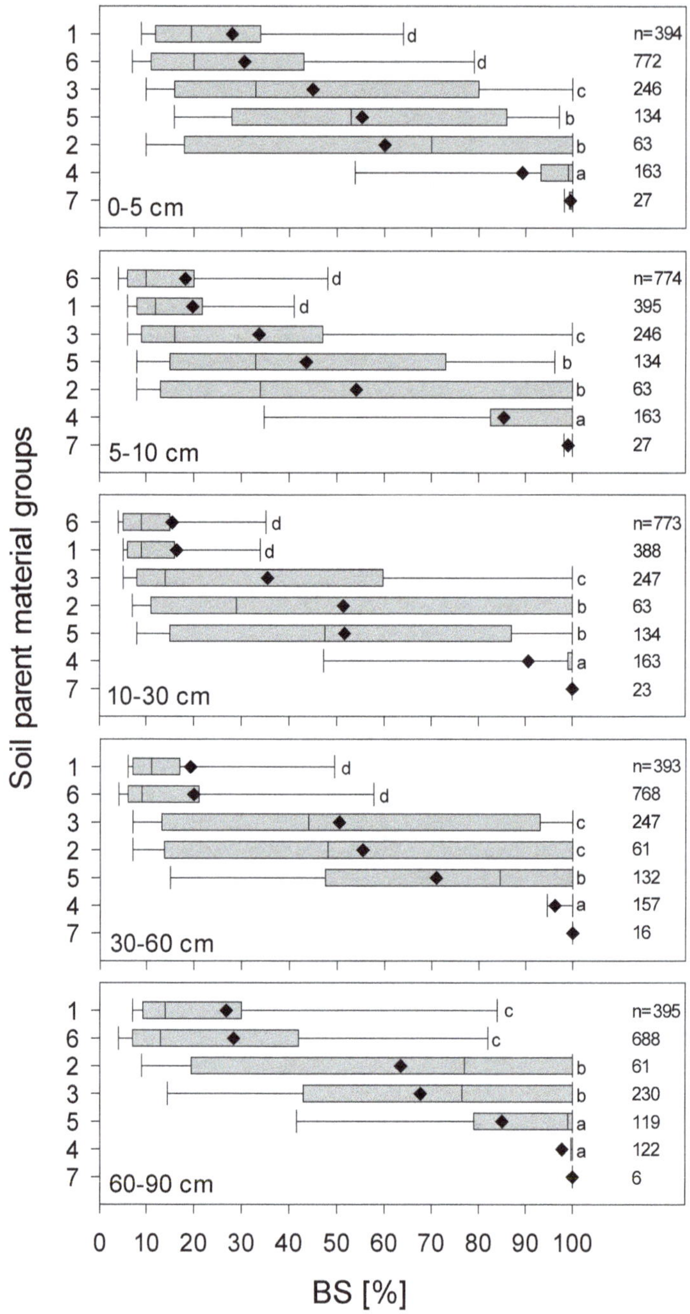

Fig. 4.6 Base saturation for different soil parent material groups in the mineral soil for NFSI II (1, soils from base-poor unconsolidated sediments; 2, soils of alluvial plains; 3, loamy soils of the lowlands; 4, soils from weathered carbonate bedrock; 5, soils from basic-intermediate bedrock; 6, soils from base-poor consolidated bedrock; 7, soils from the Alps)

consolidated bedrock (0–10 cm depth) and at soils from weathered carbonate bedrock (0–10 cm depth, Fig. 4.9). A significant decrease of base saturation was observed at all depths for soils from base-poor unconsolidated sediments, in 10–90 cm depth at soils of alluvial plains, at loamy soils of the lowlands and at soils from base-poor consolidated bedrock in 30–60 cm depth and at soils from basic-intermediate bedrock in 10–30 cm depth (Fig. 4.9). The decrease of base saturation only took place at unlimed plots of all substrates, whereas liming gave rise to base saturation at soils from base-poor unconsolidated sediments (0–10 cm depth), at soils from basic-intermediate bedrock (0–5 cm depth) and at soils from

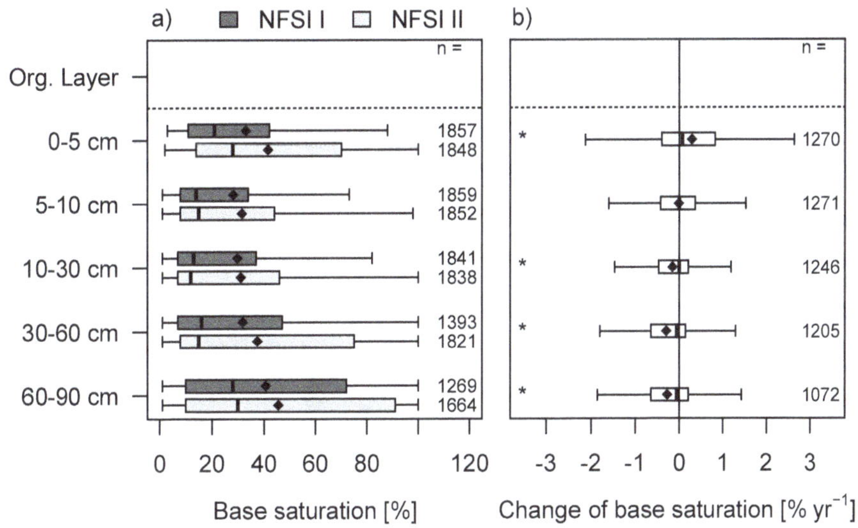

Fig. 4.7 Base saturation in the organic layer and in the mineral soil layers for NFSI I and NFSI II (**a**, complete sample including bogs and organic soils) and differences of base saturation between NFSI I and NFSI II (**b**, paired sample without bogs and organic soils)

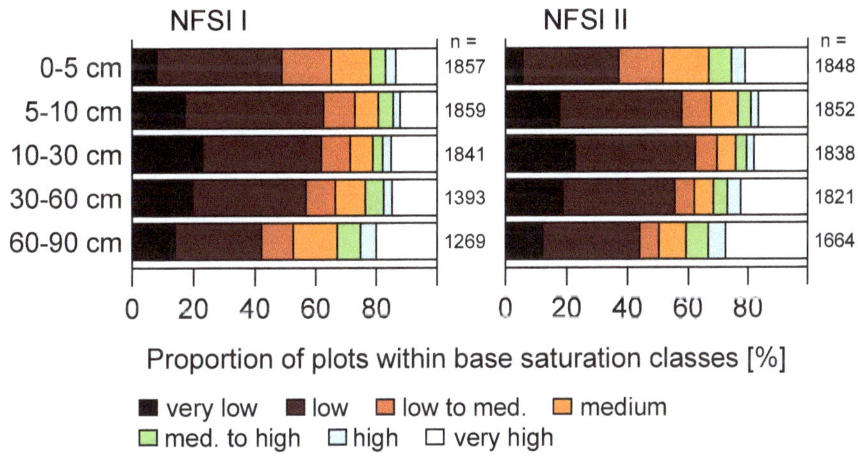

Fig. 4.8 Fractions of plots within different base saturation classes in the mineral soil for complete samples of NFSI I and NFSI II (base saturation classes according to AK Standortskartierung (2003); very low <7%, low 7–20%, medium to low 20–30%, medium 30–50%, medium to high 50–70%, high 70–85%, very high >85%)

base-poor consolidated bedrock (0–30 cm depth). Hence, liming seems to override effects of different substrates.

The effect of liming and other environmental and soil variables on the change of base saturation was analysed with a generalized additive model (GAM) (Hastie and Tibshirani 1990). A mass-weighted mean base saturation for the depth interval 0–90 cm of NFSI I and NFSI II was derived according to Spranger et al. (2015). The change of base saturation (BS) between NFSI I and NFSI II ($_{\Delta}BS = BS_{NFSI\ II} - BS_{NFSI\ I}$) was analysed with various potential predictor variables (Table 4.2).

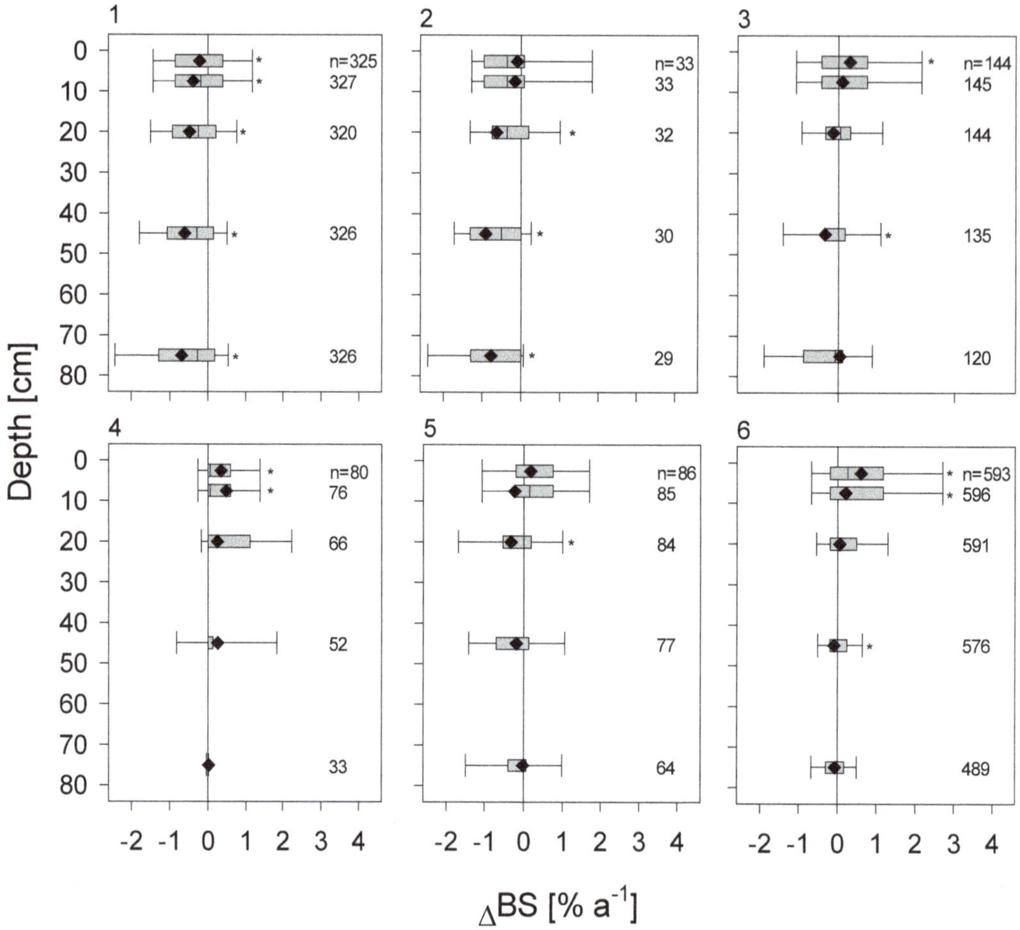

Fig. 4.9 Change of base saturation ($_{\Delta}$BS [% a^{-1}]) between NFSI I and NFSI II in the mineral soil for selected soil parent material groups (1, soils from base-poor unconsolidated sediments; 2, soils of alluvial plains; 3, loamy soils of the lowlands; 4, soils from weathered carbonate bedrock; 5, soils from basic-intermediate bedrock; 6, soils from base-poor consolidated bedrock)

For the development of the GAM, the general methodology of Wood (2006) was followed using the R-library mgcv (1.8–17). $_{\Delta}$BS and $_{\Delta}$CEC values outside the triple interquantile range (3*IQR) were excluded from the analysis. Furthermore one outlier detected from the residual plots of the final model was also excluded. For model building and variable preselection, the R add-on package mboost (2.8-1) was used (Hofner et al. 2011). To detect the optimal mstop values for application of the mboost function, a tenfold cross-validation was applied. Continuous variables were integrated as penalized splines in the model, each specific for limed and unlimed plots, respectively. Predictor variables significant at $p \leq 0.05$ were included in the model. The Ca stock in the organic layer and the mineral soil (0–90 cm), base saturation at NFSI I, the carbon stock in the organic layer, the carbon stock in the mineral soil (0–90 cm) and the K weathering rate (0–90 cm) were included in the final model (see Eq. 4.1, Table 4.3). Except the K weathering rate, each variable exhibits significantly different partial effects for limed and unlimed plots (Fig. 4.10a–e). The selected predictors account for 67% of the total variance:

Table 4.2 Predictor variables used in the generalized additive model (GAM) for the prediction of change of base saturation (0–90 cm) between NFSI I and NFSI II ($_{\Delta}$BS)

Predictor	Unit
S deposition	keq ha^{-1} a^{-1}
N deposition	keq ha^{-1} a^{-1}
BC deposition	keq ha^{-1} a^{-1}
ACpotnet deposition	keq ha^{-1} a^{-1}
Net biomass BC uptake	keq ha^{-1} a^{-1}
Calcium weathering rate	keq ha^{-1} a^{-1}
Magnesium weathering rate	keq ha^{-1} a^{-1}
Potassium weathering rate	keq ha^{-1} a^{-1}
BC release by silicate weathering	keq ha^{-1} a^{-1}
Cation exchange capacity	$mmol_c$ kg^{-1}
Base saturation mineral soil 0–90 cm	%
Carbon stock organic layer	t ha^{-1}
Carbon stock mineral soil 0–90 cm	t ha^{-1}
Carbon stock organic layer + mineral soil 0–90 cm	t ha^{-1}
Nitrogen stock organic layer + mineral soil 0–90 cm	t ha^{-1}
C/N ratio	–
Calcium stock organic layer + mineral soil 0–90 cm	kg ha^{-1}
Magnesium stock organic layer + mineral soil 0–90 cm	kg ha^{-1}
Potassium stock organic layer + mineral soil 0–90 cm	kg ha^{-1}
BC stock organic layer + mineral soil 0–90 cm	keq ha^{-1}
Liming	Yes/no
Carbonates in soil	Yes/no
Stand type	Categorial
Soil type	Categorial
Soil class	Categorial
Soil parent material group	Categorial
Seepage flux	mm
Soil water content	m^3 m^{-3}
Mean air temperature	°C
Mean annual precipitation	mm
Stand type	Categorial
Federal state	Categorial
Latitude	GK4
Longitude	GK4

Table 4.3 Estimated coefficients and statistical characteristics of the model

	Est.	SE	edf
Parametric coefficients			
Intercept	−0.3173	0.2511	
LI = YES	1.7872*	0.6924	
Approximate significance of smooth terms			
$f_{LI,NO}$(logCa)			4.428***
$f_{LI,YES}$(logCa)			2.712***
$f_{LI,NO}$(BS)			4.044***
$f_{LI,YES}$(BS)			1.000***
$f_{LI,NO}$(CO)			2.997***
$f_{LI,YES}$(CO)			1.000***
$f_{LI,NO}$(CM)			1.000***
$f_{LI,YES}$(CM)			2.458***
f_3(KW)			2.352***

Est estimated parameter value, *SE* standard error, *edf* effective degrees of freedom, *Ca* calcium stock, *BS* base saturation, *CO* carbon stock in organic layer, *CM* carbon stock in mineral soil, *KW* K weathering rate, *LI* limed, Signif. codes: <0.001***, <0.05*

$$\begin{aligned}\Delta\mathrm{BS}_i = \beta_0 &+ I_{\{\mathrm{LI}i=\mathrm{NO}\}} f_{\mathrm{LI,NO}}(\log\mathrm{Ca}_i) + I_{\{\mathrm{LI}i=\mathrm{YES}\}} f_{\mathrm{LI,YES}}(\log\mathrm{Ca}_i)\\ &+ I_{\{\mathrm{LI}i=\mathrm{NO}\}} f_{\mathrm{LI,NO}}(\mathrm{BS}_i) + I_{\{\mathrm{LI}i=\mathrm{YES}\}} f_{\mathrm{LI,YES}}(\mathrm{BS}_i)\\ &+ I_{\{\mathrm{LI}\_i=\mathrm{NO}\}} f_{\mathrm{LI,NO}}(\mathrm{CO}_i) + I_{\{\mathrm{LI}i=\mathrm{YES}\}} f_{\mathrm{LI,YES}}(\mathrm{CO}_i)\\ &+ I_{\{\mathrm{LI}i=\mathrm{NO}\}} f_{\mathrm{LI,NO}}(\mathrm{CM}_i) + I_{\{\mathrm{LI}i=\mathrm{YES}\}} f_{\mathrm{LI,YES}}(\mathrm{CM}_i)\\ &+ f_1(\mathrm{KW}) + I_{\{\mathrm{LI}i=\mathrm{YES}\}}\beta_1 + \varepsilon_i\end{aligned} \quad (4.1)$$

with

$_{\Delta}\mathrm{BS}_i$ = change in base saturation between NFSI I and NFSI II at NFSI plot i [%]
β_0 = intercept
I = indicator function, denoted by $I_{\{\mathrm{condition}\}} = \{(1, \text{if LI} = \text{YES})/(0, \text{else})\}$
f_{LI},p, (p = NO,YES) = to describe liming specific one-dimensional penalized regression splines
β_1 = parameter vector corresponding to LIME (LI)
Ca = calcium stock in organic layer + mineral soil 0–90 [kg ha^{-1}]
BS = base saturation at NFSI I in mineral soil 0–90 [%]
CO = carbon stock in organic layer [t ha^{-1}]
CM = carbon stock in mineral soil 0–90 [t ha^{-1}]
KW = K weathering rate in mineral soil 0–90 [kg ha^{-1} a^{-1}]
LI = lime classes (NO, unlimed; YES, limed)
f_1 = one-dimensional penalized regression splines
ε_i = random error term, $\varepsilon_i \sim N(0, \sigma_\varepsilon^2)$

The Ca stock displays a positive partial effect on $_{\Delta}$BS, which is stronger for unlimed as compared to limed plots (Fig. 4.10a). Contrariwise, for base saturation at NFSI I, a negative effect with higher base saturation values is apparent (Fig. 4.10b).

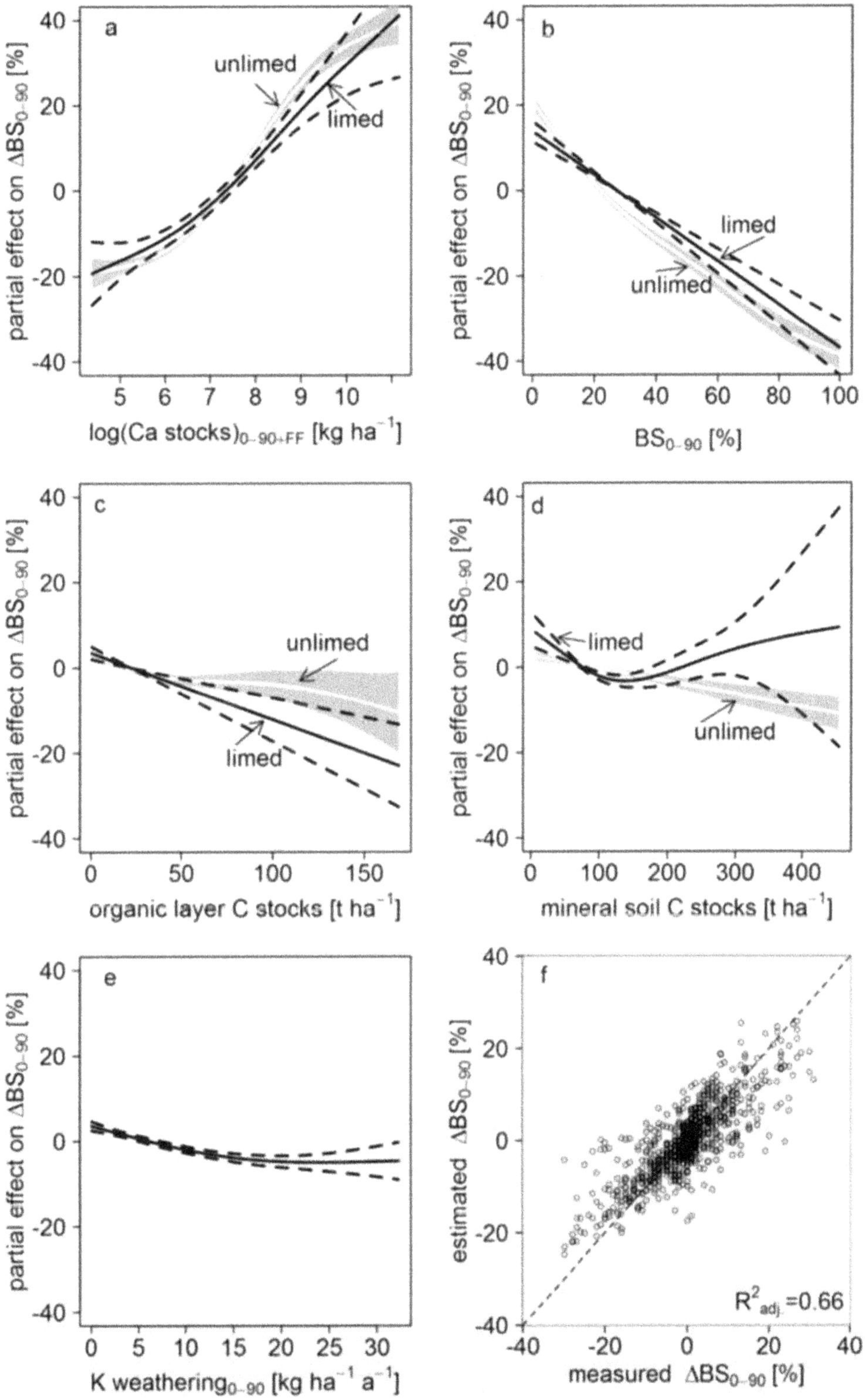

Fig. 4.10 Partial effects of Ca stock in organic layer + mineral soil 0–90 (**a**), base saturation 0–90 (**b**), carbon stocks in organic layer (**c**), carbon stocks in mineral soil (**d**), K weathering rates (**e**) and the relationship between measured and estimated base saturation changes in the soil 0–90 ($_{\Delta}$BS) (**f**). Dashed lines indicate 95% pointwise prognosis intervals

For the interpretation of these effects, a strong correlation between base saturation and calcium stock in soil ($r_{spear} = 0.7^{***}$) should be considered. The negative relation of base saturation with $_{\Delta}$BS may indicate that sites with very low base saturation had reached a low level, where the flux of base cations was restricted (Horn et al. 1989). Sites with higher base saturation have a higher potential for decrease of base saturation especially at sites where sulphate is released into soil solution. In addition, sites affected by deposition of fly ashes (see Chap. 3) show a high base saturation at NFSI I, hence a disproportional high reduction of base saturation (Riek et al. 2012). The effect size of organic layer and mineral soil carbon stock is lower than that of Ca stock and base saturation. On sites with a higher organic layer carbon stock, there is a higher retention of base cations in the organic layer and consequently a lower availability of base cations in the mineral soil (Fig. 4.10c). The model identified also an effect of mineral soil carbon stocks on $_{\Delta}$BS which is stronger on limed plots for very low carbon stocks (Fig. 4.10d). Although the effect of K weathering on $_{\Delta}$BS is only weak, it may be interpreted as a disproportional high decrease of base saturation on sandy sites with a high proportion of illite-like clay minerals, e.g. soils from base-poor unconsolidated sediments (Fig. 4.10e). From Fig. 4.10f no systematic deviation of the relation between estimated and observed change in base saturation is visible.

The sensitivity of $_{\Delta}$BS for limed and unlimed plots with respect to base saturation between 10 and 25% for different levels of Ca stock (600 and 900 kg ha^{-1}) and organic layer carbon stocks (20 and 40 t ha^{-1}) is visible in Fig. 4.11. The other variables included in the GAM were held constant at the median of NFSI II. Liming generally increases $_{\Delta}$BS especially where a large reduction of base saturation occurs indicating an efficient mitigation of further soil acidification after liming. With regard to a different depth interval (0–90 cm) considered in this study, the effect size is comparable to the findings of Guckland et al. (2012) for sites in northwest Germany, who found differences in base saturation of 20%, 5% and 3% for the depth intervals 0–10 cm, 10–20 cm and 20–40 cm of limed as compared to unlimed plots, respectively. A strong effect of organic layer carbon stocks on Ca and Mg retention and base saturation changes was also described by Guckland et al. (2012). In summary, model results suggest that $_{\Delta}$BS is mainly dependent on the soil pools of basic substances at NFSI I, which is modified with respect to liming and soil organic matter status.

4.2.3 *Aqua Regia Extractable Ca Stocks*

The mean stocks of aqua regia extractable Ca in the organic layer amounted to 288 ± 9 kg ha^{-1} for the NFSI II. High amounts of Ca were found predominantly at sites with carbonate-containing bedrock and at limed plots (Fig. 4.12). Low Ca amounts were observed in soils from base-poor substrates at low mountain ranges and in the North German lowlands. Evers et al. (2016) found for the state of Hesse a decrease of Ca stocks in the organic layer and an increase in the mineral topsoil

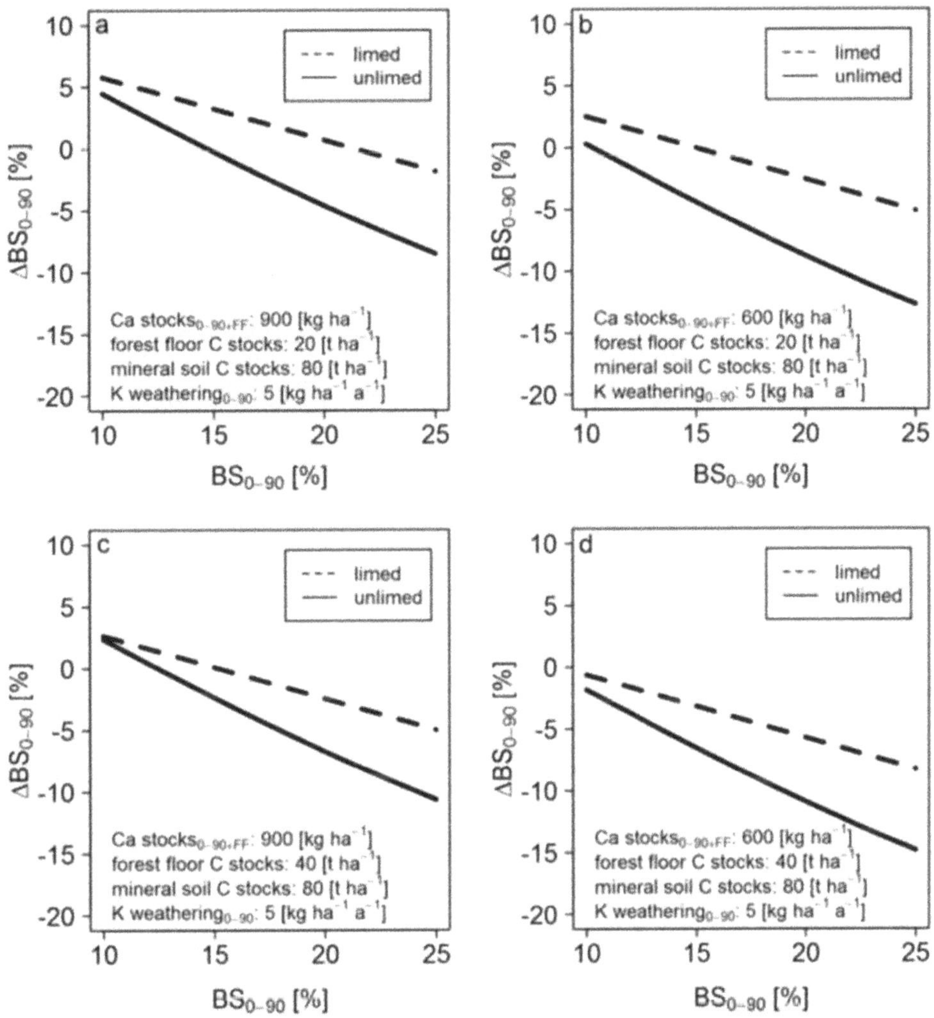

Fig. 4.11 Partial effects of base saturation on base saturation changes between NFSI I and NFSI II ($_\Delta$BS) for different levels of Ca stocks and organic layer carbon stocks. Other variables held constant

between NFSI I and NFSI II. The spatial and temporal pattern of Mg stocks was similar to Ca yet at a lower level (not shown).

4.2.4 Comparison with Long-term Studies on Soil Acidification

The time period between NFSI I and NFSI II is characterized by strong reductions of acid deposition. However, even stronger reductions occurred prior to NFSI I beginning in the 1980s. In order to compare changes in the acid-base status of forest soils between NFSI I and NFSI II to a longer time period, long-term studies from intensive

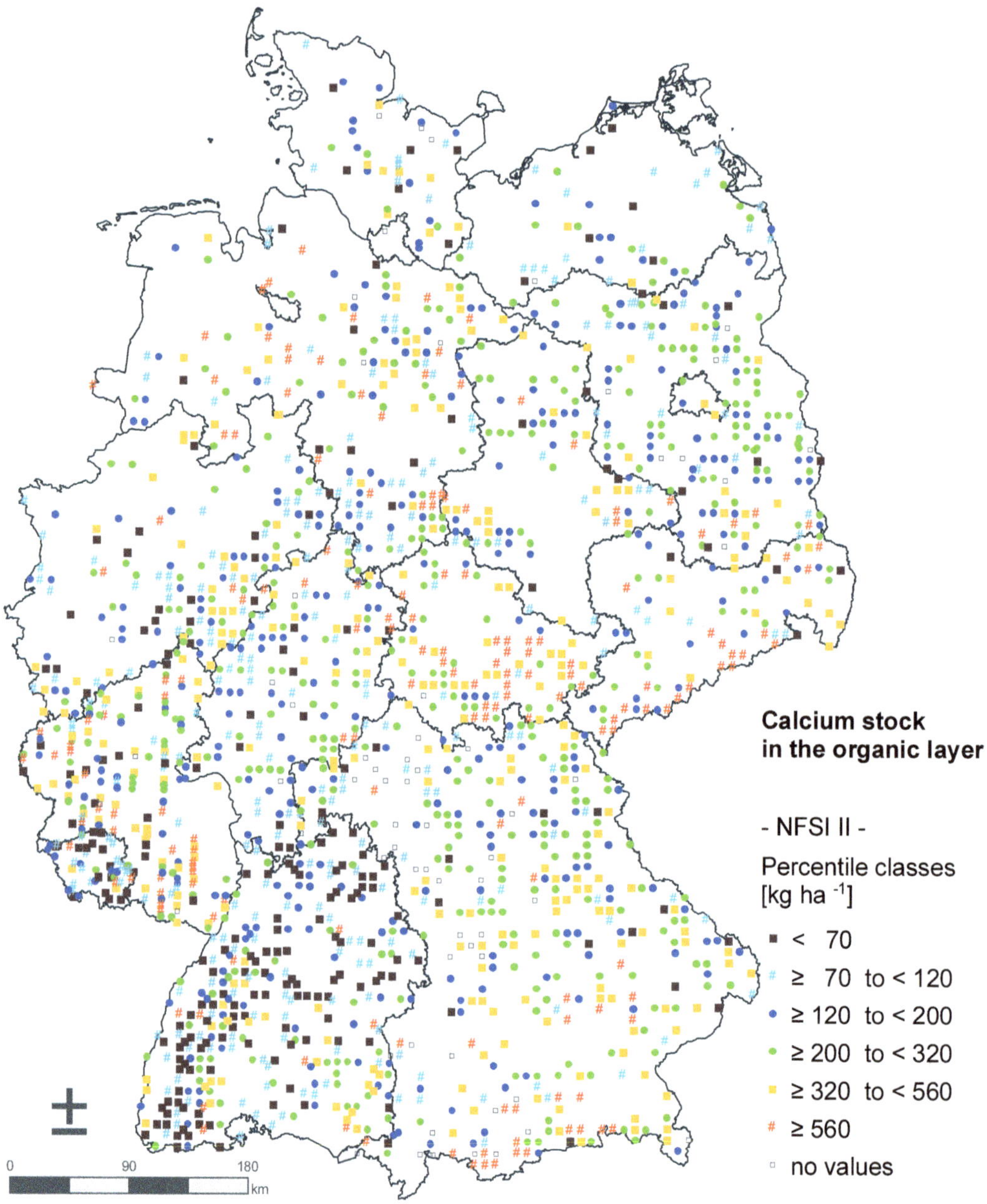

Fig. 4.12 Calcium stock in kg ha^{-1} in the organic layer of NFSI II

monitoring plots from northwestern Germany were evaluated with respect to soil acidity.

At two intensive monitoring plots of the ICP Forests Level II programme at Solling, northwestern Germany, stocked with European beech (Level II plot 304) and Norway spruce (Level II plot 305) 11 and 12 inventories of the mineral soil, respectively, have been conducted between 1966 and 2010. At both plots, base saturation showed a decreasing trend between the late 1960s and the beginning of the 2000s, whereas a slight recovery was observed within this century (Figs. 4.13 and 4.14). Recovery was more distinctive in the upper soil layers, whereas the subsoil

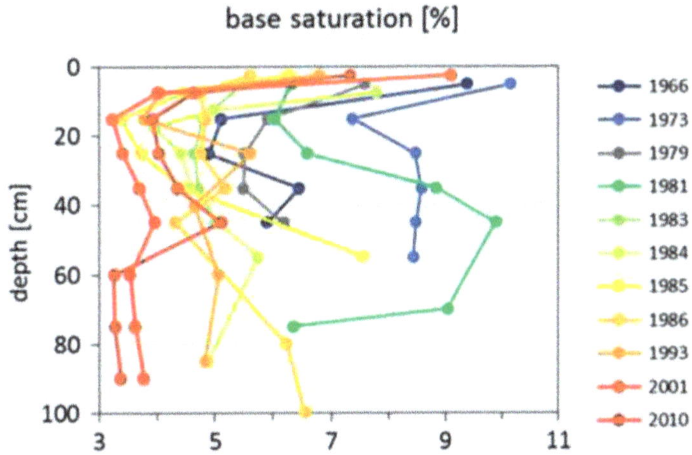

Fig. 4.13 Long-term development of base saturation over depth at Level II plot 304 Solling beech

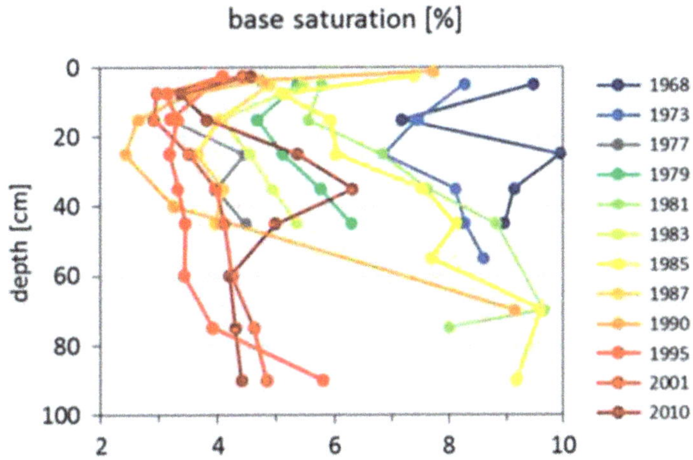

Fig. 4.14 Long-term development of base saturation over depth at Level II plot 305 Solling spruce

experienced a further acidification. The loss of base cations from the mineral soil until the 1990s was partly compensated by an accumulation in the organic layer (Meiwes et al. 2009). Since then, base cations were released into the mineral soil due to enhanced decomposition of the organic layer (Meesenburg et al. 2016).

At long-term monitoring plots at substrates with low buffer capacity and soil samplings before 1985, strong decreases of base saturation in the 0–30 cm layer of the mineral soil occurred before 1990 (Fig. 4.15). The decrease ceased at those plots approximately between 1995 and 2005, and a slight increase of base saturation can be assumed since then. The time period, when soil acidification was reversed at these plots, falls in between NFSI I and NFSI II, which might explain opposite trends of soil acidification found at the NFSI plots.

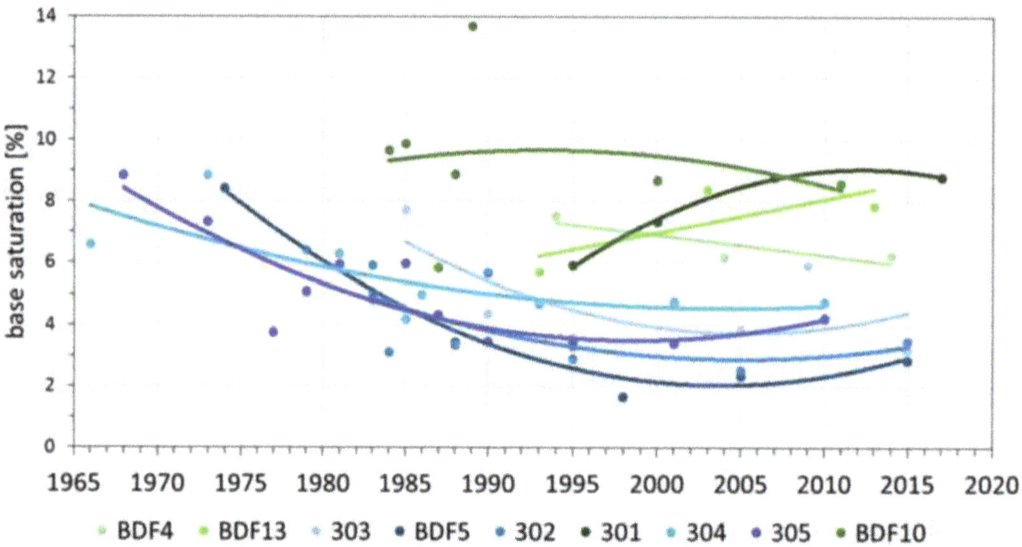

Fig. 4.15 Long-term development of base saturation at 0–30 cm depth of the mineral soil at intensive monitoring plots in northwest Germany. 301 to 305 refer to ICP Forests Level II plots, BDF4 to BDF13 to permanent soil monitoring plots; plots with soils from base-poor consolidated bedrock are in blue colour; plots with soils from base-poor unconsolidated sediments are in green colour; trend lines are second-order polynomial functions, except for plots with only three samplings, where a linear trend was applied

4.2.5 Case Study on Soil Acidification in the State of Brandenburg

A special pollution situation in the northeastern German lowlands became evident in the NFSI of the federal state of Brandenburg (Riek et al. 2015). An important outcome of this survey was that acidification of many forest soils in Brandenburg and neighbouring areas proceeded at an above-average rate in the period between NFSI I (1992) and NFSI II (2007). A decrease in the pH(KCl) values was recorded in the organic layer and the mineral topsoil to 30 cm depth, whereas pH(KCl) changes were insignificant in the subsoil. The base saturation as a sensitive indicator of soil acidification decreased significantly in all depths of the examined soil body (0–140 cm depth), and the Ca and Mg stocks have changed from predominantly low-medium to low (Ca) and from low to very low (Mg), respectively (valuation levels by AK Standortskartierung 2016).

The leaching of Ca and Mg ions, which had previously been deposited by flying ash from brown coal power plants, is regarded as a major cause for this obviously strong decrease of pH and base saturation in a relatively short period. A pronounced loss of base cations was recorded especially for those soils that had unusually high base saturations with respect to these usually nutrient-poor sandy soils at the time of NFSI I (Fig. 4.16).

Regionally occurring high pH(KCl) values in the mineral topsoil of the NFSI II sample, with simultaneously low values in the subsoil, also provide a clear indication for a still persisting effect of the former atmospheric deposition of basic dust. In

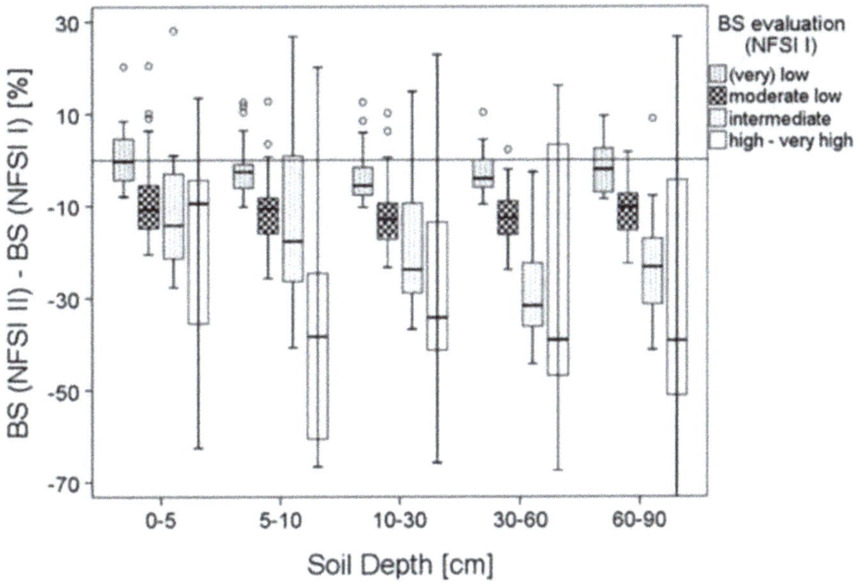

Fig. 4.16 Difference between base saturation (BS) from NFSI II and NFSI I in the state of Brandenburg, stratified by soil depth and evaluation level of base saturation at the time of NFSI I (valuation levels according to AK Standortskartierung 2003), without hydromorphic and carbonate-containing soils ($n = 126$)

addition, local maxima of the base cation stocks are currently still found in the former dust deposition areas in southern Brandenburg. Noteworthy are the natural nutrient-poor inventory plots of this old moraine area, which still have very high Ca and Mg fractions of the cation exchange capacity despite a general reduction of base saturation between NFSI I and NFSI II. The depth profiles of base saturation at NFSI plots affected by fly ashes do not fit well to any of the depth profile types of base saturation defined in Chap. 2.

The former GDR brown coal production amounted to 85.2 million tons and peaked at 312.2 million tons in the mid-1980s (Buck 1996). The use of lignite in power plants grew steadily until the year 1989. It was increasingly burned as raw brown coal in the industrial agglomeration areas of the GDR. According to studies in the Bitterfeld area by Koch et al. (2001), the fly ashes consisted of high proportions of Ca and Mg oxides. The dusts were partly deposited in the vicinity of emission sources and due to emission with high chimneys also transported over long distances settling particularly at forest ecosystems because of their high surface roughness.

On the basis of deposition measurements since the 1960s (Möller and Lux 1992), coal mining statistics (Statistik der Kohlenwirtschaft e.V. 2015) and extensive analyses of bark samples on the grid of the forest condition survey of the former GDR in the years 1985 and 1988 (Stöcker and Gluch 1992; Kallweit et al. 1985), both the temporal and the spatial distribution of the base cation deposition were roughly calculated (Riek et al. 2015). According to these estimations approximately 50 kg ha^{-1} $year^{-1}$ of Ca was deposited on average into pine stands in the state of Brandenburg at the time of the highest incidence of atmospheric pollution in the 1980s. A mean cumulative Ca deposition in the order of 2.2 t ha^{-1} for the period from 1945 to 1990 is estimated according to this approach. This corresponds to an amount of lime of 6 t ha^{-1} ($CaCO_3$ content: 90%) over a period of 45 years, with which a large part of the acid sulphur deposition could be buffered. After cessation of emission of these specific atmospheric pollutants during the early 1990s, the accumulated mobile anions of sulphuric acid were leached, taking corresponding amounts of Ca and Mg ions with them.

Thus, it is assumed that the reduced base saturation recorded in the NFSI II represents a more natural state than the artificially elevated values at the time of the NFSI I in the regions previously affected by dust deposition.

4.3 Conclusions

The results of NFSI II demonstrate first achievements of environmental policy and forest management for the recovery of forest soils from acidification. The widespread acidification of topsoils and depletion of bases, which was observed during NFSI I, has been reduced. The recovery from acidification was highest in the organic layer and decreased with depth of the mineral soil. However, pH(KCl) decreased further especially at sites in the northeastern German lowlands. Where forest sites were limed, a significant increase of pH(KCl) until 10 cm depth was observed. The pH increase can be attributed to decreased deposition of acidity and to liming measures. A larger increase of pH(H_2O) as compared to pH(KCl) is due to the decrease of the ionic strength of soil solution after deposition reduction.

According to the main buffer mechanisms the sampled soil intervals can be assigned to buffer ranges. Due to changes of soil pH, the distribution of NFSI plots among the buffer ranges changed as well. The rise of pH in the topsoil resulted in an increase of plots within the exchange to $CaCO_3$ buffer range, whereas plots within the Fe to Al buffer range decreased. In the 60–90 cm depth interval the proportion of plots in the Al and exchange buffer range decreased in favour of the silicate and $CaCO_3$ buffer range. The shift in the distribution of plots among buffer ranges was mainly caused by liming.

Base saturation increased in the topsoil between NFSI I and NFSI II, which was significant for the paired sample in the 0–5 cm interval. However, a significant decrease of base saturation was observed below 10 cm depth. Limed plots showed a significant increase of base saturation in the upper 30 cm of the mineral soil, and base saturation in this depth range was higher as compared to unlimed plots.

The Al saturation of the NFSI plots showed a depth gradient, which was inverse to the depth gradient of base saturation. Aluminium saturation was on average 40.2 $\pm$ 0.62% in the 0–5 cm depth achieving a maximum of 62.3 $\pm$ 0.79% in 10–30 cm depth and decreasing values with increasing depth. Average Mn, Fe and H saturation was highest in the uppermost depth interval (0–5 cm) with values of 2.5 $\pm$ 0.08%, 3.1 $\pm$ 0.08% and 12.8 $\pm$ 0.31%, respectively, and decreased with increasing depth.

Mean stocks of aqua regia extractable Ca and Mg in organic layers of German forest soils were 288 ± 8.5 kg ha^{-1} and 73 ± 2.7 kg ha^{-1}, respectively, for NFSI II. They were increasing with increasing depth achieving 1895 ± 177 kg ha^{-1} and 1403 ± 62 kg ha^{-1} of Ca and Mg, respectively, in 5–10 cm depth. The spatial distribution of nutrient cations was mainly related to the parent material of the soils with high stocks in soils from weathered carbonate bedrock and in soil from basic-intermediate bedrock, whereas low stocks were found in soils from base-poor substrates. Mean stocks of exchangeable nutrient cations (Ca, K, Mg) for the whole soil profile (0–90 cm) were 6670 ± 274 kg ha^{-1}, 609 ± 15.7 kg ha^{-1} and 1206 ± 62.3 kg ha^{-1}, respectively.

Liming between NFSI I and NFSI II exert a major effect on the acid-base status and the nutrient availability of forest soils. Limed sites reveal higher pH and base saturation as compared to unlimed sites. In addition, stocks of exchangeable Ca and Mg are significantly higher up to 30 cm depth.

At NFSI II plots, lower pH(KCl) values were observed in coniferous stands than in deciduous stands in the organic layer and in the mineral soil up to 10 cm depth. These differences were less distinctive in deeper soil layers.

Generally, the acid-base status of forest soils in Germany improved between NFSI I and NFSI II due to a strong reduction of acid deposition. The statistical evaluation showed a major impact of liming and the acid-base status at NFSI I on the base saturation change until NFSI II. In addition, carbon stocks in the organic layer and mineral soil exhibit a modifying influence. Soil acidification was decelerated or even reversed inducing a decreased Al concentration in soil solution with corresponding reduced toxic stress to tree roots and microorganisms. Especially those sites, which are limed and vegetated with deciduous tree species, showed indications of recovery. At limed sites the regeneration extended even to the subsoil. However, unlimed acid-sensitive sites revealed an ongoing acidification of the subsoil with corresponding losses of base cations. Major causes are the still substantial deposition of nitrogen species with subsequent generation of acidity through uptake and nitrification processes and the remobilization of previously retained sulphur. In addition, the loss of alkalinity through the uptake of nutrient cations gained increasing importance during the last decades because of increased growth rates and decreased importance of other sources of acidity. Owing to the nationwide reduction of acid inputs, the relative spatial variability of soil acidity increased. A regional peculiarity are the deposits of basic dust in parts of the northeastern German lowlands, which were subject to a strong leaching from the soils but still cause a higher level of alkalinity at some sites.

With respect to an aspired recovery of the acid-base status of forest soils in Germany, the following recommendations can be deduced from the results of NFSI:

- Further reduction of deposition of acidity to a level below critical loads of acidity
- Liming of acidified soils in order to accelerate the regeneration process and to compensate nutrient losses due to introduced mobile anions
- Limitation of nutrient removal with biomass extraction at nutrient-poor and acid-sensitive sites

- Promotion of deep-rooting tree species in order to enhance the zone, where mineral weathering contributes to the biogeochemical cycling of base cations

The above-mentioned adaptation options should be combined based on a site-specific evaluation of the current acid-base status and expected pathways of forest development.

References

Aber JD (1992) Nitrogen cycling and nitrogen saturation in temperate forest ecosystems. Trends Ecol Evol 7:220–224

Aber JD, Nadelhoffer KJ, Steudler P, Melillo JM (1989) Nitrogen saturation in northern forest ecosystems. Bioscience 39(6):378–386

Aber JD, McDowell W, Nadelhoffer K, Magill A, Berntson G, Kamakea M, McNulty S, Currie W, Rustad L, Fernandez I (1998) Nitrogen saturation in temperate forest ecosystems – hypotheses revisited. Bioscience 48(11):921–934. https://doi.org/10.2307/1313296

Ahrends B, Klinck U, Klinck C, Weis W (2018) Herleitung flächiger Verwitterungsraten. Freiburger Forstl. Forschung 101:113–149

AK Standortskartierung (2003) Forstliche Standortsaufnahme: Begriffe, Definitionen, Einteilungen, Kennzeichnungen, Erläuterungen, 6th edn. IHW-Verlag, Eiching near Munich

AK Standortskartierung (2016) Forstliche Standortsaufnahme: Begriffe, Definitionen, Einteilungen, Kennzeichnungen, Erläuterungen, 7th edn. IHW-Verlag, Eiching near Munich

Augusto L, Ranger J, Binkley D, Rothe A (2002) Impact of several common tree species of European temperate forests on soil fertility. Ann For Sci 59(3):233–253. https://doi.org/10.1051/forest:2002020

Blake L, Goulding K, Mott C, Johnston A (1999) Changes in soil chemistry accompanying acidification over more than 100 years under woodland and grass at Rothamsted Experimental Station, UK. Eur J Soil Sci 50(3):401–412

BMU (2007) National strategy on biological diversity. BMU, Berlin

Brumme R, Meesenburg H, Bredemeier M, Jacobsen C, Schönfelder E, Meiwes K, Eichhorn J (2009) Changes in soil solution chemistry, seepage losses, and input–output budgets at three beech forests in response to atmospheric depositions. In: Brumme R, Khanna P (eds) Functioning and management of European beech ecosystems. Ecological studies, vol 208. Springer, Berlin, pp 303–336

Buck HF (1996) Umweltpolitik und Umweltbelastung. In: Die wirtschaftliche und ökologische Situation der DDR in den 80er Jahren. Springer, pp 223–266

Corre MD, Brumme R, Veldkamp E, Beese FO (2007) Changes in nitrogen cycling and retention processes in soils under spruce forests along a nitrogen enrichment gradient in Germany. Glob Chang Biol 13(7):1509–1527

Davidson E, Hart S, Shanks C, Firestone M (1991) Measuring gross nitrogen mineralization, and nitrification by 15 N isotopic pool dilution in intact soil cores. J Soil Sci 42(3):335–349

de Vries W, Dobbertin MH, Solberg S, Van Dobben HF, Schaub M (2014) Impacts of acid deposition, ozone exposure and weather conditions on forest ecosystems in Europe: an overview. Plant Soil 380(1–2):1–45

de Wit HA, Eldhuset TD, Mulder J (2010) Dissolved Al reduces mg uptake in Norway spruce forest: results from a long-term field manipulation experiment in Norway. For Ecol Manag 259(10):2072–2082. https://doi.org/10.1016/j.foreco.2010.02.018

Evers J, Paar U, Schönfelder E (2016) Bodenchemische Kenngrößen in Hessen. In: Paar U, Evers J, Dammann I, König N, Eichhorn J (eds) Waldbodenzustandsbericht für Hessen: Ergebnisse der zweiten Bodenzustandserhebung im Wald (BZE II). Beitr Nordwestdeutsche Forstliche Versuchsanstalt, vol 15, pp 123–215

Falkengren-Grerup U, Linnermark N, Tyler G (1987) Changes in acidity and cation pools of south Swedish soils between 1949 and 1985. Chemosphere 16(10–12):2239–2248

Fölster H (1985) Proton consumption rates in Holocene and present-day weathering of acid forest soils. In: Drever JI (ed) The chemistry of weathering. Springer, Dordrecht, pp 197–209

Guckland A, Ahrends B, Paar U, Dammann I, Evers J, Meiwes KJ, Schönfelder E, Ullrich T, Mindrup M, König N, Eichhorn J (2012) Predicting depth translocation of base cations after forest liming: results from long-term experiments. Eur J For Res 131(6):1869–1887. https://doi.org/10.1007/s10342-012-0639-0

Hansen K, Vesterdal L, Bastrup-Birk A, Bille-Hansen J (2007) Are indicators for critical load exceedance related to forest condition? Water Air Soil Pollut 183(1–4):293–308

Hastie T, Tibshirani R (1990) Generalized additive models: monographs on statistics and applied probability. Chapman and Hall, Boca Ratan, FL

Hofner B, Müller J, Hothorn T (2011) Monotonicity-constrained species distribution models. Ecology 92(10):1895–1901

Höhle J, Bielefeldt J, Dühnelt P-E, König N, Ziche D, Eickenscheidt N, Grüneberg E, Hilbrig L, Wellbrock N, Kompa T (2018) Bodenzustandserhebung im Wald – Dokumentation und Harmonisierung der Methoden. Thünen Working Paper. Johann Heinrich von Thünen Institute, Braunschweig. https://doi.org/10.3220/WP1526989795000

Horn R, Schulze E-D, Hantschel R (1989) Nutrient balance and element cycling in healthy and declining Norway spruce stands. In: Schulze ED, Lange OL, Oren R (eds) Forest decline and air pollution. Springer, Berlin, pp 444–455

Jacobsen C, Rademacher P, Meesenburg H, Meiwes K (2002) Gehalte chemischer Elemente in Baumkompartimenten – Literaturstudie und Datensammlung. Berichte des Forschungszentrums Waldökosysteme, vol 69. Göttingen, Germany, 80 pp

Johnson A, Andersen S, Siccama T (1994) Acid rain and soils of the Adirondacks. I. Changes in pH and available calcium, 1930–1984. Can J For Res 24(1):39–45

Kallweit R, Kaatzsch S, Strube M, Keller E (1985) Bioindikation über Kiefern und Fichtenborken. ZUG, Umweltlabor, Wittenberg

Klinck U, Rademacher P, Scheler B, Wagner M, Fleck S, Ahrends B, Meesenburg H (2012) Ökosystembilanzen auf forstwirtschaftlich genutzten Flächen. In: Höper H, Meesenburg H (eds) Tagungsband 20 Jahre Bodendauerbeobachtung in Niedersachsen, GeoBerichte, vol 23. Landesamt für Bergbau, Energie und Geologie, Hannover, pp 163–174

Koch J, Klose S, Makeschin F (2001) Bioelementverteilung und Humusmorphologie in flugaschebelasteten Oberböden der Dübener Heide. Berichte Freiburger Forstliche Forschung, vol 33. Freiburg, pp 183–192

Lawrence GB, Hazlett PW, Fernandez IJ, Ouimet R, Bailey SW, Shortle WC, Smith KT, Antidormi MR (2015) Declining acidic deposition begins reversal of forest-soil acidification in the northeastern US and eastern Canada. Environ Sci Technol 49(22):13103–13111

Mareschal L, Turpault M-P, Bonnaud P, Ranger J (2013) Relationship between the weathering of clay minerals and the nitrification rate: a rapid tree species effect. Biogeochemistry 112 (1–3):293–309

Meesenburg H, Brumme R, Jacobsen C, Meiwes K, Eichhorn J (2009) Soil properties. In: Functioning and management of European beech ecosystems. Ecological studies, vol 208. Springer, Berlin, pp 33–47

Meesenburg H, Ahrends B, Fleck S, Wagner M, Fortmann H, Scheler B, Klinck U, Dammann I, Eichhorn J, Mindrup M (2016) Long-term changes of ecosystem services at Solling, Germany: recovery from acidification, but increasing nitrogen saturation? Ecol Indic 65:103–112

Meiwes K, Meesenburg H, Eichhorn J, Jacobsen C, Khanna P (2009) Changes in C and N contents of soils under beech forests over a period of 35 years. In: Functioning and management of European beech ecosystems. Ecological studies, vol 208. Springer, Berlin, pp 49–63

Möller D, Lux H (1992) Deposition atmosphärischer Spurenstoffe in der ehemaligen DDR bis 1990: Methoden und Ergebnisse. In: Kommission Reinhaltung der Luft im VDI und DIN, vol 18. Düsseldorf, 308 pp.

Nieder R, Benbi DK, Scherer HW (2011) Fixation and defixation of ammonium in soils: a review. Biol Fertil Soils 47(1):1–14

Piña RG, Cervantes C (1996) Microbial interactions with aluminium. Biometals 9(3):311–316

Reuss JO, Johnson DW (1985) Effect of soil processes on the acidification of water by acid deposition. J Environ Qual 14(1):26–31

Reuss JO, Johnson DW (1986) Acid deposition and the acidification of soils and waters. In: Ecological studies, vol 59. Springer, Berlin

Rich C (1968) Hydroxy interlayers in expansible layer silicates. Clay Clay Miner 16(1):15–30

Riek W, Russ A, Martin J (2012) Soil acidification and nutrient sustainability of forest ecosystems in the northeastern German lowlands – results of the national forest soil inventory. Folia For Pol Ser A 54(3):187–195. https://doi.org/10.5281/zenodo.30835

Riek W, Russ A, Kühn D (2015) Waldbodenbericht Brandenburg – Zustand und Entwicklung der brandenburgischen Waldböden. Ergebnisse der landesweiten Bodenzustandserhebungen BZE-2 und BZE-2a. Band 1. Eberswalder Forstliche Schriftenreihe. Landesbetrieb Forst Brandenburg, Landeskompetenzzentrum Forst Eberswalde, Eberswalde

Schöpp W, Posch M, Mylona S, Johansson M (2003) Long-term development of acid deposition (1880–2030) in sensitive freshwater regions in Europe. Hydrol Earth Syst Sci 7(4):436–446

Schwarz MT, Bischoff S, Blaser S, Boch S, Grassein F, Klarner B, Schmitt B, Solly EF, Ammer C, Michalzik B (2016) Drivers of nitrogen leaching from organic layers in central European beech forests. Plant Soil 403(1–2):343–360

Seip H (1980) Acidification of freshwater-sources and mechanisms. In: International conference on the ecological impact of acid precipitation. Sandefjord (Norway). 11–14 Mar 1980

Spranger T, Lorenz U, Gregor H-D (2015) ICP modelling & mapping: manual on methodologies and criteria for modelling and mapping critical loads & levels and air pollution effects, risks and trends – amended version. UBA Texte, vol 52/2004. Federal Environmental Agency (UBA), Berlin

Statistik der Kohlenwirtschaft e.V. (2015) http://www.kohlenstatistik.de/19-0-Braunkohle.html. Accessed 9 Feb 2015

Stöcker G, Gluch W (1992) Depositionscharakterisierung auf der Grundlage der Borkenindikation (Biomonitoring). In: Möller D, Lux H (eds) Deposition atmosphärischer Spurenstoffe in der ehemaligen DDR bis 1990: Methoden und Ergebnisse. Schriftenreihe der Kommission Reinhaltung Luft im VDI und DIN, vol 18. Düsseldorf, pp 272–287

Stoddard JL, Jeffries D, Lükewille A, Clair T, Dillon P, Driscoll C, Forsius M, Johannessen M, Kahl J, Kellogg J (1999) Regional trends in aquatic recovery from acidification in North America and Europe. Nature 401(6753):575

Sucker C, von Wilpert K, Puhlmann H (2011) Acidification reversal in low mountain range streams of Germany. Environ Monit Assess 174(1–4):65–89. https://doi.org/10.1007/s10661-010-1758-z

Tarrah J, Meiwes KJ, Meesenburg H (2000) Normative calculation of minerals in north German loess soils using the modified CIPW norm. J Plant Nutr Soil Sci 163(3):307–312

Ulrich B (1981) Ökologische Gruppierung von Böden nach ihrem chemischen Bodenzustand. Z Pflanzenernähr Bodenkd 144(3):289–305. https://doi.org/10.1002/jpln.19811440308

Ulrich B (1983) Soil acidity and its relations to acid deposition. In: Ulrich B, Pankrath J (eds) Effects of accumulation of air pollutants in forest ecosystems. Springer, Dordrecht, pp 127–146

Ulrich B (1987) Stability, elasticity, and resilience of terrestrial ecosystems with respect to matter balance. In: Schulze E-D, Zwölfer H (eds) Potentials and limitations of ecosystem analysis. Ecological studies, vol 61. Springer, Berlin, pp 11–49

Ulrich B (1994) Nutrient and acid-base budget of central European forest ecosystems. In: Godbold DL, Hüttermann A (eds) Effects of acid rain on forest processes. Wiley-VCH, New York, NY, pp 1–50

Ulrich B, Mayer R, Khanna PK (1980) Chemical changes due to acid precipitation in a loess derived soil in Central Europe. Soil Sci 130:193–199

UNECE (1979) Convention on long-range transboundary air pollution (CLRTAP). Geneva

Waldner P, Marchetto A, Thimonier A, Schmitt M, Rogora M, Granke O, Mues V, Hansen K, Karlsson GP, Zlindra D, Clarke N, Verstraeten A, Lazdins A, Schimming C, Iacoban C, Lindroos AJ, Vanguelova E, Benham S, Meesenburg H, Nicolas M, Kowalska A, Apuhtin V, Napa U, Lachmanova Z, Kristoefel F, Bleeker A, Ingerslev M, Vesterdal L, Molina J, Fischer U, Seidling W, Jonard M, O'Dea P, Johnson J, Fischer R, Lorenz M (2014) Detection of temporal trends in atmospheric deposition of inorganic nitrogen and sulphate to forests in Europe. Atmos Environ 95:363–374. https://doi.org/10.1016/j.atmosenv.2014.06.054

Wellbrock N, Bolte A, Flessa H (eds) (2016) Dynamik und räumliche Muster forstlicher Standorte in Deutschland: Ergebnisse der Bodenzustandserhebung im Wald 2006 bis 2008. Thünen Report, vol 43. Johann Heinrich von Thünen Institute, Federal Research Institute for Rural Areas, Forestry and Fisheries, Braunschweig

Wood SN (2006) Generalized additive models: an introduction with R. Chapman & Hall, Boca Raton, FL

Wright RF, Larssen T, Camarero L, Cosby BJ, Ferrier RC, Helliwell R, Forsius M, Jenkins A, Kopácek J, Majer V, Moldan F, Posch M, Rogora M, Schöpp W (2005) Recovery of Acidified European Surface Waters. Environ Sci Technol 39(3):64A–72A. https://doi.org/10.1021/es0531778

Chapter 5
Nitrogen Status and Dynamics in German Forest Soils

Stefan Fleck, Nadine Eickenscheidt, Bernd Ahrends, Jan Evers, Erik Grüneberg, Daniel Ziche, Juliane Höhle, Andreas Schmitz, Wendelin Weis, Paul Schmidt-Walter, Henning Andreae, and Nicole Wellbrock

S. Fleck (✉)
Thünen Institute of Forest Ecosystems, Eberswalde, Germany

North West German Forest Research Institute, Göttingen, Germany
e-mail: stefan.fleck@nw-fva.de

N. Eickenscheidt
State Agency for Nature, Environment and Consumer Protection of North Rhine-Westphalia, Recklinghausen, Germany
e-mail: nadine.eickenscheidt@lanuv.nrw.de

B. Ahrends · J. Evers · P. Schmidt-Walter
North West German Forest Research Institute, Göttingen, Germany
e-mail: bernd.ahrends@nw-fva.de; jan.evers@nw-fva.de; paul.schmidt-walter@nw-fva.de

E. Grüneberg · D. Ziche · A. Schmitz · N. Wellbrock
Thünen Institute of Forest Ecosystems, Eberswalde, Germany
e-mail: erik.grueneberg@thuenen.de; daniel.ziche@thuenen.de; andreas.schmitz@thuenen.de; nicole.wellbrock@thuenen.de

J. Höhle
Thünen Institute of Forest Ecosystems, Eberswalde, Germany

Public Enterprise Sachsenforst, Pirna, Germany
e-mail: juliane.hoehle@smul.sachsen.de

W. Weis
Bavarian State Institute of Forestry, Freising, Germany
e-mail: wendelin.weis@lwf.bayern.de

H. Andreae
Public Enterprise Sachsenforst, Pirna, Germany
e-mail: henning.andreae@smul.sachsen.de

N. Wellbrock, A. Bolte (eds.), *Status and Dynamics of Forests in Germany*, Ecological Studies 237, https://doi.org/10.1007/978-3-030-15734-0_5

5.1 Introduction

Nitrogen (N) in forest soils is a key variable to assess the state of forest ecosystems due to the large effects it has on forest growth, ecosystem integrity, and human health. High amounts of N are needed for biomass production in forests as N is one of the four elements that are structural components of most organic molecules forming biomass. In contrast to the other main elements essential for biomass production (carbon (C), oxygen (O), and hydrogen (H)), forest trees could over the longest part of their evolution not get these amounts of N directly by using ubiquitous media like water or gases from the atmosphere. Since atmospheric nitrogen (N_2) is a very stable molecule, requiring high amounts of energy for the transformation into reactive N species that may contribute to plant-available N, forest trees always depended on the recycling of organic N from the decomposition of dead organic matter in the soil, where organic molecules are with 95% the prevailing form of N (Rohmann and Sontheimer 1985). Their mineralization and the subsequent nitrification by microbes are exergonic and provide the inorganic molecules ammonium (NH_4^+) and nitrate (NO_3^-) that may both be taken up by plant roots. NO_3^- and NH_4^+ assimilation in the roots and leaves of trees require energy from photosynthesis, which provides the energy source keeping this forest internal N recycling running. In terms of ecosystem services, nutrient recycling is considered the economically most valuable ecosystem service of forests (Costanza et al. 1997).

Without consideration of human activities, there is only limited exchange of this ecosystem-internal cycle with the environment (Larcher 2001): The natural sources for additional plant-available N (lightnings and N_2 fixing microbes, redistribution of N by moving water or animals) are scarce; thus tree species show mechanisms for minimizing N losses from the forest ecosystem due to leaching or gaseous emission. These include to build up dense fine root and mycorrhizal networks close to the origin of newly mineralized N compounds or immediate uptake of any plant-available N, preferentially NH_4^+ (Ek et al. 1994; Posch et al. 2015). In this sense, forest ecosystems have evolved towards increased capability for N storage, with the upper parts of the soil as major N stock in temperate forests, as a response to the low reliability of N supply in nature.

In modern times, these conditions have dramatically changed due to N deposition from industrial processes with high energy consumption that lead deliberately (fertilizer production) or as a by-product (combustion processes) to the transformation of atmospheric N_2 to reactive N species. Atmospheric N deposition to forests in Germany had continuously risen since pre-industrial times until a maximum was reached between 1980 and 1995, when NFSI I took place. Partly due to the introduction of catalytic converters for Otto engines (in Germany since 1989), N deposition started to decrease and was already much lower during NFSI II (see Sect. 2.2), though historically still on a high level (Fagerli et al. 2007; Fagerli and Aas 2008). The temporal course of the calculated N deposition (Fig. 5.1) agrees well with other models of deposition history in Europe (cf. Engardt et al. 2017).

The high additional amount of NO_3^- and NH_4^+ currently reaching forest ecosystems through atmospheric deposition strongly affects the recycling of N: Forest growth is no longer limited by N availability such that other minimum factors gain

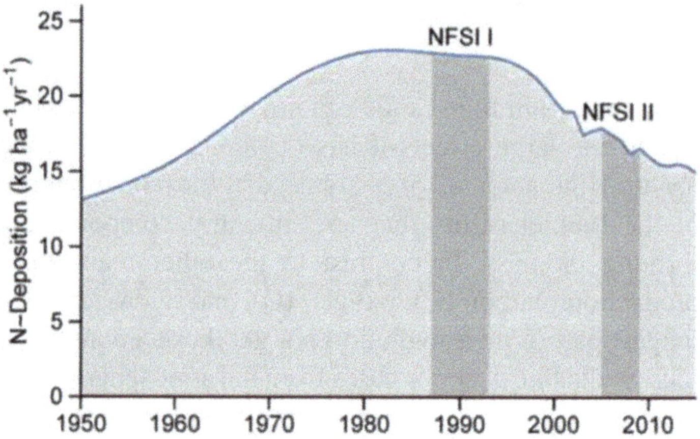

Fig. 5.1 Temporal development (1950–2015) of total (dry + wet + occult) nitrogen deposition to NFSI plots in kg ha^{-1} year^{-1}. For details on methodology, see Sects. 5.4.3 and 1.16

importance for controlling growth (Spiecker et al. 1996; Talkner et al. 2015). The increased amounts of plant-available N in the soil lead to changes in the species composition of ground vegetation, favouring nitrophilous species (Förster et al. 2017; Strengbom and Nordin 2008). If N is deposited over longer time periods, the N retention capacity of forest soils may be exceeded (N saturation, Aber et al. 1989, 1998; Lovett and Goodale 2011; Meesenburg et al. 2016), and excess N is leached in the form of NO_3^-, which leads to increased NO_3^- concentrations in the groundwater commonly used as drinking water and in lakes, where algal bloom and fish-poisonous conditions may be the consequence (Erisman et al. 2015; Oulehle et al. 2015). Lakes and brooks are also affected by acidity that is generated by plant uptake of NH_4^+ or its nitrification and may among other effects cause fish- and root-toxic aluminium ions (Al^{3+}) to dissolve, which can be leached together with NO_3^- (Vitousek et al. 1997). Due to electroneutrality, NO_3^- leaching also leads to a loss of base cations from the soil (de Vries et al. 2014). Also gaseous emissions of N_2O and other N compounds may increase with increasing N input (Eickenscheidt et al. 2011; Eickenscheidt and Brumme 2012; Krupa 2003), thereby contributing to the rising concentrations of the greenhouse gas N_2O in the atmosphere.

Since forests are typically much less affected by fertilizer application compared to agricultural areas, the groundwater under forested areas is often used for drinking water supply. Therefore, the quality of groundwater under forests is of direct relevance for human health. While a maximum of 50 mg NO_3^- per litre is tolerated according to the water framework directive of the EU (EU Directive 2006/118/EC), it is known that even much lower concentrations over longer periods may induce health risks like bowel cancer (Ward et al. 2005).

The NFSI, as the only spatially representative inventory of forest soils in Germany, may provide evidence for the effects of decreasing, but still high N deposition on N availability and N retention in the soil. It may show whether there are yet signs of recovery from the most extreme deposition rates and help to find out the time scale for such recovery. However, the soils were not only affected by changes in deposition of NO_3^- and NH_4^+: Between NFSI I and NFSI II, a strong reduction in sulphur deposition (Engardt et al. 2017), a climate change induced rise in temperatures, higher availability of carbon dioxide (CO_2), and changed availability of water may have altered the relationship between N transformation processes in the soil (Fleck et al. 2017). For example, during the vegetation periods, days with extremely low soil water availability (<40% of plant available water capacity) were much less frequent in the 10 years before NFSI I (21.6 days) than in the 10 years before NFSI II (30.4 days, Fig. 3.11), while annual precipitation remained unchanged, thereby potentially increasing the amount of wetting and drying cycles that increase the availability of soil organic matter to decay processes (Borken and Matzner 2009). Management effects of liming and conversion of tree species composition provide additional influencing factors. The measurement results must, therefore, be interpreted with care, since they are the result of several interacting environmental and management factors.

The following sub-chapters show first the status of N stocks and C/N ratios and highlight some of the most important impact factors for these patterns. The status results rely on NFSI II measurements (complete sample including peatland plots). Results from NFSI I and Intensive Forest Monitoring plots (IFM plots, i.e. Level II plots and beyond them other plots sampled according to the same methodology) are occasionally given for comparison. Nitrogen stock changes from NFSI I to NFSI II are then derived with two alternative approaches: (1) differential measurement of N stocks at two points in time (NFSI I and NFSI II), based on the paired sample without peatland plots or on IFM plot measurements, and (2) N balance of the modelled input and output rates for the period between NFSI I and NFSI II. These trend calculations are followed by the discussion of methods and a final discussion of the results.

5.2 Nitrogen Stocks in Forest Soils

Nitrogen stocks in the soil profile were assessed from the organic layer to a maximum depth of 90 cm. Because the soil depth was lower than 90 cm at a considerable proportion of sites and because of lower data availability at deeper depth, soil profile data are in most cases only shown for organic layer—60 cm. The typical gradient of N stocks with soil depth is, however, best visible for organic layer—90 cm. Considering the skewness of N stock distributions from NFSI I (see Sect. 5.5.3), medians are presented for each layer instead of means wherever distributions are not normal. All medians and means are area-weighted (see Sect. 1.17).

5.2.1 Gradient of Nitrogen Stocks with Depth in the Soil Profile

In NFSI II, 11% of N were kept in the organic layer, 52% in the uppermost 30 cm of the mineral soil, 21% in 30–60 cm depth, and 12% in 60–90 cm depth (layer medians as percentages of the organic layer—90 cm median). This gradient was similarly

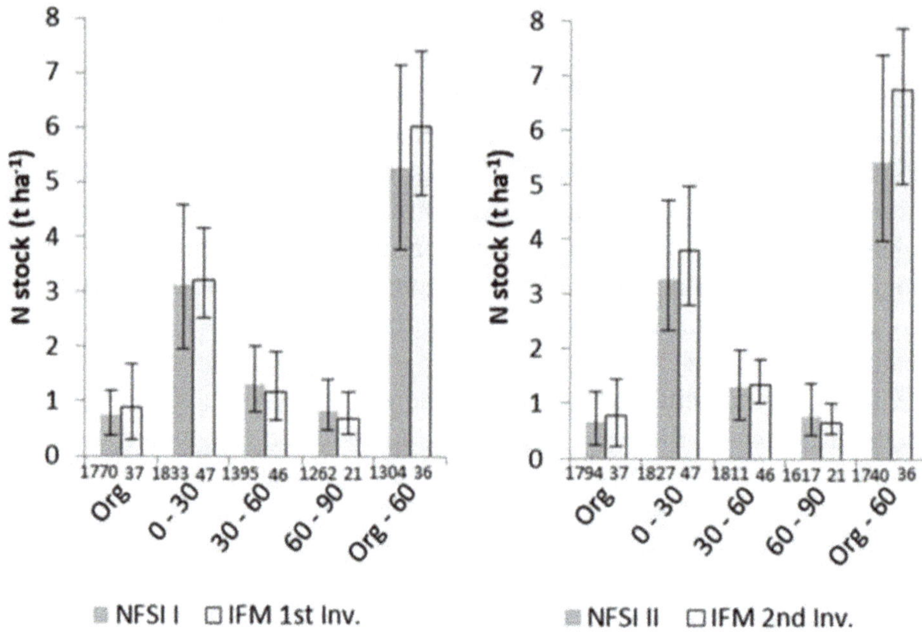

Fig. 5.2 Depth gradient of nitrogen stocks in t ha^{-1} for NFSI I (left, dark columns) and NFSI II (right, dark columns) related to the total sample of each inventory. Results from the first and second inventory on Intensive Forest Monitoring plots (IFM, light columns) measured within a time shift of maximum 5 years to each of the NFSIs (see "Results from Intensive Forest Monitoring Plots" under Sect. 5.4.2) are given for comparison. All values are medians with interquartile ranges as error bars and sample sizes given at the horizontal axis. As representative result for German forest soils, medians and interquartile ranges of NFSIs are area-weighted (see Sect. 1.17 and Fig. 5.11)

observed in NFSI I (12%, 50%, 21%, and 13%, respectively) and did not significantly change (Figs. 5.2 and 5.12a). Even the results from the much lower number of IFM plots with available inventory data show a very similar gradient (see Sect. 5.4.2). The observed gradient reflects the flow of organic material and its descendants through the ecosystem: Dead biomass is predominantly integrated via the organic layer, where the fresh litter is in the prevailing case of aerobic conditions not stored, but subject to high turnover rates and bioturbation, leading to its integration into soil organic matter. Total soil N stocks are, therefore, typically several times higher than those of only the organic layer and depend mainly on the quality and amount of litter entering the soil at its upper border or from fine roots in the upper mineral soil. Other important determinants of soil N stock are the climatic and soil-chemical conditions for litter decomposition. The sharpest decrease in N stocks occurs between the uppermost 30 cm and the next 30 cm of soil, since the aboveground organic material mainly adds to the uppermost part of the mineral soil. The decrease between 30–60 cm depth and 60–90 cm depth is rather moderate and apparently reflects the decreasing fine root density, which is mostly considered irrelevant in even deeper layers, where N concentrations often fall below the detection limit.

5.2.2 Nitrogen Stocks in the Organic Layer

Nitrogen stock in the organic layer was with 0.67 t ha^{-1} and an interquartile range (iqr) of {0.24; 1.22} t ha^{-1} (n = 1794) smaller than the amount of N stored in the upper 5 cm of the mineral soil (1.09 t ha^{-1}, iqr{0.76; 1.43}, n = 1846). Since the delineation between both compartments is to a certain extent subjective (Jansen et al. 2005), the joint N stock in organic layer—5 cm (1.81 t ha^{-1}, iqr {1.43; 2.37}, n = 1784) is a more robust estimate.

5.2.3 Nitrogen Stocks in the Soil Profile: Organic Layer—Maximum 90 cm

While the whole rooting zone may—especially on sandy soils—often go deeper than 90 cm of the mineral soil (cf. Czajkowski et al. 2009), about 6% of NFSI II plots had soil profiles shallower than 90 cm. As a representative number for German forest soils, the area-weighted median of N stocks for all soil profiles from the organic layer down to maximum depth of 90 cm of the mineral soil was 6.3 t ha^{-1} (iqr {4.5; 8.6}, n = 1647). The empirical national rating for N stocks in the rooting zone of forests (AK Standortskartierung 2016) was confirmed to some extent by the scatter of these results, since high (>10 t ha^{-1}) and very high (>20 t ha^{-1}) N stocks in the soil profile were observed on only 14% and 1% of the plots, respectively. These highest values were mainly reached in organic soils on actual or former peatland area. Very low nitrogen stocks (≤2.5 t ha^{-1}) were found on only 3% of the plots.

Nitrogen stocks of the soil profile (Fig. 5.3) were highest on peatland plots and organic soils in the western part of the North-German Lowland, the Alps and their foothills, and the Bavarian Forest. Other regions with very high N stocks are the Rhine Rift Valley (here the N redistribution on alluvial plains may play a role), and mountain ranges surrounding the Thuringian Basin, Upper Franconia, large parts of the Rhenish Slate Mountains, the Southern Swabian Alb, Harz and nearby mountain ranges in Lower Saxony, and the Saarland. Out of these regions, the western part of the North-German Lowland, the Bavarian Forest, Upper Franconia, Harz, and the Rhenish Slate Mountains received comparatively high amounts of N deposition between 1990 and 2007. Plots in the Alps, the Bavarian Forest, and plots on high altitude in other mountain ranges have low annual mean temperature, which hampers decomposition in all stages and leads to an accumulation of N (see Sect. 5.3.3). Upper Franconia, the southern Swabian Alb, the Alps and their foothills, and a part of the Saarland provide soils from carbonate bedrock.

The lowest N stocks of the soil profile are found in the Eastern part of the North-German Lowland with a focus on Brandenburg and adjacent areas in Saxony, Saxony-Anhalt, Lower Saxony, and Mecklenburg-West-Pomerania, where nutrient-poor, sandy soils in a dryer climate prevail and pine is the most abundant forest tree species (compare Sect. 5.3). Other regions with very low N stocks are parts of the mountain ranges Black Forest, Palatinate Forest, the Northern part of the

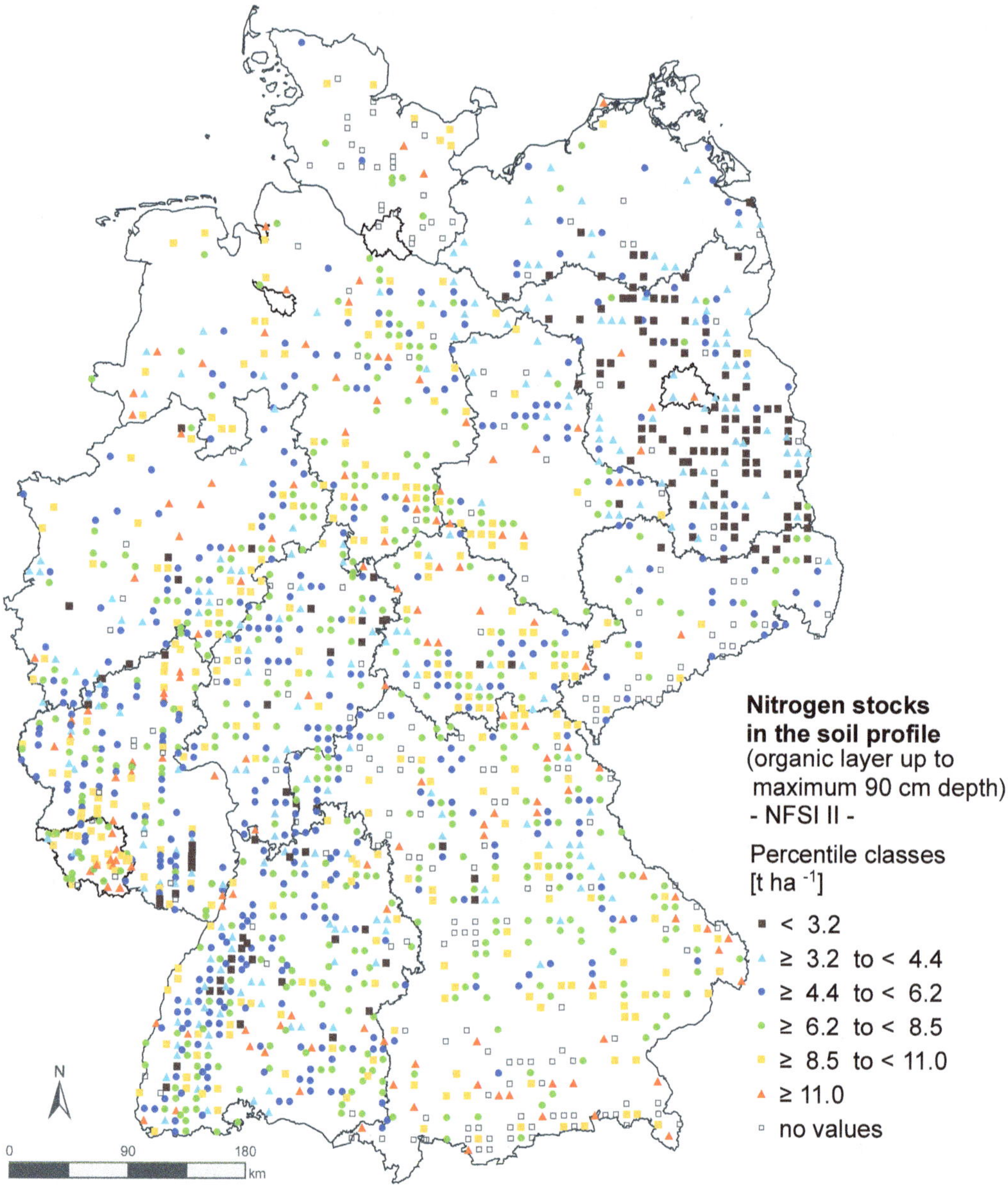

Fig. 5.3 Nitrogen stocks of the soil profile as measured during NFSI II ($n = 1647$)

Swabian Alb, as well as Odenwald and the lower mountain ranges of North-Hesse, which are (except Odenwald) among the mountain ranges with relatively low N deposition. Black Forest and Odenwald possess mostly soils from acidic bedrock.

5.2.4 *C/N Ratios in the Top Soil*

C/N ratios as indicators of soil fertility and degradability of organic material in the upper horizons are traditionally calculated for the Ah horizon of the mineral soil when the humus form is mull or mull-like moder, while the Oa horizon of the organic

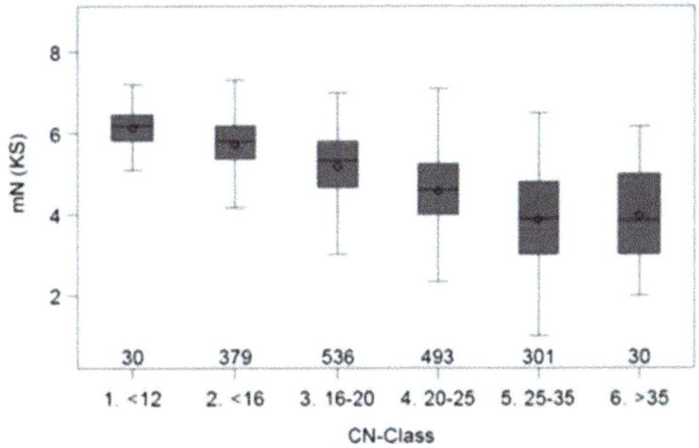

Fig. 5.4 Nutrient indicator values of the herb layer (mN (KS)) after Ellenberg et al. (2003) as dependent on the C/N ratio of the upper mineral soil (0–5 cm depth, sample sizes at the horizontal axis)

layer is used in all other cases (AK Standortskartierung 2003). In accordance with Cools et al. (2014), we separated between forest floor (organic layer) and top soil (here: 0–5 cm).

The significance of C/N ratios for soil fertility gets visible when they are related to the nutrient indicator values of the ground vegetation present at each plot. The nutrient indicator values (from nutrient poor (1) to nutrient rich (9), Ellenberg et al. 2003) of plants in the herb layer show a decrease with increasing C/N ratio class (Fig. 5.4) and thereby confirm earlier findings on plant species adaptation.

C/N ratios in the organic layer reached a mean value of 25.2 ± 0.1 (iqr {22.5; 27.3}, $n = 1792$). C/N ratios below 20.7 (10%-quantile, q10) would be considered very low values, and C/N ratios above 30.4 (90%-quantile, q90) are very high in the NFSI II dataset for the organic layer, which includes undecomposed as well as partly decomposed litter and humus. In the upper 5 cm of the mineral soil, the mean of C/N ratios was 20.6 ± 0.14 (iqr {16.5; 23.7}, $n = 1850$). Here, very low C/N ratios were the ones below 14.2 (q10), and very high C/N ratios start from 27.9 (q90). Due to the operator-dependent separation between both layers, it may sometimes be interesting to also know these values for the continuum from the organic layer to 5 cm: The mean was 24.3 ± 0.09 (iqr {21.8; 29.3}, $n = 1787$), and very low values would be smaller than 20 (q10), while values above 29.3 (q90) would be considered very high.

The mean values for the organic layer and 0–5 cm, respectively, were similar to the previously reported mean values for the forest floor and the mineral topsoil for the whole of Europe (Cools et al. 2014).

C/N ratios in the organic layer were similar among humus forms (Table 5.1); in this layer mull and mull-like moder are factually represented by merely the O_l-layer, such that C/N ratios are strongly influenced by the C/N ratios of fresh litter. In layer 0–5 cm, where the different velocities of decomposition and the bioturbation play a role, C/N ratios varied between 16.5 for mull and 26.6 for mor.

As will be discussed in Sect. 5.3, the main determinant for the pattern of C/N ratios in Germany is the occurrence of tree species, with pine as the species with highest C/N ratios and all other coniferous forest trees on the second rank (compare Sect. 5.3 and Cools et al. 2014). On the level of forest growth regions, there is also an

Table 5.1 C/N ratios of humus forms in the organic layer and 0–5 cm depth of the mineral soil

	Mull	Mull-like moder	Moder	Mor-like moder	Mor
Organic layer	25.7 ± 0.2	24.2 ± 0.2	**24.7 ± 0.2**	**25.5 ± 0.3**	**26.7 ± 0.4**
0–5 cm	**16.5 ± 0.1**	**19.4 ± 0.3**	21.9 ± 0.2	24.3 ± 0.4	26.6 ± 0.7

The humus form-specific definition of reference horizons (AK Standortskartierung 2016) requires to use the bold numbers for a comparison between humus forms

influence of land use proportions: The average C/N ratios (0–5 cm) of NFSI plots are significantly lower in growth regions with more than 50% agricultural land use, compared to growth regions with less than 50% agricultural area ($p < 0.001$). Geographically (Fig. 5.5), the highest C/N ratios in the uppermost 5 cm of the mineral soil of forests occurred mainly in the pine-dominated North-German Lowland. Coniferous tree species are also the determining factor for high C/N ratios in Thuringian Forest, Palatinate Forest, and Erzgebirge, as well as Northern Bavaria, while the lowest C/N ratios are mainly found in beech-dominated landscapes like the Rhenish Slate Mountains and in regions with soils from carbonate bedrock, such as the Swabian-Franconian Alb and the foothills of the Alps.

5.2.5 Comparison to C/N Ratios of NFSI I

For C/N ratios, comparisons to NFSI I are more adequate for organic layer—5 cm than for the both sub-layers separately, since humus forms with their different reference horizons may have changed between NFSI I and NFSI II. The C/N ratios were on average lower in NFSI I (22.4, $n = 1155$) than those from NFSI II (24.0). Also, the few organic soils not included in the paired sample had lower C/N ratios in NFSI I (16.6, $n = 9$) than in NFSI II (19.2). Considering that N stocks in the organic layer down to 5 cm of the mineral soil increased between NFSI I and NFSI II, the results show that the accumulation of N in this layer was accompanied by an even stronger and disproportionately high accumulation of C that led to a marked increase of C/N ratios. The resulting change rate of C/N ratios was +0.09 year^{-1}. While an increase of C/N ratios would generally fit to the decrease of N deposition or an increase of nitrogen uptake by the growing stand, some studies indicate that N deposition as a growth-triggering condition contributes to the variability of C stocks in organic and mineral soil layers of forests—in combination with other growth conditions, e.g. climatic factors (Bedison and Johnson 2009). Nitrogen stocks have already been proposed as an indicator for the C sequestration potential of soils (Vesterdal et al. 2008). A time shift between N deposition and the subsequently increased C input to the soil could well explain the observed increase of C/N ratios during a phase of decreasing N-deposition rates (Fig. 5.1).

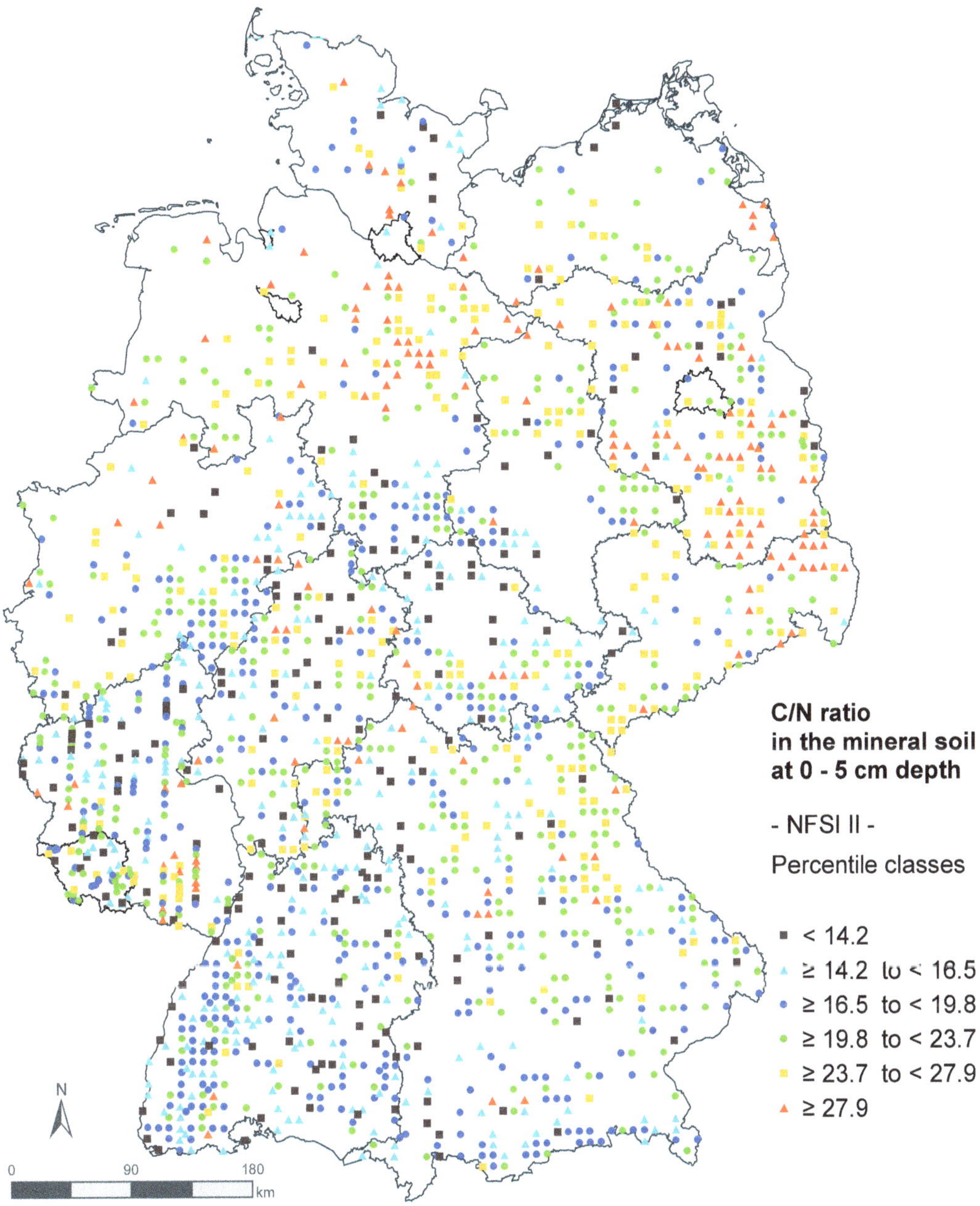

Fig. 5.5 C/N ratio in the upper 5 cm of the mineral soil during NFSI II ($n = 1850$)

5.3 Impact Factors

5.3.1 Forest Type

The different amounts and qualities of litter from the stand forming tree species have a strong influence on C/N ratios and N stocks not only of the organic layer. Due to the preferred use of certain tree species for certain soil substrates, the effect of tree species may not fully be disentangled from the one of soil substrates.

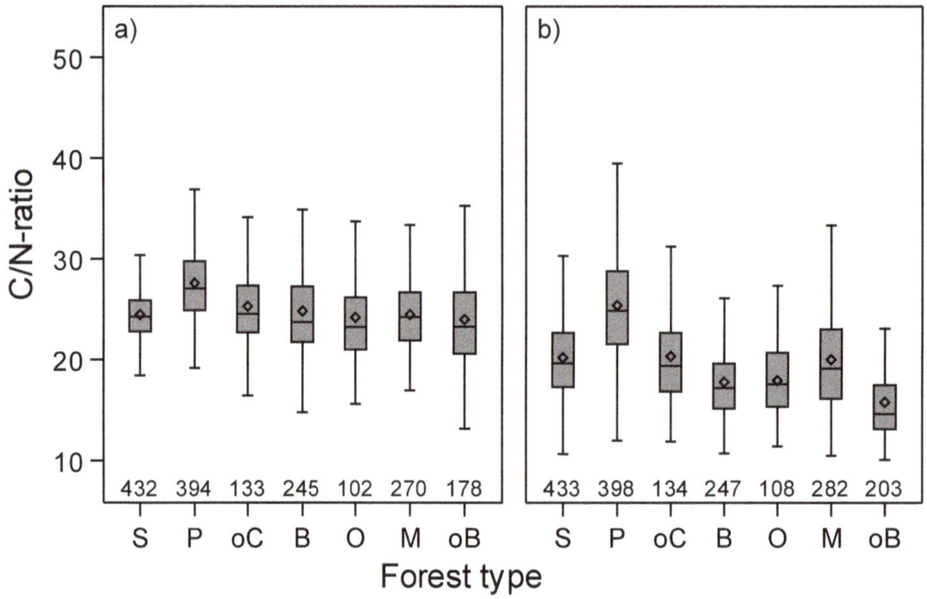

Fig. 5.6 C/N ratios of the forest types beech-dominated broadleaved forest (B), oak-dominated broadleaved forest (O), spruce-dominated coniferous forest (S), pine-dominated coniferous forest (P), other broadleaved forest (oB), mixed coniferous and broadleaved forest (M), and other coniferous forest (oC) in the organic layer (a) and in 0–5 cm of the mineral soil (b). Sample sizes are given at the horizontal axis. The other broadleaved forests are mainly forests with high proportions (>70%) of specialty hardwood trees like ash and sycamore but also hornbeam, alder, and birch. Other coniferous forests predominantly comprise silver fir, Douglas fir, and larch forests. The mixed forests are typically mixtures of beech with spruce, pine, larch, and silver fir or forest stands with pine or spruce and high proportions of birch or oak

Stands dominated by pine had with 27.6 and 25.3, respectively, the highest organic layer and 0–5 cm C/N ratios of all forest types (Fig. 5.6). This rank is occupied by pine in all individual soil substrate groups. High C/N ratios generally indicate both: poor decomposability in the initial stage of decomposition, where mainly cellulose is degraded, and good decomposability in the later stages, where lignified litter components are degraded over much longer time intervals and where the presence of too high amounts of N would hamper decomposition through the suppression of lignolytic enzyme production and the creation of recalcitrant compounds (Berg 2014). Corresponding to both, N stocks of pine (Fig. 5.7) were the highest of all forest types in the organic layer (1.14 ± 0.03 t ha^{-1}), where the initial stage of decomposition takes place, while they were lowest for organic layer—60 cm (4.47 ± 0.12 t ha^{-1}), where the decomposition of lignin is included, which agrees well with the decomposition model of Berg (2014). On the other end of the scale, the lowest C/N ratio (which means highest initial, but lowest total decomposability) in the organic layer or in 0–5 cm of the mineral soil (C/N = 24 or 15.8, respectively) is found under the canopy of broadleaved forests that are not dominated by beech or oak, and this rank is also valid across soil substrate groups. Accordingly, the organic layer N stocks of broadleaved forests are very low (0.4 ± 0.03 t ha^{-1}), while they have the highest N stocks of all forest types in organic layer—60 cm (8.82 ± 0.37 t ha^{-1}).

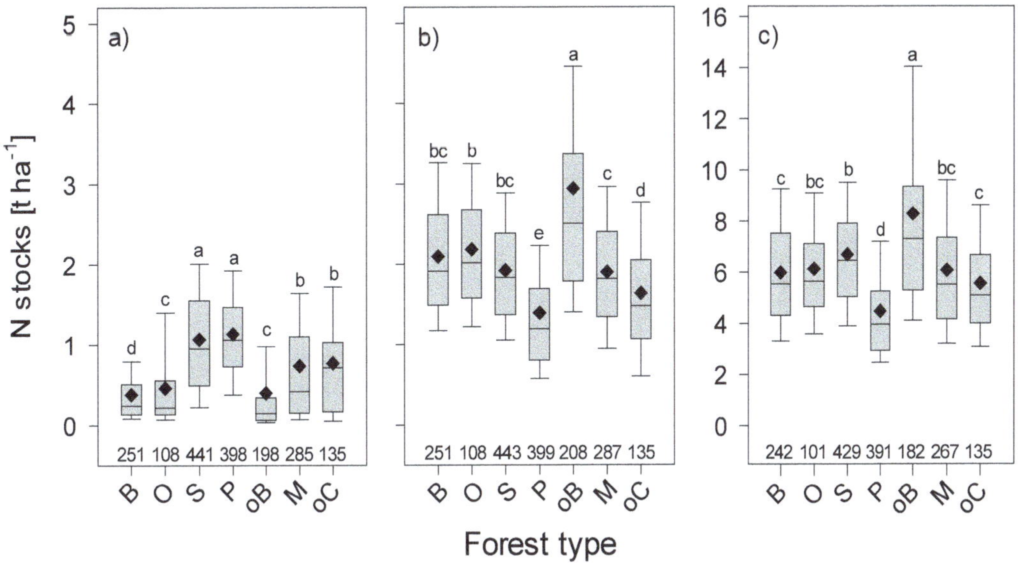

Fig. 5.7 Nitrogen stocks of the different forest types (see caption of Fig. 5.6) in the organic layer (a), in 0–10 cm of the mineral soil (b), and in the soil profile from organic layer to 60 cm of the mineral soil (c). Significant differences are indicated by minor case letters and sample sizes at the horizontal axis

Beech and oak stands were similar to the other broadleaved forests in terms of relatively low C/N ratios and lowest N stocks in the organic layer, whereof the latter is confirmed across soil substrate groups. Their N stocks in organic layer—60 cm were intermediate, comparable to those of other forest types (mixed forests, other coniferous forests).

The C/N ratios of spruce stands were the second highest of all forest types and similar to those of pine stands. The low initial decomposability of their litter agrees well with the rank as the second highest of all N stocks in the organic layer. Complying with Berg (2014), N stocks of spruce stands in organic layer—60 cm were higher than those of pine, but they were even higher than those of beech and oak stands. The effect of tree species-specific litter quality may be less clear in this case due to other factors like the permanently high leaf area index of spruce stands and, consequently, reduced exposition of their soils to sun, rain, snow, wind, and temperature changes. Spruce litter is also better protected by its acidity against decomposition. The higher productivity of spruce compared to pine is generally due to the selection of sites with better nutrient and water availability.

5.3.2 *Parent Material and Soil Acidity*

Soil parent material mainly affects N stocks through its influence on soil acidity: Soils from weathered carbonate bedrock or from basic-intermediate bedrock provide the least acidic conditions for nutrient uptake, decomposition by microbes, and redistribution of organic material by bioturbation. They have the lowest C/N ratio (16.3 or 17.7, respectively) in 0–5 cm of the mineral soil (Fig. 5.8). As an effect of a

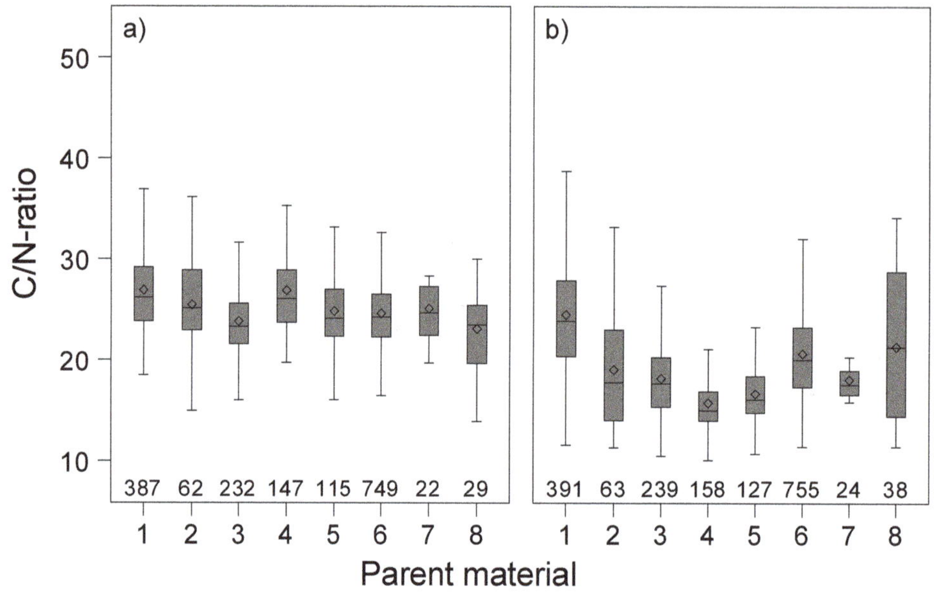

Fig. 5.8 C/N ratios of the different soil substrate groups in organic layer (**a**) and 0–5 cm depth of the mineral soil (**b**). (1) Soils from base-poor unconsolidated sediments, (2) soils of alluvial plains, (3) loamy soils of the lowlands, (4) soils from weathered carbonate bedrock, (5) soils from basic-intermediate bedrock, (6) soils from base-poor consolidated bedrock, (7) soils of the Alps, (8) organic soils from peatland. Sample sizes are given at the horizontal axis

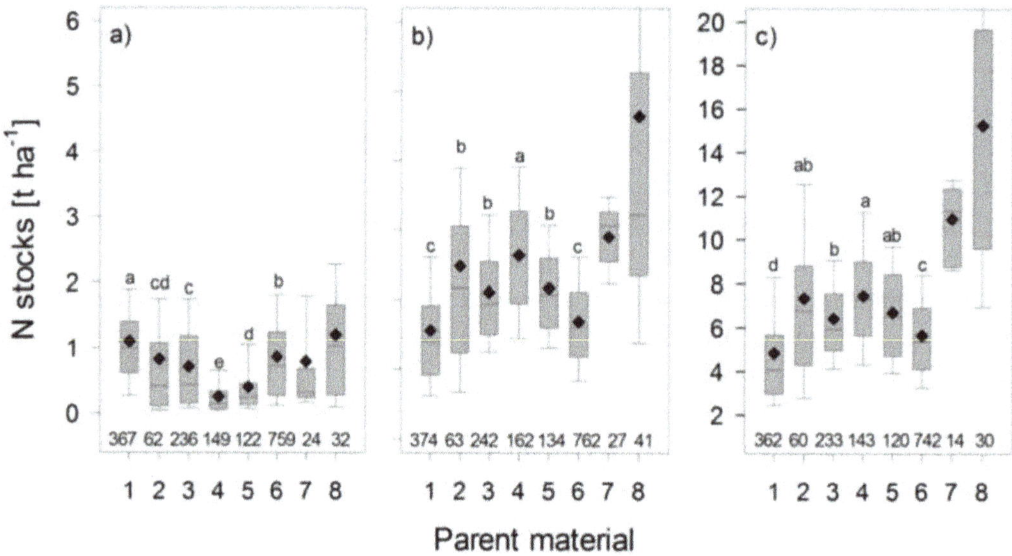

Fig. 5.9 Nitrogen stocks of the eight soil substrate groups (compare caption of Fig. 4.8) in the organic layer (**a**), 0–10 cm of the mineral soil (**b**), and the soil profile from organic layer to 60 cm of the mineral soil (**c**). Sample sizes are given at the horizontal axis

soil property, the C/N ratio differences between soil substrates are more pronounced in the uppermost part of the mineral soil than in the organic layer. In accordance with the effect of C/N ratios postulated by Berg (2014), both parent material groups have significantly lower N stocks (Fig. 5.9) in the organic layer than all other parent material groups and the highest N stocks in 0–10 cm of the mineral soil (2.64 t ha^{-1} and 2.17 t ha^{-1}, respectively) and in the soil profile from organic layer to 60 cm (7.51 t ha^{-1} and 6.71 t ha^{-1}, respectively).

Additional features that protect organic material in the mineral soil from decomposition are relevant in the three smaller substrate groups: Organic soils from peatland (15.7 t ha^{-1} in organic layer—60 cm, $n = 38$) and the partly submerged soils of alluvial plains (7.35 t ha^{-1}, $n = 63$) provide anaerobic zones through the influence of water and may be supplied with organic material from moving water, while soils from the Alps (10.97 t ha^{-1}, $n = 27$) may provide organic matter protecting enclosures between rocks (compare Sabatini et al. 2015). Organic soils from peatland are the parent material group with the highest variability in N stocks but also the highest average N stocks in all layers. This is due to the fact that the most diverse peatland types like fens and bogs are summarized in one material group.

Apart from organic soils, soils from base-poor consolidated bedrock or base-poor unconsolidated sediments have the highest C/N ratios in 0–5 cm of the mineral soil and also the highest N stocks in the organic layer. Consistent with highest C/N ratios, their total N stocks in organic layer—60 cm are lower than in all other parent material groups (4.77 and 5.67 t ha^{-1}, respectively).

Soil acidity may also be influenced by liming, which is only a relevant management option on acid-sensitive plots (see Chap. 4), and only this subsample of plots is considered in the liming evaluation. Liming did show no direct effect on C/N ratios in NFSI II. But similar to the pattern observed in soils from carbonate bedrock, limed plots showed a decrease of N stocks between NFSI I and NFSI II in the organic layer (-7.8 kg ha^{-1} $year^{-1}$ $\pm$ 2.4), but an increase of N stocks by +11.8 kg ha^{-1} $year^{-1}$ $\pm$ 4.2 in the upper mineral soil (0–30 cm). The not limed plots of the acid-sensitive subsample showed instead more or less constant N stocks in the organic layer (-0.7 kg ha^{-1} $year^{-1}$ $\pm$ 3.2) and a decrease of -11.6 kg ha^{-1} $year^{-1}$ $\pm$ 4.5 in the upper 30 cm of the mineral soil. Both trends can be explained by the acidifying effect of N-deposition: Acidity inhibits (micro-)biological activity such that particulate organic matter accumulates in the organic layer, while the formation of mineral-associated organic matter and N-storage in the mineral soil is reduced, thereby reversing the usually opposite effect of N-addition on microbial growth and decomposition occurring when acidification is counteracted with liming (Averill and Waring 2018). Supporting this explanation, a multifactorial covariance analysis (ANCOVA) on the factors liming, forest type, and clay content reveals that only the combined effect of clay content and liming had a significant influence on N accumulation rates in 0–30 cm of the mineral soil. This points to a potentially enhanced formation of N containing clay-organic matter complexes in limed soil, which contributes to the recalcitrant fraction of N compounds (Preston et al. 2009) in the mineral soil. Soil biological activity is vulnerable to acidification: Earthworm activity generally increases under the changed chemical conditions in limed soils (Homan et al. 2016), and also microbial activity increases with increases of soil pH (Anderson and Domsch 1993).

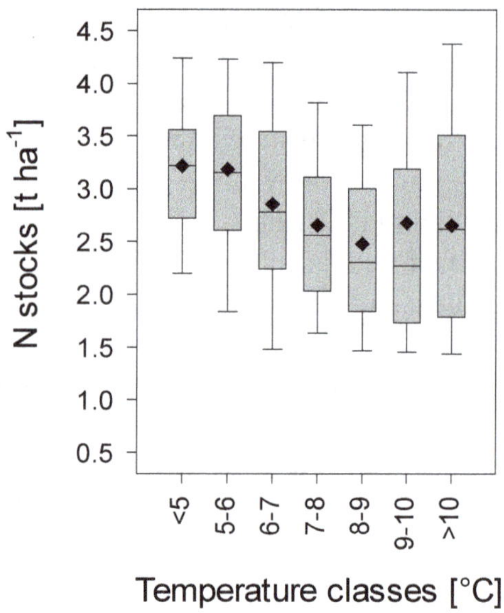

Fig. 5.10 Nitrogen stocks in the organic layer and upper 10 cm of the mineral soil for different annual mean temperature classes

5.3.3 Annual Mean Temperature

Low temperatures are an additional condition that may slow down the process of decomposition of soil organic matter. Concordantly, N stocks in the organic layer and the upper mineral soil up to 10 cm are highest on the plots with the lowest annual mean temperatures (Fig. 5.10) and increase stepwise to the higher temperature classes. The highest temperature classes above 9 °C, however, have again somewhat higher N stocks than the neighbouring class, which may be due to the relevance of dry conditions on these plots that have a prohibitive effect on decomposition of soil organic matter.

5.3.4 Agricultural Land Use

The influence of the regional distribution of agricultural land use was identified based on CORINE land cover data for the forest growth districts in Germany (Gauer and Kroiher 2012). The NHx/NOy ratio of total N deposition generally increases with the proportion of agricultural land use in growth districts ($r^2 = 0.42$, $p < 0.05$), but total N deposition to forests in the same district is not generally increased (here meteorological conditions like precipitation, wind speed, and fog rate are also important factors). However, dividing the German forest growth districts into classes of agricultural land use reveals that C/N ratios in 0–5 cm of the mineral soil are lower in growth districts with high or very high proportion of agricultural area, while they are highest in districts with less agriculture (Fig. 5.11a). This apparent effect of agricultural land use is directly due to higher N stocks in 0–5 cm of the mineral soil. In accordance with this finding, also N stocks in agricultural districts are higher than in those with less agriculture (Fig. 5.11b).

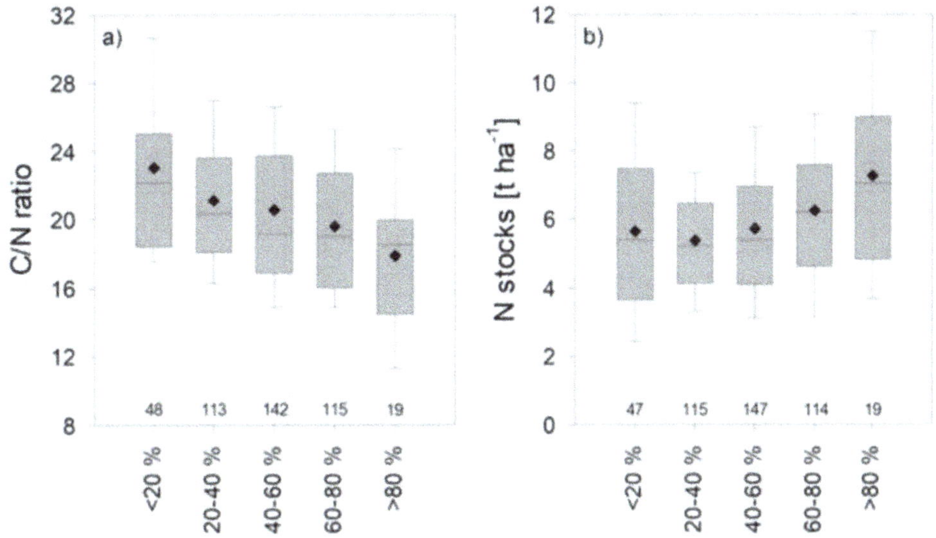

Fig. 5.11 Average C/N ratio in 0–5 cm of the mineral soil (**a**) and average N stocks in layer Org-60 cm (**b**) of NFSI plots in forest growth districts with variable proportion of agricultural land use (percentages)

As a summary, forest type, parent material, soil acidity, annual mean temperature, and agricultural land use all contribute to the variability of N stocks (Fig. 5.3) and C/N ratios (Fig. 5.5) of German forests and may explain much of the geographical pattern observed. From these factors, parent material and annual mean temperature are naturally given factors that have affected stand and soil condition for centuries, while forest type, soil acidification and agricultural land use are comparatively younger and mainly man-made factors with a shorter, but apparently long enough impact period to affect the N status of forest soils. In the case of agricultural land use, it is probable that the duration of the impact period is linked to N emissions from agriculture that started with the production of N containing mineral fertilizer since about 1910 and the increased imports of forage for more intensive livestock production. Due to multiple interactions in the processes leading to N accumulation or release, the relative contribution of these and other factors may not directly be derived and will need a more complete multivariate model where also other variables (e.g. local exposition, soil wetness, soil texture, podzolation, rooting depth, forest growth, and harvest practice) are expected to play a role.

5.4 Nitrogen Stock Changes

The result of N stock changes over more than a decade is visible in the N stock difference between NFSI I and NFSI II as differential measurement of a state variable at two points in time (Sect. 5.4.1). Apart from NFSI plots, an N stock difference was similarly also detected on IFM plots (Sect. 5.4.2). Due to the

Table 5.2 Main results of status and trends of nitrogen stocks in the different soil layers

	Complete sample			Paired sample		
	Status (kg ha^{-1})			Trend (kg ha^{-1} $year^{-1}$)		
	Median	IQR	*n*	Median	IQR	*n*
Org. layer	671	{239; 1222}	1794	−2.4	{−21.0; +20.0}	1213
0–5 cm	1087	{756; 1425}	1846	+6.5	{−10.0; +23.2}	1281
5–10 cm	644	{421; 962}	1846	+2.0	{−7.1; +11.8}	1282
10–30 cm	1559	{979; 2364}	1832	−3.5	{−22.0; +18.7}	1271
0–30 cm	3280	{2333; 4721}	1827	+7.1	{−34.1; +52.9}	1265
30–60 cm	1302	{702; 1981}	1811	−14.6	{−39.9; +6.2}	1239
Org.—5 cm	1810	{1430; 2367}	1784	+5.3	{−26.3; +35.3}	1205
Org.—30 cm	4100	{3096; 5479}	1767	+6.2	{−44.6; +53.6}	1191
Org.—60 cm	5430	{3970; 7387}	1740	−8.2	{−74.0; +48.7}	1152

integration over several years, this approach does not directly investigate the actually occurring change rates on specific plots. It may, however, indicate the average annual change rate as a trend that would explain the difference between both points in time, thereby integrating over change rate variability between years as well as seasonal and spatial variability of the processes contributing to the observed difference. A more directed approach investigating the processes leading to N stock changes is provided by N balance estimations, describing the flow of nitrogen in and out of an ecosystem with sub-models based on the known course of meteorological variables between NFSI I and NFSI II (Sect. 5.4.3).

5.4.1 Nitrogen Stock Difference on NFSI Plots

Actual status (Sect. 5.2) and trend of N stocks between both NFSIs rely on different sampling approaches: While the presented status results are always based on measurements of the complete NFSI II sample, the calculated trends may only be based on those plots that were investigated two times (paired sample). In the case of N, the paired sample lacks data from Saarland and Bavaria. Organic soils were not included in the paired sample. In all cases, area-weighted medians are reported for N stocks of NFSI instead of means due to the skewness of distributions, and trends are only integrated down to a depth of 60 cm due to the increased uncertainty of low concentration N measurements relevant in deeper layers (see Sect. 5.5.3 and Table 1.2). An overview of the main results is given in Table 5.2.

Trends in Organic Layer: 5 cm

The comparison with NFSI I values yields a non-significant trend: While N stocks in the organic layer showed a slight decrease relative to NFSI I values (−2.4 kg ha^{-1} $year^{-1}$), they increased by +6.5 kg ha^{-1} $year^{-1}$ in the uppermost 5 cm of the mineral soil, leading to a total accumulation of +5.3 kg ha^{-1} $year^{-1}$ in organic layer—5 cm.

The apparent N stock accumulation between NFSI I and NFSI II in organic layer—5 cm tends to even higher values, when a statewise comparison of the complete sample of NFSI I and NFSI II is performed (weighted medians of federal states: +0.2 kg ha^{-1} $year^{-1}$ in the organic layer and +12.2 kg ha^{-1} $year^{-1}$ in 0–5 cm). Since the measured concentrations in these layers are high, they are less affected by potential analytical errors, which underpins the reliability of this result. The N stock accumulation in the organic and uppermost soil layer coincides with the measured increase of N concentrations in tree leaves and needles between both points in time (see Chap. 9), which might indicate higher N availability for trees as well as higher input to the organic layer. It is questionable, however, whether this increase in leaf N concentrations reflects a general trend or just interannual variability (Riek et al. 2016). It has to be considered, though, that the accumulation of dead organic matter on the forest floor over several years is less sensitive to such interannual variability and might thereby help to recognize longer-term trends in nutrient input and turnover. Under conditions of rising N concentrations in leaves and in the upper mineral soil, the apparent slight decrease of N stocks in the organic layer could be interpreted as a hint towards increased turnover rates of organic material (see Chap. 6).

Trend in Organic Layer: 60 cm

The median of N stock changes in organic layer—60 cm equals −8.2 kg ha^{-1} $year^{-1}$ (Table 5.2, Fig. 5.12c). The tendency given by this number for the whole country is mainly due to very high differences between NFSI I and NFSI II in the layer 30–60 cm, while the other layers remained nearly unchanged or showed the opposite tendency (Fig. 5.12c). The tendency is regionally very diverse and may not be seen as typical for the whole country, since large parts of Germany show only negligible changes and others are not included in the paired sample. In contrast, the median for organic layer—30 cm was +6.2 kg ha^{-1} $year^{-1}$. This contrasting trend between upper and lower soil compartments comes along with a slight recovery from acidification in the upper soil compartments, while still increasing acidity is observed in the subsoil (see Chap. 4), which could explain higher N losses from deeper soil layers.

5.4.2 *Nitrogen Stock Difference on IFM Plots*

Alternative insight into the status and dynamics of N stocks in German forest soils between NFSI I and NFSI II may only be derived from Intensive Forest Monitoring plots (IFM plots) of the ICP Forests Level II network and additional plots operated according to the same methodology (UNECE ICP Forests Programme Co-ordinating Centre 2016), as the only available independent dataset from approximately the same area and within nearly the same time frame. The soil surveys conducted at these densely instrumented plots are combined with numerous other surveys (Fleck et al. 2017) and aim to provide the basis for the analysis of cause-effect relationships in forest ecosystems.

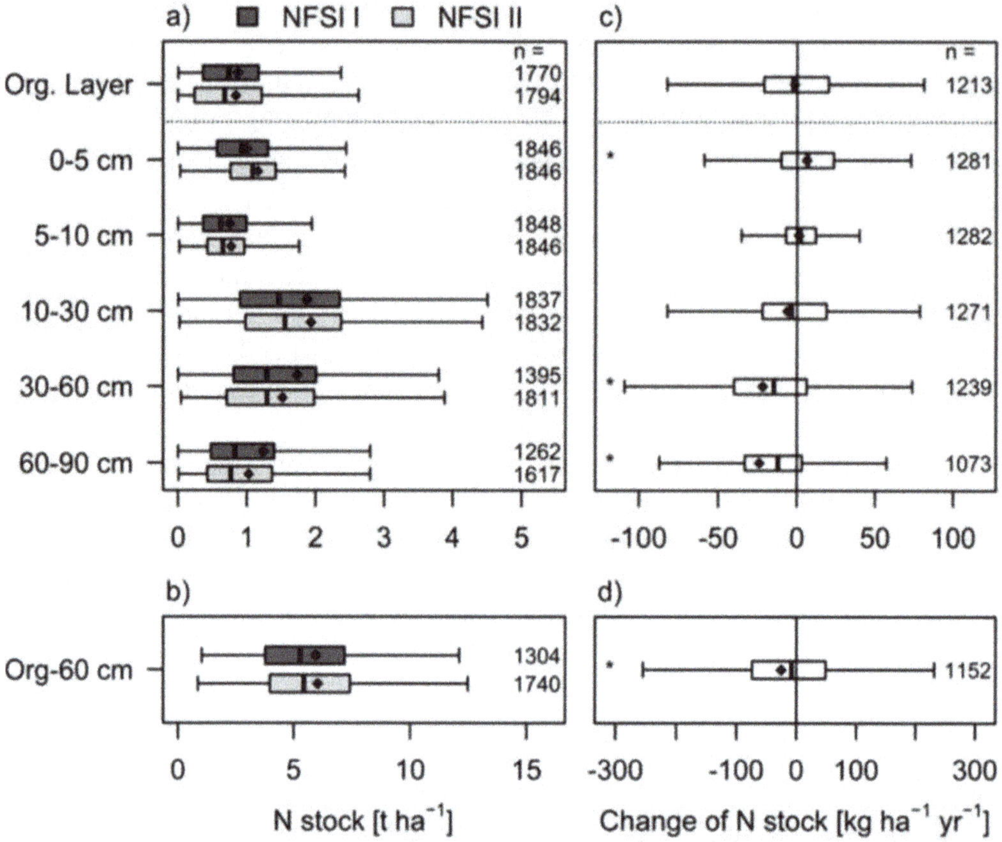

Fig. 5.12 Depth gradient of nitrogen stocks in NFSI I and NFSI II related to the total sample of each inventory (**a**) and respective sums of nitrogen stocks for the soil profile from organic layer up to 60 cm of the mineral soil (**b**). The annual change rates of nitrogen stocks (**c** and **d**) were calculated based on the paired sample of both inventories except peatland. The shown statistical significance of differences between medians (indicated as stars) does not reflect the influence of measurement uncertainties (see Sect. 5.5.2). Outliers are not shown

Intensive Forest Monitoring plots are located in forest stands covering regionally typical combinations of soil type, tree species, climate, and pollution levels (e.g. deposition rates of N and S). Due to their relatively small number and irregular spatial arrangement (Fig. 5.13), results from IFM plots are considered indicative for their respective forest ecosystem types but are not suited to derive spatially representative results (medians reported are, thus, not area-weighted). Nevertheless, ranges of measured N stocks, N deposition, and calculated N leaching rates (see Sect. 5.4.3) from these plots yield an independent source of information for the time span between NFSI I and NFSI II. Only IFM plots with a maximum temporal shift of ±5 years compared to the two NFSIs have been selected for analysis.

The depth gradient of N stocks on the intensive monitoring sites shows a similar pattern compared to NFSI plots for both the first and second inventory (Fig. 5.2). In terms of trends, N stocks of IFM plots (Fig. 5.14) show a marked increase between the first and second inventory (median: +28.0 kg N ha^{-1} $year^{-1}$, iqr{−23.9; +79.4}, $n = 36$). The direction of this overall trend is opposite to the findings from NFSI plots, but the variability of all results from IFM plots is higher than on NFSI plots.

Fig. 5.13 Selected Intensive Forest Monitoring plots (IFM plots) with repeated soil inventories with a time shift of maximum 5 years relative to each of the NFSI inventories

However, the NFSI pattern of decreasing N stocks in the organic layer and increasing N stocks in the upper layers of the mineral soil is also reflected in the results from IFM plots (−4.5 kg ha^{-1} year^{-1}, iqr{−24.7; +8.8}, n = 37 in the organic layer and +43.1 kg ha^{-1} year^{-1}, iqr{−6.6; +71.3}, n = 47 in 0–30 cm).

In contrast to NFSI findings, the N stocks in the lower layers of the mineral soil did only slightly change on IFM plots (+1.9 kg ha^{-1} year^{-1}, iqr {−23.7; +40.9}, n = 46 in 30–60 cm and −1.2 kg N ha^{-1} year^{-1}, iqr {−30.9; +24.9}, n = 21 in 60–90 cm).

Thus, while the two monitoring networks show different trends in total for organic layer—60 cm, they both indicate a marked difference between the increasing

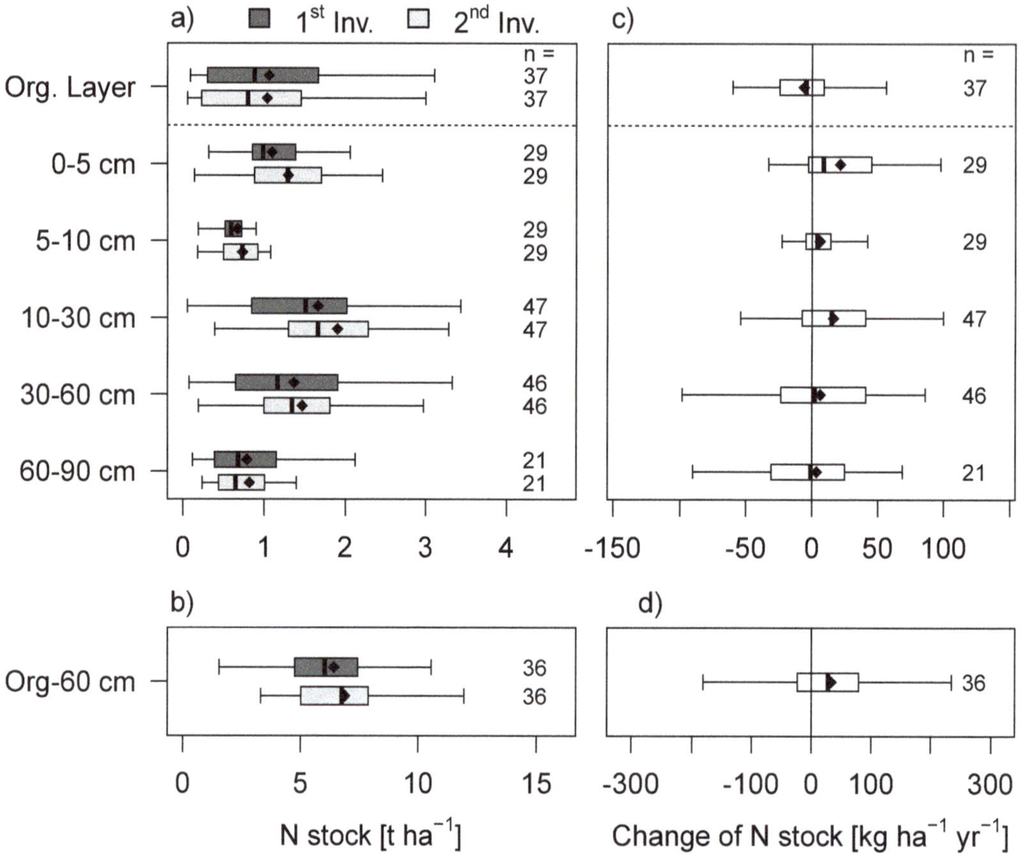

Fig. 5.14 Depth gradient of nitrogen stocks in the first and second inventory on the Intensive Forest Monitoring (IFM) plots (**a**) and respective sums of nitrogen stocks for the soil profile from the organic layer down to 60 cm of the mineral soil (**b**). The annual change rates of nitrogen stocks (**c** and **d**) were calculated based on the paired sample of both inventories. Since layers 0–5 cm and 5–10 cm were not separately sampled on many plots, the sample size of these layers is lower than, e.g. of organic layer—60 cm

trend in the soil compartments above 30 cm and the negative or constant trend in the layers deeper than 30 cm of the mineral soil. The relevance and interpretation of this finding are further discussed in Sects. 5.5 and 5.6.

5.4.3 Nitrogen Balance Estimation

The additional measurement of nitrate concentration from soil water extracts within NFSI II allows for N balance estimations as an alternative approach to derive N stock change rates.

N budgets of forest ecosystems have been used to evaluate and quantify the effects of deposition and other impacts on forest soil dynamics (Korhonen et al. 2013; Meesenburg et al. 2016; Palviainen et al. 2017). Under the assumption that N

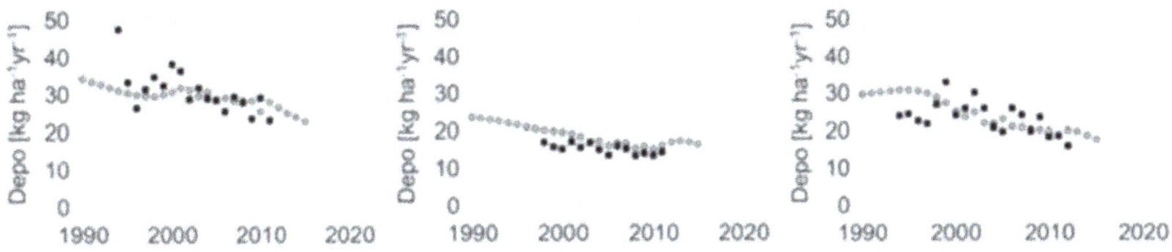

Fig. 5.15 Evaluation of reconstructed nitrogen deposition (open circle) against values calculated with a canopy budget model (Ulrich 4) from deposition measurement (filled square) of three ICP Forests Intensive Forest Monitoring plots in Germany: Augustendorf/Lower Saxony (left), Klötze/Saxony-Anhalt (centre), and Klingenthal/Saxony (right)

fixation in most forest ecosystems is negligible (Posch et al. 2015), the calculation of N pool changes in the soil was simplified to:

$$\Delta S = N_{\mathrm{D}} - N_{\mathrm{E}} - N_{\mathrm{L}} - N_{\mathrm{nU}} \tag{5.1}$$

with change of N pools in the soil (ΔS) being based on N_{D} (total N input by atmospheric deposition—wet + dry + occult), N_{E} (gaseous emission of N_2, N_2O, and NO), N_{L} (leaching of N below the rooting zone), and net N uptake of the vegetation (N_{nU}, the balance of N gains due to root uptake from the soil and N losses due to litter fall and harvest residues remaining on the floor): The assumptions regarding the calculation of each term of the N balance are given in the following sections since they are not measured but modelled values.

5.4.3.1 Atmospheric Nitrogen Deposition

Long-term trends of N deposition were calculated following Thiele et al. (2017) with the model MAKEDEP (Alveteg et al. 1998). The model was run with grid-based estimates for Germany for the year 2009 (Schaap et al. 2015). We reconstructed N deposition before 2009 using the regional trend from the EMEP database (Tarrasón and Nyiri 2008) and standard time series from Alveteg et al. (1998) as published in Ahrends et al. (2010), Fleck et al. (2017), and Hauck et al. (2012). Reconstructed deposition was evaluated with measurements from Level II plots (Fig. 5.15).

5.4.3.2 Gaseous Nitrogen Emissions

The gaseous emissions N_{E} were considered as linearly related to net N input (de Vries et al. 1994):

$$N_E = \begin{cases} f_E \cdot (N_D - N_{nU} - N_I) & \text{if} \quad N_D > N_{nU} + N_I \\ 0 & \text{else} \end{cases} \tag{5.2}$$

with N_{nU} for net removal of N in harvested trees, N_I for long-term net immobilization of N in soil organic matter, and f_E for the site-specific emission fraction ($0 \leq f_E < 1$). N_E is thus considered a second-order process after immobilization and uptake needs are fulfilled. f_E as a fraction based on soil drainage (Reinds et al. 2001) was approximated using clay content (C) (Ahrends et al. 2010; Murray et al. 2017) in an exponential function between the minimum for sandy soil (0.05; $C = 0\%$) and the maximum for clay soils (0.7; $C = 45\%$) (Rihm and Achermann 2016).

$$f_E = 0.0501 \cdot e^{0.0581 \cdot C} \tag{5.3}$$

This approximation gets very close to published texture-specific classes (Park and Shim 2001; Reinds and De Vries 2010; Nagel and Gregor 1999). The basic function was supplemented by rules taking into account peat soils ($f_E = 0.8$; Rihm and Achermann 2016), depth to groundwater level, and stagnic soil conditions. N_I was quantified after Nagel and Gregor (1999) being 1 kg ha^{-1} year^{-1} for most plots (Posch et al. 2015).

5.4.3.3 Nitrogen Leaching

Nitrogen leaching rates were estimated by multiplying the amount of annual seepage water with the mean nitrate concentration of the lowest soil layer. Plot-specific soil water fluxes were estimated with the physically based hydrological model LWF-BROOK90 (Version 3.4, Hammel and Kennel 2001). Nitrate concentration in seepage water ($NO3_{seepage}$) was estimated from soil water extracts, assuming that leaching of ammonium (NH_4^+) can be neglected in forest ecosystems due to its preferential uptake and complete nitrification within the root zone (Posch et al. 2015). Several authors confirm the possibility to empirically derive $NO3_{seepage}$ from soil water extracts, using regression approaches (Schlotter et al. 2012; Kohlpaintner et al. 2012; Evers et al. 2002; Ludwig et al. 1999). The applied methodologies, however, differ with respect to (1) the ratio between soil and water, (2) utilization of fresh or dried soil, (3) the facultative correction of nitrate concentrations measured in the soil water extract ($NO3_{extract}$) to either the actual water content (m^3 m^{-3}) at the time of soil sampling (Θ) or the calculated water content at field capacity (Θ_{FC}), and (4) the assumed functional relationship in the regression between $NO3_{seepage}$ and $NO3_{extract}$. In the following, we describe the methods used to derive estimates of $NO3_{seepage}$ for the data presented here.

The method applied during NFSI II to measure $NO3_{extract}$ is to mix one mass part of dry soil with two mass parts of deionized water. The suspension is left for 24 h at room temperature and stirred frequently. After filtration $NO3_{extract}$ is analysed (GAFA 2009).

Table 5.3 Linear regression between $NO3_{seepage}$ (mmol$_c$ l^{-1}) and $NO3_{extract}$ (mmol$_c$ l^{-1}, corrected to water content at field capacity Θ_{FC} or to actual water content Θ)

Corr.	Intercept	Slope	R^2	adj. R^2	n	RMSE	AIC
None	−0.03495	3.802***	0.388	0.381	95	0.1208	−125.93
Θ_{FC}	−0.03937*	0.5785***	0.491	0.486	94	0.1074	−146.69
Θ	−0.05500**	0.4271***	0.512	0.507	92	0.0943	−167.37

Significances are coded as $*p < 0.05$, $**p < 0.01$, $***p < 0.001$

Table 5.4 Generalized additive regression between $NO3_{seepage}$ and $NO3_{extract}$ including further soil parameters with $NO3_{extract}$ corrected to Θ_{FC} or Θ (Corr.)

Corr.	R^2	Adj. R^2	n	RMSE	AIC
None	0.511	0.480	90	0.0887	−166.13
	Intercept***	s($NO3_{extract}$)***	s(G)*	s(SI)***	
Θ_{FC}	0.747	0.726	89	0.0606	−228.58
	Intercept***	s($NO3_{extract\text{-}\Theta FC}$)***	s(S)***	s(SI)***	s(C)**
Θ	0.727	0.708	90	0.0663	−217.63
		s($NO3_{extract\text{-}\Theta}$)***	s(G)**	s(SI)***	

Significance of model variables ($*p < 0.05$, $**p < 0.01$, $***p < 0.001$) is given in the second line for each model

$NO3_{extract}$, nitrate concentration in soil water extract [mmol$_c$ l^{-1}]; G, gravel content [% mass]; S, SI, C, sand, silt, clay content in fine earth [% mass]

The statistical model used is based on water extracts from soil samples and direct measurements of $NO3_{seepage}$ in tight spatial and temporal proximity. The data was taken from Evers et al. (2002) comprising 10 plots from five federal states with 4 investigation points each and from 21 plots with 5 investigation points in Bavaria (Lutz 2015). Due to an elimination of extreme values, problems with seepage water sampling in dry soils, and lack of data, a total of 95 (85 with additional soil properties) pairs of values was available for statistical modelling. The linear regression between $NO3_{extract}$ and $NO3_{seepage}$ yielded R^2 values between 0.39 and 0.51 (Table 5.3). Best results were obtained when $NO3_{extract}$ was corrected to Θ, closely followed by the correction to Θ_{FC}, where field capacity was estimated from soil texture according to Ad-HocAG_Boden (2005). Model performance was enhanced by the consideration of additional soil properties. Significant impacts of soil bulk density, gravel content, and the content of sand, silt, and clay in the fine earth were identified using generalized additive models (GAM) (Table 5.4).

The best performing GAM (Corr. Θ_{FC}) was selected and replaced by a multilinear function to facilitate calculation (Equation 5.4), including the variables in their linear or quadratic form only in the case of significance ($p < 0.001$) and forcing the results through zero if negative:

Table 5.5 Statistical properties of model building and validation for nitrate concentrations in seepage water as calculated with Eq. (5.4)

	n	R^2	RMSE	$\text{Mean}_{\text{measured}}$	$\text{Mean}_{\text{modelled}}$	RMSE ÷ mean
Model building	85	0.722	0.0647	0.0698	0.0698	0.9272
Validation	85	0.219	0.1277	0.0933	0.0906	1.3681

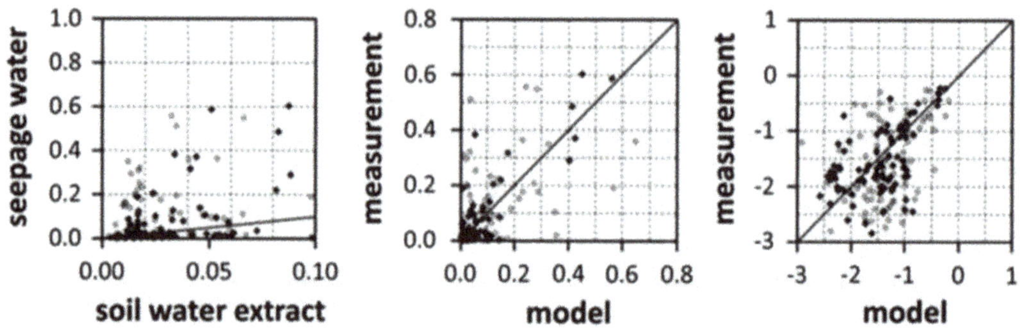

Fig. 5.16 Dispersion of measured values taken for model building (filled square) and for validation (open circle) around the 1:1-line in the relationships between $NO3_{extract}$ and $NO3_{seepage}$ (left), modelled (Eq. 5.4) and measured $NO3_{seepage}$ (centre), and modelled and measured $NO3_{seepage}$ in logarithmic display (right), $NO3_{extract}$ and $NO3_{seepage}$ in $\text{mmol}_c\ l^{-1}$

$$NO3_{seepage} = \begin{cases} a * NO3^2_{extract} + b * S + c * S^2 + d * SI^2 + e * C^2 + f \\ g * NO3_{extract} \text{ if } NO3_{seepage} < 0 \end{cases} \quad (5.4)$$

with $NO3_{extract}$ for nitrate concentration in soil water extract corrected to water content at field capacity [$\text{mmol}_c\ l^{-1}$] and variables S, SI, and C for sand, silt, and clay content in fine earth [% mass]. The coefficients are $a = 0.5911$, $b = 0.03252$, $c = -0.0001516$, $d = 0.000366$, $e = 0.000317$, $f = -1.702$, and $g = 0.3144$.

The resulting model was applied to a set of validation data from Level II plots in Germany and two smaller studies in Baden Württemberg and Bavaria. As expected, due to the high spatial and temporal variability of nitrate concentrations in soil solution, the 85 pairs of values could not compete with the dataset used for model building. As a consequence the quality of statistical properties for validation was low (Table 5.5). However, the modelled values corresponded well to measurements with respect to magnitude and value distribution, especially if compared to the direct relationship between $NO3_{extract}$ and $NO3_{seepage}$ (Fig. 5.16). This correspondence was even preserved in double logarithmic display applied due to low nitrate concentrations in a large part of the seepage water samples.

5.4.3.4 Net Nitrogen Uptake for Different Harvest Scenarios

For managed forest, the long-term average net N uptake of vegetation (NnU) is equal to the amount of N exported with harvested tree compartments and can be calculated

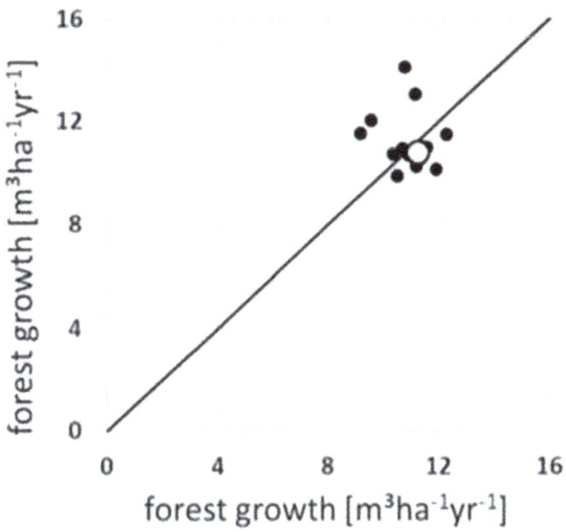

Fig. 5.17 Yearly increment of solid volume at the NFI3 plots (https://bwi.info; 77Z1Jl_L101of_2012) in the different federal states of Germany and the estimated yearly increment for the NFSI plots with modified yield tables for the period 2002–2012. Filled circle, values from the federal states of Germany; open circle, area weighed mean for Germany

by multiplying the average annual growth rate of the respective compartments with their element concentrations (Posch et al. 2015):

$$N_{\mathrm{nU}} = k_{\mathrm{gr,sv}} \cdot p_{\mathrm{sv}} \cdot \mathrm{ctN}_{\mathrm{sv}} + k_{\mathrm{gr,ba}} \cdot p_{\mathrm{ba}} \cdot \mathrm{ctN}_{\mathrm{ba}} + k_{\mathrm{gr,br}} \cdot p_{\mathrm{br}} \cdot \mathrm{ctN}_{\mathrm{br}} \tag{5.5}$$

with k_{gr} for the average annual growth rate (m^3 ha^{-1} year^{-1}), p for density (kg m^{-3}), and ctN (kg kg^{-1}) for N concentration in the compartments solid volume under bark, bark, and branches (subscripts sv, ba, and br, respectively). Dependent on the harvesting practice, the contribution of branches has to be included (whole-tree harvest) or not (stem only). Forest growth for each NFSI plot was estimated from a forest inventory in 2011–2012 using yield tables (Schober 1995) for the reconstruction up to 1990. Correction factors were applied to consider the higher growth rates due to environmental factors and mixed forests (Pretzsch 2016). The results were evaluated with growth data from the Third National Forest Inventory (BMEL 2016) for about 50,000 plots in Germany (Fig. 5.17).

The parameters for the biomass expansion factors, the density of the compartments, and the nitrogen concentration in biomass were taken from Ahner et al. (2013) and Jacobsen et al. (2002). As there is a great uncertainty regarding harvesting practice on NFSI plots and the representativeness of the long-term average for the shorter timescales relevant between NFSI I and II, we calculated N exports for both whole-tree and stem-only harvest, to consider and represent the high uncertainty in uptake estimations. It should be noted that tree stumps as harvest residues (Pretzsch 2009) were not included so that net N uptake of the vegetation may be slightly overestimated. As an approximation for realistic harvest exports, we used the averages between the two variants in the N balance.

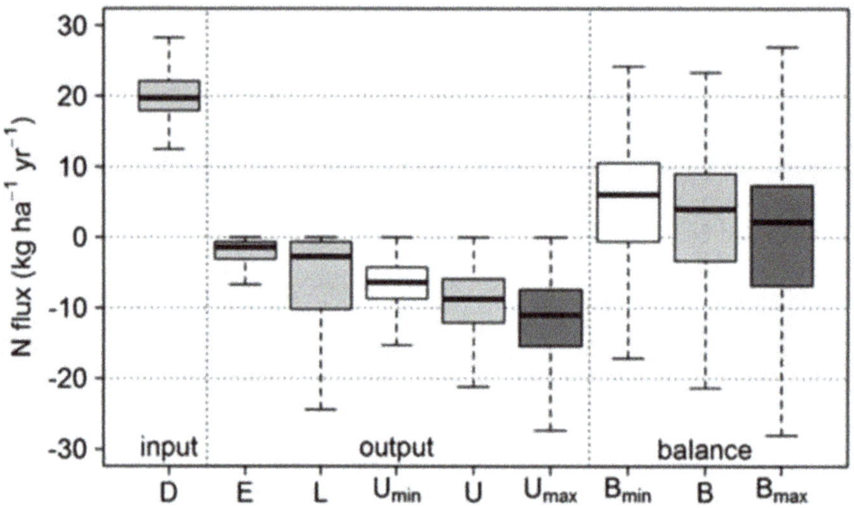

Fig. 5.18 Composition and fluxes from the nitrogen balance for forest ecosystems at German NFSI plots. D, deposition; E, emission; L, leaching; U, net uptake of the vegetation; B, balance; min, minimum harvest scenario (solid volume over bark); and max, maximum harvest scenario (whole tree harvesting without leaves and needles)

5.4.3.5 Discussion of Estimated Balances

Figure 5.18 shows the nitrogen balance and the different components for the NFSI plots in Germany. The balance was calculated for three different harvesting practice scenarios: stem only (B_{min}), whole tree without needles and leaves (B_{max}), and the average of both (B).

Between 1990 and 2007, the average annual N input for the NFSI plots by atmospheric deposition ranged from 10.6 to 40.7 kg ha^{-1} year^{-1} (median 19.7 kg ha^{-1} year^{-1}), which is in good accordance to measured deposition rates for 57 Level II plots in Germany (Borken and Matzner 2004). The generally decreasing trend for nitrogen deposition during the investigated period (Fig. 5.1) is supported by other investigations (Waldner et al. 2014). One should note that the accurate quantification of total nitrogen deposition, especially for dry deposition, for forest ecosystems is very difficult and uncertain, because of the lack of measurements, species variability, and the different deposition processes and fluxes that need to be simplified on a landscape scale (compare Harrison et al. (2000)). Simpson et al. (2011) estimated an error of 30% for the different regional models and methods in Europe. The choice of a certain deposition method depends on the purpose of the study and the availability of throughfall measurements. Because the latter are not available for NFSI plots, we used the state-of-the-art modelling approach for critical load calculations in Germany (see also Schaap et al. 2017) in combination with a reconstruction procedure. The evaluation of the estimated trend in the calculated deposition data was done by throughfall measurement in combination with a canopy budget model.

The quantified gaseous N flux to the atmosphere ranged between 0 and 16.8 kg ha^{-1} $year^{-1}$ (median 1.4 kg ha^{-1} $year^{-1}$). The results are in good agreement with the range (0.2–2 kg ha^{-1} $year^{-1}$ NO-N) given by Molina-Herrera et al. (2017). Upper values are mostly found for groundwater-affected plots or clear-cut stands (Dutch and Ineson 1990; Gundersen 1991; Tietema et al. 1991). For upland forest soils, typical rates are between 1 and 3 kg ha^{-1} $year^{-1}$ (Dutch and Ineson 1990).

Sixty-one percent of all NFSI plots leached less than 5 kg ha^{-1} $year^{-1}$ N with seepage water. The resulting median of 2.6 kg ha^{-1} $year^{-1}$ matches with an earlier study covering 57 Level II plots in Germany (Borken and Matzner 2004). There, the leaching rates ranged between 0 and 26.5 kg ha^{-1} $year^{-1}$ (median 1.4 kg ha^{-1} $year^{-1}$). About 71% of all plots released less than 5 kg ha^{-1} $year^{-1}$. Additional more recently evaluated data of the federal forest authorities using published reports on IFM plots (Barth et al. 2016; FAWF n.d.; Hammel and Kennel 2001; Hannemann et al. 2016; Karl et al. 2012; Klinck et al. 2012; Morgenstern 2015; Russ et al. 2017; Steinert and Feger 2010) confirm this range and median. Mellert et al. (2005) classified nitrate leaching rates for Bavarian forests with 66% showing values of 0–5 kg ha^{-1} $year^{-1}$, 20% of 5–15 kg ha^{-1} $year^{-1}$, and 14% higher than 15 kg ha^{-1} $year^{-1}$. In comparison, our estimates for Bavaria were 79.3% (0–5 kg ha^{-1} $year^{-1}$), 15.4% (5–15 kg N ha^{-1} $year^{-1}$), and 5.3% (>15 kg ha^{-1} $year^{-1}$). Kiese et al. (2011) estimated the nitrate leaching from German forest ecosystems by coupling Forest-DNDC to a GIS. Their results varied between 0 and 85 kg NO3-N ha^{-1} $year^{-1}$ with an area-averaged mean of 5.5 kg ha^{-1} $year^{-1}$. The higher N output is probably explained by the higher deposition rates used as input for the DNDC model based on values from Gauger et al. (2002) where problems with the nitrogen mass balance occurred in the applied version of the LOTOS-EUROS deposition model (Schaap et al. 2015).

The medians of net N uptake were between 6.6 kg ha^{-1} $year^{-1}$ (scenario stem only) and 11.9 kg ha^{-1} $year^{-1}$ (scenario whole tree without needles/leaves). Averaging both scenarios resulted in 8.7 kg ha^{-1} $year^{-1}$ and thus on a similar scale as in other studies (Ahner et al. 2013; Korhonen et al. 2013; Meesenburg et al. 2005; Nagel and Gregor 1999; Rademacher et al. 2009).

Because of the high atmospheric N input, the median of the N balance was always positive for all harvest scenarios (B_{min}: +7.0 kg ha^{-1} $year^{-1}$, B: +4.8 kg ha^{-1} $year^{-1}$, B_{max}: +2.9 kg ha^{-1} $year^{-1}$). Figure 5.19 shows the geographical distribution of the nitrogen balance for NFSI plots in Germany. Positive N balances are mainly found in the Bavarian Danube Plain, Upper Palatinate and Upper Franconia, the Ore Mountains, and nearly the whole North-German Lowland except Schleswig-Holstein. An accumulation of negative N balances is visible in the Black Forest and nearby mountain ranges, the Rhenish Slate Mountains, lower mountain ranges of North-Hesse, the Saarland, and Schleswig-Holstein.

In many regions, N stock decreases of the differential measurement approach are confirmed by the balance approach, and the federal statewise geographical distribution of change rates from differential N stock measurements on NFSI plots is rank correlated to the calculated balances (Spearman's Rho = 0.8, $n = 13$, $p < 0.005$). A plotwise comparison of modelled (balances) and measured (N stock difference) changes in N stocks showed lower rank correlation (Spearman's Rho = 0.21, $n = 979$, $p < 10^{-10}$).

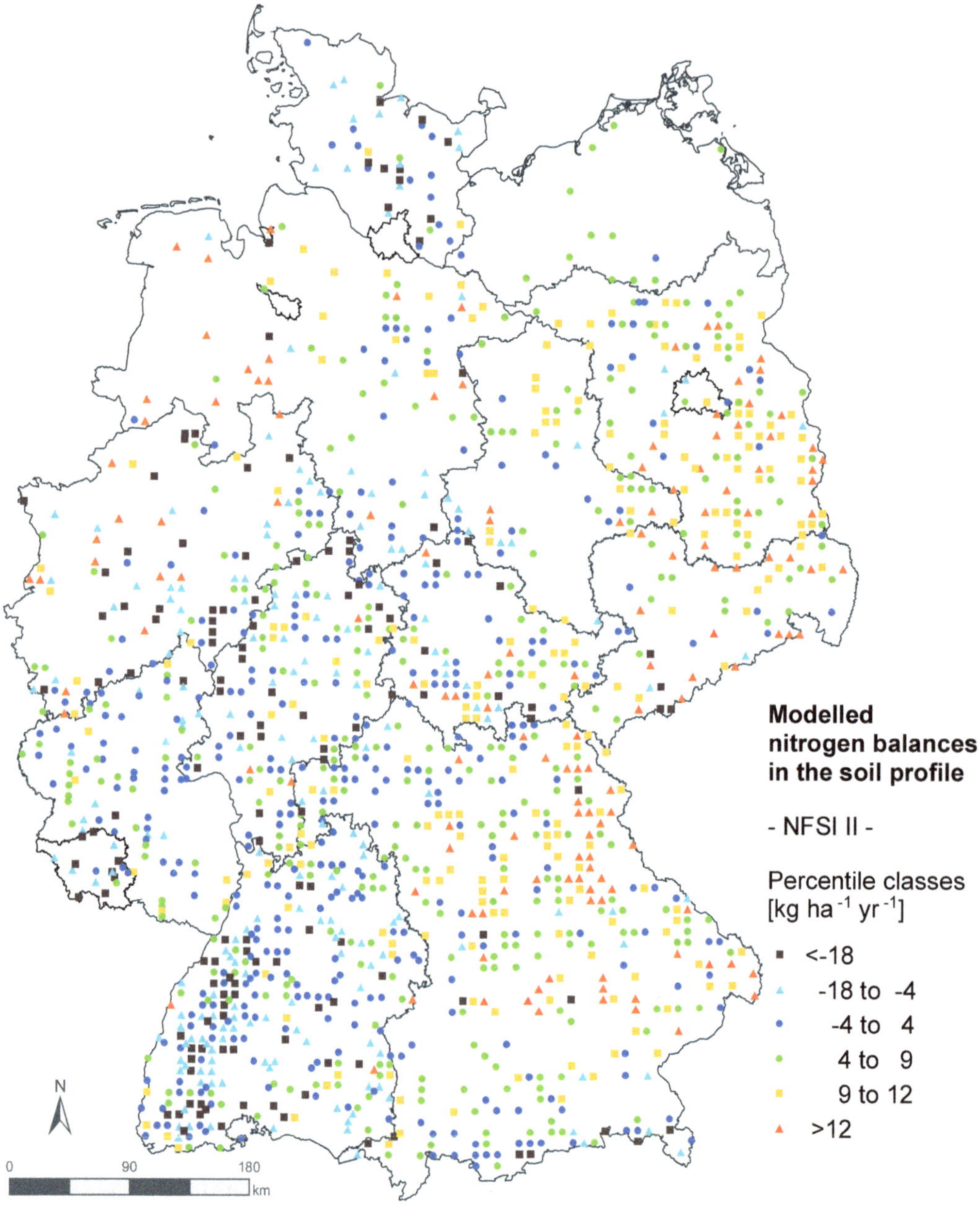

Fig. 5.19 Spatial distribution of nitrogen balances of the NFSI plots in Germany in kg ha^{-1} $year^{-1}$ (period 1990–2007)

5.5 Discussion of Methods

Nitrogen measurements are a special case in NFSI, since their repeatability in ring tests was much lower than for most other parameters, especially when concentrations were low (compare Table 1.2). It is the goal of this discussion to better understand the sources of uncertainty for these data and to highlight their special properties in order to improve the conclusions that can be drawn on their basis.

5.5.1 Spatial Variability

Nitrogen stock is one of the parameters with highest small-scale spatial variability. Next to the distribution of fine roots and earthworm activity (Meier and Leuschner 2010; Andriuzzi et al. 2016), the spatial variability of litterfall, parent material, pH, and microclimate (Sabatini et al. 2015), the distribution of stagnant conditions on the plot (Bekele et al. 2013), and the presence of N fixing bacteria in the root system of a few tree species (Rodriguez et al. 2011) were identified as causes for this variability. Mellert et al. (2008) systematically investigated the variability of NFSI soil parameters on 33 Bavarian plots along a sampling grid of 3 m distance and discovered the least spatial autocorrelation in N stocks, while other parameters like pH or C/N ratio showed lower spatial variability on small scales. While Kirwan et al. (2005) recommend to sample N at least at 36 locations per plot, Mellert et al. (2008) require total N stock changes of 37% or more as precondition for a proof of N stock changes between NFSI I and NFSI II. This degree of change was by far not reached in the presented trend results.

The representativeness of the soil subsamples (see Sect. 1.17.3) put into analysis from the bulk samples is very important under high small-scaled variability, since also the bulk samples could be more spatially variable in this parameter than in others. This is especially valid for NFSI I, where only 0.01–0.07 g of soil material were used in C/N analysers to derive N concentration of the whole sample, while the newer devices during NFSI II could analyse 1 or 2 g at once.

5.5.2 Uncertainty from Analytical Errors

In the following, we compile the magnitude of potential errors in laboratory analyses in order to estimate the uncertainty of N stocks as the sum of uncertainties of the quantities participating in N stock calculation (JCGM 2008). Since N stocks are calculated as the product of N concentration and fine-earth stock, the uncertainty (Var) of both measurands was estimated as (IPCC 2013):

$$\text{Var (measurand)} = \text{Var (repeatability)} + \text{Var(all means)} \tag{5.6}$$

The uncertainty of bulk density estimates was not considered separately since they are included in the observed uncertainty of fine-earth stocks. Uncertainty of depth delineations and the partial derivatives of quantities measured to each other were assumed to be negligible in a first approximation.

For chemical analysis of N concentration, Var (repeatability) was estimated from the pooled intra-laboratory standard deviation of N analyses within the labs (ring-test results, compare Sect. 1.8), and Var (all means) was based on the inter-laboratory standard deviation.

The uncertainty of N concentrations of the NFSI I ring tests (König and Wolff 1993) summed to 0.033 mg g^{-1} or 12% of sample means. For NFSI II, the interlaboratory ring tests (König et al. 2013) revealed an uncertainty of 0.027 mg g^{-1}, equaling ca. 4% of the sample means.

Var (fine-earth stock) was estimated based on Grüneberg et al. (2014). In their study, data from NFSI plots was used where fine-earth stocks were measured during both inventories. The fine-earth stock of NFSI I was on average 193 ± 35 t ha^{-1} higher than that of NFSI II. It was assumed that fine-earth stocks are constant between both inventories. Therefore the mean deviation plus the deviation (standard error) of the fine-earth stocks represents a certain degree of measurement inaccuracy of fine-earth stocks. This uncertainty amounted to 8% of the fine-earth stocks.

The reported relative uncertainties (percentage values) were used to calculate the uncertainties of the fine-earth stocks of each depth layer. The variance of the annual N stock changes was then summed up with inaccuracies of the measurement technique to obtain an estimate of the total uncertainty. The comparison of observed N stock differences in each layer with the total uncertainty of each layer resulted in all cases to non-significance, such that the observed N stock changes from differential measurements on NFSI plots must be considered scientifically non-significant tendencies, though statistical significance was shown (compare Wasserstein and Lazar 2016; Lemoine et al. 2016).

5.5.3 *Treatment of Very Low Concentrations*

Apart from the relative uncertainties given as percentages, absolute uncertainties were taken into account when judging the accuracy of N concentrations in the lowest soil layer (interquartile range 0.2–0.5 mg g^{-1}, 60–90 cm). As the given limits of quantification for the different devices used during NFSI I were in a range up to 0.24 mg g^{-1}, the 44% of measured values which were below or equal to this threshold were considered potentially inaccurate, especially with regard to the fact that potential measurement errors would always occur in positive direction under these circumstances, since negative values are not possible. The whole layer was excluded from trend calculation since the potential influence of measurement errors on the calculated stock differences was not negligible due to high fine-earth stocks (interquartile range 2–4.5 Gg ha^{-1}) that could easily lead to distortions when calculating stock differences.

The relevance of such a distortion is difficult to verify in the final dataset of calculated N stocks; however, there was a noticeable high proportion of exceptionally high N stocks observed especially in the data from deeper layers of NFSI I, which is reflected in the very high coefficients of skewness for these datasets (Table 5.6). While the high levels of N deposition preceding and during the NFSI I period might have caused elevated N concentrations also in deeper layers, such high coefficients of skewness have neither been observed in NFSI II nor in the inventories on IFM plots at both points in time. Coefficients of skewness were also regionally very diverse.

Table 5.6 Coefficients of skewness for each distribution of nitrogen stocks (t ha^{-1}) in the different layers, determined for NFSI I, NFSI II, and the first and second inventory of Intensive Forest Monitoring plots

Coefficient of skewness	NFSI I	NFSI II	IFM 1st	IFM 2nd
Organic layer	1.9	2.3	1.3	0.9
0–5 cm	2.4	1.1	0.5	0.0
5–10 cm	3.7	1.6	2.2	1.6
10–30 cm	8.5	2.1	1.4	1.8
30–60 cm	28.1	2.9	1.1	1.2
60–90 cm	28.3	6.4	0.8	2.2
OrgL—30 cm	6.6	1.6	1.2	1.0
OrgL—60 cm	20.0	1.7	0.7	1.0

The numerous exceptionally high N stocks from deeper layers of NFSI I are the main cause for a number of plots with extreme changes in N stocks. As usual in other long-term investigations (Johnson et al. 2007; Kiser et al. 2009), some of these stored samples were subsampled with a sample divider and reanalysed with a C/N analyser in 2017. First analyses show that a part of the high N concentrations from NFSI I may be reproduced; however, it is unclear to what extent the re-analyses are influenced by storage conditions.

In NFSI I, also the layer 30–60 cm of the mineral soil shows a high relative proportion of exceptionally high N stocks (coefficient of skewness = 28.1). The measured N concentrations were higher than in 60–90 cm (interquartile range 0.3–0.9 mg g^{-1}) and were, thus, in their majority expected to be less affected by the laboratory limit of quantification. However, since their influence is high in stock calculations (fine-earth stocks interquartile range: 2.2–4.3 Gg ha^{-1}) and also other error sources may have contributed to the skewed distribution, only weighted median values of trends derived from this layer in NFSI I are presented, thereby reducing the influence of very high or low absolute numbers on the results. For better comparability, all results of the paired and the complete sample are presented as weighted medians.

5.5.4 *Plot Selection Effects*

Most methodological problems complicating the evaluation of temporal changes between the two NFSIs equally apply for NFSI plots and IFM plots and are not likely to be the cause for the apparently opposite direction of the trends derived from both networks and from balance calculations. Of course the number of IFM plots with two N stock inventories is much smaller than the number of NFSI plots, such that outliers have the potential to affect the overall pattern of temporal changes. While NFSI plots are systematically selected along a spatial grid, IFM plots were selected in order to represent the regionally typical forest types. May sample size and plot selection be

Table 5.7 Percentage of NFSI plot subsamples of size n that would produce at least the same deviation of nitrogen stock medians relative to NFSI results as it is reported for the available number of Intensive Forest Monitoring plots for each layer

	n	First inventory (%)	Second inventory (%)	Trend (%)
Organic layer	37	9.1	22.5	37.7
0–5 cm	29	34.2	1.2	20.7
5–10 cm	29	58.5	94.6	12.8
10–30 cm	47	38.9	11.4	0.02
0–30 cm	47	36.7	1.1	0.02
30–60 cm	46	10.3	12.9	0.05
60–90 cm	21	16.5	61.2	5.7
Org—60 cm	36	15.3	0.1	0.8

Sample sizes in 0–5 cm and 5–10 cm are lower than, e.g. in organic layer—60 cm due to non-separate sampling of both layers on many plots

the causes for an opposite direction of trends derived for the time between both NFSIs?

A bootstrap analysis was performed to investigate the plot selection effect. 100 million subsamples of NFSI plots (paired sample) were randomly chosen for each layer with a sample size identical to the corresponding number of IFM plots available. If IFM plots were just another selection of plots from the same population, there would probably also be plot combinations from the NFSI plots generating the same deviation to the whole NFSI result or an even larger deviation. The results are given in Table 5.7: For example, for the organic layer, the analysis revealed that an N stock decrease ≤ -4.5 kg ha^{-1} year^{-1} (the value measured on IFM plots) would also have been the outcome of 37.7% of all samples with the same sample size ($n = 37$) drawn from NFSI plots, showing that IFM plot results are not generally different from the results of NFSI in the organic layer. On the other hand, for the mineral soil between 0 and 30 cm depth, only 0.02% of the equal-sized subsamples of the NFSI plots would lead to an N stock increase $\geq +43.1$ kg ha^{-1} year^{-1}—here NFSI plots and IFM plots are generally different.

The bootstrap analysis shows that the IFM plot results of the first inventory could also have been derived from a reasonable number of combinations from NFSI plots. Especially in the upper part of the mineral soil and to a lesser extent in the other layers, IFM plots had in the first inventory N stocks typical for NFSI plots. In the second inventory, N stocks of layers deeper than 30 cm of the mineral soil and of the organic layer of IFM plots were again quite typical results for subsamples of the NFSI, but the upper 30 cm of the mineral soil had results that could hardly be generated based on NFSI plots: Only 1.1% of the random NFSI plot subsamples would produce such a result. Also the trends derived from the difference between first and second inventory may not be generated based on subsamples of the NFSI in 0–30 cm and in 30–60 cm of the mineral soil.

It may be concluded from this analysis that it is not just the low number of IFM plots that is responsible for the apparently opposite direction of their trend results: Some IFM plot results especially of the second inventory appear to be not from the same population as the results of NFSI plots. It may also be concluded that the initial plot selection is not the primary cause for the diverging results in terms of absolute N stocks, since the first inventory after plots were selected could reasonably be reproduced by subsamples of NFSI plots. Only with respect to the occurring change rates, the selected IFM plots are not comparable to the NFSI dataset.

Several alternative explanations for the different trend results from both networks are possible: (1) Since the inventories were not all performed in exactly the main year of NFSI I or NFSI II, it could be that N stock changes are more variable in time than expected, such that the results of both networks represent different stages of N stock development. Other long-term monitoring programmes on N stocks in forests also found subsequent N stock changes with reversed direction that were difficult to explain as a long-term development (Johnson et al. 2007; Johnson and Turner 2014; Kiser et al. 2009; Binkley et al. 2000). (2) The originally more typical state of N stocks on IFM plots may have changed after the first inventory due to their use as IFM plots. Most IFM plots are highly instrumented plots with less regular thinning, with the consequence that the stands are on average older than the mean age of NFSI plots and, thus, could accumulate more N in the soil during their lifetime. Many of the plots are fenced, which lowers bioturbation, browsing, and predation by larger mammals. (3) The small methodological differences between both networks may have contributed to the deviation in trend results. Differences exist, e.g. with regard to the number of soil samples taken in the forest, such that the high spatial variability of N stocks may better be accounted for by IFM plots. (4) No IFM plots with sufficient data are located in those areas in the southwest of Germany where predominantly negative change rates of soil N stocks have been observed. Thus, a systematic underrepresentation of geographic areas with negative change rates within the set of IFM plots contributes to the results from the bootstrapping approach.

Summarizing the methods discussion, it may not be excluded that the results of differential N stock measurements are subject to relevant over- or underestimations. Namely the measurements in deeper layers are influenced by the mentioned error sources, since the employed techniques especially in NFSI I operated close to their limit of quantification. However, they are the only representative benchmark values we can relate to from the 1990s, and there is no proof that would justify to discard them. It is of course important to take the uncertainties mentioned into account when drawing conclusions especially on long-term trends.

5.6 Summary and Conclusions

Nitrogen stocks in German forest soils are with 6.3 t ha^{-1} on the most frequently observed level in European forest soils (5–10 t ha^{-1}, Fleck et al. 2016), which is considered a medium level according to the empirical rating for German forest soils (AK Standortskartierung 2016). This result was similarly observed in NFSI I and NFSI II and agrees in general terms with the measurements on IFM plots (Fig. 5.2).

More than 50% of N in forest soils is generally stored in the uppermost 30 cm of the mineral soil. The next 30 cm contains more than 20%, and roughly another 15% is stored in 60–90 cm depth of the mineral soil. The organic layer contains usually the remaining 10–15% of N in the whole soil profile up to 90 cm. This result is confirmed by NFSI I and NFSI II measurements as a representative result for Germany, and it is in the same order of magnitude also observed on the selected IFM plots.

A general property of N stocks in German forest soils is their high regional variability, which is visible in all evaluations (Figs. 5.3, 5.5, and 5.19). The observed pattern may largely be explained as the long-term impact of factors like tree species, parent material, soil acidification, annual mean temperature, and agricultural land use with the related N deposition. The depth gradient of N stocks observed on plots from the different tree species and parent materials is well explained by the decomposability approach of Berg (2014), whereby high C/N ratios in organic layer—5 cm indicate organic material with low initial decomposability (decomposition of cellulose), but high total decomposability (including the decomposition of lignin), leading to high N stocks in the organic layer, but low N stocks in total for organic layer—60 cm. Low C/N ratios in turn were nearly always associated with low N stocks in the organic layer and high N stocks in organic layer—60 cm. These results demonstrate the central relevance of C/N ratios of the organic material provided for its decomposability in the soil.

A potential explanation for the relationship between C/N ratio and the depth gradient of N stocks may be found in the observation that C/N ratios initially decrease during decomposition due to N immobilization taking place in the first 9 months of decomposition (Hasegawa and Takeda 1996). Parton et al. (2007) also showed that immobilization during the first phases of decomposition is highest in litter with high C/N ratios (when microbial biomass growth is N limited) and very low when C/N ratios are small and microbial biomass growth is C limited. The effect of C/N ratios on microbial biomass growth and subsequent decomposition is, however, twofold: While an increase of microbial biomass reduces the amount of easily accessible particulate organic matter, it increases the amount of mineral-associated organic matter, which mainly originates from microbial necromass and exudates (Averill and Waring 2018).

The effect of soil pH on decomposition processes is visible from the liming effect on acid-sensitive plots, where N stocks in the organic layer decreased in the years between NFSI I and NFSI II, while N stocks increased in the mineral soil (0–30 cm): Apparently, this effect is independent from C/N ratios, since there was no direct effect on C/N ratios, while the directly stimulating effect of increased soil pH on microbial activity is known (Anderson and Domsch 1993). Increased decomposition of particulate organic matter by microbes in the organic layer may here again be associated with an increase of mineral-associated organic matter in the mineral soil, eventually enhanced by the decreasing influence of liming on soil pH with depth.

In the years between NFSI I and NFSI II, an N accumulation took place in organic layer—5 cm. More precisely, the N accumulation was concentrated in the uppermost 5 cm of the mineral soil, while there appeared to be a slight decrease of N stocks in the organic layer. This result is derived from differential N stock measurements on NFSI plots as well as on IFM plots. C/N ratios in organic layer—5 cm increased in the same time from a moderate level in NFSI I to a somewhat higher level in NFSI II. Such an increase of C/N ratios could be the first visible effect of the slowly decreasing N deposition rates, since uptake of N compounds into the organic layer is reduced with decreasing deposition. This interpretation is consistent with the fact that the formerly observed decrease of C/N ratios in the years around NFSI I had been attributed to the increasing N deposition rates at that time (Geissen and Brümmer 1999). The expected inhibitory effect of increasing C/N ratios on microbial biomass growth in the organic layer (low initial decomposability) did not occur and may have been overridden in this case by the increase of pH values in the organic layer (compare Chap. 4) associated with decreasing N deposition. Also the climatic changes in the period before each of the NFSIs may have contributed to higher organic matter decomposition by microbes in the organic layer: Increasing temperatures and the potentially increasing frequency of drought-rewetting cycles in the 10 years before sampling (compare Chap. 3) may yet in these years have led to accelerated decomposition and increased microbial activity, thereby decreasing the amount of particulate organic matter N in the organic layer and increasing the N stocks in mineral-associated organic matter in the upper mineral soil. A direct link between N deposition and C/N ratios appears to be likely also from the observed impact of agricultural land use.

While N accumulation in organic layer—5 cm may be derived from NFSI as well as IFM plots, increasing N stocks in the mineral soil above 30 cm depth and decreasing N stocks below this depth were only found on NFSI plots. IFM plots confirm the increase in the mineral soil above 30 cm, but yield constant N stocks in the layers below. However, a discrepancy in the change rates between mineral soil layers above and below 30 cm depth seems to be a common feature of both networks' results on N stock change rates from differential N stock measurements. Regardless of the uncertainties associated with low concentration measurements in deeper layers, a potential explanation for such a discrepancy would most likely be connected to the acidifying effect of N deposition, which has been found to be independent of the effect of C/N ratios. After N (and S) deposition had slowly decreased, an increase of soil pH and base saturation was only observed in the uppermost layers (compare Chap. 4). This reduction of acidity together with climatic changes induced an increase of (micro-) biological activity, reducing the amount of stored N in the organic layer by decomposition, but increasing the amount of mineral-associated organic matter in the upper mineral soil, while the ongoing acidification in the deeper mineral soil of acid-sensitive plots would be expected to reduce microbial activity and potentially even the amount of N stored in this compartment.

The approaches to derive N stock changes between NFSI I and NFSI II show a diverse picture. While the modelled N balances lead to the expectation of on average increasing N stocks (+2.8 to +7.0 kg ha^{-1} $year^{-1}$, depending on harvest scenario), they also estimate N stock decreases for a substantial proportion of the sites, even if only a minimum amount of harvesting is assumed. The differential N stock measurement results from IFM plots, which partly follow the scenario of no harvest, indeed show a clear increase of N stocks, which is with +28.0 kg ha^{-1} year^{-1} even higher than the range given from modelled N balances. While N stocks on NFSI plots in organic layer—30 cm slightly increased (+6.2 kg ha^{-1} year^{-1}), N stock decreases are reported for the deeper layers with their generally low N concentrations that turn the result for organic layer—60 cm to a median decrease of −8.2 kg ha^{-1} year^{-1}. Since both NFSI and IFM plot results are outside the range given by N balance estimations, it may not be excluded that they are over- or underestimations. The representativeness of modelled N balances from NFSI plots is considered higher than their differential N stock measurements, since N stock differences of the latter, coined by the extremely high spatial and temporal variability of N concentrations in soil samples and resulting measurement uncertainties, were scientifically not significant.

Summarizing, the effect of decreasing N deposition rates yet appears to be visible as increasing C/N ratios and pH values (Chap. 4) in the organic layer, whereof the latter is partly achieved by liming and the reduction of S deposition and contributes to decreasing N stocks in the organic layer. Changes in climate, litter quality, and soil pH appear to be responsible for the observed shift of N stocks towards the upper mineral soil. The forest soils contain still a medium high amount of N stocks and further accumulate N, preferentially in the uppermost 30 cm of the mineral soil, while there are indications for large N stock decreases in the deeper layers of the mineral soil that are potentially influenced by continuing acidification in these layers. While no general increase of N in groundwater from forest catchments has been reported for Germany in the years between NFSI I and NFSI II (Sucker et al. 2011), the risk of such a development requires more intensive monitoring of processes in the deeper mineral soil. Modelling studies are needed especially on acid-sensitive forest sites to better predict the development of N-leaching under forest soils recovering from acidification. The data show, how vulnerable biological activity and N-storage in forest soils may be under changing conditions, and losses of N by leaching from forests need to be considered as a serious possibility. Further reductions of N deposition are, thus, still needed as a preventive measure for environmental protection.

References

Aber JD, Nadelhoffer KJ, Steudler P, Melillo JM (1989) Nitrogen saturation in northern forest ecosystems. BioScience 39(6):378–386

Aber JD, McDowell W, Nadelhoffer K, Magill A, Berntson G, Kamakea M, McNulty S, Currie W, Rustad L, Fernandez I (1998) Nitrogen saturation in temperate forest ecosystems—hypotheses revisited. BioScience 48(11):921–934. https://doi.org/10.2307/1313296

Ad-HocAG_Boden (ed) (2005) Bodenkundliche Kartieranleitung (KA 5), vol 5. Schweizerbart'sche Verlagsbuchhandlung, Stuttgart, Germany

Ahner J, Ahrends B, Engel F, Hansen J, Hentschel S, Hurling R, Meesenburg H, Mestermacher U, Meyer P, Möhring B, Nagel J, Nagel R, Pape B, Rohde M, Rumpf H, Schmidt M, Schmidt M,

Spellmann H, Sutmöller J (2013) Waldentwicklungsszenarien für das Hessische Ried. Entscheidungsunterstützung vor dem Hintergrund sich beschleunigt ändernder Wasserhaushalts-und Klimabedingungen und den Anforderungen aus dem europäischen Schutzgebietssystem Natura 2000. Beiträge aus der Nordwestdeutschen Forstlichen Versuchsanstalt, vol 10. Nordwestdeutsche Forstliche Versuchsanstalt, Göttingen, Germany

Ahrends B, Meesenburg H, Döring C, Jansen M (2010) A spatio-temporal modelling approach for assessment of management effects in forest catchments. In: Status and perspectives of hydrology in small basins, IAHS Publ. 336, pp 32–37

AK Standortskartierung (2003) Forstliche Standortsaufnahme: Begriffe, Definitionen, Einteilungen, Kennzeichnungen, Erläuterungen, 6th edn. IHW-Verlag, Eiching near Munich, Germany

AK Standortskartierung (2016) Forstliche Standortsaufnahme: Begriffe, Definitionen, Einteilungen, Kennzeichnungen, Erläuterungen, 7th edn. IHW-Verlag, Eiching near Munich, Germany

Alveteg M, Walse C, Warfvinge P (1998) Reconstructing historic atmospheric deposition and nutrient uptake from present day values using MAKEDEP. Water Air Soil Pollut 104 (3–4):269–283. https://doi.org/10.1023/a:1004958027188

Anderson T, Domsch K (1993) The metabolic quotient for CO_2 (qCO_2) as a specific activity parameter to assess the effects of environmental conditions, such as pH, on the microbial biomass of forest soils. Soil Biol Biochem 25:393–395

Andriuzzi WS, Ngo PT, Geisen S, Keith AM, Dumack K, Bolger T, Bonkowski M, Brussaard L, Faber JH, Chabbi A, Rumpel C, Schmidt O (2016) Organic matter composition and the protist and nematode communities around anecic earthworm burrows. Biol Fertil Soils 52(1):91–100. https://doi.org/10.1007/s00374-015-1056-6

Averill C, Waring B (2018) Nitrogen limitation of decomposition and decay: how can it occur? Global Change Biol 24:1417–1427

Barth N, Tannert R, Kurzer H-J, Kolber H, Andreae H, Haferkorn U, Rust M, Grunert M (2016) Stickstoffmonitoring sächsicher Böden. Landesamt für Umwelt, Landwirtschaft und Geologie, Dresden, Germany

Bedison JE, Johnson AH (2009) Controls on the spatial patterns of carbon and nitrogen in Adirondack forest soils along a gradient of nitrogen deposition. Soil Sci Soc Am J 73 (6):2105–2117. https://doi.org/10.2136/sssaj2008.0336

Bekele A, Kellman L, Beltrami H (2013) Plot level spatial variability of soil organic carbon, nitrogen, and their stable isotopic compositions in temperate managed forest soils of Atlantic Canada. Soil Sci 178(8):400–416. https://doi.org/10.1097/ss.0000000000000003

Berg B (2014) Decomposition patterns for foliar litter—a theory for influencing factors. Soil Biol Biochem 78:222–232. https://doi.org/10.1016/j.soilbio.2014.08.005

Binkley D, Son Y, Valentine DW (2000) Do forests receive occult inputs of nitrogen? Ecosystems 3:321–331

BMEL (2016) Der Wald in Deutschland—Ausgewählte Ergebnisse der dritten Bundeswaldinventur, 2nd edn. Bundesministerium für Ernährung und Landwirtschaft (BMEL), Berlin, Germany

Borken W, Matzner E (2004) Nitrate leaching in forest soils: an analysis of long-term monitoring sites in Germany. J Plant Nutr Soil Sci 167(3):277–283. https://doi.org/10.1002/jpln.200421354

Borken W, Matzner E (2009) Reappraisal of drying and wetting effects on C and N mineralization and fluxes in soils. Global Change Biol 15(4):808–824. https://doi.org/10.1111/j.1365-2486.2008.01681.x

Cools N, Vesterdal L, de Vos B, Vanguelova E, Hansen K (2014) Tree species is the major factor explaining C:N ratios in European forest soils. Forest Ecol Manag 311:3–16. https://doi.org/10.1016/j.foreco.2013.06.047

Costanza R, d'Arge R, de Groot R, Farber S, Grasso M, Hannon B, Limburg K, Naeem S, O'Neill RV, Paruelo J, Raskin RG, Sutton P, van den Belt M (1997) The value of the world's ecosystem services and natural capital. Nature 387(6630):253–260. https://doi.org/10.1038/387253a0

Czajkowski T, Ahrends B, Bolte A (2009) Critical limits of soil water availability (CL-SWA) for forest trees—an approach based on plant water status. Landbauforschung Volkenrode 59 (2):87–93

de Vries W, Kros J, van der Salm C (1994) Long-term impacts of various emission deposition scenarios on Dutch forest soils. Water Air Soil Pollut 75:1–35

de Vries W, Dobbertin MH, Solberg S, Van Dobben HF, Schaub M (2014) Impacts of acid deposition, ozone exposure and weather conditions on forest ecosystems in Europe: an overview. Plant Soil 380(1–2):1–45

Dutch J, Ineson P (1990) Denitrification of an upland forest site. Forestry 63(4):363–377. https://doi.org/10.1093/forestry/63.4.363

Eickenscheidt N, Brumme R (2012) NO_x and N_2O fluxes in a nitrogen-enriched European spruce forest soil under experimental long-term reduction of nitrogen depositions. Atmos Environ 60:51–58

Eickenscheidt N, Brumme R, Veldkamp E (2011) Direct contribution of nitrogen deposition to nitrous oxide emissions in a temperate beech and spruce forest—a ^{15}N tracer study. Biogeosciences 8:621–635

Ek H, Andersson S, Arnebrant K, Söderström B (1994) Growth and assimilation of NH4+ and NO3- by *Paxillus involutus* in association with *Betula pendula* and *Picea abies* as affected by substrate pH. New Phytol 128:629–637

Ellenberg H, Weber HE, Düll R, Wirth V, Werner W (2003) Zeigerwerte von Pflanzen in Mitteleuropa. Scripta Geobotanica XVIII, Datenbank. Erich Goltze Göttingen, Germany

Engardt M, Simpson D, Schwikowski M, Granat L (2017) Deposition of sulphur and nitrogen in Europe 1900-2050. Model calculations and comparison to historical observations. Tellus Ser B Chem Phys Meteorol 69(1):1328945. https://doi.org/10.1080/16000889.2017.1328945

Erisman J, Dammers E, van Damme M, Soudzilovskaia N, Schaap M (2015) Trends in EU nitrogen deposition and effects on ecosystems. Air Waste Manag Assoc Mag 65:31–35

Evers J, König N, Wolff B, Meiwes KJ (2002) Vorbereitung der Zweiten Bodenzustandserhebung im Wald (BZE II): Untersuchungen zur Laboranalytik, Stickstoffbestimmung und zeitlichen Variabilität bodenchemischer Parameter. Federal Ministry of Food, Agriculture and Consumer Protection, Northwest German Forest Research Institute, Bonn, Germany

Fagerli H, Aas W (2008) Trends of nitrogen in air and precipitation: model results and observations at EMEP sites in Europe, 1980-2003. Environ Pollut 154(3):448–461. https://doi.org/10.1016/j.envpol.2008.01.024

Fagerli H, Legrand M, Preunkert S, Vestreng V, Simpson D, Cerqueira M (2007) Modeling historical long-term trends of sulfate, ammonium, and elemental carbon over Europe: a comparison with ice core records in the Alps. J Geophys Res Atmos 112:D23. https://doi.org/10.1029/2006jd008044

FAWF (n.d.) Forschung an Dauerbeobachtungsflächen. http://www.fawf.wald-rlp.de/fileadmin/website/fawfseiten/fawf/FUM/index.htm?umweltmonitoring/DBFL/forschung.html. Accessed 12/15/2017

Fleck S, Cools N, Vos BD, Meesenburg H, Fischer R (2016) The Level II aggregated forest soil condition database links soil physicochemical and hydraulic properties with long-term observations of forest condition in Europe. Ann Forest Sci 73(4):945–957. https://doi.org/10.1007/s13595-016-0571-4

Fleck S, Ahrends B, Sutmöller J, Albert M, Evers J, Meesenburg H (2017) Is biomass accumulation in forests an option to prevent climate change induced increases in nitrate concentrations in the North German Lowland? Forests 8(6):219

Förster A, Becker T, Gerlach A, Meesenburg H, Leuschner C (2017) Long-term change in understorey plant communities of conventionally managed temperate deciduous forests: effects of nitrogen deposition and forest management. J Veg Sci 28(4):747–761. https://doi.org/10.1111/jvs.12537

GAFA (ed) (2009) Handbuch Forstliche Analytik (HFA). Grundwerk und 1. - 4. Ergänzung des Gutachterausschuss Forstliche Analytik (GAFA). Bundesministerium für Verbraucherschutz, Ernährung und Landwirtschaft, Bonn, Germany

Gauer J, Kroiher F (2012) Waldökologische Naturräume Deutschlands—Forstliche Wuchsgebiete und Wuchsbezirke—Digitale Topographische Grundlagen—Neubearbeitung Stand 2011. Landbauforschung vTI Agriculture and Forestry Research, Braunschweig, Germany

Gauger T, Anshelm F, Schuster H, Draaijers GPJ, Bleeker A, Erisman JW, Vermeulen AT, Nagel H-D (2002) Kartierung ökosystembezogener Langzeittrends atmosphärischer Stoffeinträge und Luftschadstoffkonzentrationen in Deutschland und deren Vergleich mit Critical Loads und Critical Levels. Forschungsvorhaben im Auftrag des BMU/UBA. Institut für Navigation, Universität Stuttgart, Stuttgart

Geissen V, Brümmer GW (1999) Decomposition rates and feeding activities of soil fauna in deciduous forest soils in relation to soil chemical parameters following liming and fertilization. Biol Fertil Soils 29:335–342

Grüneberg E, Ziche D, Wellbrock N (2014) Organic carbon stocks and sequestration rates of forest soils in Germany. Global Change Biol 20(8):2644–2662. https://doi.org/10.1111/gcb.12558

Gundersen P (1991) Nitrogen deposition and the forest nitrogen cycle—role of denitrification. Forest Ecol Manag 44(1):15–28. https://doi.org/10.1016/0378-1127(91)90194-z

Hammel K, Kennel M (2001) Charakterisierung und Analyse der Wasserverfügbarkeit und des Wasserhaushalts von Waldstandorten in Bayern mit dem Simulationsmodell BROOK90. Forstliche Forschungsberichte München, vol 185. Technische Uni München Wissenschaftszentrum Weihenstephan, Munich, Germany

Hannemann J, Russ A, Kallweit R, Riek W (2016) Betrachtungen zu den Stoffbilanzen von Level II-Flächen in Brandenburg. In: Ministerium für Ländliche Entwicklung Umwelt und Landwirtschaft des Landes Brandenburg (ed) 30 Jahre Forstliches Umweltmonitoring in Brandenburg. Eberswalder Forstliche Schriftenreihe. LFE, Eberswalde, Germany

Harrison A, Schulze E-D, Gebauer G, Bruckner G (2000) Canopy uptake and utilization of atmospheric pollutant nitrogen. In: Carbon and nitrogen cycling in European forest ecosystems. Springer, Berlin, pp 171–188

Hasegawa M, Takeda H (1996) Carbon and nutrient dynamics in decomposing pine-needle litter in relation to fungal and faunal abundances. Pedobiologia 40:171–184

Hauck M, Zimmermann J, Jacob M, Dulamsuren C, Bade C, Ahrends B, Leuschner C (2012) Rapid recovery of stem increment in Norway spruce at reduced SO_2 levels in the Harz Mountains, Germany. Environ Pollut 164:132–141. https://doi.org/10.1016/j.envpol.2012.01.026

Homan C, Beier C, McCay T, Lawrence G (2016) Application of lime ($CaCO_3$) to promote forest recovery from severe acidification increases potential for earthworm invasion. Forest Ecol Manag 368:39–44. https://doi.org/10.1016/j.foreco.2016.03.002

IPCC (2013) Good practice guidance for land use, land-use change and forestry. Institute for Global Environmental Strategies for the Intergovernmental Panel on Climate Change (IPCC), Hayama, Japan

Jacobsen C, Rademacher P, Meesenburg H, Meiwes K (2002) Gehalte chemischer Elemente in Baumkompartimenten—Literaturstudie und Datensammlung. Berichte des Forschungszentrums Waldökosysteme. Niedersächsische Forstliche Versuchsanstalt Göttingen, Göttingen, Germany

Jansen M, Chodak M, Saborowski J, Beese F (2005) Determination of humus stocks and qualities of forest floors in pure and mixed stands of spruce and beech (Erfassung von Humusmengen und -qualitaten in organischen Auflagen in Rein- und Mischbestanden von Buchen und Fichten unterschiedlichen Alters). Allgemeine Forst- und Jagdzeitung 176(9/10):176–186

JCGM (2008) Evaluation of measurement data—guide to the expression of uncertainty in measurement

Johnson DW, Turner J (2014) Nitrogen budgets of forest ecosystems: a review. Forest Ecol Manag 318:370–379

Johnson DW, Todd DE Jr, Trettin CF, Sedinger JS (2007) Soil carbon and nitrogen changes in forests of walker branch watershed, 1972 to 2004. Soil Sci Soc Am J 71(5):1639–1646. https://doi.org/10.2136/sssaj2006.0365

Karl S, Block J, Schüler G, Schultze B, Scherzer J (2012) Wasserhaushaltsuntersuchungen im Rahmen des forstlichen Umweltmonitorings und bei waldbaulichen Versuchen in Rheinland-Pfalz. Mitteilungen aus der Forschungsanstalt für Waldökologie und Forstwirtschaft Rheinland-Pfalz (FAWF), vol 71/12. Trippstadt, Germany

Kiese R, Heinzeller C, Werner C, Wochele S, Grote R, Butterbach-Bahl K (2011) Quantification of nitrate leaching from German forest ecosystems by use of a process oriented biogeochemical model. Environ Pollut 159(11):3204–3214. https://doi.org/10.1016/j.envpol.2011.05.004

Kirwan N, Oliver MA, Moffat AJ, Morgan GW (2005) Sampling the soil in long-term forest plots: the implications of spatial variation. Environ Monit Assess 111(1–3):149–172. https://doi.org/10.1007/s10661-005-8219-0

Kiser LC, Kelly JM, Mays PA (2009) Changes in forest soil carbon and nitrogen after a thirty-year interval. Soil Sci Soc Am J 73(2):647–653. https://doi.org/10.2136/sssaj2008.0102

Klinck U, Rademacher P, Scheler B, Wagner M, Fleck S, Ahrends B, Meesenburg H (2012) Ökosystembilanzen auf forstwirtschaftlich genutzten Flächen. In: Höper H, Meesenburg H (eds) GeoBerichte 23—Tagungsband 20 Jahre Bodendauerbeobachtung in Niedersachsen. Landesamt für Bergbau, Energie und Geologie, Hannover, Germany, pp 163–174

Kohlpaintner M, Huber C, Göttlein A (2012) Improving the precision of estimating nitrate (NO_3^-) concentration in seepage water of forests by prestratification with soil samples. Eur J Forest Res 131(5):1399–1409. https://doi.org/10.1007/s10342-012-0606-9

König N, Wolff B (1993) Abschlussbericht über die Ergebnisse und Konsequenzen der im Rahmen der bundesweiten Bodenzustandserhebung im Wald (BZE) durchgeführten Ringanalysen. Reports of the Research Centre for Forest Ecosystems/Forest Decline—Series B. Göttingen University, Göttingen, Germany

König N, Schönfelder E, Blum U (2013) Auswertung der Standardmessungen und der Ringversuche im Rahmen der BZE II. Federal Ministry of Food, Agriculture and Consumer Protection, Berlin, Germany

Korhonen JFJ, Pihlatie M, Pumpanen J, Aaltonen H, Hari P, Levula J, Kieloaho AJ, Nikinmaa E, Vesala T, Ilvesniemi H (2013) Nitrogen balance of a boreal Scots pine forest. Biogeosciences 10 (2):1083–1095. https://doi.org/10.5194/bg-10-1083-2013

Krupa SV (2003) Effects of atmospheric ammonia (NH_3) on terrestrial vegetation: a review. Environ Pollut 124(2):179–221. https://doi.org/10.1016/s0269-7491(02)00434-7

Larcher W (2001) Ökophysiologie der Pflanzen, 6th edn. Ulmer, Stuttgart

Lemoine NP, Hoffman A, Felton AJ, Baur L, Chaves F, Gray J, Yu Q, Smith MD (2016) Underappreciated problems of low replication in ecological field studies. Ecology 97 (10):2554–2561

Lovett GM, Goodale CL (2011) A new conceptual model of nitrogen saturation based on experimental nitrogen addition to an oak forest. Ecosystems 14(4):615–631. https://doi.org/10.1007/s10021-011-9432-z

Ludwig B, Meiwes KJ, Khanna P, Gehlen R, Fortmann H, Hildebrand EE (1999) Comparison of different laboratory methods with lysimetry for soil solution composition—experimental and model results. J Plant Nutr Soil Sci 162(3):343–351. https://doi.org/10.1002/(sici)1522-2624(199906)162:3<343::aid-jpln343>3.0.co;2-e

Lutz F (2015) Methodenvergleich zur Ableitung der Ionenkonzentrationen im Sickerwasser aus Analysen der Bodenfestphase. Technische Universität München, München

Meesenburg H, Mohr K, Dämmgen U, Schaaf S, Meiwes KJ, Horváth B (2005) Stickstoff-Einträge und -Bilanzen in den Wäldern des ANSWER-Projektes—Eine Synthese. Landbauforschung Völkenrode 279:95–108

Meesenburg H, Ahrends B, Fleck S, Wagner M, Fortmann H, Scheler B, Klinck U, Dammann I, Eichhorn J, Mindrup M (2016) Long-term changes of ecosystem services at Solling, Germany: recovery from acidification, but increasing nitrogen saturation? Ecol Indicat 65:103–112

Meier IC, Leuschner C (2010) Variation of soil and biomass carbon pools in beech forests across a precipitation gradient. Global Change Biol 16(3):1035–1045. https://doi.org/10.1111/j.1365-2486.2009.02074.x

Mellert KH, Gensior A, Kolling C (2005) Nitrogen saturation in Bavarian forests—results of the nitrate inventory (Stickstoffsättigung in den Wäldern Bayerns—Ergebnisse der Nitratinventur.). Forstarchiv 76(2):35–43

Mellert KH, Kölling C, Rücker G, Schubert A (2008) Kleinräumige Variabilität von Waldboden-Dauerbeobachtungsflächen in Bayern—Ein Beitrag zur Unsicherheitsabschätzung der BZE II. Waldökologie, Landschaftsforschung und Naturschutz 6:43–61

Molina-Herrera S, Haas E, Grote R, Kiese R, Klatt S, Kraus D, Kampffmeyer T, Friedrich R, Andreae H, Loubet B, Ammann C, Horvath L, Larsen K, Gruening C, Frumau A, Butterbach-Bahl K (2017) Importance of soil NO emissions for the total atmospheric NO_X budget of Saxony, Germany. Atmos Environ 152:61–76. https://doi.org/10.1016/j.atmosenv.2016.12.022

Morgenstern Y (2015) Wasserhaushaltsmodellierung der Intensivmessstellen Level-II unter Buche und Fichte mit dem 1D-Standortsmodell LWF-Brook90. FVA Freiburg, Freiburg

Murray CA, Whitfield CJ, Watmough SA (2017) Uncertainty-based terrestrial critical loads of nutrient nitrogen in northern Saskatchewan, Canada. Boreal Environ Res 22:231–244

Nagel H-D, Gregor H-D (1999) Ökologische Belastungsgrenzen—Critical Loads & Levels. Ein internationales Konzept für die Luftreinhaltepolitik. Springer, Berlin

Oulehle F, Cosby BJ, Austnes K, Evans CD, Hruska J, Kopacek J, Moldan F, Wright RF (2015) Modelling inorganic nitrogen in runoff: seasonal dynamics at four European catchments as simulated by the MAGIC model. Sci Total Environ 536:1019–1028. https://doi.org/10.1016/j.scitotenv.2015.05.047

Palviainen M, Pumpanen J, Berninger F, Ritala K, Duan B, Heinonsalo J, Sun H, Koster E, Koster K (2017) Nitrogen balance along a northern boreal forest fire chronosequence. PLoS One 12(3): e0174720. https://doi.org/10.1371/journal.pone.0174720

Park S-U, Shim JM (2001) Estimation of critical loads of sulfur and nitrogen for the Korean ecosystem. Int J Soc Mater Eng Resour 10(2):121–129

Parton W, Silver WL, Burke IC, Grassens L, Harmon ME, Currie WS, King JY, Adair EC, Brandt LA, Hart SC, Fasth B (2007) Global-scale similarities in nitrogen release patterns during long-term decomposition. Science 315(5810):361–364

Posch M, de Vries W, Sverdrup HU (2015) Mass balance models to derive critical loads of nitrogen and acidity for terrestrial and aquatic ecosystems. In: de Vries W, Hettelingh JP, Posch M (eds) Critical loads and dynamic risk assessments: nitrogen, acidity and metals in terrestrial and aquatic ecosystems. Environmental pollution, vol 25. Springer, Dordrecht, pp 171–205. https://doi.org/10.1007/978-94-017-9508-1_6

Preston CM, Nault JR, Trofymow JA, Smyth C, Cidet Working Group (2009) Chemical changes during 6 years of decomposition of 11 litters in some Canadian forest sites. Part 1. Elemental composition, tannins, phenolics, and proximate fractions. Ecosystems 12(7):1053–1077. https://doi.org/10.1007/s10021-009-9266-0

Pretzsch H (2009) Forest dynamics, growth and yield: from measurement to model. Springer, Berlin

Pretzsch H (2016) Ertragstafel-Korrekturfaktoren für Umwelt- und Mischungseffekte. AFZ-Der Wald 14:47–50

Rademacher P, Khanna PK, Eichhorn J, Guericke M (2009) Tree growth, biomass, and elements in tree components of three beech sites. In: Brumme R, Khanna PK (eds) Functioning and management of European beech ecosystems. Ecological studies, vol 208. Springer, Berlin, pp 105–136

Reinds GJ, De Vries W (2010) Uncertainties in critical loads and target loads of sulphur and nitrogen for European forests: analysis and quantification. Sci Total Environ 408(8):1960–1970. https://doi.org/10.1016/j.scitotenv.2009.12.001

Reinds GJ, Posch M, de Vries W (2001) A semi-empirical dynamic soil acidification model for use in spatially explicit integrated assessment models for Europe. Alterra Report. Alterra Green World Research, Wageningen, Netherlands

Riek W, Russ A, Hannemann J, Kallweit R (2016) Bodenzustand und Baumernährung: Kennwerte aus BZE und Level II-Programm. In: 30 Jahre forstliches Umweltmonitoring in Brandenburg. Eberswalder Forstliche Schriftenreihe, vol 63. Ministerium für Ländliche Entwicklung Umwelt und Landwirtschaft des Landes Brandenburg, Eberswalde, Germany, pp 40–60

Rihm B, Achermann B (2016) Critical loads of nitrogen and their exceedances. Swiss contribution to the effects-oriented work under the Convention on Long-range Transboundary Air Pollution (UNECE). Environmental Studies. Federal Office for the Environment, Bern, Switzerland

Rodriguez A, Duran J, Covelo F, Fernandez-Palacios JM, Gallardo A (2011) Spatial pattern and variability in soil N and P availability under the influence of two dominant species in a pine forest. Plant Soil 345(1–2):211–221. https://doi.org/10.1007/s11104-011-0772-4

Rohmann U, Sontheimer H (1985) Nitrat im Grundwasser. Ursachen - Bedeutung - Lösungswege. DVGW-Forschungsstelle am Engler-Bunte-Institut der Universität Karlsruhe, Karlsruhe, Germany

Russ A, Riek W, Martin J (2017) Forstliches Umweltmonitoring Mecklenburg-Vorpommern—Ergebnisse der Untersuchungen auf den Intensivmonitoringflächen (Level II). Mitteilungen aus dem Forstlichen Versuchswesen Mecklenburg-Vorpommern. Landesforst Mecklenburg-Vorpommern (AöR), Schwerin, Germany

Sabatini FM, Zanini M, Dowgiallo G, Burrascano S (2015) Multiscale heterogeneity of topsoil properties in southern European old-growth forests. Eur J Forest Res 134(5):911–925. https://doi.org/10.1007/s10342-015-0899-6

Schaap M, Wichink Kruit R, Hendriks C, Kranenburg R, Segers A, Builtjes P, Banzhaf S, Scheuschner T (2015) Atmospheric deposition to German natural and semi-natural ecosystems during 2009. Interim Report to UFOPLAN Project 3712 63 240-1—1st PINETI Federal Environment Agency (UBA), Dessau-Roßlau, Germany

Schaap M, Banzhaf S, Scheuschner T, Geupel M, Hendriks C, Kranenburg R, Nagel H-D, Segers AJ, von Schlutow A, Wichink Kruit R, Builtjes PJH (2017) Atmospheric nitrogen deposition to terrestrial ecosystems across Germany, Biogeosciences Discuss. Dessau-Roßlau, Germany. https://doi.org/10.5194/bg-2017-491

Schlotter D, Schack-Kirchner H, Hildebrand EE, von Wilpert K (2012) Equivalence or complementarity of soil-solution extraction methods. J Plant Nutr Soil Sci 175(2):236–244. https://doi.org/10.1002/jpln.201000399

Schober R (1995) Ertragstafeln wichtiger Baumarten bei verschiedener Durchforstung. 4th edn. Sauerländer, Frankfurt a. M.

Simpson D, Aas W, Bartnicki J, Berge H, Bleeker A, Cuvelier K, Dentener F, Dore T, Erisman JW, Fagerli H (2011) Atmospheric transport and deposition of reactive nitrogen in Europe

Spiecker H, Mielikäinen K, Köhl M, Skovsgaard JP (1996) Growth trends in European forests: studies from 12 countries. Springer, Berlin

Steinert M, Feger KH (2010) Modellierung und Interpretation des Wasser- und Stoffhaushaltes für die Thüringer Hauptmessstationen Großer Eisenberg, Holzland, Possen und Lehesten. Dresden Technical University, Department for Soil Science and Site Ecology, Dresden, Germany

Strengbom J, Nordin A (2008) Commercial forest fertilization causes long-term residual effects in ground vegetation of boreal forests. Forest Ecol Manag 256(12):2175–2181. https://doi.org/10.1016/j.foreco.2008.08.009

Sucker C, von Wilpert K, Puhlmann H (2011) Acidification reversal in low mountain range streams of Germany. Environ Monit Assess 174(1–4):65–89. https://doi.org/10.1007/s10661-010-1758-z

Talkner U, Meiwes KJ, Potocic N, Seletkovic I, Cools N, De Vos B, Rautio P (2015) Phosphorus nutrition of beech (*Fagus sylvatica* L.) is decreasing in Europe. Ann Forest Sci 72(7):919–928. https://doi.org/10.1007/s13595-015-0459-8

Tarrasón L, Nyiri Á (2008) Transboundary acidification, eutrophication and ground level ozone in Europe in 2006. EMEP status report. The Meteorological Institute, Oslo, Norway

Thiele JC, Nuske RS, Ahrends B, Panferov O, Albert M, Staupendahl K, Junghans U, Jansen M, Saborowski J (2017) Climate change impact assessment—a simulation experiment with Norway spruce for a forest district in Central Europe. Ecol Model 346(2017):30–47

Tietema A, Bouten W, Wartenbergh PE (1991) Nitrous oxide dynamics in an oak beech forest ecosystem in the Netherlands. Forest Ecol Manag 44(1):53–61. https://doi.org/10.1016/0378-1127(91)90197-4

Ulrich B (1994) Nutrient and acid-base budget of central european forest ecosystems. In: Godbold D, Hüttermann A (eds) Effects of acid rain on forest processes. Wiley-Liss, New York, pp 1–50

UNECE ICP Forests Programme Co-ordinating Centre (2016) Manual on methods and criteria for harmonized sampling, assessment, monitoring and analysis of the effects of air pollution on forests. Eberswalde, Germany

Vesterdal L, Schmidt IK, Callesen I, Nilsson LO, Gundersen P (2008) Carbon and nitrogen in forest floor and mineral soil under six common European tree species. Forest Ecol Manag 255 (1):35–48. https://doi.org/10.1016/j.foreco.2007.08.015

Vitousek PM, Aber JD, Howarth RW, Likens GE, Matson PA, Schindler DW, Schlesinger WH, Tilman D (1997) Human alteration of the global nitrogen cycle: sources and consequences. Ecol Appl 7(3):737–750. https://doi.org/10.2307/2269431

Waldner P, Marchetto A, Thimonier A, Schmitt M, Rogora M, Granke O, Mues V, Hansen K, Karlsson GP, Zlindra D, Clarke N, Verstraeten A, Lazdins A, Schimming C, Iacoban C, Lindroos AJ, Vanguelova E, Benham S, Meesenburg H, Nicolas M, Kowalska A, Apuhtin V, Napa U, Lachmanova Z, Kristoefel F, Bleeker A, Ingerslev M, Vesterdal L, Molina J, Fischer U, Seidling W, Jonard M, O'Dea P, Johnson J, Fischer R, Lorenz M (2014) Detection of temporal trends in atmospheric deposition of inorganic nitrogen and sulphate to forests in Europe. Atmos Environ 95:363–374. https://doi.org/10.1016/j.atmosenv.2014.06.054

Ward MH, deKok TM, Levallois P, Brender J, Gulis G, Nolan BT, VanDerslice J (2005) Workgroup report: drinking-water nitrate and health - recent findings and research needs. Environ Health Perspect 113(11):1607–1614. https://doi.org/10.1289/ehp.8043

Wasserstein RL, Lazar N (2016) The ASA's statement on p-values: context, process, and purpose. Am Statistician 70(2):129–133

Chapter 6
Carbon Stocks and Carbon Stock Changes in German Forest Soils

Erik Grüneberg, Ingo Schöning, Winfried Riek, Daniel Ziche, and Jan Evers

6.1 Introduction

Forests play an important role in the global carbon (C) cycle. Through photosynthesis plants convert atmospheric carbon dioxide (CO_2) into plant biomass. About half of the C stored in the plant biomass is subsequently released into the atmosphere by respiration in plants. Most of the remaining C enters the soil, e.g., as dead leaf or root biomass. The amount of soil C is determined by the net balance of the rate of organic C input and its decomposition. Soils are the largest reservoir of organic C in the active C cycle of terrestrial ecosystems.

Globally, they contain about 1.5–2.0 × 10^{15} tons of organic C to a depth of 1 m (Amundson 2001). Forest soils store one third of the global organic C. They are a larger reservoir for C than plants and atmosphere combined (Schlesinger 1991).

With regard to climate protection policy, the sequestration of atmospheric C in soils as stable organic matter is discussed as a potential contribution to mitigate atmospheric CO_2 concentrations (Janzen 2004). A proportion of the C entering the soil is mineralized by microorganisms resulting in the release of CO_2 and/or

E. Grüneberg (✉) · D. Ziche
Thünen Institute of Forest Ecosystems, Eberswalde, Germany
e-mail: erik.grueneberg@thuenen.de; daniel.ziche@thuenen.de

I. Schöning
Max-Planck-Institute for Biogeochemistry, Jena, Germany
e-mail: Ingo.Schoening@bgc-jena.mpg.de

W. Riek
University for Sustainable Development and Eberswalde Forestry State Center of Excellence, Eberswalde, Germany
e-mail: winfried.riek@hnee.de

J. Evers
Northwest German Forest Research Institute, Göttingen, Germany
e-mail: jan.evers@nw-fva.de

N. Wellbrock, A. Bolte (eds.), *Status and Dynamics of Forests in Germany*,
Ecological Studies 237, https://doi.org/10.1007/978-3-030-15734-0_6

methane. Carbon can also be leached from soil as dissolved or particulate organic matter soil C not mineralized by soil microorganisms—ranging from young recently deposited plant material to very old organic constituents in the so-called stable organo-mineral associations—is stored in the soil as organic matter. The soil organic matter fulfills important soil functions as it serves as a nutrient exchanger and buffer for plants, improves the retention of water, and increases the buffering and exchange capacity of soils. Soil organic matter also provides energy for heterotrophic soil organisms as well as for the microflora.

An increase in C stocks is expected to be a result of environmental as well as human-induced factors (Tan et al. 2004; Liski et al. 2002; de Vries et al. 2009). The mineralization of organic matter is limited by temperature and water availability, and thus the amount and dynamic of soil C pools are substantially affected by climate (Davidson and Janssens 2006; Lorenz and Lal 2010). Currently, it is not yet fully resolved why one part of soil C is decomposed rapidly, while other parts are characterized by turnover times ranging from decades to millennia (Baisden et al. 2002). One possible reason is that the accumulation of soil C is controlled by the restricted availability of decomposing organisms and their enzymes (Jastrow et al. 2007). Furthermore, the lack of nutrients and energy sources may restrict the growth of microorganisms involved in decomposition processes.

The amount of C stored in the soil is controlled by the potential of soils to stabilize organic C. Various studies have shown that less CO_2 is released from soils where large amounts of C are associated with the mineral phase (Six et al. 2002). This is generally explained by an interaction between reactive mineral surfaces and organic C by mechanisms such as ligand exchange, polyvalent cation bridges, and hydrophobic interactions (Vogel et al. 2015; von Lutzow et al. 2006). Therefore, the size of the clay and silt fraction is important for the storage of soil C (Tan et al. 2004). Moreover, chemical soil properties such as the pH value or the cation exchange capacity can affect the microbial turnover of organic matter (Ladegaard-Pedersen et al. 2005; Leuschner et al. 2013).

Direct and indirect anthropogenic activities influence the C dynamic in forest ecosystems. Various studies showed an impact of tree species selection and accompanied forest stand composition on C stocks (Ladegaard-Pedersen et al. 2005; Prietzel and Bachmann 2012). Forest management practices such as drainage, thinning, timber harvesting, and liming can also affect soil C dynamics (Johnson et al. 2002; Nave et al. 2010).

Investigations of liming induced effects on C stock changes have revealed contradictory results. On the one hand, it has been shown that after the application of lime, the decomposition of the organic layer increased accompanied by an accumulation of C in the mineral soil (Andersson and Nilsson 2001; Evers et al. 2008). On the other hand, various studies have demonstrated that a long-lasting decrease ranging from weeks up to 1 year followed the initial increase in soil respiration after liming (Illmer and Schinner 1991; Melvin et al. 2013).

In addition, various studies suggested an impact of nitrogen (N) deposition on C sequestration in forest soils, but the results are highly uncertain and vary by two orders of magnitude (de Vries et al. 2009; Janssens et al. 2010). The increase in N deposition on forests over a longer time period may reduce the decomposition of organic matter. Increased organic matter input in soil through enhanced aboveground biomass productivity or increased recalcitrance of N-enriched litter may lead to reduced long-term decomposition rates of organic matter (de Vries et al. 2009).

Results of the 1st National Forest Inventory (NFSI) were presented by Wolff and Riek (1996), who showed spatial patterns of C stocks and morphological humus forms and found correlations to bedrock, soil texture, and main soil types. They reported effects of a partial decoupling of material cycles on C dynamics due to long-lasting acid and N deposits during the last decades. Estimates of nationwide C pools and repeated soil C inventories or soil C monitoring are still rare, even though such information is essential to fulfill national commitments under Article 3.4 of the Kyoto Protocol and under the 2016 Paris climate agreement. With the repetition of the NFSI, Germany obtained a comprehensive data base to evaluate the status quo of soil organic matter and the C sequestration rate of forested soils over time.

6.2 Carbon Stocks in German Forest Soils

6.2.1 Carbon Stocks

Forest soils (the organic layer and the mineral soil to a maximum depth of 90 cm) contained on average 117.1 ± 1.7 Mg ha^{-1} of organic C. The median of the organic C stock was 105.0 Mg ha^{-1} indicating a right-skewness with a disproportionately large number of sites with high organic C stocks (Fig. 6.1). The 10% soils with the lowest organic C stocks contained less than 64.8 Mg ha^{-1}, while the 10% soils with the highest organic C stocks contained more than 177.4 Mg ha^{-1}. Highest organic C stocks were usually found in soils where the decomposition of organic matter was restricted due to oxygen deficiency caused by water saturation and/or by per humid climate. Soils that are limited in depth by continuous rocks, gravel, or stones contained the lowest organic C stocks. In respect to the amount of C stored in the whole soil, the organic layer contained 18% of C (Fig. 6.2a). Organic C stocks in the organic layer were highly variable with C stocks ranging from 0.2 to 190 Mg ha^{-1} and a coefficient of variation of 94%. The upper 30 cm of the mineral soil contained 59% of the organic C stored in the complete profile, while the depth increments 30–60 cm and 60–90 cm contained 18% and 9% of the total organic C, respectively. This gradient is similarly to observe N stocks (see Chap. 5). In contrast to averaged C stocks which decreased with depth, the relative variability of C stocks increased with depth with a coefficient of variation ranging from 50% in depth increment 0–5 cm to 169% in depth increment 60–90 cm. The organic C stocks in different depth increments of the mineral soil varied considerably ranging from 0.4 to 142.5 Mg ha^{-1} in depth increment 0–5 cm and from 0.6 to 590.9 Mg ha^{-1} in depth increment 10–30 cm.

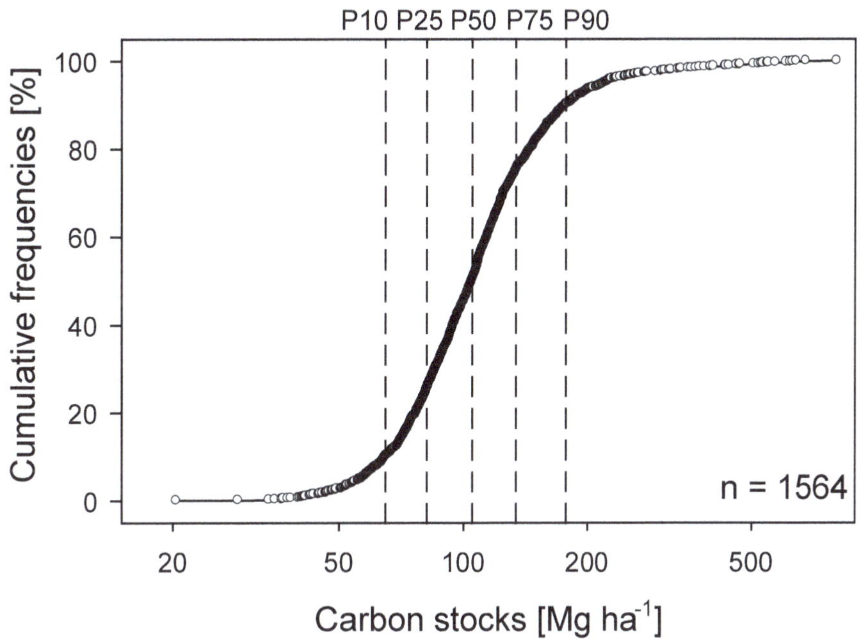

Fig. 6.1 Cumulative frequencies [in %] of organic stocks in German soils (organic layer and mineral soil down to a depth of 90 cm). 10, 25, 50, 75, and 90% quantiles are indicated (P10, P25, P50, P75, P90)

6.2.2 Organic Carbon Stock Changes in German Forest Soils

Estimates of organic C stock changes in the organic layer and mineral soil between the inventories are based on sites that were sampled at both inventories. Bogs and fens were not included in the analyses. The thickness of the humus body in wetland soils varies over time, and therefore, a reference level for a repeated measurement is nonexistent. Carbon stocks of the organic layer increased negligible by 0.06 ± 0.03 Mg ha^{-1} year^{-1} since the first inventory (Fig. 6.2b). In contrast, C stocks increased significantly in increments up to a depth of 30 cm with 0.25 ± 0.02 Mg ha^{-1} in 0–5 cm, 0.16 ± 0.01 Mg ha^{-1} in 5–10 cm, and 0.28 ± 0.03 Mg ha^{-1} in 10–30 cm ($p < 0.001$). Significant organic C stock changes in the soil below 30 cm could not be found. The averaged organic C stock in German forest soils down to a depth of 90 cm increased significantly by 0.75 ± 0.09 Mg ha^{-1} year^{-1} resulting in a total increase of 11.3 Mg ha^{-1} between the inventories (a time period of 15 years). In a previous analysis, the organic layer was identified as a stable C pool, while the mineral soil sequestered 0.41 Mg ha^{-1} year^{-1} (Grüneberg et al. 2014). The apparent tendency of an accumulation in the organic layer and the upper depth increment of the mineral soil are also reflected by N stocks (see Chap. 5). The pattern of decreasing N stocks with depth is mainly due to high differences in 30–60 cm which was not observed for C. Since the measured concentrations are much higher for C compared to N, they are less affected by potential analytical errors. On the European scale estimates of C stock changes, rates ranged from 0.1 to 1.0 Mg ha^{-1} year^{-1} (Luyssaert et al. 2010). Jonard et al. (2017) quantified changes in soil organic C stocks based on the data from two soil

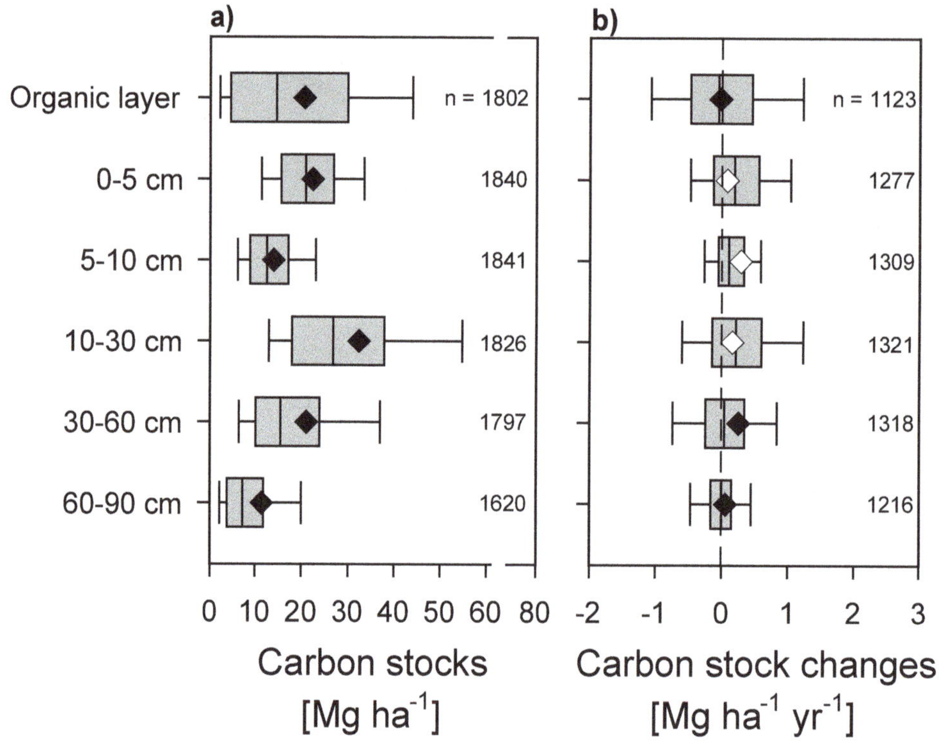

Fig. 6.2 (**a**) Carbon stocks and (**b**) carbon stock changes in the organic layer and depth increments of the mineral soil. Box plots on the carbon stocks refer to all plots considered in the 2nd National Forest Soil Inventory, while carbon stock changes refer to plots exclusively considered on both National Forest Soil Inventories. Box plots with white diamonds on carbon stock changes indicate significant differences based on a one-sample t- test with H0: 0 ($p < 0.05$). Organic soils were excluded from the analysis of carbon stock changes

surveys carried out in the French forest monitoring plots. The authors estimated for forest soils across France an accumulation of 0.35 Mg C ha^{-1} year^{-1} between 1993 and 2012.

For the first time, Grüneberg et al. (2014) estimated countrywide C stock changes of forest soils in Germany. Differences between these results and results presented here arise from different statistical approaches. In contrast to the results presented here which are based on paired samples, previous estimates of C stocks and C stock changes considered both paired and unpaired samples. They applied site-specific forest stand and soil information which was available from nationwide geographical datasets. The plots were grouped into classes, which correspond to the classification scheme of the geographical datasets. For the organic layer, inventory plots were grouped into classes of forest stand types (coniferous forest, mixed forest, and broadleaf forest) and substrate groups (dystrophic and eutrophic unconsolidated sediments as well as rocks with high and low base saturation). For the mineral soil, inventory plots were grouped according to soil and substrate information. Federal states were included in the classification to consider differences in the sampling density among German federal states. Detailed information about the methods can be found in Grüneberg et al. (2014).

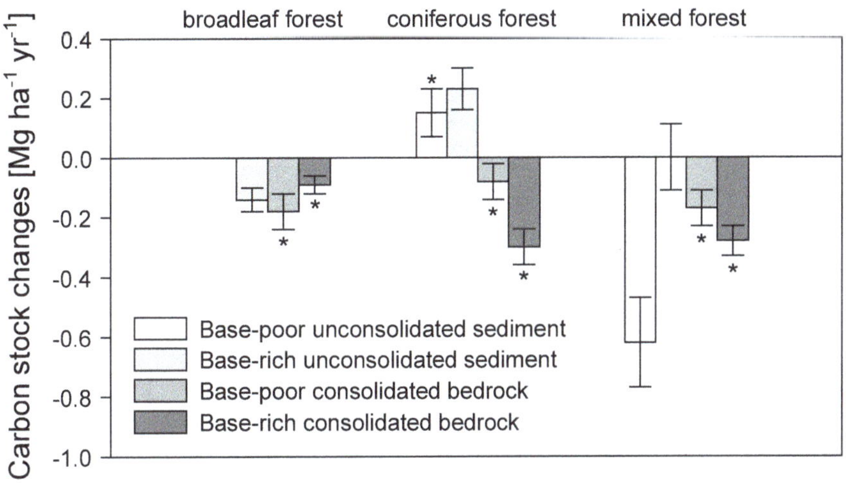

Fig. 6.3 Organic carbon stock changes in the organic layers of different substrate groups under various forest stands between the 1st and 2nd National Forest Inventory (Grüneberg et al. 2014). Asterisks indicate a 0.05 significance level, tested by a two-way ANOVA and the attended Tukey HSD test for each substrate group individually

The C stock changes of the organic layer ranged from -0.62 ± 0.15 Mg ha^{-1} year^{-1} under broadleaf forest to 0.23 ± 0.07 Mg ha^{-1} year^{-1} under coniferous forest, both in soils derived from eutrophic unconsolidated sediments (Fig. 6.3). The site-specific organic C stocks have increased since the first national inventory especially in the North German Lowland and in the Alpine foothills (Fig. 6.4). Both regions are characterized by dystrophic unconsolidated sediments and coniferous forests. In contrast, C was not sequestered but rather released under broadleaf and mixed forest. In the low mountain ranges consisting of solid rocks with low base saturation in the middle and southern parts of Germany, the organic layer is considered to be a C source as it released C. Carbon was also released from organic layers in the Alps as well as in the mountainous regions of southern Germany with solid rocks high in base saturation. The mountainous regions with high base saturation are not characterized by specific forest stands. However, organic layers in broadleaf and mixed forest lost more C than in coniferous forest.

The C stock changes in the upper 30 cm of the mineral soil ranged from -0.71 ± 0.28 Mg ha^{-1} year^{-1} in soils derived from redeposited material from lime stone, marlstone, and dolomite to 1.35 ± 0.19 Mg ha^{-1} year^{-1} in soils derived from eutrophic sandy deposits (Fig. 6.5). Similar to the organic layer, the mineral soil to a depth of 30 cm showed a strong increase in C stocks in mostly sandy soils of the North German Lowland (Fig. 6.6). The largest changes could be detected in the eastern part of the North German Lowland, where the initial C pool was low. Similar increases could be detected in the northern part of the Upper Rhine Valley and in the Rhine-Main Lowland. Moderate negative changes were found in soils derived from solid rocks of mountains and hills especially in the middle and south of Germany as well as in soils of the Alps. Furthermore, soils derived from limestone weathering or from weathered marlstone and claystone as, for example, in the Swabian and Franconian Alb and Upper Palatinate Jura showed slightly decreasing C stocks.

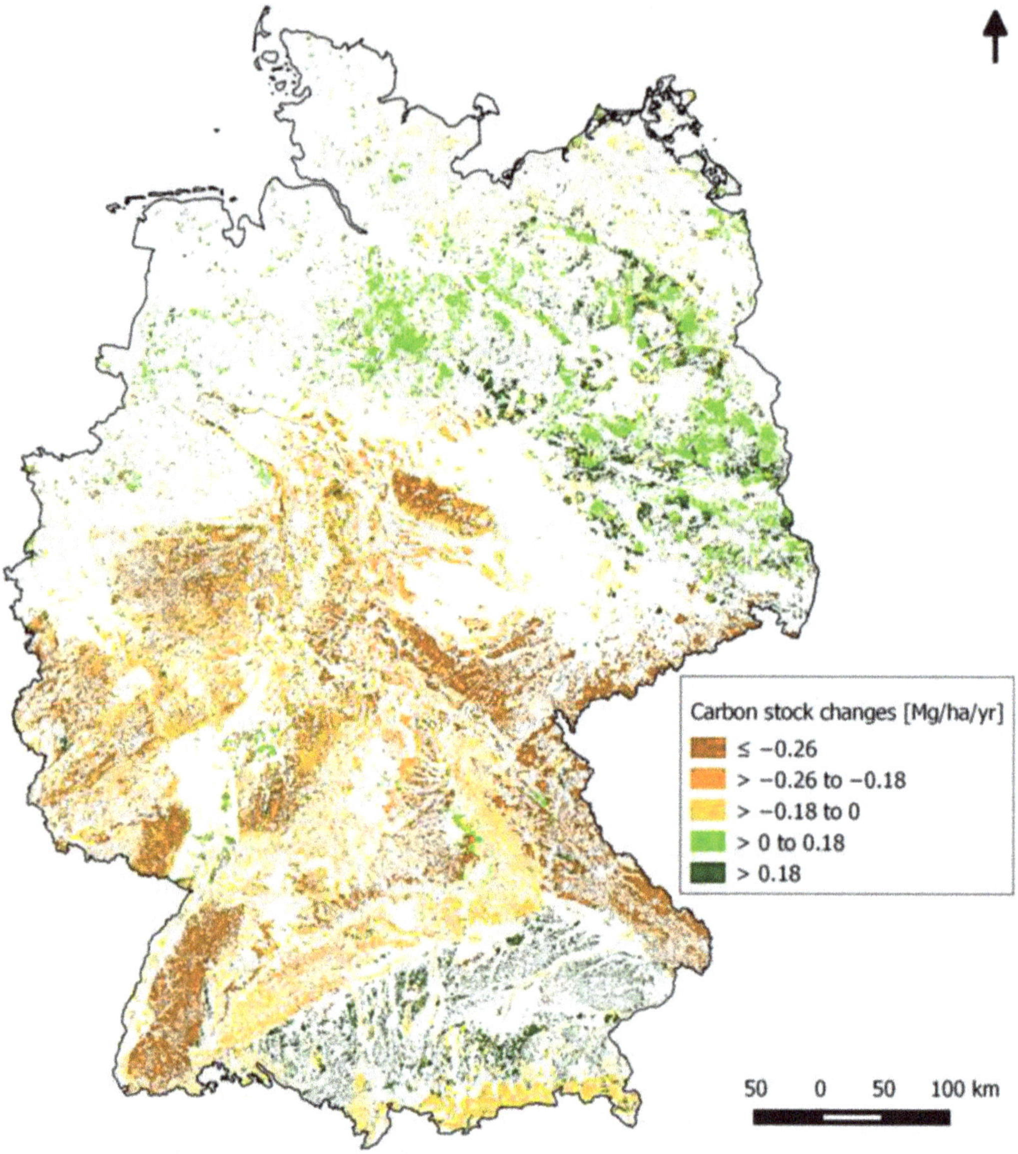

Fig. 6.4 Area-related organic carbon stock changes in the organic layer of German forest soils between the 1st and 2nd National Forest Soil Inventory

6.2.3 Effects of Forest Stands and Parent Material on Carbon Stocks

6.2.3.1 Forest Stands-Specific Carbon Stocks

Effects of forest stands on C stocks of the organic layer and mineral soil were investigated by analyses of variance (ANOVA) and a subsequent pairwise comparison between the groups by Bonferroni test ($p < 0.05$). Forest stands were distinguished in oak, beech, other broadleaf stands, pine, spruce, other coniferous stands, and mixed stands. In the organic layer, C stocks ranged from 7.8 ± 1.1 Mg ha^{-1}

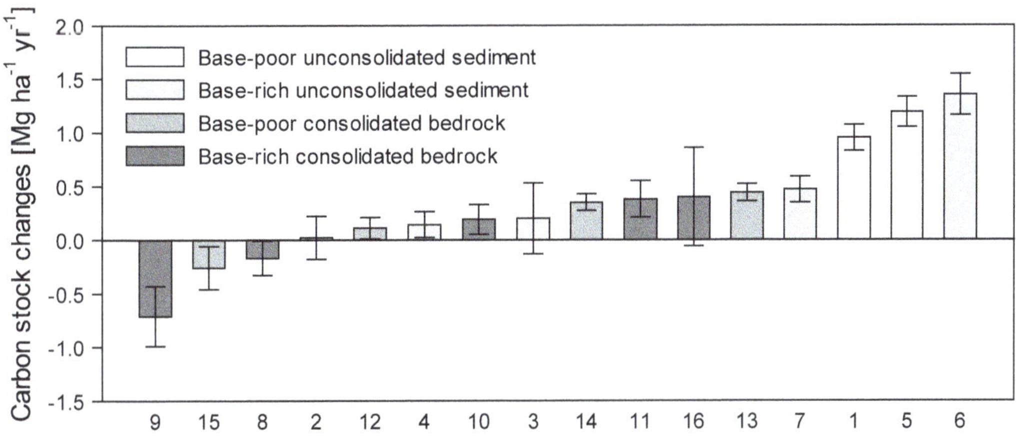

Fig. 6.5 Organic carbon stock changes in the mineral soil down to a depth of 30 cm under different soil and substrate groups between the 1st and 2nd National Forest Soil Inventory. Numbers on the *X*-axis represent soil parent material groups defined in Grüneberg et al. (2014)

under other broadleaf trees to 31.4 ± 0.9 Mg ha^{-1} under pine (Fig. 6.7a). They differed significantly among forest stands (ANOVA; $p < 0.001$). Apart from the comparison between beech with oak and between other coniferous trees with mixed stands, the C stocks between all other stands were significantly different. Broadleaf forest stands showed the lowest values, while C stocks of coniferous forest stands were an order of magnitude higher. This can be explained by an inhibited decomposition of needle litter and the associated enrichment of organic matter in the organic layer.

Broadleaf forest stands stored with 88.9 ± 5.0 Mg ha^{-1} in the upper 30 cm of the mineral soil the highest amount of organic C, while pine forest stands showed with 60.0 ± 1.9 Mg ha^{-1} the lowest stocks (Fig. 6.7b). The organic C stocks in the mineral soil to a depth of 30 cm were, though less pronounced, still significantly different between the forest stands (ANOVA; $p < 0.001$). Significantly higher C stocks were found under other broadleaf forest than under other forest stands, while pine forest stands stored significantly less C in this depth increment than other forest stands.

A significant part of the C is observed in 30–90 cm emphasizing the subsoil as a stable, long-term C pool. The amount of C stored in the subsoil was half the C stored in the 0–30 cm depth increment (Fig. 6.7c). The variation of C stocks among the groups in the subsoil (30–90 cm) was lower than in the depth increment above. Nevertheless, significant differences among the forest stands were evident (ANOVA; $p < 0.001$). The variation within the groups, however, was similar to the 0–30 cm depth increment. Broadleaf forest stands are prone to store more C in the subsoil than coniferous forest. Under pine forest stands, we detected with 24.8 ± 1.7 Mg ha^{-1} the lowest C stocks. Similar to the 0–30 cm depth increments, other broadleaf forest stands stored significantly higher C (42.6 ± 4.0 Mg ha^{-1}) in the subsoil than all other forest stands. At the same time, also the largest variation was found under other broadleaf forest stands probably due to a high variability of site conditions among the stands. However, the pattern that more C is stored under

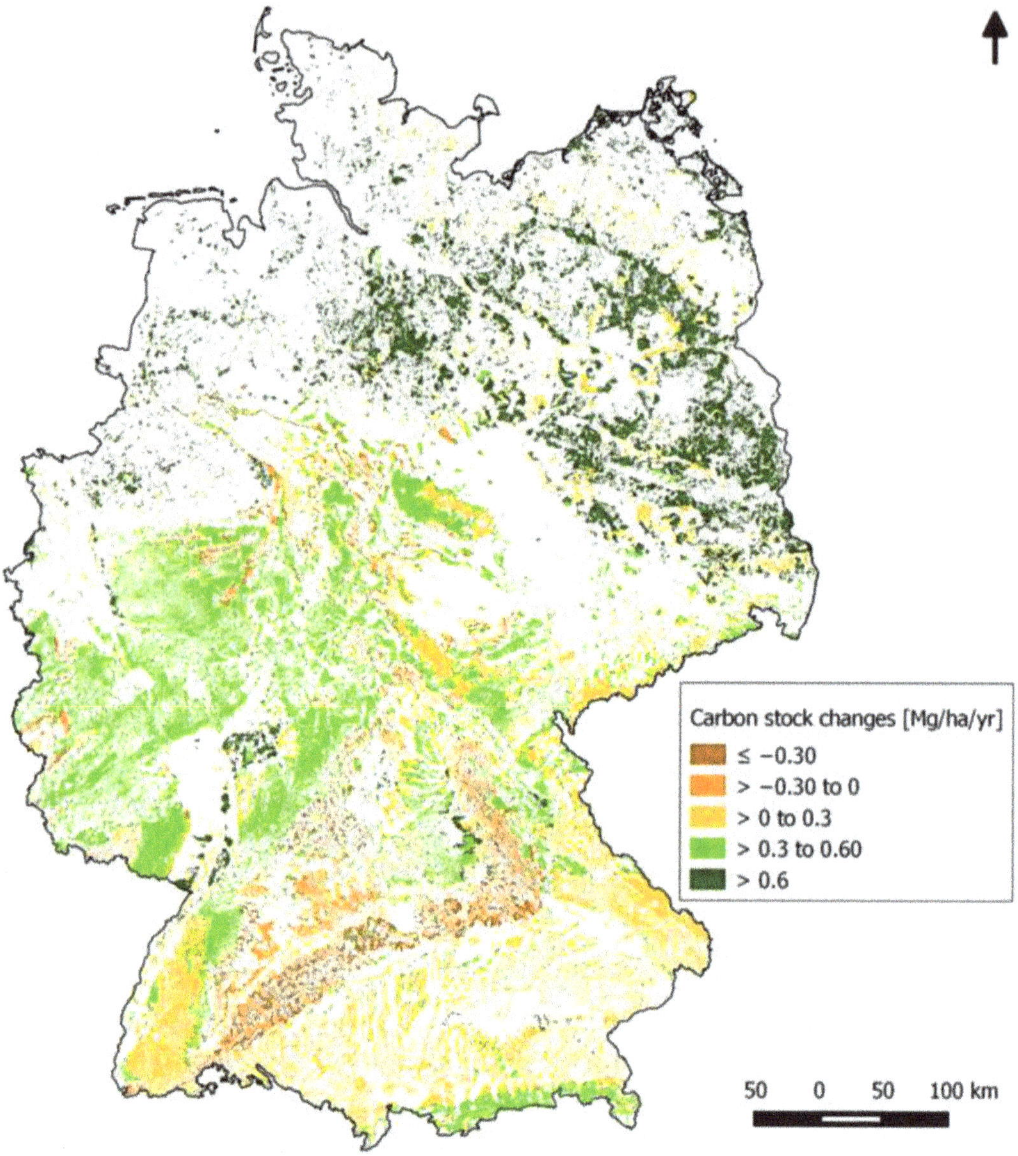

Fig. 6.6 Area-related organic carbon stock changes in the mineral soil down to a depth of 30 cm of German forest soils between the 1st and 2nd National Forest Soil Inventory

broadleaf compared to coniferous forest stands might be caused by regional and site-specific differences of soil properties.

6.2.3.2 Organic Carbon Stocks of Different Soil Parent Materials

The soil parent material groups determine to a large extend soil physical and chemical properties of a forest site that in turn affect the amount and turnover of soil organic matter. Differences of organic C stocks among soil parent material groups were analyzed by ANOVA and subsequent pairwise comparisons between

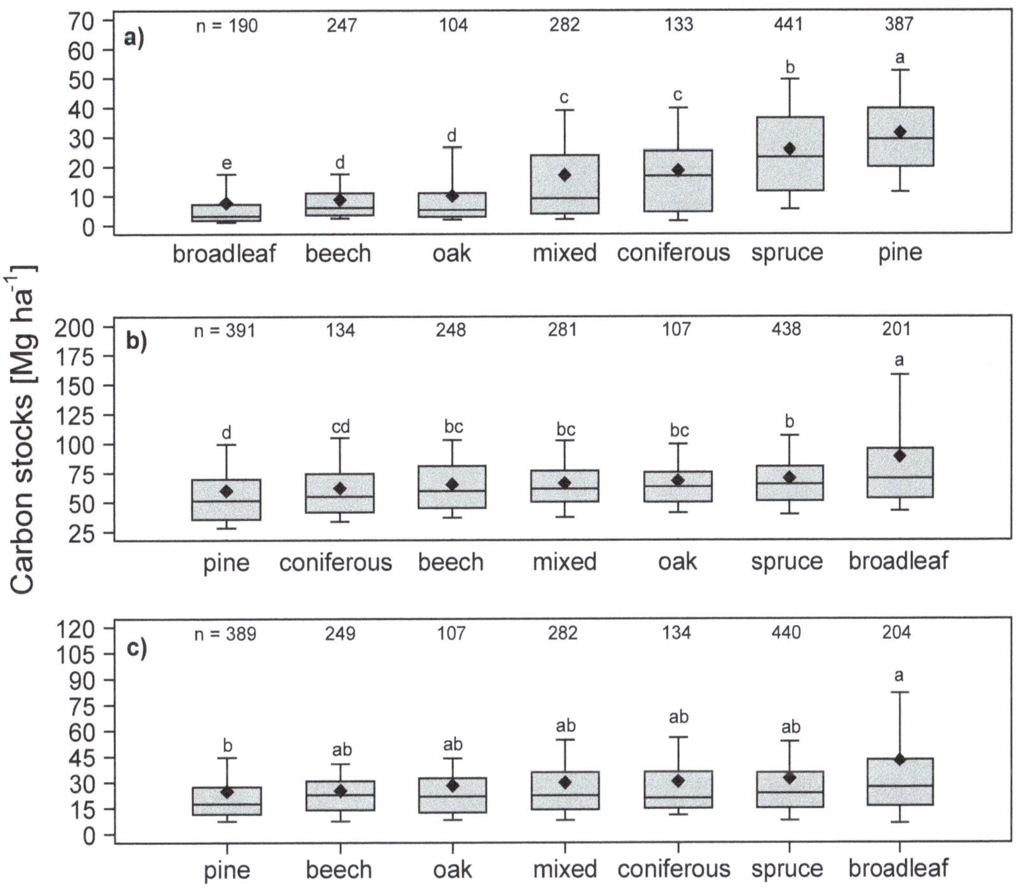

Fig. 6.7 Organic carbon stocks in (**a**) the organic layer, (**b**) the mineral soil down to a depth of 30 cm and (**c**) in the depth increment 30–90 cm under various forest types. Forest types with different letters are significantly different ($p < 0.05$). The line within a box represents the median and the diamond represents the mean. Forest types on the *X*-axis; for definition see Chap. 2

the groups by Bonferroni test ($p < 0.05$). Soils of bogs, fens, and other soils with high organic C concentrations (>15%) as well as soils from the Alps were not considered in the ANOVA. Soils of the Alps were excluded because of their low sample size, while organic soils were not studied due to their clearly distinctive development, properties, and reaction to environmental effects compared to mineral soils. In particular the thickness of the peat layer determines the amount of organic C stored in organic soils. Peat layers are characterized by notably high C stocks and a high spatial variability of organic C stocks. The classification into different soil parent material groups, however, could reduce the variability within the groups compared to the forest stand types because highly variable organic soils were analyzed separately as organic soils.

The results showed a relation of C stocks between the organic layer and the parent material. The averaged C stored of the organic layer ranged from 6.2 ± 0.5 Mg ha^{-1} in soils from weathered carbonate bedrock to 29.6 ± 1.1 Mg ha^{-1} in soils from base-poor unconsolidated sediment (Fig. 6.8a). Forest sites rich in nutrients and base cations are characterized by low organic C stocks in the organic layer due to favorable

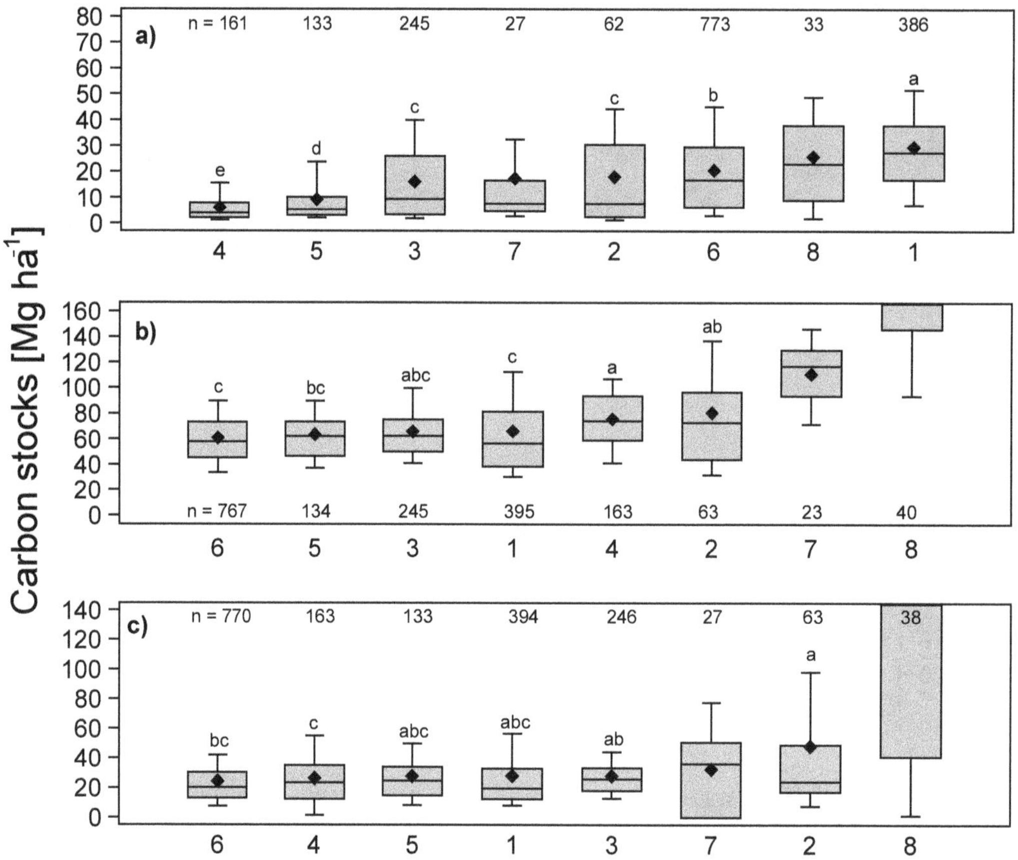

Fig. 6.8 Organic carbon stocks in (**a**) the organic layer, (**b**) the mineral soil down to a depth of 30 cm, and (**c**) the depth increment 30–90 cm in soils derived from various parent materials. Soil parent material groups with different letters are significantly different ($p < 0.05$). The line within a box represents the median, and the diamond represents the mean; numbers on the *X*-axis represent soil parent material groups with (1) soils from base-poor unconsolidated sediment, (2) soils of alluvial plains, (3) loamy soils of the lowland, (4) soils from weathered carbonate bedrock, (5) soils from basic-intermediate bedrock, (6) soils from base-poor hard bedrock, (7) soils from the Alps, and (8) organic soils

conditions for soil organisms to decompose organic matter. In contrast, the high organic C stocks in soils from base-poor unconsolidated sediment or in organic soils can be explained through reduced decomposition of organic matter due to low nutrition availability or anaerobic conditions. The Bonferroni test revealed significantly higher C stocks of the organic layer in soils from base-poor unconsolidated sediment than in all other groups. The organic layer of soils from base-poor bedrock stored significantly more C than the organic layer of soils from weathered carbonate bedrock, soil from basic-intermediate bedrock, as well as loamy soils of the lowland and soils in broad river valleys. Carbon stocks were also significantly higher in the organic layer of soils in broad river valleys and of loamy soils of the lowland compared to the organic layer of soils from weathered carbonate bedrock and of soils from basic to intermediate bedrock.

Organic layer C stocks were stronger influenced by forest stand types than by soil parent material, while C stocks of the mineral soil down to 30 cm depth were most affected by soil parent material. Soil C stocks were higher and characterized by a considerably larger variability than C stocks in the organic layer (Fig. 6.8b). Moreover, the magnitude of organic C stored in the mineral soil derived from different parent materials was contrary to the amount of C stored in the organic layer. Base cation-poor sites showed lower organic C stocks in the mineral soil than sites rich in base cations. The averaged values for the 0–30 cm depth increment ranged between 60.9 ± 0.9 t C ha^{-1} in soils from base-poor bedrock and 110.9 ± 5.7 t C ha^{-1} in soils from the Alps. The high organic C stocks in the soils from the Alps can be explained by a slow decomposition of organic matter due to effects of high altitude. The studied organic soils stored with 216.6 ± 19.0 Mg ha^{-1} much more C than mineral soils. Soils in broad river valleys and soils from weathered carbonate bedrock were significantly more enriched in C than soils from base-poor unconsolidated sediment and bedrock. Particularly soils in broad river valleys received organic matter by lateral transport during past flooding events.

Carbon stocks of the subsoil (depth increment 30–90 cm) hardly differed among the soil parent materials. Moreover, the variability within most of the groups was less pronounced compared to the 0–30 cm depth increment. The average organic C stocks ranged between 24.3 ± 0.8 Mg ha^{-1} in soils from base-poor bedrock and 48.2 ± 8.8 Mg ha^{-1} in soils in broad river valleys (Fig. 6.8c). Organic soils exhibited the highest organic C stocks in 30–90 cm (179.3 ± 24.8 Mg ha^{-1}). In contrast to the 0–30 cm depth increment, the C stocks in soils in broad river valleys were significantly higher than in soils from weathered carbonate bedrock. The differences between the 0–30 cm and 30–90 cm depth increments might be explained by higher stone contents in sites derived from carbonate rocks compared to sites derived from material transported by water as in broad river valleys. The organic C stocks of loamy soils of the lowland or of soils from basic-intermediate bedrock were not different from sites poor in cations although the mineralogical composition of those soils was expected to stabilize larger amounts of C over a long period of time.

6.2.3.3 Interactions Between Forest Stand Types and Soil Parent Material

One-dimensional considerations prevent separating interactions between forest stand types and soil parent material due to the low sample size and the high variability of any of the groups. Coniferous and broadleaf forest stands were separated insufficiently by the NFSI method because pure stands were separated in which at least 70% of the trees in the main canopy are of a single species. This means that pure forest stands can provide 30% of other tree species. To enhance the number of sample points for a sufficient statistical analysis in each class, we further distinguished between pure coniferous and broadleaf forest stands based on a proportion of 90% of single tree species in the main canopy. The following soil parent material groups were considered for the analyses: soils from base-poor consolidated bedrock,

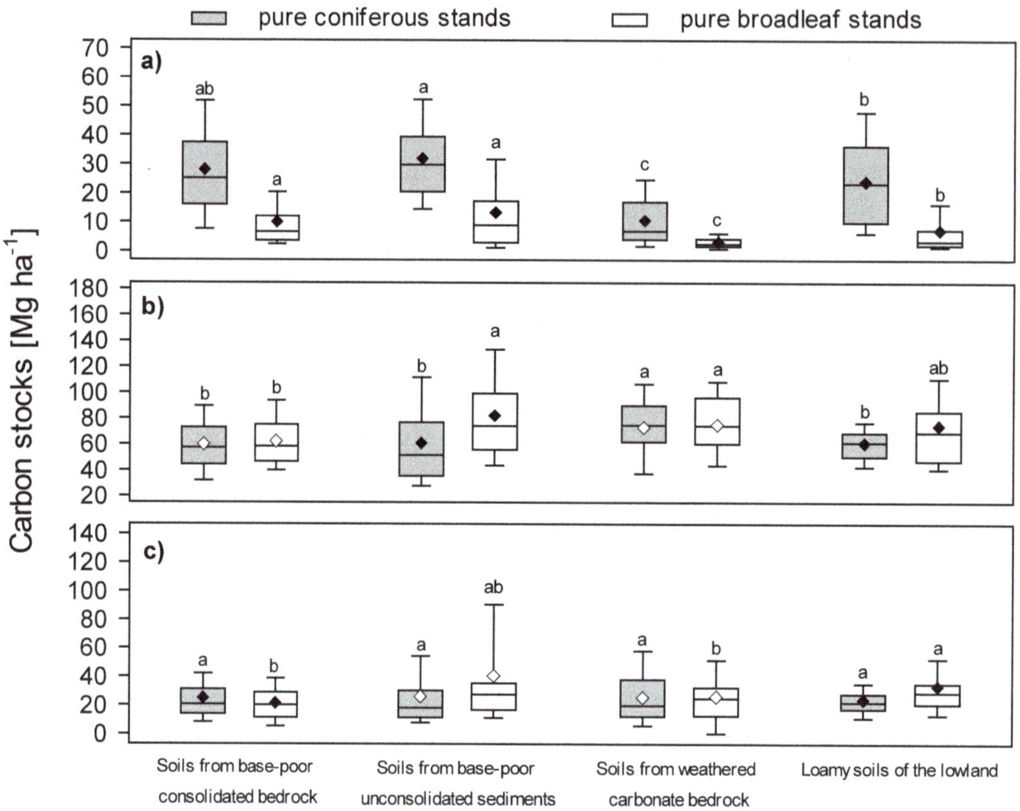

Fig. 6.9 Carbon stocks in the organic layer (**a**), as well as in the mineral soil down to a depth of 30 cm (**b**) and in depth increment 30–90 cm (**c**) under pure coniferous and broadleaf stands on soils derived from various parent materials. Black diamonds indicate significant differences between pure stands within the same soil parent material group based on a two-sample *t*-test ($p < 0.05$). Comparisons of the same pure stand among soil parent material groups are based on Bonferroni test ($p < 0.05$). Same letters indicate no significant differences ($p < 0.05$). Outliers are masked; a line within a box represents the median, and the diamond represents the mean

Table 6.1 Results of analysis of variance (ANOVA) testing effects of pure forest stands and soil parent material on carbon stocks in the organic layer

Variables	DF	MSE (type III)	*F* value	*p* value
Pure forest stand	1	260.965	371.43	<0.0001
Soil parent material	3	112.663	53.45	<0.0001
Pure forest stand*soil parent material	3	4.788	2.27	0.0786

DF degrees of freedom, *MSE* mean square error

soils from base-poor unconsolidated sediments, loamy soils of the lowland, and soils from weathered carbonate bedrock.

The organic C stocks of the organic layer ranged from 3.2 ± 0.3 Mg ha^{-1} under pure broadleaf forest in soils from weathered carbonate bedrock to 31.9 ± 1.04 Mg ha^{-1} under pure coniferous forest in soils from base-poor unconsolidated sediments (Fig. 6.9a). The ANOVA revealed a significant relation between organic C stocks in the organic layer of pure forest stands and soil parent material group (Table 6.1).

Table 6.2 Results of analysis of variance (ANOVA) testing effects of pure forest stands and soil parent material on carbon stocks in the mineral soil down to a depth of 30 cm

Variables	DF	MSE (type III)	*F* value	*p* value
Pure forest stand	1	2.729	14.93	0.0001
Soil parent material	3	4.171	7.61	<0.0001
Pure forest stand*soil parent material	3	2.323	4.23	0.0055

DF degrees of freedom, *MSE* mean square error

Similar to the classification into soil parent material groups without distinguishing forest types among forest stands, where coniferous forest showed the highest and broadleaf forest showed the lowest C stocks, here we found maximum organic C stocks under pure coniferous forest and minimum organic C stocks under pure broadleaf forest across all soil parent material groups. The differences between the pure forest stands were significant for all soil parent material groups. We found significant differences under pure coniferous forest exclusively between soils from base-poor consolidated bedrock and loamy soils of the lowland and soils from base-poor unconsolidated sediments, respectively. Taking broadleaf forest into account, significant differences were shown between most of the groups except for comparing soils from base-poor unconsolidated sediments with soils from base-poor consolidated bedrock. It could be shown that organic C stocks in organic layers are affected by soil parent material as we found higher organic C stocks on base cation-poor substrates compared to substrates low in base cations where in turn the organic C stocks were lower.

For average C stocks in the 0–30 cm depth increment, the variability within the groups was much larger than between the groups. The organic C stocks reached from 60.2 ± 1.3 t C ha^{-1} in soils from base-poor consolidated bedrock under coniferous forest to 84.9 ± 7.3 t C ha^{-1} in soils from base-poor unconsolidated sediments under broadleaf forest (Fig. 6.9b). In contrast to the organic layer, the ANOVA revealed significant higher C stocks under broadleaf than under coniferous forest (Table 6.2). Taking the soil parent material into account, soils from base-poor unconsolidated sediments and loamy soils of the lowland showed significant higher C stocks under broadleaf forest in comparison with coniferous forest. Coniferous and broadleaf forest stored significantly more C in soils from weathered carbonate bedrock than in soils from base-poor consolidated bedrock. Under coniferous forest C stocks were significantly higher in soils from weathered carbonate bedrock compared to soils from base-poor unconsolidated sediments and in loamy soils of the lowland. Furthermore, under broadleaf forest, the amount of C stored in soils from base-poor unconsolidated sediments was significantly higher than in soils from base-poor consolidated bedrock. The results of the ANOVA showed, however, significant interactions between pure stands and soil parent groups.

In the 30–90 cm depth increment (subsoil), we estimated averaged C stocks ranging from 22.4 ± 1.2 Mg ha^{-1} in soils from base-poor consolidated bedrock to 37.8 ± 6.5 Mg ha^{-1} in soils from base-poor unconsolidated sediments (Fig. 6.9c). Both the highest and lowest C stocks were found under broadleaf forest regardless of the

Table 6.3 Results of analysis of variance (ANOVA) testing effects of pure forest stands and soil parent material on organic carbon stocks in the mineral soil depth increment 30–90 cm

Variables	DF	MSE (type III)	*F* value	*p* value
Pure forest stand	1	4.524	8.67	0.0033
Soil parent material	3	9.521	6.08	0.0004
Pure forest stand*soil parent material	3	11.420	7.30	<0.0001

DF degrees of freedom, *MSE* mean square error

soil parent material. The C stock variability in the 30–90 cm depth increment is similar to that in the 0–30 cm depth increment though the C stored in the subsoil was half the C stored in the mineral soil down to 30 cm depth. The ANOVA revealed a significant relation to the forest type for the 30–90 cm depth increment (Table 6.3). The C stored in loamy soils from the lowland was significantly higher under broadleaf forest compared to coniferous forest. In contrast, soils from base-poor consolidated bedrock showed under coniferous forest higher C stocks than under broadleaf forest. Nevertheless, C stocks under broadleaf forest in the subsoil were prone to be affected by soil parent material. This can be proved by loamy soils of the lowlands which stored significant more C than soils from base-poor consolidated bedrock or soils from weathered carbonate bedrock. In contrast to broadleaf forest, the C stocks among the soil parent material groups were not significantly different. Similar to the 0–30 cm depth increment, the ANOVA revealed significant interactions between forest stands and soil parent material groups.

6.3 Effects of Natural and Anthropogenic Environmental Factors on Carbon Stocks in Forest Soils

We used structural equation modelling (SEM) to analyze direct and indirect factors affecting organic C stocks and organic C stock changes. Structural equation modelling was performed with R 3.2.2 (R Core Team 2015) by means of the work package lavaan (Rosseel 2012). A hypothetical model was created initially before pre-postulated cause-effect relationships could be verified by the data of the NFSI. Structural equation models can be used to analyze simultaneously dependent and independent variables in order to study more complex relations among variables. Independent variables are denoted as exogenous, and dependent variables are denoted as endogenous. Results of the SEM can be represented graphically by path analyses where the strength of the causal relationship is expressed by path coefficients. A direct effect revealed by the SEM and showed by the path analyses is presented with an aligned arrow, while covariance is displayed with a curved double arrow. It is supposed that on national scale the amount of C stored in the organic layer and mineral soil is mainly affected by climate, plant biomass, and related litter inputs as well as physical and chemical soil properties. Further effects can arise from forest management practices and deposition of N. For the modelled effects on

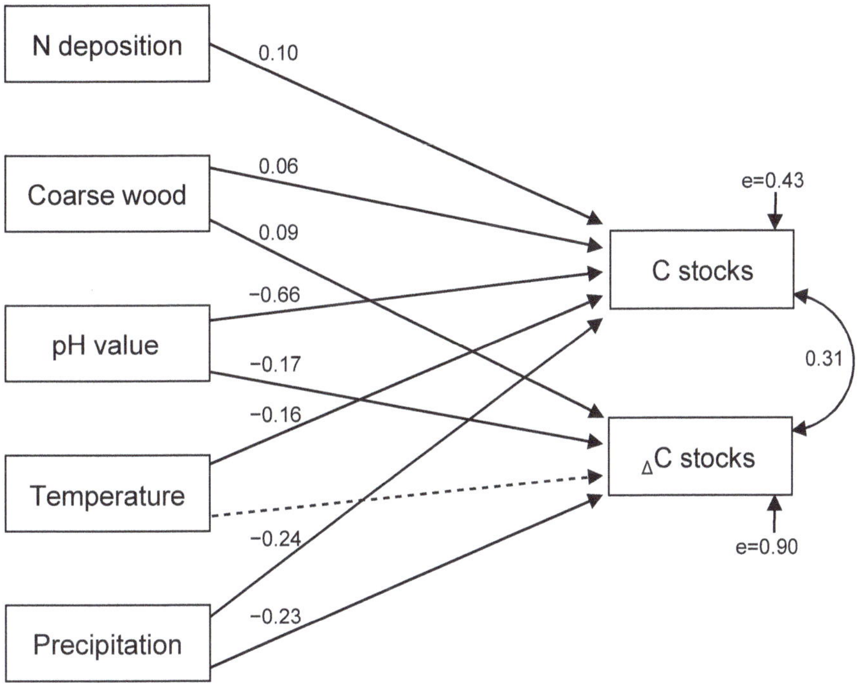

Fig. 6.10 Results of the structural equation modelling with $n = 1114$, $P(\chi 2) = 0.831$, 11 degrees of freedom, and a RMSEA (root mean square error of approximation) of 0 (90% confidence interval, [0; 0.047]) for the organic layer with exogenous (coarse wood = coarse wood volume, pH value [1 M KCl], temperature = mean annual temperature, precipitation = mean annual precipitation) and endogenous (C stocks = organic carbon stocks, ΔC stocks = annual organic carbon stock changes between NFSI I and NFSI II) variables in boxes and path coefficients with significant (continuous arrow) and not significant (dashed arrow) assignments as well as error variances (e)

organic layer C, we considered the averaged mean temperature and the annual precipitation, the coarse wood volume per hectare, the averaged mean N deposition, as well as the pH value (1 M KCl) as variables describing site properties. The more complex model for the mineral soil down to 30 cm depth additionally included the clay concentrations as an exogenous variable and the cation exchange capacity and the C concentration as endogenous variables. It should be considered here that the effective cation exchange capacity (CEC) is partially determined by the C concentration.

The results of the SEM for the organic layer showed that the magnitude of C stocks was strongest correlated to the pH value (−0.66), while relations to the climate variables temperature (−0.16) and precipitation (−0.24) were less distinct, and relations to the N deposition (0.10) as well as to the coarse wood volume (0.06) were weak (Fig. 6.10). Carbon stock changes in the organic layer were affected by coarse wood volume (0.09), pH value (−0.17), and precipitation (−0.23). The model for the mineral soil revealed weak correlations between organic C stocks and the exogenous variables N deposition (0.18) and pH value (0.06) (Fig. 6.10). In contrast, organic C stocks were strongly affected by the endogenous variable organic C concentration which in turn was largely determined by the clay concentrations (0.48). The C concentration was not considered in the model for the organic layer

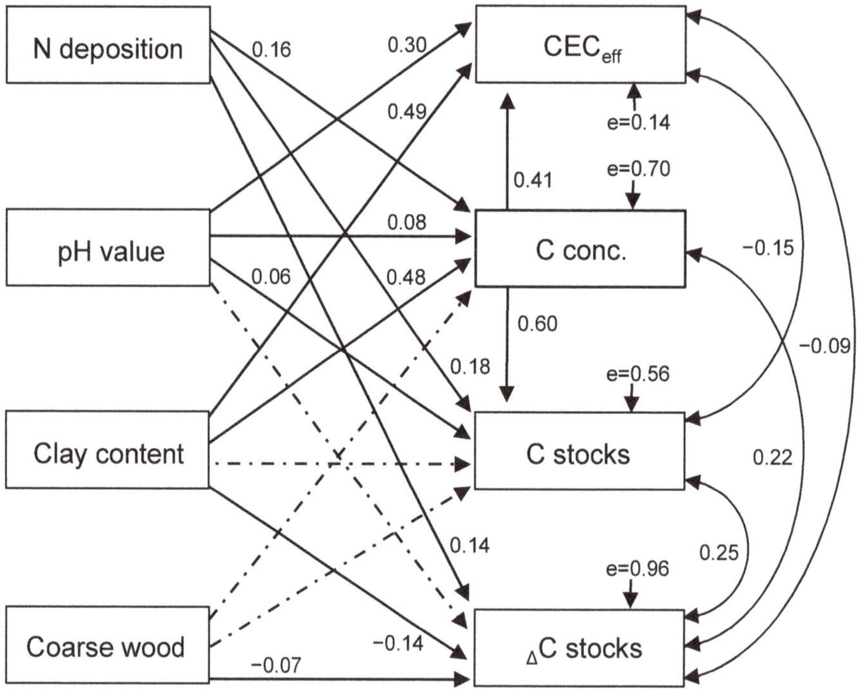

Fig. 6.11 Results of the structural equation modelling with $n = 886$, $P(\chi2) = 0.330$, 22 degrees of freedom, and a RMSEA (root mean square error of approximation) of 0.011 (90% confidence interval, [0; 0.069]) for the mineral soil down to a depth of 30 cm with exogenous (coarse wood = coarse wood volume, clay concentrations, pH value [1 M KCl], N deposition = mean annual N deposition) and endogenous (CEC_{eff} = effective cation exchange capacity, C conc. = organic carbon concentrations, C stocks = organic carbon stocks, $_{\Delta}$C stocks = annual organic carbon stock changes between NFSI I and NFSI II) variables in boxes and path coefficients with significant (continuous arrow) and not significant (dashed arrow) assignments as well as error variances (e)

because the amount of C stored in the organic layer is mainly controlled by the thickness and mass of the organic layers and not—as in the mineral soil—controlled by the C concentration. Additionally to direct effects on organic C stocks, coefficients revealed a path of N deposition (0.16) and pH value (0.08) on C concentration. Indirect effects on organic C stocks—that were weaker than direct effects—operating through C concentration were therefore caused by N deposition (0.10) and pH value (0.05). The clay concentrations indirectly affected (0.32) C stocks. The indirect effect was much larger than the direct effect, which was not significant disguised the negative relation of the direct effect between clay concentrations and C stocks.

The SEM revealed that the C stock changes of the mineral soil depend on N deposition (0.14), clay concentrations (−0.14), and coarse wood volume (−0.07) (Fig. 6.11). Additionally, a covariance (0.25) was shown between C stocks and C stock changes. Taking this covariance into account, N deposition (0.05) operated through C stocks on C stock changes. Further relations with C stock changes resulted from indirect effects of the exogenous variables N deposition (0.04), pH value (0.02), and clay concentrations (0.11) operating through the C concentrations and the respective covariance. Furthermore, clay concentrations (0.05) were also related with C stock changes through the covariance of the effective cation exchange capacity. All revealed effects operating through the covariance are less pronounced than the revealed direct effects. In both models the magnitudes of the sequestered C strongly depend on the amount of C stored in the soil. Various studies demonstrated negative relationships between initial soil C stocks and C stock changes which were explained by a change in the forest management (Goidts and van Wesemael 2007; Riley and Bakkegard 2006) or by changed environmental conditions (Bellamy et al. 2005; Saby et al. 2008). Repeated measurements on the same subject or unit of observation can lead to a statistical phenomenon called "regression to the mean" (Barnett et al. 2005). This statistical effect has been debated causing artifact losses from initially C-rich soil (Lark et al. 2006) and C gains in soils with initially low C stocks (Callesen et al. 2015). The revealed covariance between C stocks and C stock changes indicated the effect of the regression to the mean.

6.4 Effects of Natural Environmental Factors

The results of the studied soil parent material groups as well as the results of the SEM showed that site conditions are important factors affecting the storage of soil organic C. Base cation-poor sites showed for the organic layer highest C stocks, while base cation-rich sites stored low amounts of C (Fig. 6.12). Effects of soil parent material on the amount of stored C in the organic layer and mineral soil have been shown by various studies. It was demonstrated that the decomposition of litter was higher on sites rich in base cations or that the C accumulation of the organic layer was caused by reduced decomposition of the litter on sites low in base cations (Ladegaard-Pedersen et al. 2005; Vesterdal et al. 2008). In contrast to the organic layer, the magnitude of C stocks in mineral soils was higher on base cation-rich sites. Similar results were presented by other studies (Davis et al. 2004; Richardson and Stolt 2013; Conforti et al. 2016), while Bavarian forest soils showed no relation between soil parent material and C stocks in the mineral soil (Wiesmeier et al. 2013). The various groups of soil parent material are characterized by differing soil texture. It is well-known that clayey soils store more C than sandy soils. In the Good Practice Guidance for Land Use, Land-Use Change and Forestry (IPCC 2003), the texture is considered by default values for C stocks down to a depth of 30 cm for a warm temperate, moist climate. Sandy soils include all soils having >70% sand and <8% clay and are indicated to store 34 t C ha^{-1}. Soils with high activity clay minerals are lightly to moderately weathered soils, which are dominated by 2:1 silicate clay minerals. High activity soils are assumed to store 88 t C ha^{-1} (IPCC 2003). Texture-related differences in organic C stocks can be explained by a decrease in decomposing from sand to silt to clay-sized complexes. This is due to the positive relation between C concentrations and clay content, which may be attributed to the stabilization of organic matter by the formation of stable complexes with clay minerals (Six et al. 2002). The specific surface area of soils increases with decreasing

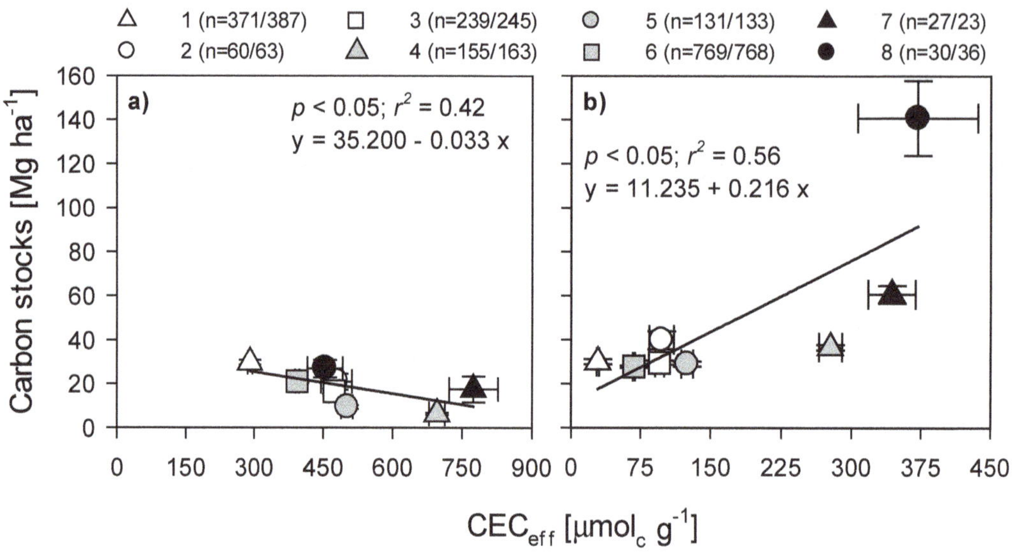

Fig. 6.12 Relation between the effective cation exchange capacity and carbon stocks (**a**) in the organic layer and (**b**) in the 10–30 cm depth increment of the mineral soil. The linear regression was performed between the averaged effective cation exchange capacity and carbon stocks of various soil parent material groups; the numbers in the legend represent soil parent material groups with (1) soils from base-poor unconsolidated sediment, (2) soils of alluvial plains, (3) loamy soils of the lowland, (4) soils from weathered carbonate bedrock, (5) soils from basic-intermediate bedrock, (6) soils from base-poor hard bedrock, (7) soils from the Alps, and (8) organic soils

particle size resulting in an increased importance of interaction between the organic matter and the mineral phase. Furthermore, an increased stability of soil organic matter might also be caused by its location within aggregates (Schrumpf et al. 2013; Golchin et al. 1997). The potential of a soil to form aggregates depends on the size distribution of the primary complexes and their characteristics which in turn depend on fundamental soil properties, such as clay mineralogy as well as on the abundance of microorganisms. Aggregates physically protect soil organic matter by forming physical barriers between microorganisms and their substrates and controlling food web interactions and consequently microbial turnover (Six et al. 2002). Moreover, the enhanced C concentration in fine-textured soils compared to soils with low clay content can probably be explained by higher soil moisture in clay-rich soils. Soils with high clay content have higher proportions of fine pores, which are less accessible for biological activities. We assume that this may have resulted in a reduced decomposition of the litter inputs at clay-rich sites and a subsequent accumulation of C (Grüneberg et al. 2013). In contrast to C stocks that correlate positively with clay content, we found increasing C sequestration with increasing particle size. The extent of decomposing soil organic matter decreases from sand- to silt- to clay-sized complexes suggesting that the binding capacity of clay and silt particles is more saturated in the finer-textured soils than in coarser- textured soils (Christensen 2001; Guggenberger et al. 1995). Schulten and Leinweber (2000) supposed that the importance of the clay fraction can become notably relevant with lower clay concentrations. Therefore, we assume that there is an enrichment of C in fine particle

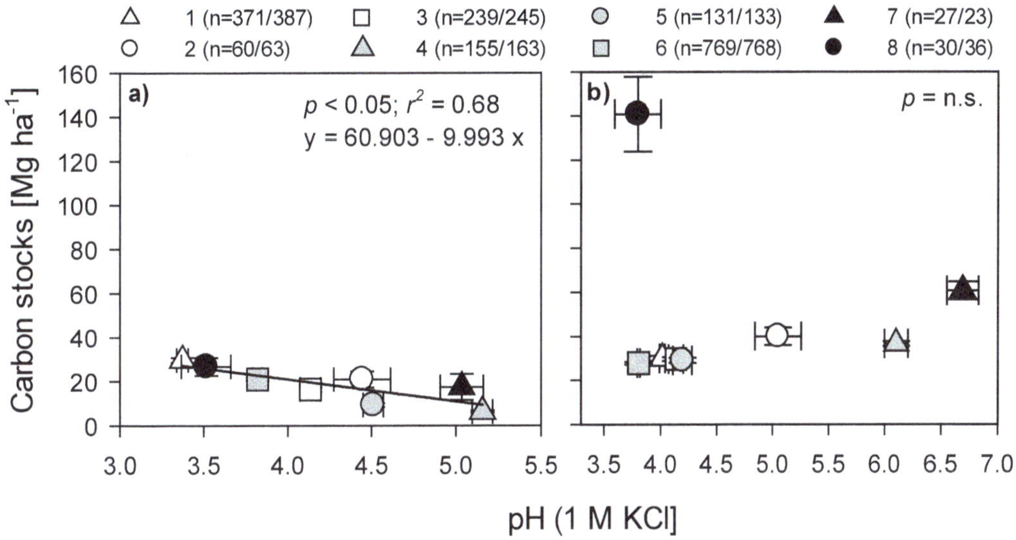

Fig. 6.13 Relation between the pH value and carbon stocks (**a**) in the organic layer and (**b**) in the 10–30 cm depth increment of the mineral soil. The linear regression was performed between the averaged pH value carbon stocks of various soil parent material groups; the numbers in the legend represent soil parent material groups with (1) soils from base-poor unconsolidated sediment, (2) soils of alluvial plains, (3) loamy soils of the lowland, (4) soils from weathered carbonate bedrock, (5) soils from basic-intermediate bedrock, (6) soils from base-poor hard bedrock, (7) soils from the Alps, and (8) organic soils

fractions in low clay sandy lowland soils. Consequently, the relative magnitude of C accumulation is elevated in low clay soils compared to high clay soils.

The SEM for the organic layer and the mineral soil revealed indications on a relation between soil acidity and C stocks. The pH value of the organic layer correlated negatively with C stocks, while the mineral soil showed a positive correlation. Furthermore, the results in respect to forest stand types and soil parent material groups especially for the organic layer indicated a relation between pH value and C stocks (Fig. 6.13). Both, soil parent material groups, representing a broad range of soil acidity, and forest stand types, representing a broad range of base and acidic cations caused by the litter of various tree species, showed a significant gradient of acidity. Moreover, C stock changes of the organic layer were negatively related to pH value. In forest soils liming is performed to compensate for acidic inputs and to improve the cation supply of the trees. Therefore, liming increases the pH value which in turn affects the organic carbon cycle in forest soils (see Sect. 6.5.3). Various studies demonstrated that nutrient-rich soils tend to be associated with higher rates of litter decomposition, whereas the accumulation of soil organic matter in less fertile soils occurs due to reduced decomposition (Oostra et al. 2006; Vesterdal et al. 2013). Vesterdal (1999) demonstrated that the accumulation of C in the organic layer was higher on less fertile sites than on fertile sites. This was explained by higher activity of organisms on fertile sites that incorporate material from the organic layer into the mineral soil. A study on Norway spruce stands on a soil poor in nutrient showed that the root mass in the forest floor was much greater

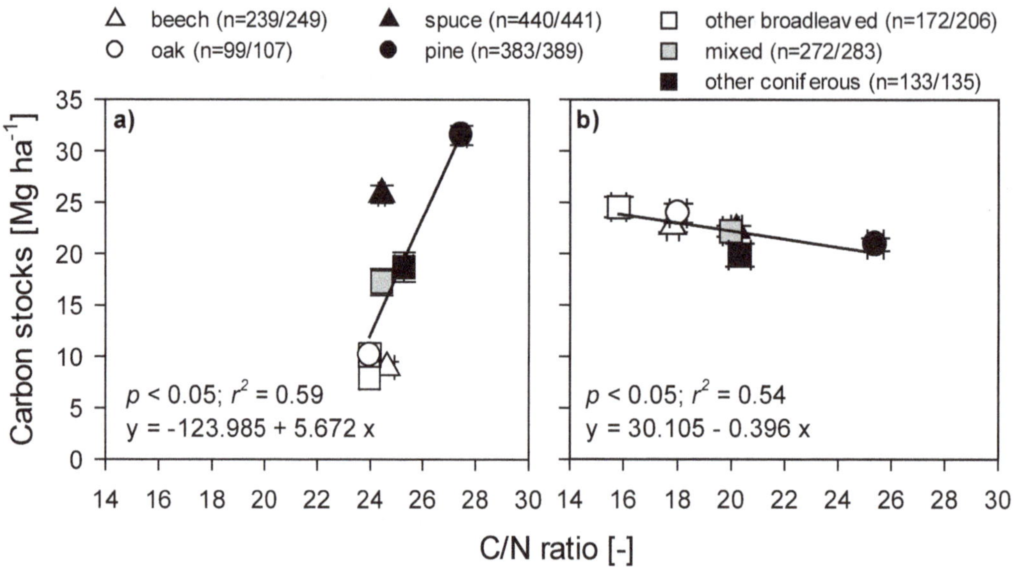

Fig. 6.14 Relation between the C/N ratio and carbon stocks (**a**) in the organic layer and (**b**) in the 0–5 cm depth increment of the mineral soil. The linear regression was performed between the averaged pH value carbon stocks of various forest stand types; the numbers in the legend represent forest stand types; for definition see Chap. 2

than in stands on more nutrient-rich soils (Vesterdal et al. 1995). A larger input of root litter to the organic layer at acidic sites poor in nutrient may thus additionally contribute to a higher sequestration of C in the organic layer.

6.5 Effects of Anthropogenic Factors

6.5.1 Forest Stand Structure

Significant differences in C stocks of the organic layer could be found for different tree species (Fig. 6.7a). Carbon stocks in the organic layer were highest under coniferous trees followed by beech and other broadleaf tree species. This order corresponds with findings compiled from various European and North American studies (Vesterdal et al. 2013). The results have in common that despite various methodological approaches, C stocks of the organic layers were affected by tree species. Differences in the accumulation of C can be explained by litter quality (Fig. 6.14). Various studies found higher C/N ratios in litter from coniferous trees compared to litter from broadleaf trees. Beech trees showed higher C/N ratios than litter from oaks and valuable broadleaved tree species (Vesterdal et al. 1995; Guckland et al. 2009). The C/N ratio of the litter can also serve as a proxy for other important chemical properties of the aboveground litter such as lignin content (Hobbie et al. 2006). The lignin/N ratio in turn is related to the turnover and accumulation of organic matter. The lignin/N ratios described in literature decreased as followed: spruce, pine, beech > oak > other broadleaved trees (Kalbitz et al. 2006; Lovett et al. 2004).

The NFSI also confirmed an influence of tree species on C stock in mineral soil. We found a trend in sequestering more C in subsoils under broadleaf trees (30–90 cm). There are only a few studies on regional or national scale reporting tree species depending C accumulation in mineral soil. Some large-scale inventories included soil sampling and thus provide information on C stocks under different forest stands or tree species. Effects of tree species, however, might be masked by site-related factors (Vesterdal et al. 2013). Nevertheless, recent studies reported a tendency to store more C under broadleaf and valuable broadleaved tree species compared to coniferous trees and beech, respectively (Langenbruch et al. 2012; Oostra et al. 2006; Vesterdal et al. 2008). Tree species-related mechanisms and processes that affect C stocks could not be verified with the data from the NFSI. We suggest that the tree species could be driven by differences in soil fauna activity and related incorporation of litter into the soil. Variations in earthworm abundance between broadleaf and coniferous tree species have been reported by various studies (de Schrijver et al. 2012; Reich et al. 2005). These studies clearly reported that earthworm abundance or biomass increased from spruce over beech and valuable broadleaved tree species. We assume that earthworm abundance or biomass is a likely explanation for differences in C stocks between forest stand types or tree species due to higher inputs of organic layer material to upper mineral soil. A further mechanism explaining differences in mineral soil C stocks is deemed to be tree species-related differences in root biomass and its turnover. Temperate climate regions are known to have a similar C input through root biomass as aboveground C input through litter fall (Kleja et al. 2008) and to have significant higher below-ground biomass of broadleaf trees compared to coniferous trees (Finer et al. 2007). Furthermore, the C released by root litter and exudates is more stable than C that entered the soil by aboveground litter input (Crow et al. 2009).

The tree biomass was of minor importance explaining differences in C stocks and C stock changes of the organic layer (Fig. 6.10). Carbon stocks in the organic layer only slightly increased with tree biomass and accompanied increase of forest stand age. In contrast to our results, most of the studies could hardly reveal differences in C-related litter input of various tree species. Therefore in their studies, direct effects on organic layer C stocks could not be proved (Binkley and Valentine 1991; Trum et al. 2011; Vesterdal et al. 2008). Organic matter decomposition and turnover depend on the activity of decomposers which in turn is controlled by climate and/or forest management (Augusto et al. 2002). Higher tree biomass likewise is accompanied by a recovery of the site since the last forest management activity. A long-lasting period without disturbances due to forest management might thus favor accumulation of C (Schulze et al. 1999). Growing and yield charts suggest a decrease of forest stand productivity of aged stands resulting in steady-state equilibrium after decades. Studies on coniferous forests also showed an increase of organic layer mass with increasing forest stand age developing steady-state equilibrium after several decades (Böttcher and Springob 2001). The age class distribution of forest stands illustrates that in our study the major part of forest stands are rather in a growth period than in steady-state equilibrium. In contrast to the organic layer, we could not find a relation between tree biomass and C stocks for the mineral soil (Fig. 6.11). Investigations about effects of forest stand age, which is proportional related to coarse wood volume, revealed marginal correlations with C stocks (Böttcher and Springob 2001). Furthermore, based on yield charts, it can be derived a higher tree biomass of stands similar in age for coniferous than for broadleaf trees indicating that revealed effects might also be induced by differing tree species. This is supported, e.g., by a higher C accumulation of organic layers in coniferous forests compared to broadleaf forest. The negative relation between coarse wood volume and C stock changes in mineral soil is attributable to the fact that sites with high C stocks are established at soils from weathered carbonate rocks, loamy soils of the lowland, or soils of broad river valleys usually planted with broadleaf trees, whereas low C stocks can be found in soils from both base-poor unconsolidated sediment and bedrock usually planted with coniferous trees. Furthermore, soils under older forest stands may be rather in steady state of C input and output than soils under younger forest stands that have greater potential to fix C, and this difference may be attributed to a loss of organic matter following management disturbance (Covington 1981).

6.5.2 *Atmospheric Nitrogen Deposition*

A previous NFSI report discussed imbalances between pH value, C/N ratio, and humus forms that were attributed to atmospheric deposition of N (Wolff and Riek 1996). By implementing NFSI data into SEM, it could be shown that organic layer C stocks were positively correlated with N deposition, while organic layer C stock changes showed no response to N deposition (Fig. 6.11). More complex relations showed that both C concentrations and C stock as well as C stock changes in the mineral soil were affected by N depositions. This can be explained by the fact that the additional available N potentially increases the forest stand growth rates which in turn are likely to cause an accumulation of organic matter by enhanced biomass input into the soil. Furthermore, the oversupply of N stimulates microorganisms decomposing less C to receive sufficient nutrients. Long-term experiments in tempered and northern hardwood forest soils revealed a decrease in soil respiration under enhanced N deposition (Bowden et al. 2004; Burton et al. 2004). The positive relation between N deposition and C stocks and C stock changes revealed in the SEM might also be caused by collinearity with the data used for modelling N deposition especially for interactions of N deposition with height above sea level, geographical position, or forest stand type.

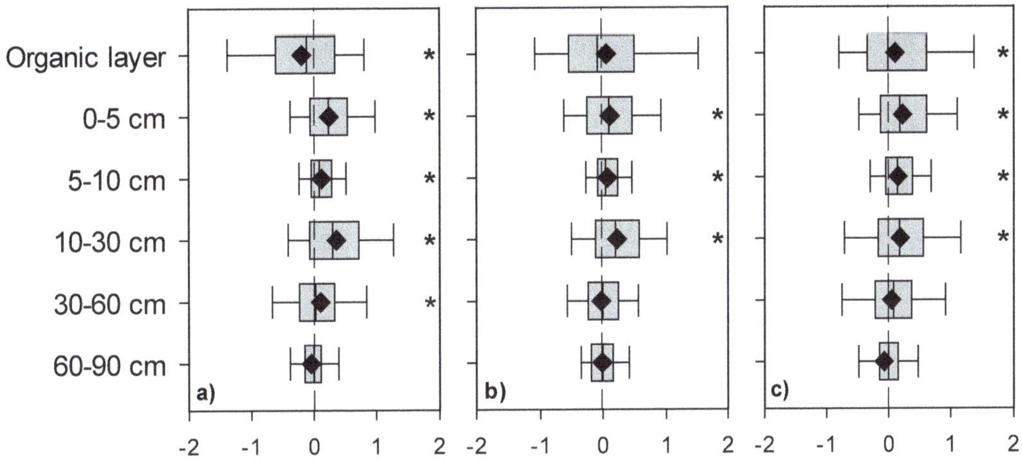

Fig. 6.15 Depth distribution of the annual organic carbon stock changes of (**a**) limed and (**b**) not limed sites sensitive to soil acidification as well as (**c**) sites both not limed and not sensitive to soil acidification. Asterisks indicate significant differences based on a one-sample *t*-test with H0: 0 ($p < 0.05$)

6.5.3 Forest Liming

The 1st NFSI report indicated a nationwide soil acidification widely independent from soil parent material (Wolff and Riek 1996). The consequences of soil acidification are manifold for eco-systematic functions of forests. Concepts for forest liming that consider specific site conditions were developed for an efficient and long-lasting regeneration of essential soil functions. The identified sites sensitive to soil acidification comprise sites that are suitable to lime according to specific liming concepts (see Chap. 2). Altogether, 749 out of 1859 survey plots of the NFSI were identified as sites sensitive to soil acidification including at least 385 survey plots limed since the 1st NFSI.

The results showed relations between liming and C stocks in the organic layer and mineral soil. We found in the organic layer at limed sites a significant decrease in C stock changes of 0.19 ± 0.05 Mg ha^{-1} year^{-1} (Fig. 6.15a). Increasing C stock changes of the organic layer were detected for not limed but sensitive sites to soil acidification with 0.06 ± 0.08 Mg ha^{-1} year^{-1} (Fig. 6.15b) and for not limed sites not sensitive to soil acidification with 0.11 ± 0.04 Mg ha^{-1} year^{-1} (Fig. 6.15c) sensitive to soil acidification. Taking the mineral soil into account, C stocks at limed sites increased significantly by 0.25 ± 0.04 Mg ha^{-1} year^{-1} in 0–5 cm, by 0.12 ± 0.02 Mg ha^{-1} year^{-1} in 5–10 cm, by 0.36 ± 0.05 Mg ha^{-1} year^{-1} in 10–30 cm, and by 0.11 ± 0.05 Mg ha^{-1} year^{-1} in 30–60 cm since the 1st NFSI. The gain of C in mineral soil thus offsets clearly the loss of C in organic layer. In respect to C stocks of the 1st NFSI, total soil organic C stocks throughout the whole soil profile showed a net increase of ~9%. A significant increase of 0.11 ± 0.05 Mg ha^{-1} year^{-1} in 0–5 cm, of 0.07 ± 0.02 Mg ha^{-1} in 5–10 cm, and of 0.22 ± 0.06 Mg ha^{-1} in 10–30 cm was also found at not limed forest sites sensitive to soil acidification. The increase, however, was less distinct as for the limed sites

Table 6.4 Results of analysis of covariance (ANCOVA) testing effects of clay content, temperature, and liming on organic carbon stock changes in the organic layer

Variables	DF	MSE (type III)	F value	*p* value
Clay content	1	18.800	17.02	<0.0001
Temperature	1	26.552	24.04	<0.0001
Liming	1	0.004	0.04	0.8433
Clay content*temperature	1	21.680	19.63	<0.0001
Clay content*liming	1	1.437	1.30	0.2544
Temperature*liming	1	0.239	0.22	0.6422
Clay content	1	18.800	17.02	<0.0001

DF degrees of freedom, *MSE* mean square error

and the C change of the soil profile since the 1st NFSI has totally increased by ~6%. Annual C stock changes of sites not sensitive to soil acidification showed significant increases in the organic layer of 0.11 ± 0.04 Mg ha^{-1} $year^{-1}$ and in the mineral soil of 0.23 ± 0.03 Mg ha^{-1} $year^{-1}$ in 0–5 cm, of 0.16 ± 0.02 Mg ha^{-1} $year^{-1}$ in 5–10 cm, and of 0.20 ± 0.04 Mg ha^{-1} $year^{-1}$ in 10–30 cm. The annual C stock changes were not significant in the subsoil (below 30 cm soil depth).

To exclude interactions with factors affecting C stock changes within the sites sensitive to soil acidification, multivariate analyses of covariance (ANCOVA) were performed for the organic layer as well as for the mineral soil (0–30 cm). We used for the organic layer the input variables liming, clay content (representing site properties), and temperature (representing climate). The ANCOVA revealed a clear influence of clay content and temperature on C stock changes in the organic layer (Table 6.4). The SEM in contrast has illustrated no temperature effects. On the other hand, the results of the ANCOVA showed an interaction between clay content and temperature. The differing results pointed out that C stock changes of the organic layer depended on soil parent material, although strength and direction of the relation were controlled by averaged mean temperature. No evidence was found for an influence of liming on C stock changes by multivariate analysis.

The SEM for mineral soil exposed effects of clay content and N deposition on C stock changes. This result was verified by ANCOVA in respect to effects of liming exclusively with the data from sites sensitive to soil acidification. The ANCOVA revealed direct effects of liming, while the influence of clay content and N deposition was not significant (Table 6.5). There were, however, significant interactions between clay content and liming as well as between clay content and N deposition. These findings indicated that liming favors the accumulation of C in mineral soils, strength, and direction, however, was depending from clay content and/or N deposition. The SEM supported the results of the ANCOVA revealing indirect positive as well as direct negative interdependences. The multivariate analysis also demonstrated that limed sites with low clay content accumulated high amounts of C, while limed sites with high clay concentrations tended to lose C. Similar results were found for N deposition where sites with high depositions sequestered C, while sites with low depositions were affected by a loss of C. This suggests that limed sites that are characterized by high N deposition are different in turnover and

Table 6.5 Results of analysis of covariance (ANCOVA) testing effects of clay content, temperature, and liming on organic carbon stock changes in the mineral soil down to a depth of 30 cm

Variables	DF	MSE (type III)	*F* value	*p* value
Clay content	1	6.213	2.77	0.0966
N deposition	1	0.107	0.05	0.8273
Liming	1	16.030	7.14	0.0077
Clay content*N deposition	1	5.887	2.62	0.1058
Clay content*liming	1	17.091	7.61	0.0059
N deposition*liming	1	11.839	5.27	0.0219

DF degrees of freedom, *MSE* mean square error

decomposing soil organic matter as sites that are either exclusively limed or affected by high N deposition. Especially not limed sites close to arable land or to road traffic might be affected by higher N inputs.

The calcium applied by the lime plays an important role in forest ecosystems and therefore can affect directly C pools and fluxes. The loss of C of limed organic layers compared to organic layers not limed may be caused by changes in the microbial and faunal composition and function. It is known that the microbial activity (Illmer and Schinner 1991; Andersson and Nilsson 2001) and the earthworm abundance (Hobbie et al. 2006; Reich et al. 2005) are stimulated when the pH value and calcium concentrations increase. The increasing C sequestration rate of the mineral soil on limed sites can be attributed to an enhanced calcium availability through the application of lime, which may reduce the dissolution and mobility of dissolved organic matter by forming cation bridges (Oste et al. 2002). These cation bridges are thought to stabilize soil organic matter and reduce decomposition which in turn causes enhanced retention of soil organic matter. Studies about effects of liming on soil respiration have observed a decline in respiration after an initial enhancement period to a few years post-liming (Illmer and Schinner 1991; Melvin et al. 2013; Groffman et al. 2006). On the other hand, observations of increased decomposition with liming have been reported in various studies (Baath and Arnebrant 1994; Priha and Smolander 1994). Nevertheless, reasons for these differential responses remain unclear. A reduction of soil respiration indicates that liming is affecting relations between microbial communities and the organic matter mineralized. Changes in microbial communities by liming may cause changes of recalcitrance of produced soil organic matter or physical stabilization of soil organic matter.

6.6 Summary and Conclusions

The analysis of the data from NFSI revealed a C sequestration rate of 0.75 Mg ha^{-1} in the organic layer and the mineral soil down to 90 cm for the time period between the inventories. Especially the organic layer and depth increments down to 30 cm depth showed an increase, while in the subsoil (below 30 cm soil depth), changes were marginal. With respect to climate protection policy, the sequestration of atmospheric

C as stable long-lasting soil organic matter is of interest due to the fact that our subsoils alone stored 25% of total C stocks. Carbon stocks of the organic layer and the mineral soil down to 90 cm totaled to 117.1 Mg ha^{-1} in 2006, of which 17% of organic C was stored in the organic layer and 58% in the upper 30 cm of the mineral soil.

The results showed that the variation of C stocks in the organic layer was especially controlled by tree species as soils under broadleaf trees stored less C in the organic layer than soils under coniferous trees. In contrast, tree species effects on C stocks of mineral soil were comparatively less pronounced. This may be explained by both differences in litter acidity and nutrient availability. Sites with broadleaf trees may store more C in subsoil compared to sites with pine and spruce trees and are therefore predestined for long-term C sequestration. This is probably due to the fact that there is under broadleaf and mixed forests a higher input of fairly stable organic matter to deeper soil horizon by deeper rooting trees, a higher biological activity, and/or less disturbance of the soil caused by more stable stands. Soil C stocks are furthermore affected by site conditions and the resulting soil texture. An effective selection of tree species combined with specific site conditions may therefore enhance C sequestration of mineral soils. This can probably be facilitated by establishing conifer mixed forests or by specific placements of broadleaf tree species in pure coniferous stands especially on soils derived from unconsolidated sediments and/or on soils derived from base-poor bedrock. Besides the analysis of C input and output (mineralization, leaching, etc.), the quantitative estimation of soil organic matter stabilization has to be investigated in more detail to understand processes of C sequestration in forest soils.

In explaining more or less large-scaled patterns of changes in soil organic C stocks, various regional factors have to be considered. Wide areas of the Northern German Lowland are characterized by weakly developed sandy soils derived from pre-Pleistocene sediments that were depleted of soil organic matter by extensive overuse. The site quality, however, has improved by reversing the degradation of conscious forest management since NFSI I. In particular cases it can be assumed that rather locally effective processes are crucial in explaining C stock changes. The increase or decrease of C stocks in the organic layer and mineral soil can be controlled by forest management such as forest liming, tree species selection, and forest conservation. Besides natural environmental effects of soil parent material, texture, and increasing climate-induced water stresses, the pathway analysis revealed additional specific effects of N deposition on C stock changes. Furthermore important is the canopy uptake or the proximity to arable land in respect to the total deposition. Nevertheless, both factors could not be integrated in the model and have to be disregarded. In general, terrestrial ecosystems respond to enhanced N deposition when they are N limited. The additional N has the potential to increase sequestering C by an increase in productivity and accumulation of soil organic matter through increased litter production. Carbon accumulation through N deposition may also be due to increased recalcitrance of N-enriched litter and decreased soil respiration, both leading to reduced long-term decomposition rates of organic matter. Furthermore, modelling N input is affected by covariance, e.g., between climate and N deposition, because the amount of N deposed at a site depends on in situ precipitation. The relationship between climate variables and N deposition could be further assessed by simultaneously measuring N stocks and fluxes when aspects of climate and deposition may be recorded separately as conducted by the intensive (Level II) monitoring.

Besides timber harvest and decommissioned forests, a further aspect discussed in forest ecology is the application of lime to reduce negative consequences of atmospheric deposition of acids. Our results showed that liming may affect C cycling in forest ecosystems as C stocks increased in mineral soil, while a loss of C was observed in the organic layer. In literature various results have been discussed in light of the impact of liming on soil organic C cycling in forest ecosystems. Recent research showed that soil respiration was both stimulated and inhibited by forest liming depending on various environmental conditions. Further research is needed to identify the most important factors affecting turnover of soil organic matter in respect to the impact of forest liming especially on dynamics of microbial communities as well as on recalcitrance and stabilization of soil organic matter. A prolonged impact on C sequestration and N retention is expected considering Germany received a long-term and area-wide, largely substrate-independent acidification and soil depletion through atmospheric inputs of acids. From this point of view, more detailed research is needed studying effects of the application of cations by liming on C pools and fluxes in forest ecosystems.

References

Amundson R (2001) The carbon budget in soils. Annu Rev Earth Planet Sci 29:535–562. https://doi.org/10.1146/annurev.earth.29.1.535

Andersson S, Nilsson SI (2001) Influence of pH and temperature on microbial activity, substrate availability of soil-solution bacteria and leaching of dissolved organic carbon in a mor humus. Soil Biol Biochem 33(9):1181–1191. https://doi.org/10.1016/s0038-0717(01)00022-0

Augusto L, Ranger J, Binkley D, Rothe A (2002) Impact of several common tree species of European temperate forests on soil fertility. Ann Forest Sci 59(3):233–253. https://doi.org/10.1051/forest:2002020

Baath E, Arnebrant K (1994) Growth-rate and response of bacterial communities to pH in limed and ash treated forest soils. Soil Biol Biochem 26(8):995–1001. https://doi.org/10.1016/0038-0717(94)90114-7

Baisden WT, Amundson R, Brenner DL, Cook AC, Kendall C, Harden JW (2002) A multiisotope C and N modeling analysis of soil organic matter turnover and transport as a function of soil depth in a California annual grassland soil chronosequence. Glob Biogeochem Cycles 16(4):82/01–82/26. https://doi.org/10.1029/2001gb001823

Barnett AG, van der Pols JC, Dobson AJ (2005) Regression to the mean: what it is and how to deal with it. Int J Epidemiol 34(1):215–220. https://doi.org/10.1093/ije/dyh299

Bellamy PH, Loveland PJ, Bradley RI, Lark RM, Kirk GJD (2005) Carbon losses from all soils across England and Wales 1978-2003. Nature 437(7056):245–248. https://doi.org/10.1038/nature04038

Binkley D, Valentine D (1991) 50-year biogeochemical effects of green ash, white-pine, and Norway spruce in a replicated experiment. Forest Ecol Manag 40(1–2):13–25. https://doi.org/10.1016/0378-1127(91)90088-d

Böttcher J, Springob G (2001) A carbon balance model for organic layers of acid forest soils. J Plant Nutr Soil Sci Zeitschrift für Pflanzenernährung und Bodenkunde 164(4):399–405. https://doi.org/10.1002/1522-2624(200108)164:4<399::aid-jpln399>3.0.co;2-6

Bowden RD, Davidson E, Savage K, Arabia C, Steudler P (2004) Chronic nitrogen additions reduce total soil respiration and microbial respiration in temperate forest soils at the Harvard Forest. Forest Ecol Manag 196(1):43–56. https://doi.org/10.1016/j.foreco.2004.03.011

Burton AJ, Pregitzer KS, Crawford JN, Zogg GP, Zak DR (2004) Simulated chronic NO_3-deposition reduces soil respiration in northern hardwood forests. Glob Chang Biol 10 (7):1080–1091. https://doi.org/10.1111/j.1365-2486.2004.00737.x

Callesen I, Stupak I, Georgiadis P, Johannsen VK, Østergaard HS, Vesterdal L (2015) Soil carbon stock change in the forests of Denmark between 1990 and 2008. Geoderma Reg 5:169–180. https://doi.org/10.1016/j.geodrs.2015.06.003

Christensen BT (2001) Physical fractionation of soil and structural and functional complexity in organic matter turnover. Eur J Soil Sci 52(3):345–353. https://doi.org/10.1046/j.1365-2389.2001.00417.x

Conforti M, Luca F, Scarciglia F, Matteucci G, Buttafuoco G (2016) Soil carbon stock in relation to soil properties and landscape position in a forest ecosystem of southern Italy (Calabria region). Catena 144:23–33. https://doi.org/10.1016/j.catena.2016.04.023

Covington WW (1981) Changes in forest floor organic-matter and nutrient content following clear cutting in northern hardwoods. Ecology 62(1):41–48. https://doi.org/10.2307/1936666

Crow SE, Lajtha K, Filley TR, Swanston CW, Bowden RD, Caldwell BA (2009) Sources of plant-derived carbon and stability of organic matter in soil: implications for global change. Glob Chang Biol 15(8):2003–2019. https://doi.org/10.1111/j.1365-2486.2009.01850.x

Davidson EA, Janssens IA (2006) Temperature sensitivity of soil carbon decomposition and feedbacks to climate change. Nature 440(7081):165–173. https://doi.org/10.1038/nature04514

Davis AA, Stolt MH, Compton JE (2004) Spatial distribution of soil carbon in southern new England hardwood forest landscapes. Soil Sci Soc Am J 68(3):895–903

de Schrijver A, de Frenne P, Staelens J, Verstraeten G, Muys B, Vesterdal L, Wuyts K, van Nevel L, Schelfhout S, de Neve S, Verheyen K (2012) Tree species traits cause divergence in soil acidification during four decades of postagricultural forest development. Glob Chang Biol 18 (3):1127–1140. https://doi.org/10.1111/j.1365-2486.2011.02572.x

de Vries W, Solberg S, Dobbertin M, Sterba H, Laubhann D, van Oijen M, Evans C, Gundersen P, Kros J, Wamelink GWW, Reinds GJ, Sutton MA (2009) The impact of nitrogen deposition on carbon sequestration by European forests and heathlands. Forest Ecol Manag 258 (8):1814–1823. https://doi.org/10.1016/j.foreco.2009.02.034

Evers J, Dammann I, Noltensmeier A, Nagel R-V (2008) Auswirkungen von Bodenschutzkalkungen auf Buchenwälder (*Fagus sylvatica L.*). In: Ergebnisse angewandter Forschung zur Buche. Beiträge aus der NW-FVA, vol 3. Nordwestdeutsche Forstliche Versuchsanstalt (NW-FWA), Göttingen, pp 21–50

Finer L, Helmisaari HS, Lohmus K, Majdi H, Brunner I, Borja I, Eldhuset T, Godbold D, Grebenc T, Konopka B, Kraigher H, Mottonen MR, Ohashi M, Oleksyn J, Ostonen I, Uri V, Vanguelova E (2007) Variation in fine root biomass of three European tree species: Beech (*Fagus sylvatica L.*), Norway spruce (*Picea abies L. Karst.*), and Scots pine (*Pinus sylvestris L.*). Plant Biosyst 141(3):394–405. https://doi.org/10.1080/11263500701625897

Goidts E, van Wesemael B (2007) Regional assessment of soil organic carbon changes under agriculture in Southern Belgium (1955-2005). Geoderma 141(3-4):341–354. https://doi.org/10.1016/j.geoderma.2007.06.013

Golchin A, Baldock JA, Oades JM (1997) A model linking organic matter decomposition, chemistry and aggregate dynamics. In: Lal R, Kimble JM, Follett RF, Stewart BA (eds) Soil processes and the carbon cycle. CRC, Boca Raton, FL, pp 245–266

Groffman PM, Fisk MC, Driscoll CT, Likens GE, Fahey TJ, Eagar C, Pardo LH (2006) Calcium additions and microbial nitrogen cycle processes in a northern hardwood forest. Ecosystems 9 (8):1289–1305. https://doi.org/10.1007/s10021-006-0177-z

Grüneberg E, Schöning I, Hessenmöller D, Schulze ED, Weisser WW (2013) Organic layer and clay content control soil organic carbon stocks in density fractions of differently managed German beech forests. Forest Ecol Manag 303:1–10. https://doi.org/10.1016/j.foreco.2013.03.014

Grüneberg E, Ziche D, Wellbrock N (2014) Organic carbon stocks and sequestration rates of forest soils in Germany. Glob Chang Biol 20(8):2644–2662. https://doi.org/10.1111/gcb.12558

Guckland A, Jacob M, Flessa H, Thomas FM, Leuschner C (2009) Acidity, nutrient stocks, and organic-matter content in soils of a temperate deciduous forest with different abundance of European beech (*Fagus sylvatica L.*). J Plant Nutr Soil Sci Zeitschrift für Pflanzenernährung und Bodenkunde 172(4):500–511. https://doi.org/10.1002/jpln.200800072

Guggenberger G, Zech W, Haumaier L, Christensen BT (1995) Land-use effects on the composition of organic-matter particle-size separates of soil. 2. CPMAS and solution C-13 NMR analysis. Eur J Soil Sci 46(1):147–158. https://doi.org/10.1111/j.1365-2389.1995.tb01821.x

Hobbie SE, Reich PB, Oleksyn J, Ogdahl M, Zytkowiak R, Hale C, Karolewski P (2006) Tree species effects on decomposition and forest floor dynamics in a common garden. Ecology 87 (9):2288–2297. https://doi.org/10.1890/0012-9658(2006)87[2288:tseoda]2.0.co;2

Illmer P, Schinner F (1991) Effects of lime and nutrient salts on the microbiological activities of forest soils. Biol Fertil Soils 11(4):261–266. https://doi.org/10.1007/bf00335845

IPCC (2003) Good practice guidance for land use, land-use change and forestry. Intergovernmental Panel on Climate Change (IPCC), Kanagawa Prefecture, Japan

Janssens IA, Dieleman W, Luyssaert S, Subke JA, Reichstein M, Ceulemans R, Ciais P, Dolman AJ, Grace J, Matteucci G, Papale D, Piao SL, Schulze ED, Tang J, Law BE (2010) Reduction of forest soil respiration in response to nitrogen deposition. Nat Geosci 3(5):315–322. https://doi.org/10.1038/ngeo844

Janzen HH (2004) Carbon cycling in earth systems—a soil science perspective. Agric Ecosyst Environ 104(3):399–417. https://doi.org/10.1016/j.agee.2004.01.040

Jastrow JD, Amonette JE, Bailey VL (2007) Mechanisms controlling soil carbon turnover and their potential application for enhancing carbon sequestration. Clim Change 80(1–2):5–23. https://doi.org/10.1007/s10584-006-9178-3

Johnson DW, Knoepp JD, Swank WT, Shan J, Morris LA, Van Lear DH, Kapeluck PR (2002) Effects of forest management on soil carbon: results of some long-term resampling studies. Environ Pollut 116:S201–S208. https://doi.org/10.1016/s0269-7491(01)00252-4

Jonard M, Nicolas M, Coomes DA, Caignet I, Saenger A, Ponette Q (2017) Forest soils in France are sequestering substantial amounts of carbon. Sci Total Environ 574:616–628. https://doi.org/10.1016/j.scitotenv.2016.09.028

Kalbitz K, Kaiser K, Bargholz J, Dardenne P (2006) Lignin degradation controls the production of dissolved organic matter in decomposing foliar litter. Eur J Soil Sci 57(4):504–516. https://doi.org/10.1111/j.1365-2389.2006.00797.x

Kleja DB, Svensson M, Majdi H, Jansson PE, Langvall O, Bergkvist B, Johansson MB, Weslien P, Truusb L, Lindroth A, Agren GI (2008) Pools and fluxes of carbon in three Norway spruce ecosystems along a climatic gradient in Sweden. Biogeochemistry 89(1):7–25. https://doi.org/10.1007/s10533-007-9136-9

Ladegaard-Pedersen P, Elberling B, Vesterdal L (2005) Soil carbon stocks, mineralization rates, and CO_2 effluxes under 10 tree species on contrasting soil types. Can J Forest Res 35 (6):1277–1284. https://doi.org/10.1139/x05-045

Langenbruch C, Helfrich M, Flessa H (2012) Effects of beech (*Fagus sylvatica*), ash (*Fraxinus excelsior*) and lime (*Tilia spec.*) on soil chemical properties in a mixed deciduous forest. Plant Soil 352(1–2):389–403. https://doi.org/10.1007/s11104-011-1004-7

Lark RM, Bellamy PH, Kirk GJD (2006) Baseline values and change in the soil, and implications for monitoring. Eur J Soil Sci 57(6):916–921. https://doi.org/10.1111/j.1365-2389.2006.00875.x

Leuschner C, Wulf M, Bäuchler P, Hertel D (2013) Soil C and nutrient stores under Scots pine afforestations compared to ancient beech forests in the German Pleistocene: the role of tree species and forest history. Forest Ecol Manag 310:405–415. https://doi.org/10.1016/j.foreco.2013.08.043

Liski J, Perruchoud D, Karjalainen T (2002) Increasing carbon stocks in the forest soils of western Europe. Forest Ecol Manag 169(1–2):159–175. https://doi.org/10.1016/s0378-1127(02)00306-7

Lorenz K, Lal R (2010) Carbon sequestration in forest ecosystems. Springer, Dordrecht

Lovett GM, Weathers KC, Arthur MA, Schultz JC (2004) Nitrogen cycling in a northern hardwood forest: do species matter? Biogeochemistry 67(3):289–308. https://doi.org/10.1023/B:BIOG.0000015786.65466.f5

Luyssaert S, Ciais P, Piao SL, Schulze ED, Jung M, Zaehle S, Schelhaas MJ, Reichstein M, Churkina G, Papale D, Abril G, Beer C, Grace J, Loustau D, Matteucci G, Magnani F, Nabuurs GJ, Verbeeck H, Sulkava M, van der Werf GR, Janssens IA, Team C-IS (2010) The European carbon balance. Part 3: forests. Glob Chang Biol 16(5):1429–1450. https://doi.org/10.1111/j.1365-2486.2009.02056.x

Melvin AM, Lichstein JW, Goodale CL (2013) Forest liming increases forest floor carbon and nitrogen stocks in a mixed hardwood forest. Ecol Appl 23(8):1962–1975. https://doi.org/10.1890/13-0274.1

Nave LE, Vance ED, Swanston CW, Curtis PS (2010) Harvest impacts on soil carbon storage in temperate forests. Forest Ecol Manag 259(5):857–866. https://doi.org/10.1016/j.foreco.2009.12.009

Oostra S, Majdi H, Olsson M (2006) Impact of tree species on soil carbon stocks and soil acidity in southern Sweden. Scand J Forest Res 21(5):364–371. https://doi.org/10.1080/02827580600950172

Oste LA, Temminghoff EJM, Van Riemsdijk WH (2002) Solid-solution partitioning of organic matter in soils as influenced by an increase in pH or Ca concentration. Environ Sci Technol 36 (2):208–214. https://doi.org/10.1021/es0100571

Prietzel J, Bachmann S (2012) Changes in soil organic C and N stocks after forest transformation from Norway spruce and Scots pine into Douglas fir, Douglas fir/spruce, or European beech stands at different sites in Southern Germany. Forest Ecol Manag 269:134–148. https://doi.org/10.1016/j.foreco.2011.12.034

Priha O, Smolander A (1994) Fumigation-extraction and substrate-induced respiration derived microbial biomass-C, and respiration rate in limed soil of Scots pine sapling stands. Biol Fertil Soils 17(4):301–308. https://doi.org/10.1007/bf00383986

R Core Team (2015) R: A language and environment for statistical computing. R Foundation for Statistical Computing, Vienna, Austria

Reich PB, Oleksyn J, Modrzynski J, Mrozinski P, Hobbie SE, Eissenstat DM, Chorover J, Chadwick OA, Hale CM, Tjoelker MG (2005) Linking litter calcium, earthworms and soil properties: a common garden test with 14 tree species. Ecol Lett 8(8):811–818. https://doi.org/10.1111/j.1461-0248.2005.00779.x

Richardson M, Stolt M (2013) Measuring soil organic carbon sequestration in aggrading temperate forests. Soil Sci Soc Am J 77(6):2164–2172. https://doi.org/10.2136/sssaj2012.0411

Riley H, Bakkegard M (2006) Declines of soil organic matter content under arable cropping in southeast Norway. Acta Agric Scand Sec B Soil Plant Sci 56(3):217–223. https://doi.org/10.1080/09064710510029141

Rosseel Y (2012) lavaan. An {R} package for structural equation modeling. J Stat Softw 48(2):1–36

Saby NPA, Bellamy PH, Morvan X, Arrouays D, Jones RJA, Verheijen FGA, Kibblewhite MG, Verdoodt A, Berenyiuveges J, Freudenschuss A, Simota C (2008) Will European soil-monitoring networks be able to detect changes in topsoil organic carbon content? Glob Chang Biol 14(10):2432–2442. https://doi.org/10.1111/j.1365-2486.2008.01658.x

Schlesinger WH (1991) Biogeochemistry: an analysis of global change. Academic, San Diego

Schrumpf M, Kaiser K, Guggenberger G, Persson T, Kögel-Knabner I, Schulze ED (2013) Storage and stability of organic carbon in soils as related to depth, occlusion within aggregates, and attachment to minerals. Biogeosciences 10(3):1675–1691. https://doi.org/10.5194/bg-10-1675-2013

Schulten H-R, Leinweber P (2000) New insights into organic-mineral particles: composition, properties and models of molecular structure. Biol Fertil Soils 30(5):399–432. https://doi.org/10.1007/s003740050020

Schulze ED, Lloyd J, Kelliher FM, Wirth C, Rebmann C, Luhker B, Mund M, Knohl A, Milyukova IM, Schulze W, Ziegler W, Varlagin AB, Sogachev AF, Valentini R, Dore S, Grigoriev S, Kolle O, Panfyorov MI, Tchebakova N, Vygodskaya NN (1999) Productivity of forests in the Eurosiberian boreal region and their potential to act as a carbon sink—a synthesis. Glob Chang Biol 5(6):703–722. https://doi.org/10.1046/j.1365-2486.1999.00266.x

Six J, Conant RT, Paul EA, Paustian K (2002) Stabilization mechanisms of soil organic matter: implications for C-saturation of soils. Plant Soil 241(2):155–176. https://doi.org/10.1023/a:1016125726789

Tan ZX, Lal R, Smeck NE, Calhoun FG (2004) Relationships between surface soil organic carbon pool and site variables. Geoderma 121(3-4):187–195. https://doi.org/10.1016/j.geoderma.2003.11.003

Trum F, Titeux H, Ranger J, Delvaux B (2011) Influence of tree species on carbon and nitrogen transformation patterns in forest floor profiles. Ann Forest Sci 68(4):837–847. https://doi.org/10.1007/s13595-011-0080-4

Vesterdal L (1999) Influence of soil type on mass loss and nutrient release from decomposing foliage litter of beech and Norway spruce. Can J Forest Res 29(1):95–105. https://doi.org/10.1139/cjfr-29-1-95

Vesterdal L, Dalsgaard M, Felby C, Raulundrasmussen K, Jorgensen BB (1995) Effects of thinning and soil properties on accumulation of carbon, nitrogen and phosphorus in the forest floor of Norway spruce stands. Forest Ecol Manag 77(1–3):1–10. https://doi.org/10.1016/0378-1127(95)03579-y

Vesterdal L, Schmidt IK, Callesen I, Nilsson LO, Gundersen P (2008) Carbon and nitrogen in forest floor and mineral soil under six common European tree species. Forest Ecol Manag 255(1):35–48. https://doi.org/10.1016/j.foreco.2007.08.015

Vesterdal L, Clarke N, Sigurdsson BD, Gundersen P (2013) Do tree species influence soil carbon stocks in temperate and boreal forests? Forest Ecol Manag 309:4–18. https://doi.org/10.1016/j.foreco.2013.01.017

Vogel C, Heister K, Buegger F, Tanuwidjaja I, Haug S, Schloter M, Kögel-Knabner I (2015) Clay mineral composition modifies decomposition and sequestration of organic carbon and nitrogen in fine soil fractions. Biol Fertil Soils 51(4):427–442. https://doi.org/10.1007/s00374-014-0987-7

von Lützow M, Kögel-Knabner I, Ekschmitt K, Matzner E, Guggenberger G, Marschner B, Flessa H (2006) Stabilization of organic matter in temperate soils: mechanisms and their relevance under different soil conditions—a review. Eur J Soil Sci 57(4):426–445. https://doi.org/10.1111/j.1365-2389.2006.00809.x

Wiesmeier M, Prietzel J, Barthold F, Sporlein P, Geuss U, Hangen E, Reischl A, Schilling B, von Lützow M, Kögel-Knabner I (2013) Storage and drivers of organic carbon in forest soils of southeast Germany (Bavaria)—implications for carbon sequestration. Forest Ecol Manag 295:162–172. https://doi.org/10.1016/j.foreco.2013.01.025

Wolff B, Riek W (1996) Deutscher Waldbodenbericht 1996—Ergebnisse der bundesweiten Bodenzustandserhebung im Wald von 1987-1993. Bundesministerium für Ernährung, Landwirtschaft und Forsten, Bonn, Germany

Chapter 7
Heavy Metal Stocks and Concentrations in Forest Soils

Jens Utermann, Cihan Tarih Aydın, Norbert Bischoff, Jürgen Böttcher, Nadine Eickenscheidt, Joachim Gehrmann, Nils König, Birte Scheler, Florian Stange, and Nicole Wellbrock

J. Utermann (✉)
Ministry for Environment, Agriculture, Conservation and Consumer Protection of the State of North Rhine-Westphalia (MULNV), Düsseldorf, Germany
e-mail: jens.utermann@mulnv.nrw.de

C. T. Aydın
State Agency of Agriculture, Environment and Rural Areas of Schleswig-Holstein, Flintbek, Germany
e-mail: cihan-tarih.aydin@LLUR.Landsh.de

N. Bischoff
Thünen Institute for Climate Smart Agriculture, Braunschweig, Germany
e-mail: norbert.bischoff@thuenen.de

J. Böttcher
Gottfried Wilhelm Leibniz University, Hannover, Germany
e-mail: boettcher@ifbk.uni-hannover.de

N. Eickenscheidt
State Agency for Nature, Environment and Consumer Protection of North Rhine-Westphalia, Recklinghausen, Germany

Thünen Institute of Forest Ecosystems, Eberswalde, Germany
e-mail: nadine.eickenscheidt@lanuv.nrw.de

J. Gehrmann
State Agency for Nature, Environment and Consumer Protection of North Rhine-Westphalia, Recklinghausen, Germany

N. König · B. Scheler
Northwest German Forest Research Institute, Göttingen, Germany
e-mail: nils.koenig@nw-fva.de; birte.scheler@nw-fva.de

F. Stange
Federal Institute for Geosciences and Natural Resources, Hannover, Germany
e-mail: florian.stange@bgr.de

N. Wellbrock
Thünen Institute of Forest Ecosystems, Eberswalde, Germany
e-mail: nicole.wellbrock@thuenen.de

N. Wellbrock, A. Bolte (eds.), *Status and Dynamics of Forests in Germany*, Ecological Studies 237, https://doi.org/10.1007/978-3-030-15734-0_7

7.1 Introduction

From a scientific perspective, heavy metals are metals with a density >5 g cm^{-3}. In this section, the term "heavy metals" includes the elements arsenic (As), lead (Pb), cadmium (Cd), chromium (Cr), copper (Cu), nickel (Ni), mercury (Hg) and zinc (Zn); chemically, arsenic is considered a semimetal.

Heavy metals occur naturally in the geosphere at highly variable concentrations. Concentrations vary primarily as a function of the mineral composition of the parent rock. Elements with geogenically elevated concentrations are principally found where there is solid bedrock. In Germany, geogenically elevated element concentrations are present in particular where soils are found on periglacial sites over basic magmatic and metamorphic rock (Cr, Ni, Cu, Zn), clay (especially Liassic clay) (As, Cd, Cr, Cu, Ni, Pb, Zn), acidic metamorphic rock (Cu, Ni) and mica slate (Cr, Cu, Ni, Zn), as well as carbonate rocks as a result of accumulation of residual clay in terra fusca-type soils. The subsoil and bedrock layers that have little loess concentration contain higher concentrations of the elements compared to the topsoil; in particular for Cr and Ni in areas with basic magmatic and metamorphic bedrock, these elements then significantly trace into the topsoil (Utermann et al. 2006).

Due to the action of trees in filtering out airborne pollutants (the "Auskämmeffekt"), heavy metal loading from the atmosphere is especially high in the soils under forest. On both regional and national scales, this effect can lead to a significant change of the natural heavy metal concentrations in soils. Regionally high concentrations of heavy metals are found in particular in areas leeward of large centres of heavy industry, such as in North Rhine-Westphalia, due to long-term emissions from industry (Gehrmann 2013). Although pollutant emissions from factories and steel processing plants have now been reduced considerably, the resulting loads in the soils will persist for many decades because heavy metals are not biodegradable and migrate relatively slowly through soil. Patterns of topsoil loading across the country are apparent especially for lead, as this element was released diffusely and ubiquitously through burning of leaded fuels for decades up until 1988. As a result, there are elevated levels of lead in the soils under forests of the low mountain ranges throughout Germany (BGR 2016).

In addition to inputs of heavy metals from industrial emissions and traffic exhaust, mining of metal ores and the associated metallurgy have also deposited significant loads in the soils of regions such as the Harz Mountains, Mansfeld Land, the Rhenish Slate Mountains or the Erzgebirge (Wippermann 2000). In many cases, former waste and slag heaps in these regions are protected, as heavy metal sites are listed in Annex 1 of the FFH Directive (Habitats Directive) (habitat type 6130, metallophyte grasslands of the Violetalia calaminariae) and represent a federally protected biotype according to § 30 BNatSchG (Nature Protection Act). If these sites had no metallophytes and were left to themselves, over time and through processes of natural succession, they would develop into forested sites with substantial heavy metal contamination (Knolle et al. 2011).

At low concentrations, some of the heavy metals (e.g. Cu, Zn) can be essential nutrients and therefore have nutritional value. At higher concentrations, the metals have toxic effects for both ecosystems and humans (Ohnesorge and Wilhelm 1991; Wilke et al. 2003). Other heavy metals (e.g. Cd, Pb, Hg) are only known to have toxic effects (Litz et al. 2015). In the geosphere, the ecotoxicological effects of the heavy metals are present only when the metal is in a bioavailable form, either dissolved in soil solution or bound to the solid phase of the soil. The heavy metal concentrations discussed in this section have been extracted from soil using a strong acid ("aqua regia") and hence represent the total heavy metal concentration. This includes the fraction that is not bioavailable or cannot be displaced; consequently, conclusions based on this database are limited with respected statements about effects. In general, with the exception of the elements that are more likely present in the soil in anionic form (e.g. As), the heavy metals tend to be more mobile/bioavailable at lower pH values in the soil. Hence, higher heavy metal concentrations in forest soils, where pH is typically significantly lower compared to agricultural sites, present a higher risk for both the biosphere and in terms of displacement into groundwater. Different heavy metals are absorbed to different extents by mineral surfaces and, depending on pH, can also precipitate as low solubility oxides, hydroxides or carbonates or in other forms. The degree of absorption tends to follow the sequence Cd, Zn < Cu < Ni, Cr < Pb. The ability of heavy metals to form complexes with humic substances is also variable: Cd and Zn form only weak complexes, while Pb, Cr, Ni and Cu form stable complexes, with Cu binding more strongly to low molecular weight humic substances and Pb to higher molecular weight substances (König et al. 1986). Thus, both absorption and the tendency to form complexes depend on pH, and both factors affect the rate of displacement of heavy metals into groundwater.

The datasets from the NFSI I are limited nationally to analyses of heavy metal concentrations in the organic layer. In the NFSI II, heavy metal concentration was measured in the first two depth levels (0–5, 5–10 cm) of the mineral soil at all sample plots and over the entire depth profile at selected sites. Comparative analyses between the NFSI I and NFSI II are thus possible at a national level only for the organic layer. For this reason, data from North Rhine-Westphalia are used by way of example for the mineral soil, because in this state heavy metal concentrations were measured in the first two depth increments of the mineral soil in both inventories.

7.2 Heavy Metal Stocks in the Organic Layer and Mineral Soil

7.2.1 Status and Depth Gradients

Figures 7.1 and 7.2 present boxplots illustrating the depth gradients (organic layer, 0–5, 5–10 cm) for the stocks of all heavy metals analysed for NFSI II as well as in

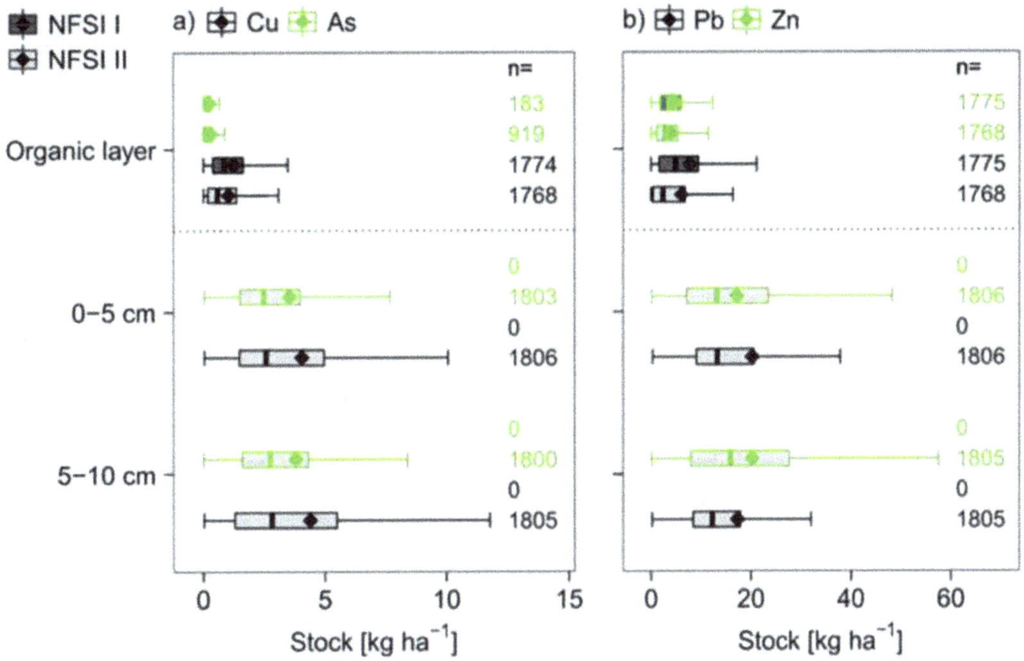

Fig. 7.1 Depth gradients of stocks for the elements (**a**) copper (Cu) and arsenic (As), (**b**) lead (Pb) and zinc (Zn) for the organic layer and the first two mineral soil increments (0–5, 5–10 cm)

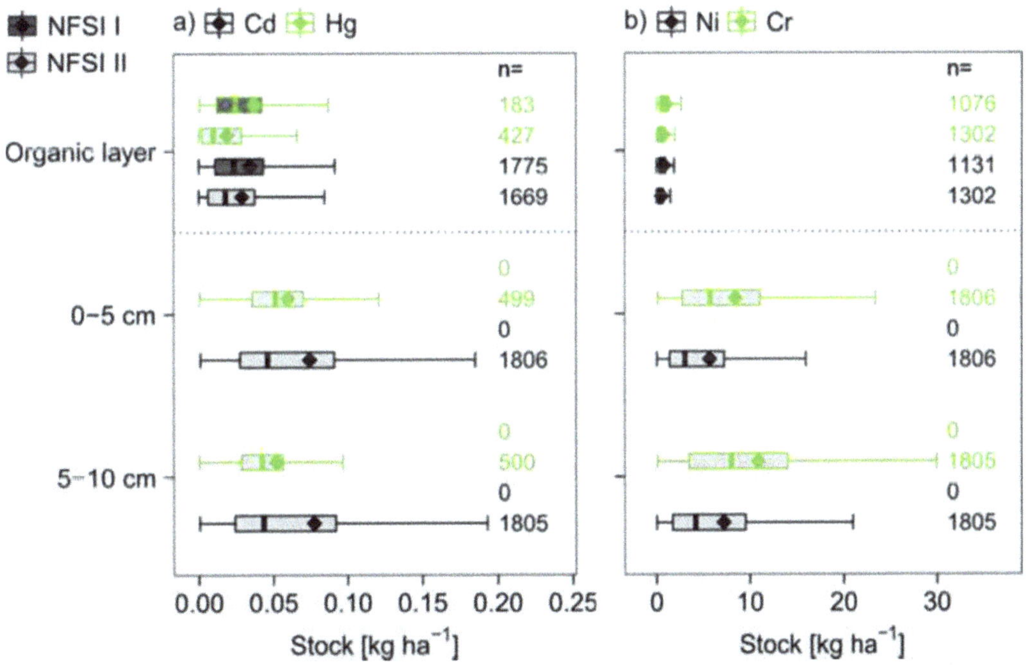

Fig. 7.2 Depth gradients of stocks for the elements (**a**) cadmium (Cd) and mercury (Hg), (**b**) nickel (Ni) and chromium (Cr) for the organic layer and the first two mineral soil increments (0–5, 5–10 cm)

case of organic layer the stocks of NFSI I for the elements Pb, Cd, Cu, Cr, Ni and Zn. Unless otherwise specified, means or comparisons of middle values are in relation to the median. The elements with data from the NFSI II encompassed all depth levels. In the NFSI I, only the data for the organic layer can be considered as this inventory did not measure heavy metals for the mineral soil layer throughout Germany. Two heavy metals are shown in each figure; the pairings are based on comparable stocks for the elements.

In terms of stocks, all the heavy metals studied have a right-skewed distribution since all elements have a few very high measurements. The upper quartile and upper whisker are therefore considerably farther from the median than are the lower quartile and lower whisker. For the same reason, the arithmetic mean for all heavy metals is greater than the median (Figs. 7.1 and 7.2). The magnitudes of the values and the depth profiles differ for each element; usually, the stocks in the organic layer are lower than that in the mineral soils because the organic layer is lacking the geogenic components. The lowest stocks both in the organic layer and in the mineral soil layers were found for Hg and Cd (NFSI II; organic layer; median; Cd, 0.02 kg ha^{-1}; Hg, 0.01 kg ha^{-1}) (Fig. 7.2), followed by As and Cu (NFSI II; organic layer; median; As, 0.14 kg ha^{-1}; Cu, 0.63 kg ha^{-1}) (Fig. 7.1) and Ni and Cr (NFSI II; organic layer; median; Ni, 0.34 kg ha^{-1}; Cr, 0.43 kg ha^{-1}) (Fig. 7.2). The highest stocks were documented for Pb and Zn (NFSI II; organic layer; median; Pb, 2.58 kg ha^{-1}; Zn, 2.86 kg ha^{-1}) (Fig. 7.1). The elements Ni, Cr and As have significantly lower stocks in the organic layer compared to the mineral soil, which points to a source that is primarily geogenic. For As, it is important to note that this element is present in an anionic form and therefore is only weakly bound in the organic layer, and rapid deposition into the mineral subsoil is possible. The elements Ni, Cr, As and Zn have the highest stocks at 5–10 cm, which (except in the case of As) is further evidence in support of an origin that is primarily geogenic. In contrast, the elements Pb and Hg have the highest stocks at 0–5 cm and decrease at the 5–10 cm depth, which suggests a significant anthropogenic impact (atmospheric sources) on the heavy metal stocks arising from geogenic processes. This trend is less obvious for Cd and Cu, which on average have similar inventories in the 0–5 cm and 5–10 cm layers.

In general, the stocks in the organic layer decreased for all heavy metals between NFSI I and NFSI II; there was particularly clear decline for Pb. The mean (arithmetic mean) decrease is significant for all the heavy metals analysed (Pb, Cd, Cu, Cr, Ni, Zn) (see Sect. 7.2.3).

In the following sections, the results for Pb and Ni are used by way of example to highlight and elucidate one element that is characterised by a more anthropogenic influence (Pb) and one that is more geogenic (Ni). Nevertheless, all heavy metals are discussed in the analyses. If an element features a deviation from the typical pattern exemplified by Pb or Ni, this case is explained in more detail.

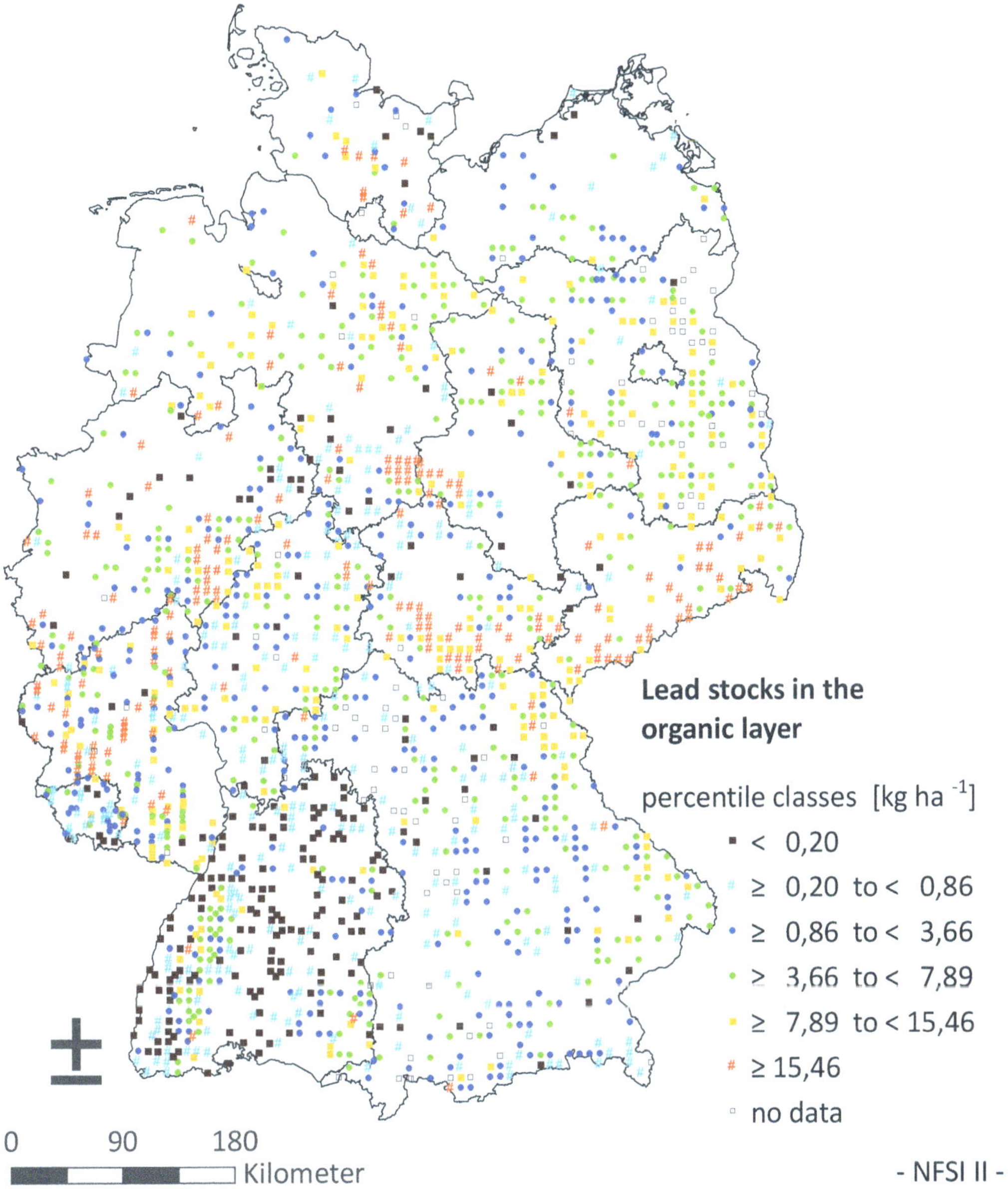

Fig. 7.3 Distribution of specific heavy metal stocks in the organic layer into six percentile classes for the forest soils of Germany using lead as an example

7.2.2 Spatial Distributions

7.2.2.1 The Organic Layer

The spatial distribution of heavy metal stocks in the organic layer is shown in Fig. 7.3 using Pb as an example. In the map, specific stock values are classified into one of six percentile groups. The difference between the stocks for the individual elements is comparatively low, as the levels of heavy metals are determined to a large extent by the stock levels of organic matter in the organic layer. Section 7.3

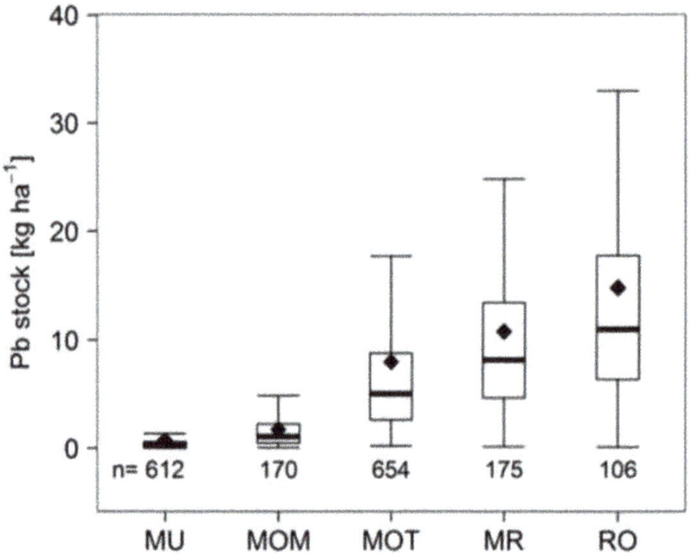

Fig. 7.4 Lead stocks [kg Pb ha^{-1}] in the topsoil (0–10 cm), differentiated according to humus type. *MU* mull, *MOM* mull-like moder, *MOT* typical moder, *MR* mor humus-like moder, *RO* mor humus

therefore addresses the differences in heavy metal concentrations in the organic layer.

The relationship between the amount of humus and heavy metal stock also becomes apparent in comparing with humus stocks (see Chap. 5). Hence, the thick organic layer of the acidic soils in the low mountainous regions (e.g. the Harz, Hunsrück, Thuringian and Bavarian forests) has significantly higher heavy metal stocks than do the calcareous regions with relatively thin organic layers (e.g. the Swabian and Franconian mountains and the hills of Lower Saxony). The soils of the lowlands of northern Germany have a greater small-scale variability of heavy metal stocks in the organic layer compared to sites in the low mountainous regions. Figure 7.4 shows the relationship between heavy metal stocks in the organic layer and the humus stocks using Pb as an example. The figure highlights the difference between regions with thinner organic layers and lower heavy metal levels in contrast to regions with thicker layers of humus and higher heavy metals stocks. For all heavy metals, the increase in the stocks thus follows the sequence MU $<$ MOM $<$ MOT $<$ MR $<$ RO.

7.2.2.2 Mineral Soil Layers

Compared to the levels in the organic layer, the heavy metal stocks in the mineral topsoil are much more strongly characterised by the heavy metal concentrations. This is understandable since the dry bulk density, with the exception of the organic soils, varies within a much narrower range compared to the concentrations of heavy metals. This can be clearly seen by comparing Figs. 7.5 and 7.17 with Ni as the example element.

The Ni stocks in the upper mineral soil layer (0–5 cm), like the stocks for Cr and Zn, which show a very similar distribution, feature a distinct dichotomy across Germany. While the lowlands of northern Germany have a clear predominance of

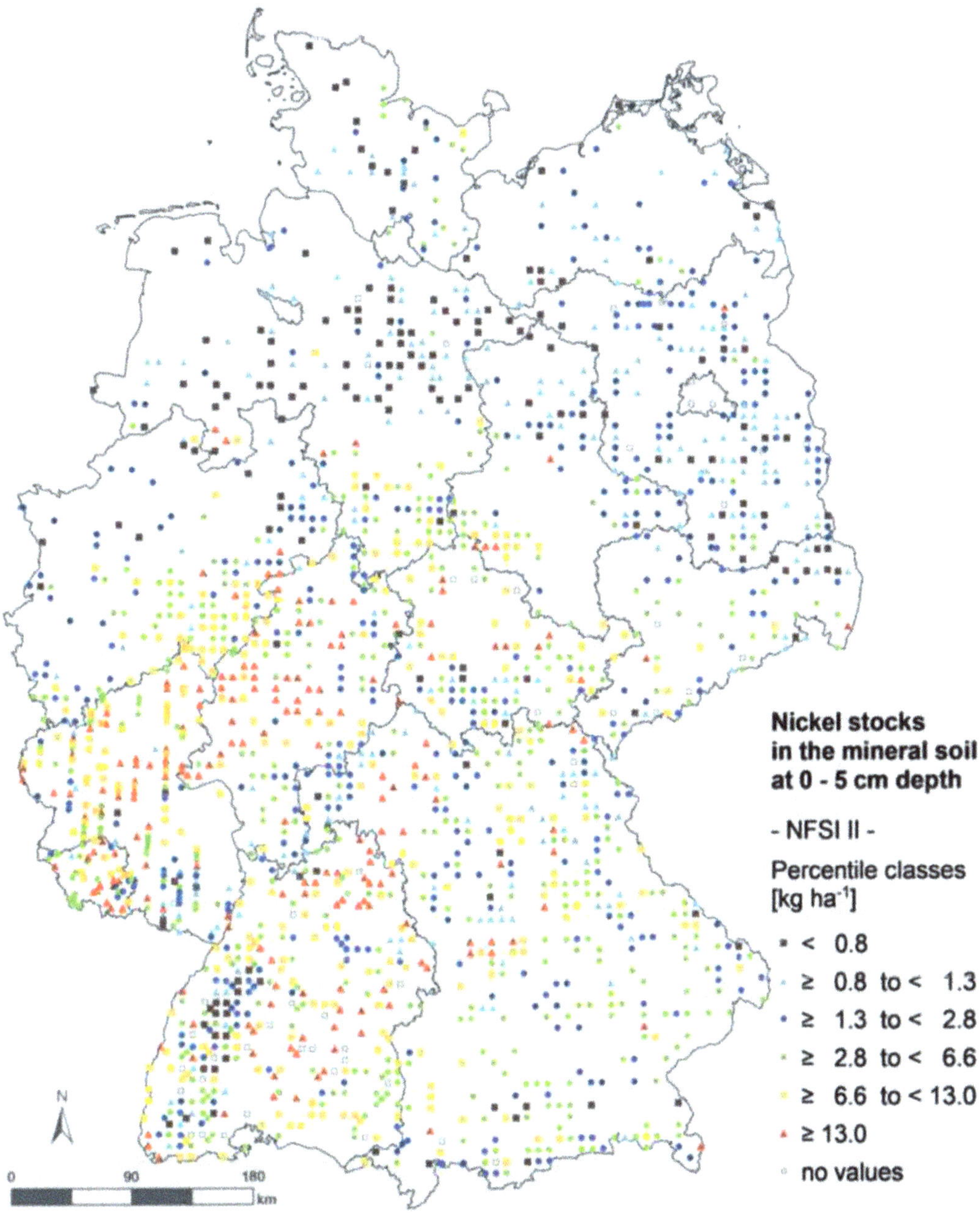

Fig. 7.5 Distribution of specific heavy metal stocks in the first layer of the mineral soil into six percentile classes (map volume Table II-6-50) for the forest soils of Germany using nickel as an example

sites with distribution of the element in the "lower" classes (0–10%, 10–25% and 25–50%), locations in southern Germany (low mountain ranges, alpine foothills, mountains) are mostly in the higher classes (50–75%, 75–90% and 90–100%). The mineral soils of forested areas in the Palatinate Forest, in Odenwald/Spessart and in the northern regions of the Black Forest have lower stocks (and also concentrations) than other low mountain forests (Fig. 7.17). This is consistent with the distribution of background values of Ni in the topsoil documented in the Soil Atlas of

Germany (BGR 2016), although with higher classes for locations in the Erzgebirge, in the Süderbergland and in places in the hills of Schleswig-Holstein.

7.2.3 Changes in Heavy Metal Stocks

The differences in heavy metal stocks in the organic layer between NFSI I and NFSI II are calculated based on the collective of paired samples. Because no paired measurements are available for As and Hg, differences cannot be determined for these elements. The time span between NFSI I and NFSI II varies between the respective sites. Whereas in most cases the time span accounts for about 17 years, it is less than 10 years for 94 sites.

Table 7.1 shows the change in heavy metal stocks in the organic layer between NFSI I and NFSI II for the entire country. It is apparent that the stocks of all the heavy metals measured (Pb, Cd, Cr, Cu, Ni, Zn) and the humus stock have declined. The largest decrease was documented for Pb, at −33%, while Cd, Cr, Cu and Ni are all at about −20%. Relative to the other heavy metals, the decline for Zn is lowest (−11%). Overall, the declines of the heavy metals exceed the decrease in the humus stock, which on average is only −2%.

Figure 7.6 illustrates the change in heavy metal stocks between NFSI I and NFSI II in g ha^{-1} $year^{-1}$ differentiated by humus type, using lead as the example. The pattern shown is similar for all the heavy metals (Cd, Cr, Cu, Ni, Zn). The scatter of absolute differences in heavy metal stocks between NFSI I and NFSI II diminishes with decreasing thickness of the humus type (MU < MOM < MOT < MR < RO). This result is based on the fact that heavy metal stocks are generally higher in thicker humus types. On average, the heavy metal stocks in the organic layer decreased most strikingly from NFSI I to NFSI II in the humus type MOM (Pb: −96.3 g ha^{-1} $year^{-1}$). In general, thinner sections of the organic layer (MU, MOM) showed decreases in heavy metals, while thicker organic layers (RO) instead showed an increase in heavy metals.

The difference in heavy metal stocks and the organic layer per year relative to the total stock for each humus type is shown in Fig. 7.6 using Pb as the example. This

Table 7.1 Average stocks (kg ha^{-1}) of the heavy metals in the organic layer and the humus stocks in the paired sample collective dataset for NFSI I and NFSI II as well as differences and changes (%)

	Pb	Cd	Cr	Cu	Ni	Zn	Organic layer
Median NFSI I	4.49	0.024	0.67	0.83	0.48	3.22	47,909
Median NFSI II	3.01	0.019	0.52	0.68	0.39	2.86	47,024
NFSI II – NFSI I	−1.47	−0.005	−0.14	−0.15	−0.09	−0.36	−885
Change (%)	−33	−20	−22	−18	−20	−11	−2
n	1183	1091	520	1183	551	1183	1224

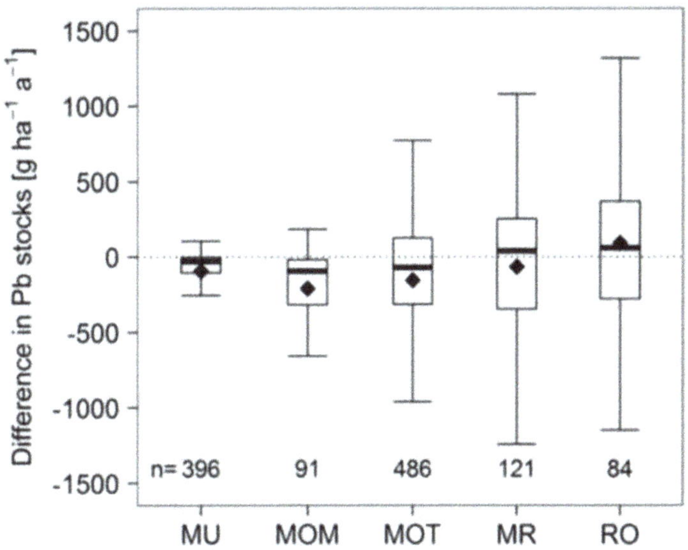

Fig. 7.6 Change in lead stocks [g Pb ha^{-1} $year^{-1}$] in the organic layer between NFSI I and NFSI II in relation to humus type. *MU* mull, *MOM* mull-like moder, *MOT* typical moder, *MR* mor humus-like moder, *RO* mor humus

figure clearly illustrates that separation according to humus type makes the scatter of the differences comparable. The differences in stocks correlated with humus type, with greater declines in stock in thinner humus types (Pb, MU −5.0%, MOM −4.2%) and slight increases in stock in thicker humus types (Pb, RO 0.9%). The decrease in heavy metal stocks in the humus types MU and MOM indicates a decline of atmospheric inputs of heavy metals in the period between NFSI I and NFSI II, since these humus types (as a result of their rapid mineralisation) have been completely reconditioned and accordingly reveal the current (at the time just prior to NFSI II) status of heavy metal input. As a whole, the decrease in heavy metal stock is more marked than the decrease in humus stock, and the increase in heavy metal stocks is less than the increase in humus stock. This result is further evidence to support the observation that the change in heavy metal stocks cannot be attributed solely to the change in humus stock; instead, an additional translocation of heavy metals into the mineral soil and/or a reduction of atmospheric deposition have taken place. This is most apparent for Pb (Fig. 7.7), Cr and Ni and to a lesser extent for Cu and Zn. This result is weakest for Cd (Fig. 7.8).

The decrease in atmospheric deposition of heavy metals in the period between NFSI I and NFSI II apparent in the change in heavy metal stocks in the organic layer is confirmed in Figs. 7.9 and 7.10 for Pb and Cd. These figures illustrate the clear decrease in inputs of Pb and Cd, differentiated by tree species and area, in 20 sample sites in Lower Saxony and Hessen between 1982 and 2014. The total inputs (canopy drip + litterfall + stemflow) are between 30 and 50% higher than inputs from canopy drip alone (Keuffel-Türk et al. 2012).

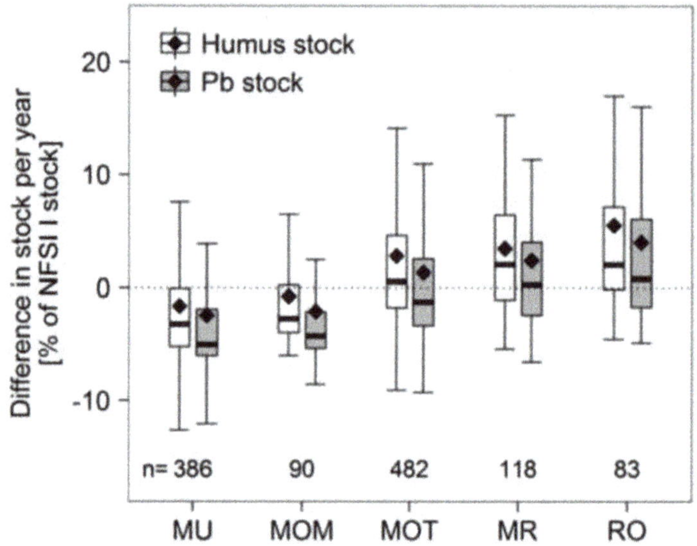

Fig. 7.7 Change in stock between NFSI I and NFSI II per year [% of the NFSI I content] for the humus content and lead in relation to humus type. Nine sites with a change in lead stock per year >100% were removed as outliers. *MU* mull, *MOM* mull-like moder, *MOT* typical moder, *MR* mor humus-like moder, *RO* mor humus

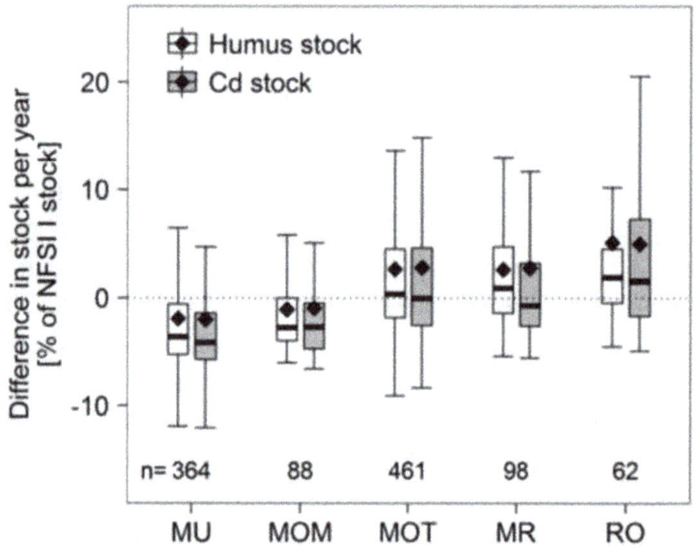

Fig. 7.8 Change in stock between NFSI I and NFSI II per year [% of the NFSI I content] for the humus content and cadmium in relation to humus type. Three sites with a change in cadmium stock per year > 100% were removed as outliers. *MU* mull, *MOM* mull-like moder, *MOT* typical moder, *MR* mor humus-like moder, *RO* mor humus

7.2.3.1 Impacts of Liming on Changes in Heavy Metal Stocks in the Organic Layer

Figures 7.11 and 7.12 illustrate the effect of liming on the heavy metal stocks in the organic layer. Depending on the humus type, liming the organic layer causes the humus stocks to decompose or negatively impacts the formation of humus by moving the organic substances into the upper layers of the mineral soil (see Chap. 6). The effect of heavy metal depletions is somewhat enhanced by liming. Hence, liming tends to reduce the heavy metal stocks in the organic layer. This applies to almost all the elements to the same extent but is somewhat more pronounced in thicker humus types such as mor humus. For lead, the effect of liming is weakest in mor humus, which may be related to the strong affinity of lead to form complexes and the simultaneous increase in pH.

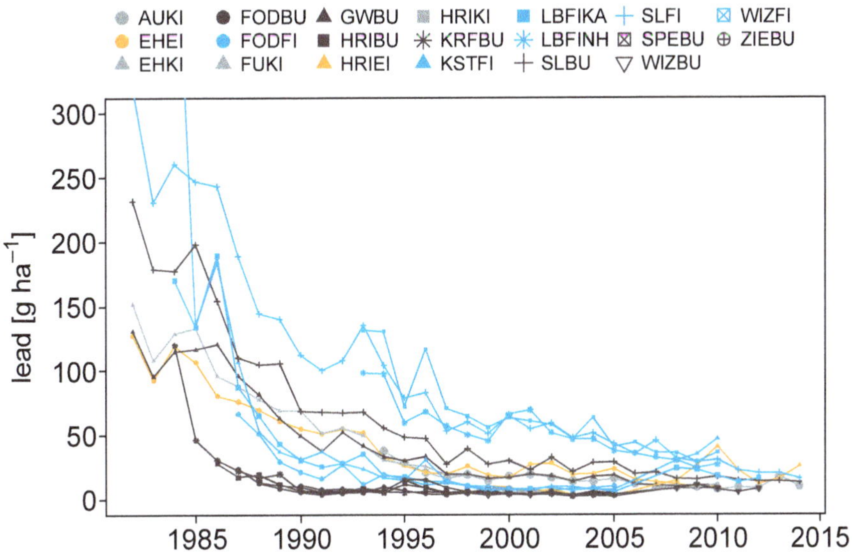

Fig. 7.9 Lead input with canopy drip for 20 sites in Lower Saxony and Hessen: red, beech stands (BU); blue, spruce stands (FI); grey, pine stands (KI); yellow, oak stands (EI); locations: *AU* Augustendorf, *FOD* Fürth/Odw, *GW* Göttingen Forest, *HRI* Hessian Ried, *LB* Lange Bramke (KA, ridge; NH, northern slope), *WIZ* Witzenhausen, *EH* Ehrhorn, *SPE* Spessart, *ZIE* Zierenberg, *KR* Krofdorf, *KST* Königstein

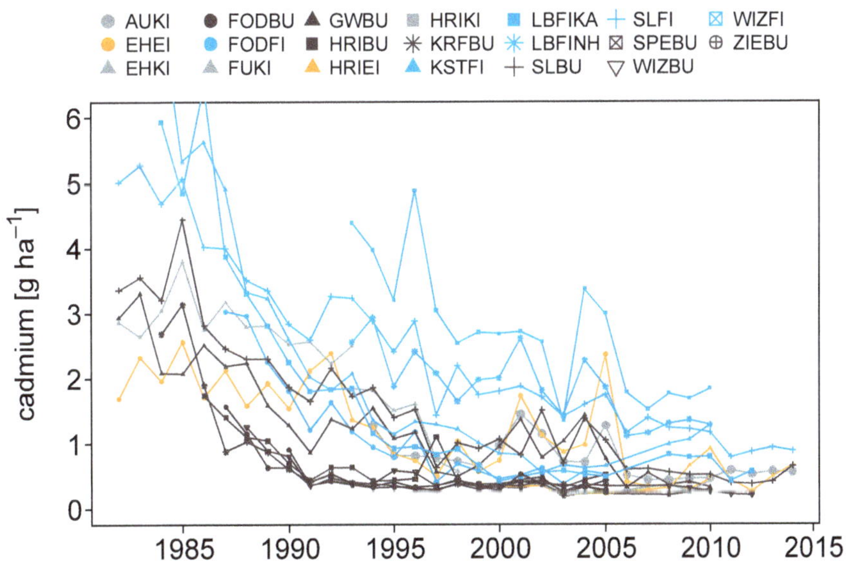

Fig. 7.10 Cadmium input with canopy drip for 20 sites in Lower Saxony and Hessen: red, beech stands (BU); blue, spruce stands (FI); grey, pine stands (KI); yellow, oak stands (EI); locations: *AU* Augustendorf, *FOD* Fürth/Odw, *GW* Göttingen Forest, *HRI* Hessian Ried, *LB* Lange Bramke (KA, ridge; NH, northern slope), *WIZ* Witzenhausen, *EH* Ehrhorn, *SPE* Spessart, *ZIE* Zierenberg, *KR* Krofdorf, *KST* Königstein

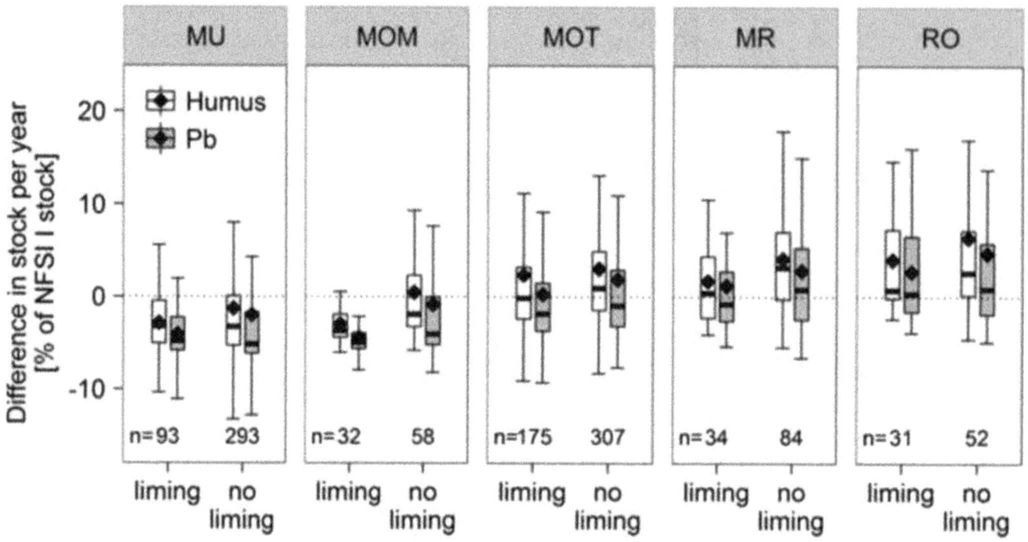

Fig. 7.11 Change in stock between NFSI I and NFSI II per year [% of the NFSI I stock] for the humus content and lead in relation to humus type and separated by sites that were limed or not. Nine sites with a change in lead stock per year > 100% were removed as outliers. *MU* mull, *MOM* mull-like moder, *MOT* typical moder, *MR* raw humus-like moder, *RO* raw humus

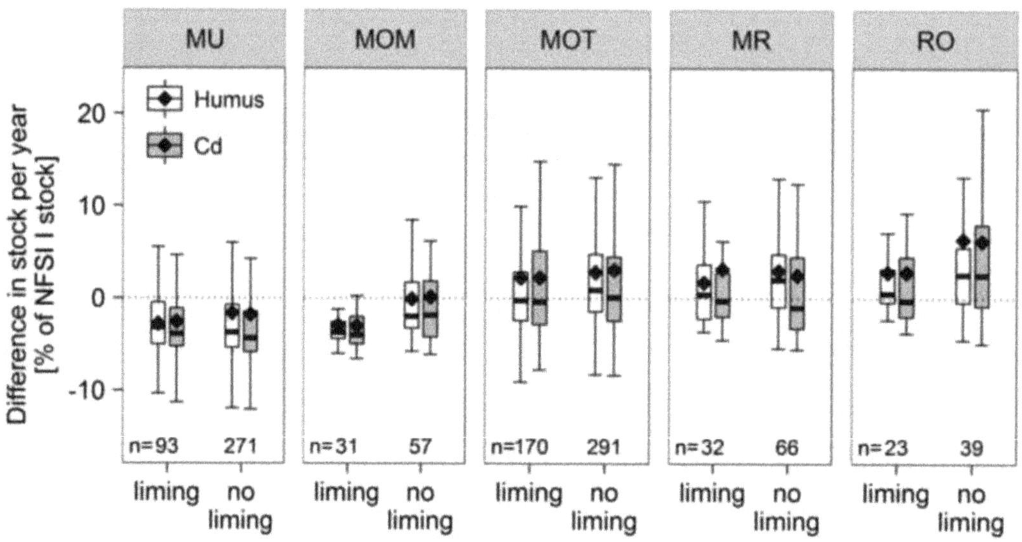

Fig. 7.12 Change in stock between NFSI I and NFSI II per year [% of the NFSI I stock] for the humus content and cadmium in relation to humus type and separated by sites that were limed or not. Three sites with a change in cadmium stock per year > 100% were removed as outliers. *MU* mull, *MOM* mull-like moder, *MOT* typical moder, *MR* mor humus-like moder, *RO* mor humus

7.2.3.2 Inventory Changes in the Soil (Organic Layer and Topsoil) Using the NFSI Plots in North Rhine-Westphalia

Interpretation of the changes in heavy metal stocks between NFSI I and NFSI II must also consider the changes in the mineral soil layers. Although data on heavy metal stocks in the mineral soil for the entire country are available only for NFSI II, in

North Rhine-Westphalia, these analyses were conducted for both NFSI I and NFSI II. Therefore changes in the heavy metal stocks between NFSI I and NFSI II can be explored for the mineral soil in this state.

Sampling for NFSI II in North Rhine-Westphalia consisted of a nationwide 8 × 8 km grid consolidated to 4 × 4 km squares in the lowland forests and in the Egge forested hills. Oriented on the middle inventory year of NFSI I and NFSI II, the comparisons below relate to a span of 17 years.

The humus stocks in the organic layer and in the mineral soil layers that had been measured directly in NFSI I and NFSI II were used as the basis for stock calculations. However, the quantity of fine soil in the mineral soil was determined only once for each inventory plot (NFSI I) and was then assumed as a constant in the comparisons of stocks.

Heavy metal stocks in the organic layer declined in the period between inventories for all elements (Table 7.2). The most distinct changes were observed for Pb and Hg. Mean stocks compared to NFSI I dropped by 59% for Pb and 51% for Hg. Stocks of the remaining heavy metals dropped over the same period by 33% for Cd and 46% for Cr, while the humus stock decreased by only 17% over this period.

Statistically, the changes in heavy metal stocks (arithmetic mean) were compared in two-way Gauss's test for stratified samples to better assess the signal strength of the changes after the first repeat of the inventory. With this test, the decrease in heavy metal stocks in the organic layer can be confirmed with a significance level of 1% as can the dry substance quantity in the organic layer.

In the mineral soil depth of 0–10 cm (Table 7.2), the stocks of a few of the heavy metals have changed in patterns deviating from the trend in the organic layer. In some cases, depletion in the organic layer is tied to an accumulation in the upper layers of the mineral soil. This type of shift in concentration to the mineral soil is seen for As, Pb, Cu, Hg and Zn. For some, the mean stocks in the 0–10 cm depth have increased only slightly by 8% (Cu) and 9% (Pb), whereas Hg increased significantly by 41%. At the same time, the decrease seen for Cd and Cr in the organic layers was continued in the upper layer of the mineral soil, with decreases of −16% and −14%, respectively, while the mean stocks of As, Ni and Zn showed only negligible change in the mineral soil compared to the first inventory.

The changes in stocks for the mineral soil (arithmetic mean) were not statistically significant for As, Cu and Ni. This result is most likely explained by the relatively high geogenic components in North Rhine-Westphalia, especially for Ni and As. In contrast are the highly significant changes for Cd and Cr that have taken place as a result of the decreased atmospheric inputs of these elements into the upper layers of the mineral soil. The overall increase in the quantity of humus in the mineral soil has had not caused an increase in the stocks of these heavy metals. In contrast, the highly significant increase in Pb and Hg in the mineral soil is due to both the influence of humus dynamics and the close correlation between the stocks of these heavy metals and organic matter. This relationship is further strengthened by inputs of Hg that are presumably increasing. The significance level for the slight increase in concentration or stocks of Zn in the mineral soil is at 90%.

Table 7.2 Average stocks [kg ha^{-1}] for heavy metals and humus and changes between NFSI I and NFSI II in the organic layer, the mineral soil (0–10 cm) and the total compartment (organic layer + mineral soil) in North Rhine-Westphalia

	As	Pb	Cd	Cr	Cu	Ni	Hg	Zn	Organic layer
Organic layer									
Median NFSI I	0.29	8.12	0.032	0.77	1.09	0.58	0.018	4.95	43,084
Median NFSI II	0.16	3.34	0.021	0.41	0.61	0.37	0.009	2.93	35,672
NFSI II – NFSI I	−0.13	−4.78	−0.010	−0.35	−0.48	−0.21	−0.009	−2.02	−7412
Change (%)	−45	−59	−33	−46	−44	−37	−51	−41	−17
Mineral soil									
Median NFSI I	8.58	62.7	0.177	23.25	10.00	10.80	0.086	41.68	
Median NFSI II	8.90	68.3	0.148	20.05	10.76	11.16	0.121	42.53	
NFSI II – NFSI I	0.33	5.62	−0.029	−3.19	0.76	0.36	0.035	0.84	
Change (%)	4	9	−16	−14	8	3	41	2	
Total stocks									
Median NFSI I	9.00	75.41	0.221	24.33	11.35	11.56	0.115	49.18	
Median NFSI II	9.11	74.63	0.176	21.06	11.79	11.80	0.136	48.56	
NFSI II – NFSI I	0.11	−0.79	−0.045	−3.28	0.44	0.24	0.021	−0.63	
Change (%)	1	−1	−20	−13	4	2	18	−1	

Because of the relatively low quantities of dry matter in the organic layer, the total of the stocks in the mineral soil plus the organic layer (Table 7.2) are determined primarily by the stocks in the mineral soil. Considering the entire layer from topsoil to a depth of 10 cm, based on NFSI II (median), the quantities of heavy metals bound in the organic layer make up a maximum of 12% of the total stocks.

As the comparison of the average total stocks shows, stocks of Pb, Cu, Zn, Ni and As in the topsoil have not changed significantly. The quantities of heavy metals that have been released from the organic layer are mirrored in the mineral soil to a depth of 10 cm, with the exception of Hg, Cd and Cr. While the Cr stock decreased by 13%, the amount of Cd declined by as much as 20%. The change in Hg is most striking; for this heavy metal, the increase by 18% is opposite to the general trend. This increase in Hg in the mineral soils clearly exceeds the reduction that was documented for the organic layer.

While the change in the total stocks of Cd, Cr and Hg in the topsoil was significant at a 1% level, the changes in Pb, Cu, Zn, Ni and As were not significant. For these elements the changes in the organic layer and the mineral soil cancel each other out. Thus, most heavy metals have no significant shift within the depth profile into deeper mineral soil layers. A slight translocation to the deeper mineral soil layers was observed only for Cd, which is a relatively mobile heavy metal. According to the data from the NFSI, at this plots an increase in the atmospheric inputs of Hg is assumed; however, this conclusion needs to be verified with more data.

7.3 Heavy Metal Concentrations in the Organic Layer and Mineral Soil

7.3.1 Spatial Distribution

7.3.1.1 Organic Layer

The spatial distribution of heavy metals concentrations in organic layer apparently differs throughout Germany. In order to help elucidate these differences, the maps for Pb and Zn concentrations are compared exemplarily for the total set of heavy metals investigated. Unlike the heavy metal stocks in the mineral topsoil (0–5 cm), for which there is a clear division in the mapping of the data across Germany (see Sect. 7.2.2), the distribution of heavy metal concentrations in the organic layer features a smaller-scale variability and hence suggests smaller-scale input pathways. For example, for Pb both atmospheric input and direct input from mining activities are evident (Fig. 7.13). The Süderbergland located in the leeward side of the industrial centres of North Rhine-Westphalia and sites in the higher elevations of the central German lowlands (e.g. the Thuringian Forest) are examples of sites with the highest Pb concentrations. Considerable impact from mining is evident in the Harz region and for As also in the Erzgebirge. The distribution of Cu is similar to that of Pb but with a few locations in Hessen and the Nuremburg/Fürth region in higher

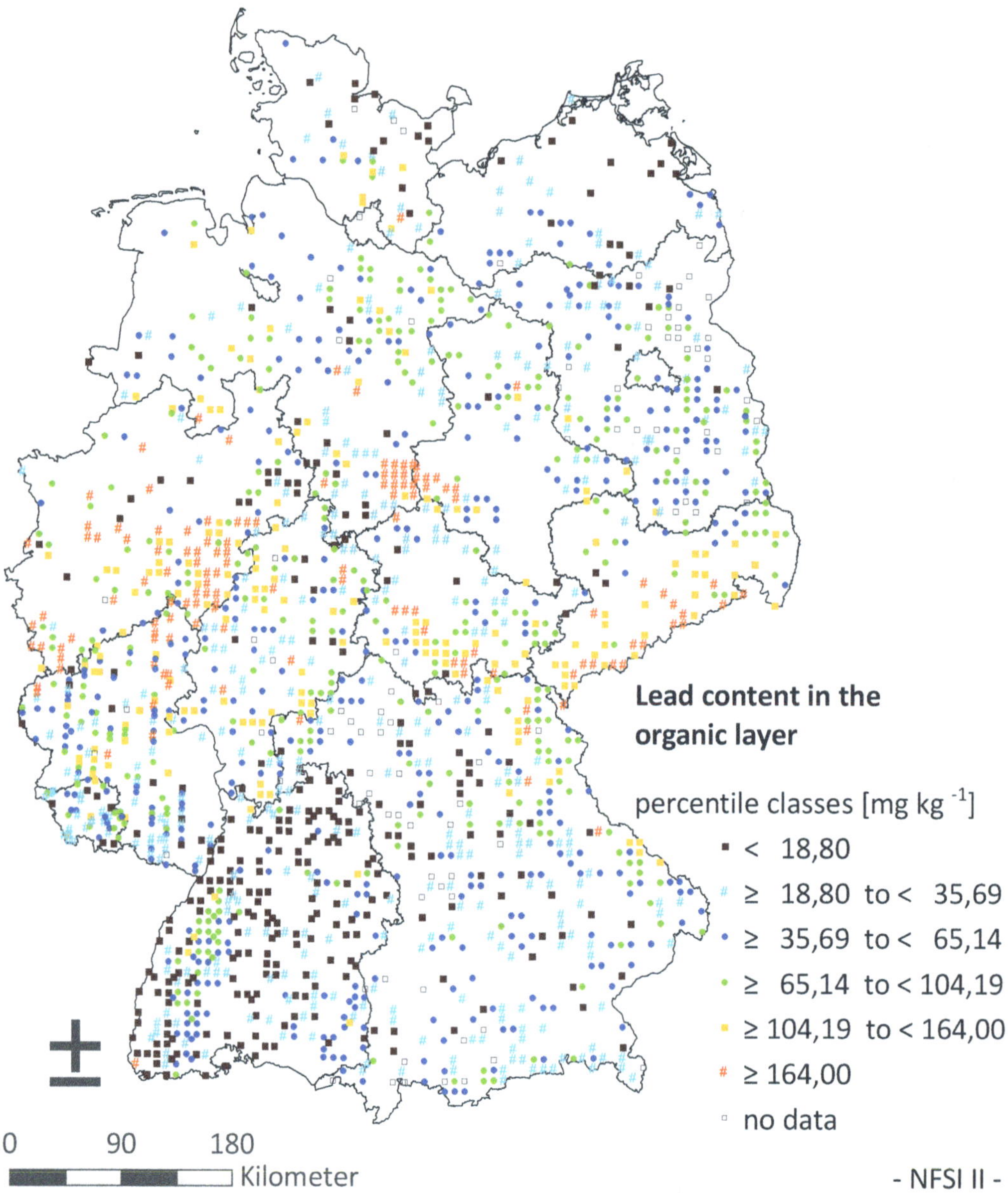

Fig. 7.13 Distribution of heavy metal content measured in the organic layer into six percentile classes for the forest soils of Germany using lead as an example

percentile classes. Figure 7.14 depicts the distribution of Zn concentrations in the organic layer. Elevated Zn concentrations are found in the regions around Nuremburg/Fürth and eastwards as far as the Upper Palatinate Forest. Similarly in the higher percentile classes are the Zn and also the Cd concentrations in the Alps; in contrast, the Cr, Ni and Pb concentrations in the organic layer of the Alps are below average. Conspicuously low Zn concentrations were documented in the organic layers in sites in north-eastern Germany. The heavy metal concentrations in the Black Forest organic layer are also relatively low.

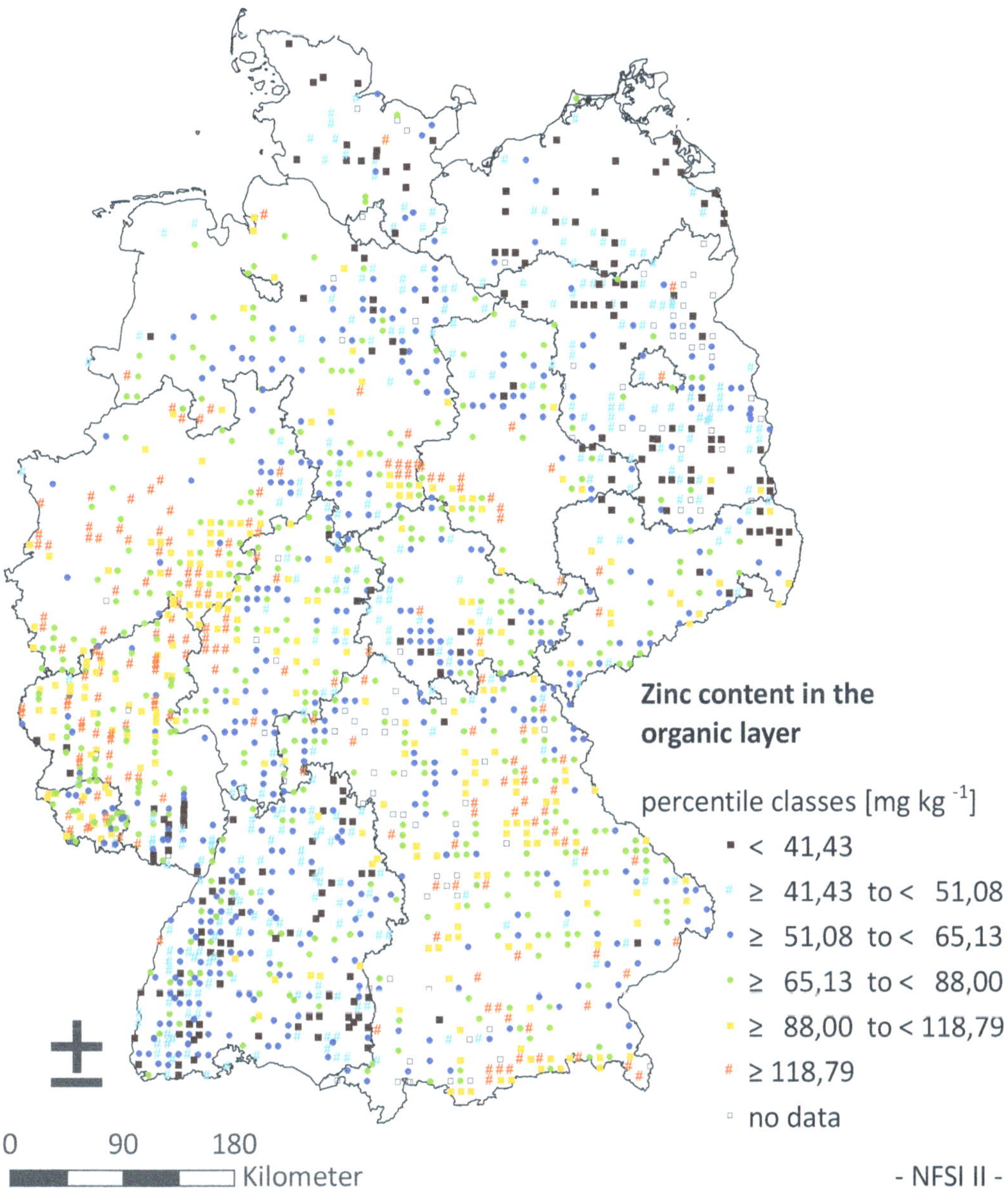

Fig. 7.14 Distribution of heavy metal content measured in the organic layer into six percentile classes for the forest soils of Germany using zinc as an example

Heavy Metal Concentrations Differentiated by Organic Horizon

Figure 7.15 differentiates the heavy metals in the organic layer according to the concentration in the organic horizons using principal component analysis. It appears that the variance in heavy metal concentrations can be represented well with just two components. Component 1 (PC1) and component 2 (PC2) alone explain 66.5% of the total variance: PC1 explains 49.0%, and PC2 explains 17.5% of the variance.

There is clear differentiation by organic layer horizon. From bottom left to top right, there is a gradient in the sequence L – L + Of ≈ Of ≈ L + Of + Oh ≈ Of + Oh – Oh, with a clear distinction between the L- and Oh-horizons and between the mixed and Of-horizons. The vectors for As and Pb and somewhat less so for Cu, Cr

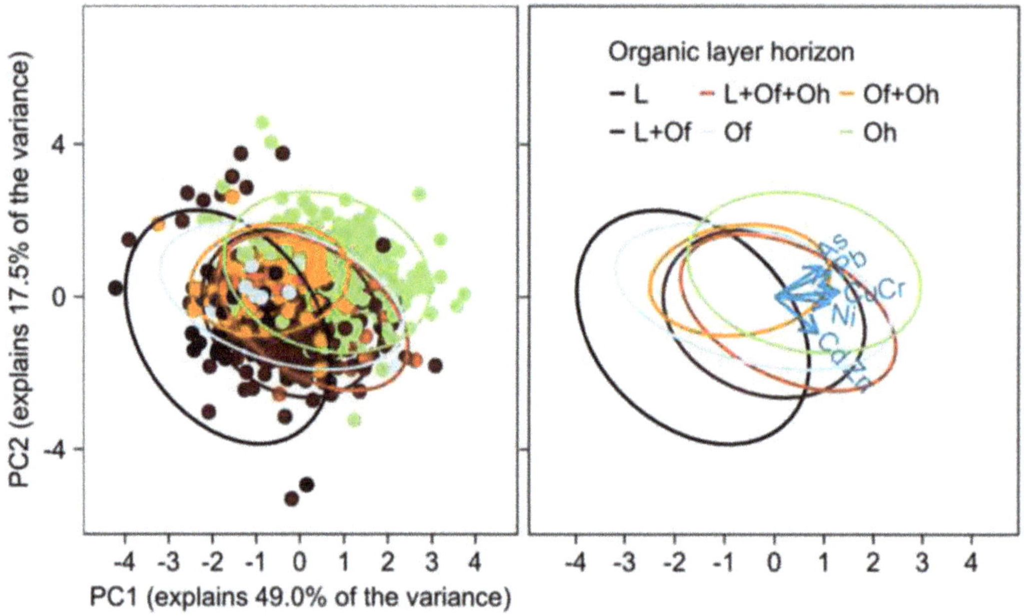

Fig. 7.15 Principal component analysis (PCA) for heavy metal contents [mg kg^{-1}] in the organic layer. The element mercury is not shown due to the low number of count sites in the organic layer. Organic layer horizons are distinguished by colour. Left: All measurements and the 95% confidence regions. Right: Vectors for the heavy metals and 95% confidence regions

and Ni correlate with these gradients, whereas the vectors for the elements Cd and Zn are orthogonal to the gradients along the organic layer horizons. This means that unlike the other elements, Cd and Zn are not differentiated by organic layer horizon. The direction of the vectors also indicates that the highest concentrations of the elements analysed, other than Cd and Zn, are found in Oh-horizons. This pattern can be explained based on the significantly weaker binding of Cd and Zn to organic substances in the organic layer.

Differentiation of the concentrations of Cd and Pb by organic layer horizon is shown in Fig. 7.16. As highlighted by the principal components analysis, no differentiation is apparent for Cd – the concentrations are all comparably low in all horizons; in contrast, Pb concentration is significantly higher in the Oh-horizon.

7.3.1.2 Mineral Soil

The distribution of concentrations in the uppermost layer of the mineral soil largely corresponds with the stocks distribution (see Sect. 7.2.2). For example, Fig. 7.17 shows the spatial distribution of concentrations for the element Ni in 0–5 cm of the mineral soil. As explained in Sect. 7.2.2, the observed distribution of Ni concentrations in the topsoil corresponds closely with the distribution of background values for Ni in the topsoil (BGR 2016). This result is clear evidence of the influence of the parent rock on Ni concentration in the mineral soil; this topic will be discussed in more detail in the next section.

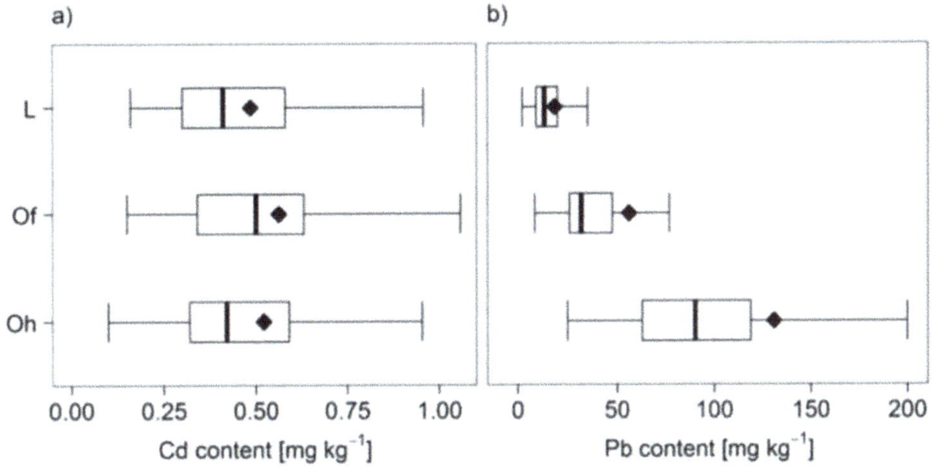

Fig. 7.16 Differentiation of the contents [mg kg^{-1}] of (**a**) cadmium and (**b**) lead by horizon of the organic layer. $N = 81$ per boxplot

7.3.2 *Influence of the Parent Rock on Heavy Metal Concentrations in the Mineral Soil*

One of the most important factors influencing heavy metal concentrations in the mineral soil is the parent rock (see Chap. 2). Heavy metal concentrations in the mineral soil vary depending on the geogenic base concentration of the parent rock. Heavy metal concentrations elevated as a result of geogenic processes occur in particular in soils derived from periglacial layers over basic magmatic and metamorphic rock (Cr, Ni, Cu, Zn), clay rock (especially Liassic clay) (As, Cd, Cr, Cu, Ni, Pb, Zn), acidic metamorphic rock (Cu, Ni) and mica slate (Cr, Cu, Ni, Zn), as well as carbonate rocks as a result of accumulation of residual clay.

Figures 7.18 and 7.19 depict the differentiation of concentrations by parent rock using Pb and Ni as examples. Differentiation of concentrations by parent rock was based on the classifications of parent rocks throughout Germany into 16 classes by the Federal Institute for Geosciences and Natural Resources (Bundesanstalt für Geowissenschaften und Rohstoffe, BGR). Sediments in littoral zones (Unit 1), sandy loess (Unit 8) and tephra (Unit 13) are three classes not shown here as they have little relevance on a nationwide basis for forested sites. Sediments and alluvial soils (Unit 2) are not listed as the heavy metal concentrations of these substrates differ considerably depending on region and the source area of the watercourses; hence, general conclusions are not valid. Because only the topsoil is considered, till and boulder clay with sandy surface layers (Unit 5) are combined with sand and thick sandy surface layers (Unit 4) for this analysis.

In general, when differentiated by parent rock type, concentrations of heavy metals that are significantly characterised by anthropogenic processes, such as Pb, are strongly overlaid by inputs from emissions. This pattern is also apparent in the result that concentrations in the 0–5 cm depth layer always exceed those in the 5–10 cm depth. This result is particularly obvious for clay soils (Unit 10), where Pb

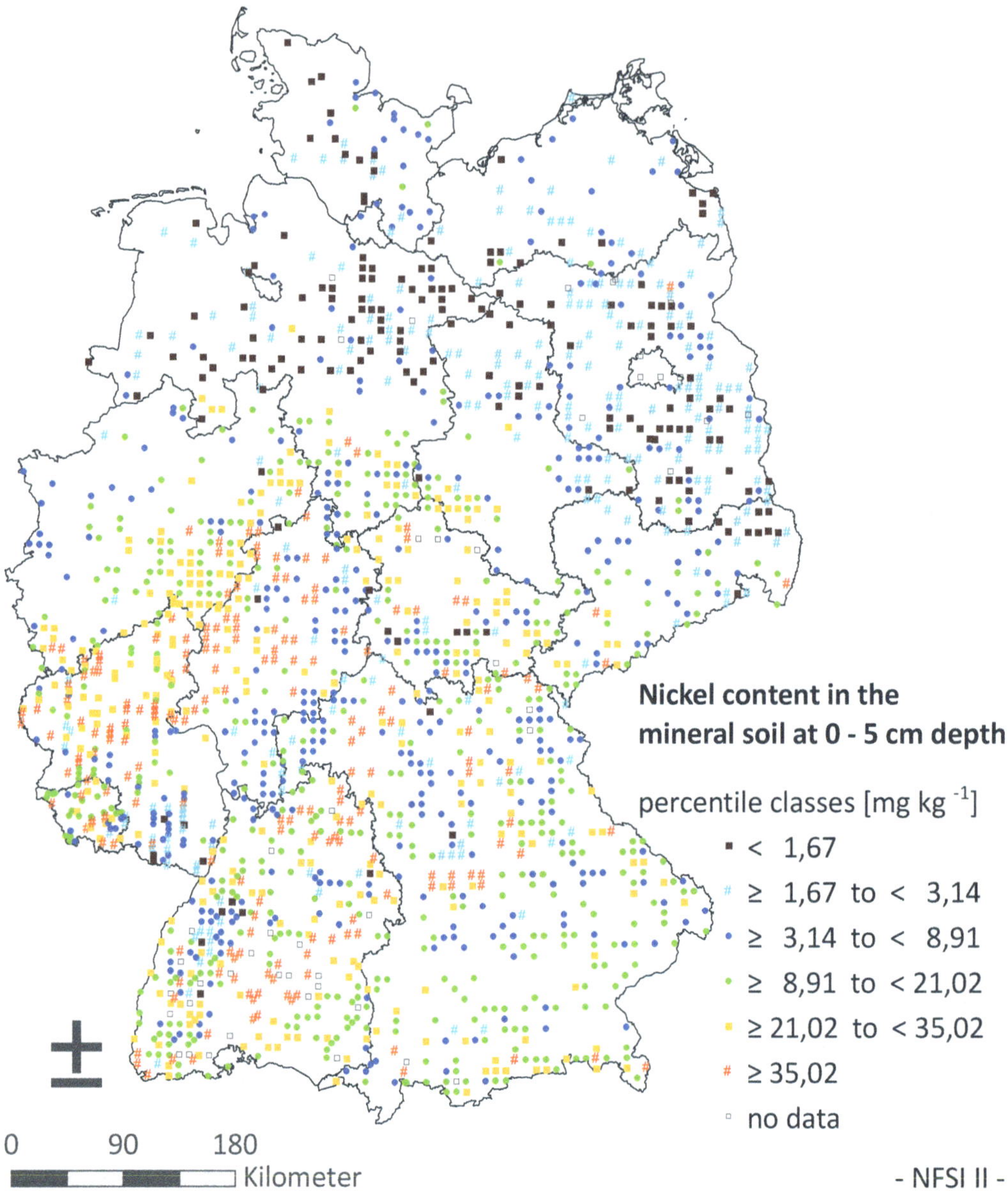

Fig. 7.17 Classification of the measured heavy metal contents in the first layer (0–5 cm) of the mineral soil into six percentile classes for the forest soils of Germany using nickel as an example

as well as As feature the highest concentrations. The explanation lies both in a high binding capacity in soils rich in clay as well as the spatial distribution of clay stones across Germany. Clay stones are particularly predominant in the Rhenish Slate Mountains leeward of the Ruhr district known as an industrial and steel production centre, and hence emissions of heavy metals are high. Concentrations that are above average on clay soils also include the elements Cu, Cr, Ni and Zn. Cd and Zn are highest on carbonate rocks (Unit 9), as these elements are naturally present in this rock type and the high soil pH due to the carbonate which favours binding of heavy

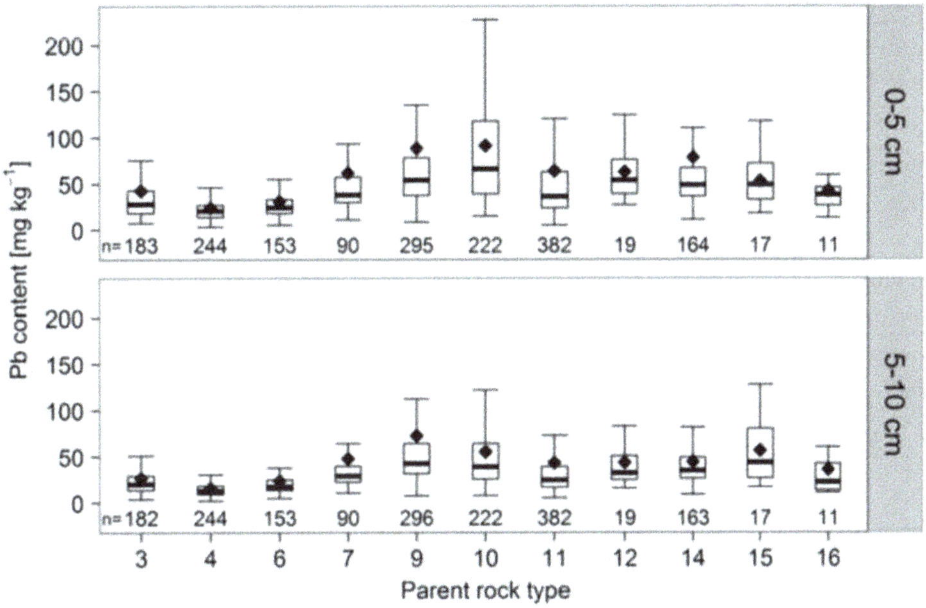

Fig. 7.18 Lead content [mg Pb kg^{-1}] differentiated by parent rock for the depth increments 0–5 cm and 5–10 cm. Numbering units: (3) terrace and gravel sediments, (4) sand and thick sandy surface layers plus till and boulder clay with sandy surface layers, (6) till and boulder clay, (7) loess and loess derivatives, (9) carbonate rock, (10) clay, (11) sandstone and other silicate rocks, (12) basic and intermediate magmatic and metamorphic rock, (14) acidic magmatic and metamorphic rock, (15) peat bog, (16) anthropogenic soil

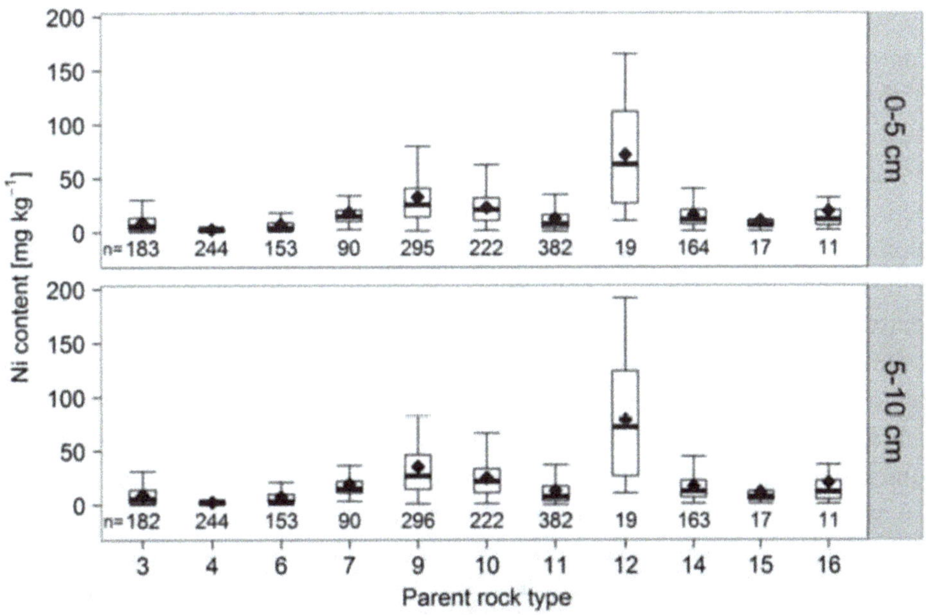

Fig. 7.19 Nickel content [mg Ni kg^{-1}] differentiated by parent rock for the depth increments 0–5 cm and 5–10 cm. An explanation of the numbering system for parent rock type is given in the caption for Fig. 7.18

Table 7.3 Proportion of the variance in heavy metal concentrations explained by the stratum parent rock type [R^2adj] and *F*-values and *p*-values. Mercury is not included in the analysis since data was available only for North Rhine-Westphalia, Rhineland Palatinate, Hessen and Saarland

	Parent rock type		
Heavy metal	R^2adj	*F*-value	*p*-value
Arsenic	0.34	126.8	<0.001
Lead	0.25	81.7	<0.001
Cadmium	0.29	100.9	<0.001
Chromium	0.47	211.0	<0.001
Copper	0.33	117.7	<0.001
Nickel	0.43	182.0	<0.001
Zinc	0.41	167.2	<0.001

metals. The elements As, Cr, Cu, Ni and Pb also have above average concentrations on carbonate rocks (Unit 9).

A different pattern emerges in the differentiation by parent rock type for the elements Ni and Cr, which are primarily geogenic in origin (Ni, Fig. 7.19). This figure indicates that the concentrations in 5–10 cm are higher than those in the 0–5 cm layer. Because of this high basic geogenic concentration, the highest concentrations of Cr and Ni, as well as Cu, are on basic and intermediate magmatic and metamorphic rocks (BAG 12). The elements Cd, Zn and Pb also have higher than average concentrations for this rock category. Consistently low heavy metal concentrations are found for terrace and gravel sediments (Unit 3), sandy and thick sandy surface layers (Unit 4), as well as till and boulder clay (Unit 6). Mid-range concentrations for all elements are found for loess and loess derivatives (Unit 7). Sandstone and other silicate rocks (Unit 11) feature low (Cd, Cr, Ni) to moderate (As, Cu, Pb, Zn) heavy metal concentrations, while the class of acidic magmatic and metamorphic rocks (Unit 14) have moderate (Cd, Cu, Zn) to high (As, Cr, Ni, Pb) concentrations. The pattern for peat bog (Unit 15) is relatively heterogeneous: there are elements with concentrations that are low (Cr, Ni), moderate (As, Cu, Zn) as well as high (Cd and Pb). Because moors are not likely to have high underlying geogenic concentrations of heavy metals, it is apparent that in particular Pb and Cd are deposited from anthropogenic sources via air pathways. Anthropogenic soils (Unit 16) have low (Cr, Cu, Ni) to mid-range (As, Pb, Cd, Zn) heavy metal concentrations.

Table 7.3 indicates the strength of the relationship between heavy metal concentrations and the parent rock type based on explained variance (R^2_{adj}) and the *F*-value. All 16 classes of parent rock were included in this analysis. The element Hg was not included due to the low sample size. The elements Pb and Cd show the lowest R^2_{adj} and *F*-values, presumably as a result of the considerable impact of anthropogenic factors for the concentrations of these elements in the topsoil. As and Cu have mid-range R^2_{adj} and *F*-values, whereas the values for Cr, Ni and Zn indicate the strongest relationship with parent rock type. The latter result is evidence of the importance of the geogenic origin of these three elements.

Fig. 7.20 Ratio of heavy metal content in the organic layer to the mineral soil (0–10 cm) for seven heavy metals. $N = 806$ for each element

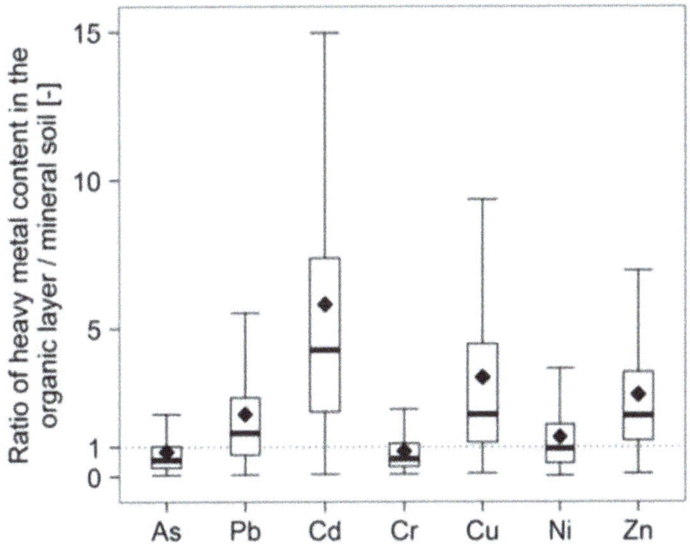

The ratio between heavy metal concentrations in the organic layer and the mineral soil (1–10 cm) is shown in Fig. 7.20. With equal stocks, the heavy metal concentrations differ due to their mass reference between the organic layer and the mineral soil based on the significantly lower dry bulk density of the organic layer, by up to a factor of 5–7; hence the heavy metal concentration in the organic layer should be five to seven times greater as the concentration in the mineral soil. In fact, the heavy metal stock in the organic layer accounts for only 12% of the stock in the mineral soil (see Sect. 7.2.3), such that the effect of the different dry bulk densities on the heavy metal concentrations in the organic layer and the mineral soil approximately cancel out. The relationship between heavy metal concentrations in the organic layer and the mineral soil depicted in Fig. 7.20 is therefore best evaluated on a comparative basis. For elements with inputs that are primarily from anthropogenic (atmospheric) processes, a significantly higher proportion (>>1) of the concentrations are expected in the organic layer compared to the mineral soil. However, other factors that affect this ratio must be considered, including the ionic charge of the heavy metal compounds present in the soil and uptake of the element by plants and thus the amount of the metal deposited through litterfall. Hence, elements that are present in the soil principally in anionic form, such as As, are less bound in the organic layer and are therefore transferred quickly into the mineral soil. In contrast, elements such as Cu and Zn are sometimes taken up from the mineral soil by plants and returned to the organic layer in litterfall ("vegetation pump"). The ratio for As is relatively low, which can be attributed to its more geogenic origin and anionic binding. Chromium and Ni also have moderate ratios < 1, confirming the predominantly geogenic origin of these elements. Cadmium, Cu, Zn and Pb have ratios considerably > 1; the "vegetation pump" may play a role for Cu and Zn as well. However, the high ratios for Pb and Cd must be attributed to input pathways that are mostly anthropogenic (atmospheric).

In Fig. 7.21, the depth profile for the concentrations of a heavy metal with a more anthropogenic influence (Pb) is contrasted with an element with a more geogenic

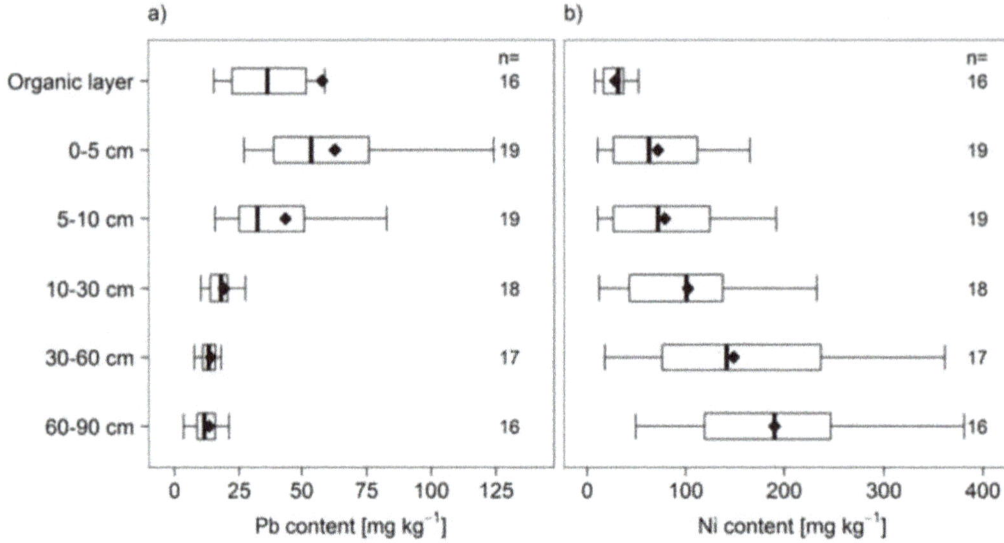

Fig. 7.21 Depth profile for contents [mg kg^{-1}] of (**a**) lead and (**b**) nickel for the parent rock class basic and intermediate magmatic and metamorphic rock (Unit 12)

origin for the entire profile on basic and intermediate magmatic and metamorphic rock (Unit 12). For Pb, concentrations are higher in the topsoil (0–5 cm) and rapidly decrease with increasing depth, whereas the opposite is true for Ni. This contrast clearly reveals the anthropogenic (atmospheric) overprinting of Pb within this parent rock class and confirms the primarily geogenic origin of the element Ni for soils on basic and intermediate magmatic and metamorphic rock.

7.3.3 Evaluation of Heavy Metal Concentrations in the Mineral Soil

The precautionary values listed in Annex 2 No. 4.1 of the Federal Soil Protection and Contaminated Sites Ordinance (Bundes-Bodenschutz- und Altlastenverordnung; (BBodSchV 1999)) can be used to evaluate the heavy metal concentrations in the mineral soil. In accordance with § 8 (2) of the Federal Soil Protection Act (Bundes-Bodenschutzgesetz; BBodSchG (1998)), exceedance of precautionary values for soil, when taking into account the harmful substance concentration from sources that are geogenic or due to extensive settlement, generally raises concern that harmful changes to the soil have taken place pursuant to § 9 BBodSchV. The precautionary values thus do not represent a danger threshold but instead indicate a threshold for concern, which, if exceeded, demands that precautionary measures such as avoiding or reducing further inputs should be taken.

So far, the BBodSchV (1999) regulates the precautionary values for seven heavy metals (Cd, Pb, Cr, Cu, Hg, Ni, Zn). An amended edition of the BBodSchV will also include precautionary values for As and thallium (Tl). The present evaluation of

Table 7.4 Precautionary values for inorganic substances (in mg kg^{-1} dry mass, fine soil aqua regia digestion)

Heavy metal	Sand	Loam/silt	Clay
Arsenic	10	20	20
Lead	40	70	100
Cadmium	0.4	1	1.5
Chromium	30	60	100
Copper	20	40	60
Nickel	15	50	70
Mercury	0.2	0.3	0.3
Zinc	60	150	200

Table 7.5 Proportion [%] of seven heavy metals in relation to class in terms of the extent to which they exceed or fall short of their precautionary value. "Other" includes missing values as well as peat bogs and sites with humus concentrations > 30 mass-%, for which no precautionary value has been defined

	Proportion [%]						
	As	Pb	Cd	Cr	Cu	Ni	Zn
≤50% of the precautionary value	54.9	32.0	82.0	65.6	77.0	70.1	68.2
>50% ≤100% of the precautionary value	26.8	38.9	8.6	21.8	12.4	16.8	20.3
>100% ≤150% of the precautionary value	6.7	11.3	1.8	3.5	2.4	3.7	3.0
>150% ≤200% of the precautionary value	2.0	4.8	0.4	0.8	0.4	1.1	1.1
>200% of the precautionary value	2.5	6.0	0.2	1.4	0.8	1.4	0.4
Others	7.0	6.9	6.9	6.9	6.9	6.9	6.9

heavy metal concentrations in the mineral soil references the precautionary values proposed in the third working draft of the revision of the BBodSchV (2015); hence, the same reference scale is used for all the heavy metal concentrations analysed. In addition, this reference scale allows all mineral soils to be included, whereas the BBodSchV (1999) is officially valid only for soils with humus concentration up to 8 mass-%. Table 7.4 lists the precautionary values from the third working draft of the BBodSchV (2015). The values are differentiated for the heavy metals according to the main soil types: sand, loam/silt and clay.

The heavy metal concentrations were then correlated with their precautionary values to evaluate the concentrations in the mineral soil. Six classes were identified, and the proportion of each element in these six classes was calculated as a percentage. Table 7.5 shows the results of this process.

The results of this analysis indicate that in the majority of forest soils in Germany, the heavy metal concentrations are below the precautionary values. For the elements As, Cd, Cr, Cu, Ni and Zn, the majority of concentrations are even below or equal to 50% of the precautionary values. This is a positive result, in that it makes clear that the forest soils are mostly in good condition in terms of heavy metal concentrations. In the country as a whole, the elements As and Pb exceed the precautionary levels by

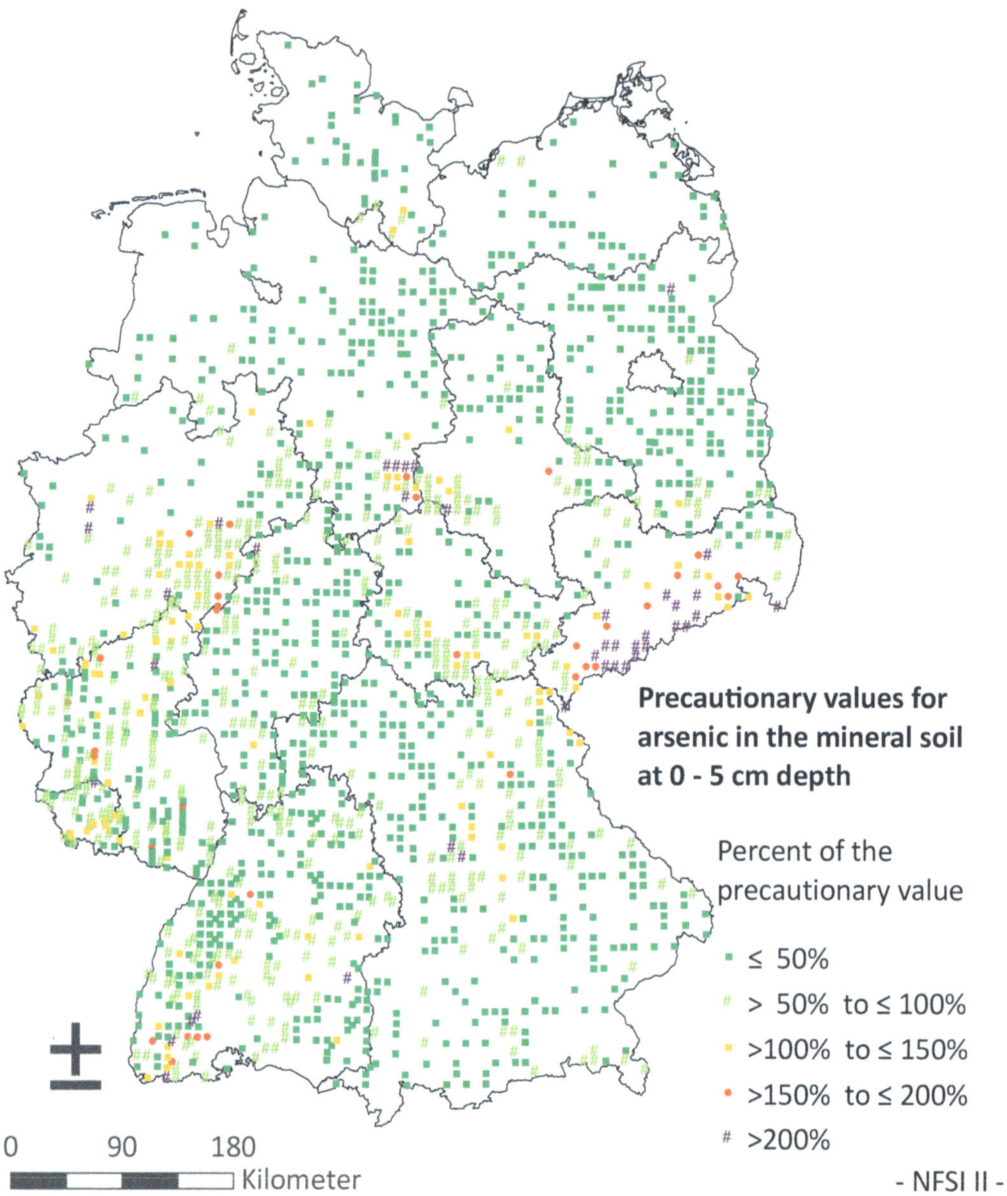

Fig. 7.22 Spatial distribution of arsenic in terms of the extent to which they exceed or fall short of their precautionary values in mineral soil (0–5 cm) for forest soils in Germany. Precautionary values are not defined for peat bogs or for sites with humus contents > 30 mass-%, so these sites are not mapped here

11.2% for As and 22.1% for Pb. Figures 7.22 and 7.23 show the spatial distribution of As and Pb in terms of this class of exceedance of the precautionary values. The most obvious results here are the loading of As in the Erzgebirge region that was described already in Sect. 7.2.2, as well as the elevated As concentrations in several locations in the Harz region, the Süder uplands and parts of the Black Forest. Although the cause for these elevated levels can be primarily attributed to mining in the Harz and Erzgebirge regions, the source of As in the Süderbergland is largely geogenic plus atmospheric inputs from the Ruhr district. In the southern Black

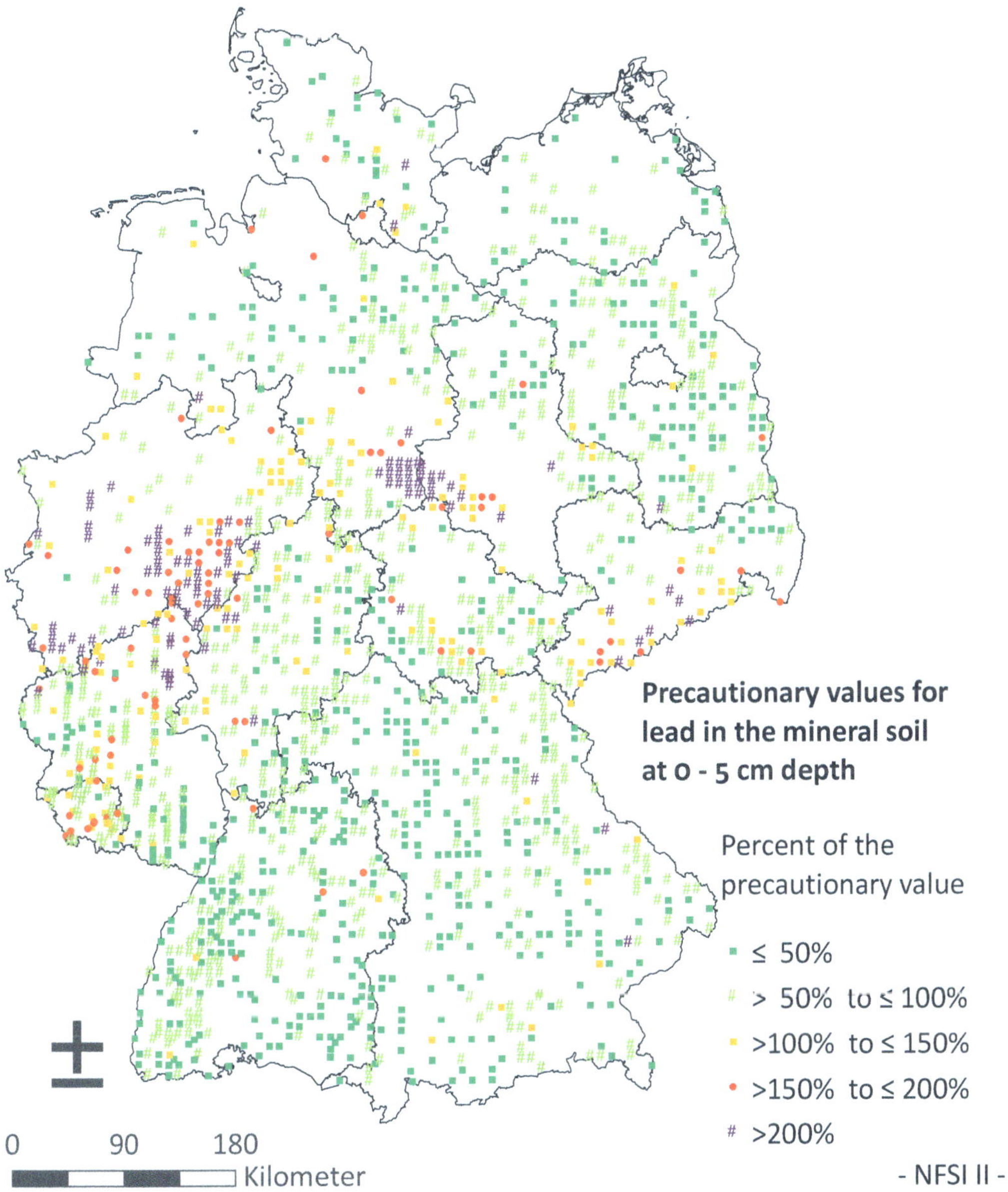

Fig. 7.23 Spatial distribution of lead in terms of the extent to which they exceed or fall short of their precautionary values in mineral soil (0–5 cm) for forest soils in Germany. Precautionary values are not defined for peat bogs or for sites with humus contents > 30 mass-%, so these sites are not mapped here

Forest, the occurrence of hydrothermal precipitates and the use of calcium arsenate for combating bark beetles have been cited as potential causes.

Lead concentrations that exceed the precautionary value are found in particular in sites in the Harz region, the Rhenish Slate Mountains, the Erzgebirge region and parts of the Thuringian Forest. In these cases too, the former mining activity can be named among the causes for the high concentrations in the Harz and Erzgebirge regions, while atmospheric inputs dominate in the Thuringian Forest and Rhenish Slate Mountains.

For Cd, 82% of sites included in the NFSI II are below or at 50% of the precautionary value, and only 2.4% of sites exceed the value (Table 7.5). Only isolated sites in the Harz region, the Westphalian Bay and the Rhenish Slate Mountains have Cd concentrations that are well above the precautionary values.

For Ni and Cr, 6.2% and 5.7% of sites, respectively, have concentrations above the precautionary values. In places, the explanation lies in geogenic inputs (basic magmatic and metamorphic rocks in Vogelsberg, limestones in the Swabian Alb), and in other sites anthropogenic causes are the source (Rhenish Slate Mountains). For Zn and Cu, <5% of sites have concentrations above the precautionary value. These sites are distributed heterogeneously throughout the country.

7.4 Conclusions/Recommendations for Action

The results of the NFSI on heavy metal burdens in forest soils suggest a need for further reduction of anthropogenic (atmospheric) input of heavy metals. This need is justified on the basis of the inability of heavy metals to biodegrade. Forest soils play an essential role in the balance of ecosystems; they must ensure that heavy metals entering the soils cannot reach the ground and surface waters unchecked but instead these substances permanently and extensively remain below levels that are harmful to plants, animals and other organisms. Thus, the nationwide NFSI is further justified by the risks that are posed by pollution of ground, spring and surface waters. In addition to the calculated heavy metal concentrations and stocks, forest management demands concrete evidence from the NFSI regarding the appropriate actions to maintain the filtering and regulatory function of forest soils. Key parameters that can favourably influence the natural retention capacity of forest soils include the humus concentration, pH and base saturation levels. These variables that describe soils can also be positively affected by forest management practices.

From the perspective of conserving soils and water systems, forest management practices must seek to counter forms of humus in the organic layer and promote the formation of humus in the mineral soil. In addition to increased cultivation and targeted maintenance of deciduous trees in mixed growth forests, forest liming is an option that creates favourable conditions for enriching the humus in the mineral soil layers. Forest liming has also proven useful for increasing the base saturation level of soils while decreasing acidity. If these measures are well coordinated, there is the potential to considerably increase stable, ecologically compatible compounds between the heavy metals and humus particles in mineral soil.

These measures can be concentrated on a regional basis, taking into account the calculated heavy metal concentrations and stocks, as well as the acidification of forest soils. Regions where forest soils have particularly high heavy metal burdens and significant soil acidification should be the focus of forest management practices that place a higher value on soil and water conservation than in regions with relatively lower heavy metal concentrations in the topsoil.

7.5 Summary

The heavy metal concentrations of forest soils in Germany are the result of both the geogenic basic concentration in the parent rock and the atmospheric inputs that can overprint this basic geogenic concentration especially in the topsoil layers. Evidence of significant atmospheric overprinting can be found for Pb, Cd and Hg, while As and Cu are less distinctly overprinted. Concentrations of the elements Ni, Cr and Zn are primarily determined by the basic geogenic concentration. The highest heavy metal concentrations are found on clay, carbonate rocks and basic and intermediate magmatic and metamorphic rocks, whereas relatively lower concentrations are documented on terrace and gravel sediments, sand and thick sandy surface layers as well as till and boulder clay.

Especially for elements with a high affinity for binding to organic substances such as Pb, Cu, Cr, Ni and Hg, heavy metal concentrations in the organic layer are clearly differentiated by humus horizon with increasing concentrations from the L- through Of- to the Oh-horizons. In contrast, Cd and Zn concentrations are not differentiated by horizon.

Over the period between the NFSI I and NFSI II, atmospheric inputs of heavy metals decreased, which, together with relocation/incorporation of the metals into the upper increments of the mineral soil, has resulted in decreased concentrations in the organic layer. Forest liming, which causes a shift of the humus from the organic layer to the upper mineral soil, has also tended to enhance this effect. Using the data from North Rhine-Westphalia by way of example, an analysis of the change in heavy metal stocks in the organic layer and the first two depth layers of the mineral soil was carried out; the results indicated that while heavy metal stocks shift from the organic layer to the mineral topsoil, the total stock of heavy metals remains largely unchanged. Special attention should be paid to Hg, as this element is the only one that has increased in soil, due to various elevated inputs through deposition.

Precautionary values based on the BBodSchV were used to assess the heavy metal concentrations in the mineral soil. For most heavy metals, concentrations in the majority of forest soils in Germany appear to be below the precautionary value; thus, in terms of heavy metal burden, forest soils can be said to be in relatively good condition in most areas. Only the elements As and Pb exceed the precautionary values in many regions, with 11.2% for As and 22.1% for Pb significantly exceeding the precautionary values.

References

BBodSchG (1998) Bundes-Bodenschutzgesetz (Federal Soil Protection Act)

BBodSchV (1999) Bundes-Bodenschutz- und Altlastenverordnung (Federal Soil Protection and Contaminated Sites Ordinance)

BBodSchV (2015) Bundes-Bodenschutz- und Altlastenverordnung (Federal Soil Protection and Contaminated Sites Ordinance).

BGR (2016) Bodenatlas Deutschland—Böden in thematischen Karten. Bundesanstalt für Geowissenschaften und Rohstoffe, Hannover

Gehrmann J (2013) Schwermetalle in nordrhein-westfälischen Waldböden auf Basis der Bodenzustandserhebung (BZE). In: Ministerium für Klimaschutz U, Natur- und Verbraucherschutz des Landes Nordrhein-Westfalen (ed) Waldzustandsbericht 2013—Bericht über den ökologischen Zustand des Waldes in NRW, Düsseldorf, Germany, pp 48–57

Keuffel-Türk A, Jankowski A, Scheler B, Rademacher P, Meesenburg H (2012) Stoffeinträge durch Deposition. In: Höper H, Meesenburg, H (eds) 20 Jahre Bodendauerbeobachtung in Niedersachsen, Hannover, Germany. Landesamt für Bergbau, Energie und Geologie, pp 19–37

Knolle F, Ernst WHO, Dierschke H, Becker T, Kison HU, Kratz S, Schnug E (2011) Schwermetallvegetation, Bergbau und Hüttenwesen im westlichen GeoPark Harz—Eine ökotoxikologische Exkursion. Braunschweiger Naturkundliche Schriften 10(1):44

König N, Baccini P, Ulrich B (1986) Der Einfluß der natürlichen organischen Substanzen auf die Metallverteilung zwischen Boden und Bodenlösung in einem sauren Waldboden. J Plant Nutr Soil Sci 149(1):68–82

Litz N, Wilcke W, Wilke B-M (2015) Bodengefährdende Stoffe: Bewertung – Stoffdaten – Ökotoxikologie – Sanierung. Wiley-VCH, Weinheim, Germany

Ohnesorge FK, Wilhelm M (1991) Zinc. In: Merian E (ed) Metals and their compounds in the environment: occurrence, analysis and biological relevance. VCH, Weinheim, Germany, pp 1309–1342

Utermann J, Düwel O, Möller A (2006) Geogen erhöhte Spurenelemente in Böden Deutschlands—Ausmaß und Bedeutung für den Vollzug der BBodSchV. In: 4. Marktredwitzer Bodenschutztage, Marktredwitz, Bavaria, Germany. City of Marktredwitz, pp 74–79

Wilke B-M, Fuchs M, See R, Skoluda R, Hund-Rinke K, Görtz T, Pieper S, Kratz W, Römbke J, Kalsch W (2003) Entwicklung ökotoxikologischer Orientierungswerte für Böden. Abschlussbericht (Förderkennzeichen: UFOPLAN 299 71 207). Umweltbundesamt (UBA), Berlin, Germany

Wippermann T (ed) (2000) Bergbau und Umwelt—langfristige geochemische Einflüsse. Springer, Berlin

Chapter 8
Occurrence and Spatial Distribution of Selected Organic Substances in Germany's Forest Soils

Marc Marx, Juliane Ackermann, Simone Schmidt, Jens Utermann, and Bernd M. Bussian

8.1 Introduction

Persistent organic contaminants are hardly decomposed in the environment. Once emitted from anthropogenic sources, these substances can accumulate in various environmental compartments, particularly in soils and sediments, and remain there for many years. In addition to their persistence, the substances can have harmful effects on humans and on the environment. They may have a cancerogenic or mutagenic effect to humans or impair the reproduction. In the environment, these substances can accumulate in organisms, have an adverse effect on them or influence their hormonal balance. Substances like polycyclic aromatic hydrocarbons (PAH), polychlorinated biphenyls (PCB), dichlorodiphenyltrichloroethane (DDT), and hexachlorobenzene (HCB) have been emitted from anthropogenic sources for several decades and are now ubiquitously distributed. The worldwide distribution of the persistent organic substances is caused by transport through air and water, leading to the detection of the substances on places far away from their emission sources, e.g. in polar regions (Kallenborn 2007; Lohmann et al. 2007; Jones and de Voogt 1999). Some persistent organic substances can naturally arise under certain conditions, but these sources are negligible compared to the anthropogenic emissions. Due to their proven harmful properties for humans and the environment, the use of these substances was governed and partly totally forbidden by the international Stockholm

M. Marx (✉) · J. Ackermann · S. Schmidt
German Environment Agency, Dessau-Roßlau, Germany
e-mail: marc.marx@uba.de; juliane.ackermann@uba.de; simone.schmidt@uba.de

J. Utermann
Ministry for Environment, Agriculture, Conservation and Consumer Protection of the State of North Rhine-Westphalia (MULNV), Düsseldorf, Germany
e-mail: jens.utermann@mulnv.nrw.de

B. M. Bussian
Formerly German Environment Agency, Dessau-Roßlau, Germany

N. Wellbrock, A. Bolte (eds.), *Status and Dynamics of Forests in Germany*, Ecological Studies 237, https://doi.org/10.1007/978-3-030-15734-0_8

Convention on Persistent Organic Pollutants. Twelve organic substances known as "dirty dozen" were banned worldwide by the Stockholm Convention (UNEP 2001; Lohmann et al. 2007).

The investigated substances are briefly described in the following:

Polycyclic Aromatic Hydrocarbons: PAH

PAHs consist of hydrocarbon rings (usually benzene rings), which are connected to one another via common sides. Due to the combination of the rings, there are approximately 10,000 different PAH. For the present study, the 16 indicator PAHs (EPA) were selected using established analytical methods for this substance group.

The PAHs have been the focus of science and the public for decades because of their harmful properties for humans and the environment. Many PAHs are assessed in European chemicals legislation (REACH) as so-called "PBT" substances because of their negative environmental behaviour. They decompose very slowly or not at all in the environment (P, persistent), accumulate in organisms and food chains (B, bioaccumulating), and are toxic to organisms (T, toxic) even at very low environmental concentrations. Once released, such substances remain very long, accumulate, and can thus develop their toxic effect in the environment over a long period. Additionally, numerous PAHs are also carcinogenic and therefore belong to the "CMR" substances (C, carcinogen; M, mutagenic; R, toxic for reproduction). The ubiquitous occurrence of PAHs in the environment makes them a particularly problematic substance group in combination with their properties (UBA 2016).

Polychlorinated Biphenyls and Dioxin-Like Polychlorinated Biphenyls: PCB and dl-PCB

PCBs are chlorinated hydrocarbons with a similar chemical structure to dioxins. On the backbone of a biphenyl (two benzene rings linked via a single bond), one or more hydrogen atoms are replaced by chlorine atoms. In total, there are 209 possible compounds (congeners). In the present study, six indicator congeners were analysed.

Polychlorinated biphenyls are thermally and chemically stable, non-flammable, non-conductive, and hydrophobic. They are technical chemicals that have been used extensively in various applications since 1929 until the early 1970s, including the use in transformers, electrical capacitors, and hydraulic systems as hydraulic fluids, as softeners in paints, sealants, insulation materials, and plastics. In addition, PCBs can originate from combustion. The use of PCBs has been banned in Germany since 1989, but polluting the environment by disposing PCB-containing waste (transformers, capacitors, and hydraulic systems) is still a major problem worldwide. Because of accidents, improper waste management, and diffuse inputs, PCBs have continued to spread throughout the environment. They accumulate as persistent and bioaccumulative substances in the food chain and can now be detected even in fish in the Antarctic, in mother's milk, and in human fat tissue. The chronic effects of PCBs on humans can be very diverse (e.g. chloracne, liver damage, teratogenicity, immune system damage). PCBs are amongst the 12 organic substances known as "dirty dozen" (Batool et al. 2016; UBA 2017).

Twelve of the possible 209 PCB congeners (see above) have a similar structure as well as properties as the dioxins and are therefore called dioxin-like PCB. The most toxic one is PCB 126 (UBA 2017).

Dichlorodiphenyltrichloroethane: DDT and Its Metabolites (DDx)

DDT is an insecticide that has been used since the early 1940s. Because of its good protection against insects, its low toxicity to mammals, and its low-cost manufacturing process, it was the most used insecticide for decades DDT is degraded, beginning with the conversion into the very long-lived metabolites dichlorodiphenyldichloroethylene (DDE) and dichlorodiphenyldichloroethane (DDD) (Aislabie et al. 1997; Boul 1995). Because of their good fat solubility, DDT and its metabolites accumulate in the fat tissue of humans and animals through the food chain (Aichner et al. 2013). The use of DDT in most industrial countries was banned in the 1970s. It also belongs to the "dirty dozen". In countries that have ratified the Stockholm Convention, the production and use of DDT are only permitted for the control of disease-transmitting insects, particularly the carriers of malaria (Wang et al. 2016; Batool et al. 2016; Lohmann et al. 2007; Boul 1995).

Hexachlorobenzene: HCB

HCB consists of an aromatic benzene ring in which all hydrogen atoms are replaced by chlorine atoms. HCB has a high vapour pressure and is stable in the air. In addition to the persistence, HCB has several harmful properties for humans and the environment. It accumulates in organisms and along the food chain because of its high fat solubility, and this might lead to harmful effects to humans.

HCB was used as a fungicide, being stopped between 1981 and 1984 due to the negative properties in both parts of Germany (Heinisch et al. 2006a). In addition, HCB also belongs to the "dirty dozen". HCB is a by-product or waste product in numerous chemical production processes, including chlorine-containing products (Fiedler et al. 1995). Therefore, many chlorinated pesticides and other chemicals contain traces of HCB contamination. HCB can also arise from combustion processes of chlorine-containing compounds.

Polychlorinated Dibenzo-*p*-Dioxins, Polychlorinated Dibenzofurans: PCDD/F

PCDD/F are composed of 75 polychlorinated dibenzo-para-dioxins (PCDD) and 135 polychlorinated dibenzofurans (PCDF). They consist as mixtures of single compounds (congeners) with different composition. For the toxic effects of the 210 dioxins and furan congeners, 17 compounds are relevant since they are strongly enriched in living organisms. Therefore, only the concentration of these 17 indicator congeners is generally analysed. The most toxic dioxin of them is the 2,3,7,8-tetrachlordibenzo-*p*-dioxin (2,3,7,8-TCDD), which is widely known since the chemical accident in Seveso, Italy, in 1976. Dioxins are already highly toxic in small amounts. They accumulate mainly in fat tissue due to their lipophilicity (fat solubility). Dioxins have been classified as cancerogenic. They can arise as unwanted by-products in the presence of chlorine and organic carbon during combustion processes and thus be present as impurities in chemicals and products, e.g. in PCB, in pentachlorophenol, and in other chloro-organic pesticides. They may also be present as impurities in materials, e.g. in asphalt, in building site rubble, or in PCB sealing compounds. Dioxins can also arise from fires of forests, coal seams, dumps, and peats and during volcanic eruptions (UBA 2017; Dopico and Gomez 2015).

For the unified risk assessment of PCDD/F and dl-PCB mixtures, the World Health Organization introduced toxicity equivalence factors (TEF) in 1998, which were modified in 2005. The concentration of a given congener is multiplied by its TEF to calculate the equivalence concentration related to 2,3,7,8-TCDD (normalized to 1). The total value of the PCDD/F and dl-PCB equivalents (TEQ) is then calculated from the sum of these concentrations (van den Berg et al. 2006).

Studies show that forests have a higher air/soil distribution coefficient due to their scavenging effect (the "Auskämmeffekt", i.e. trees filter out airborne pollutants) than agricultural or urban areas. Furthermore, soils with high total organic carbon (TOC) concentration act as a sink for persistent organic substances (Wania and Mackay 1996; Meijer et al. 2003; Posado-Baquero and Ortega-Calvo 2011; Horstmann and McLachlan 1998; Nizzetto et al. 2006; Sweetman et al. 2005). On the other hand, soils can also act as a source for some of these substances, particularly for highly volatile substances.

Numerous publications deal with the occurrence of volatile organic substances in the atmosphere and the deposition from the atmosphere (Hafner et al. 2005; Hageman et al. 2010) as well as their occurrence in agricultural and urban soils. However, there are only a few studies on forest sites, and they are often restricted to smaller areas or small sample numbers (Manz et al. 2001; Aichner et al. 2007; Wilcke 2007; Maliszewska-Kordybach et al. 2008). Exceptions to this are, for example, a study from Austria (Weiss et al. 2000) as well as studies from Bavaria, Switzerland, Great Britain, and Norway (Schmid et al. 2005; Belis et al. 2009; Offenthaler et al. 2009; Desaules et al. 2008; Schuster et al. 2011) and a recent study from Pandelova et al. (2018), dealing with PCDD/F and PCB in German forest soils. Some of these studies discuss the relationships between environmental parameters or local conditions such as altitude, population density, precipitation, or the concentration of organic matter in soil. However, a systematic comparison of the results of these studies is difficult because they differ in their sampling methods, the investigated soil horizons, and investigated substances.

A homogeneous database, which is necessary for a systematic evaluation, is not available for a large forest area. For this reason, the sampling area of the second National Forest Soil Inventory (NFSI) with about 470 sampling points of the BIOSOIL grid is ideally suited to investigate the occurrence of persistent organic substances in forest soils under climatically moderate conditions with a wide range of environmental parameters and local conditions. The aim of the present study is to show the concentrations, stocks, and spatial distribution of the investigated organic substances in forest soils across Germany.

8.2 Material and Methods

The soil sampling procedure is described in detail in BMELV (2006). In brief, soil was sampled in a grid at the BIOSOIL spots of 16 × 16 km as systematic sampling scheme within the German forests. Additionally, about 50 sampling spots of the

Table 8.1 Classification of humus levels according to Ad-HocAG_Boden (2005)

Notation (humus level)	Humus concentration (mass-%)
No humus	0
Very weak humic	<1
Weak humic	$1 - <2$
Medium humic	$2 - <5$
Strong humic	$5 - <10$
Very strong humic	$10 - <15$
Extremely strong humic	$15 - <30$
Organic, peat	≥ 30

NFSI 8 × 8 km grid were selected to consider specific regional topography or emission situations. Amongst others, polychlorinated dioxins and furans and dl-PCBs were analysed in the context of a supplementary project of the Federal Environment Agency (UBA) in the humus layer at a selection of 86 sampling spots.

For details concerning the analysis of organic substances, see Aichner et al. (2013). To calculate the humus concentration, total organic carbon concentration was multiplied by 1.72 or by 2 for organic layers and peat, respectively. For discussion purposes, humus is classified according to Ad-HocAG_Boden (2005), as is compiled in Table 8.1. At least 15 sampling sites per humus level had to be present for data evaluation.

For the evaluation of data, the lower bound assessment was chosen, i.e. values lower than the quantification limit was set to 10^{-6} μg kg^{-1} or ng TEQ05 kg^{-1} for dl-PCB and PCDD/F, respectively. Zero values, as normally required for the lower bound approach, were not used, since statistical analysis was carried out on log-transformed data and log (0) is not defined. Scales in the figures are also on a log basis. Kruskal-Wallis (H test) and post hoc Dunn test were conducted to identify significant differences on a $p < 0.05$ level. A nonlinear multiple regression (support vector machine, SVM) was used to investigate the driving factors explaining the concentrations of organic substances. The results of SVM were compared to a linear regression model. The factors examined were soil texture, humus level according to Ad-HocAG_Boden (2005) representative for total organic carbon (TOC) concentration, forest type (deciduous, mixed, and coniferous), soil depth (depth increment), and western and eastern federal states for DDx.

Stocks [g ha^{-1}] were calculated by multiplying a given concentration of a considered organic substance in soil [mg kg^{-1}] with the considered depth increment (i.e. 0.05 m; 0–5 cm or >5–10 cm) and the soil bulk density [g cm^{-3}] of the considered depth increment and a conversion factor of 10. The stocks of the organic layer were referred to the TOC stocks (see Chap. 6), since the organic matter mainly consists of TOC. The total stocks for a considered organic substance were calculated by summing up the stocks of the organic layer and the two depth increments.

All statistical analyses were performed with R 3.2.5 (R Core Team 2015).

8.2.1 *Concentrations of Organic Substances in German Forest Soils*

Data of organic substances in soils of forest soils are rare in comparison to urban or agricultural soils. Krauss et al. (2000) published a local study from northern Bavaria. They measured PAH (as sum of 20 congeners) concentrations from 24 to 15,056 μg kg^{-1} in the O horizons of 15 forest sites. The range for PCB (as sum of 12 congeners) in their study amounted to 11.6 201.5 μg kg^{-1}. In a study of Fortmann and Meesenburg (2007) from Lower Saxony, PAH concentrations (as sum of 16 congeners) from 61 18,266 μg kg^{-1} were measured in the O horizons.

Aichner et al. (2013) evaluated the same data as in the present study and classified the organic substances concentrations as generally low. They mention studies and the respective results for organic substances from other countries, e.g. Czech Republic, France, UK, and Norway. However, due to varying sampling strategies (e.g. different horizons or depth increments) and/or different methods (e.g. determination sum of 16 PAKs versus sum of 14 PAKs, sum of 6 PCBs versus sum of 12 PCBs), a comparison from the above-mentioned studies to the present study is not appropriate. Additionally, in the present study, results were not restricted to a specific region in Germany.

For some of the organic substances considered in this study, the Federal Soil Protection and Contaminated Sites Ordinance (Bundes-Bodenschutz- und Altlastenverordnung; BBodSchV (1999)) provides so-called precaution values to evaluate the organic substance concentrations in the mineral soil. In accordance with § 8 (2) of the Federal Soil Protection Act (Bundes-Bodenschutzgesetz; BBodSchG (1998)), exceedance of precaution values for soil, when taking into account the harmful substance concentration from sources that are geogenic or due to extensive settlement, generally raises concern that harmful changes to the soil have taken place pursuant to § 9 BBodSchV. The precaution values thus do not represent a danger threshold but instead indicate a threshold for concern, which, if exceeded, demands that precautionary measures such as avoiding or reducing further inputs should be taken.

PAH

The nonlinear multiple regression analysis (SVM) revealed that most of the PAH concentration can be explained by the humus level classified according to Ad-HocAG_Boden (2005). The humus level, which is representative for TOC concentration, explained 44% of the total variance followed by the forest type (5%) and soil depth (4%). In total, 53% of the variance was explained (Fig. 8.1). This was compared to a multiple linear regression, which explained 43% of the total variance, showing a nonlinear relationship between the PAH concentration and the explanatory parameters. The results for the PAH concentrations depending on the humus level, forest type, and the depth as explanatory factors are shown in Fig. 8.2.

Highest and lowest PAH concentrations with 7.5–14,889.0 μg kg^{-1} were both measured in coniferous forests. As expected, the PAH concentrations increased with

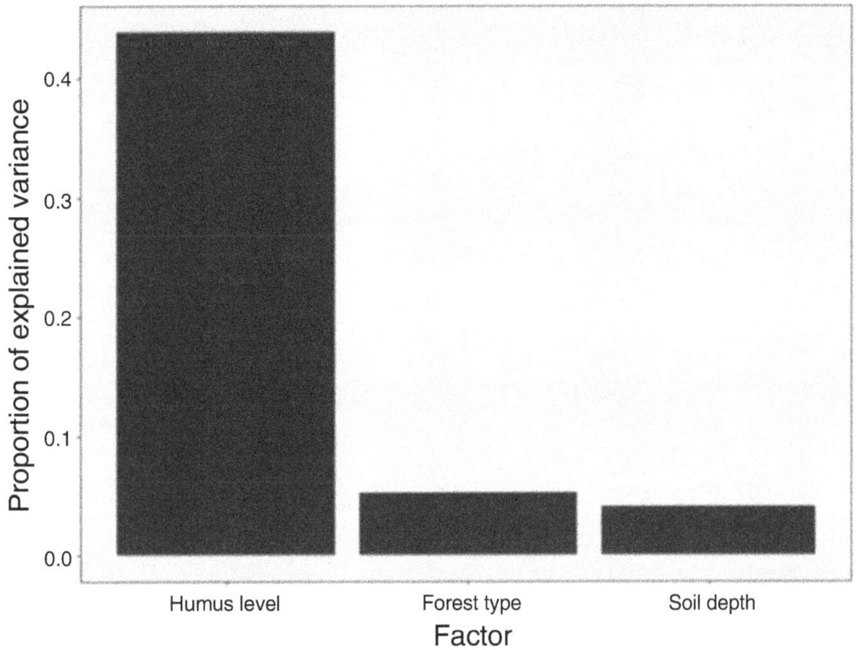

Fig. 8.1 Proportion of the variance explained by given factors calculated by support vector machines for PAH content in forest soils

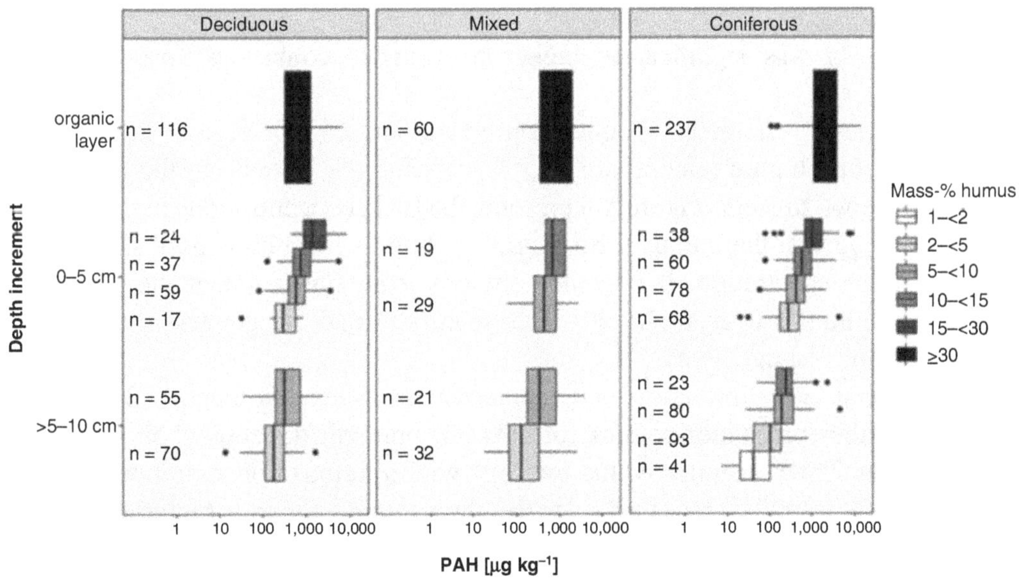

Fig. 8.2 PAH concentrations grouped by forest type, depth increment, and humus level

increasing humus level, i.e. increasing TOC concentration. In the organic layer, PAH concentration in soils under coniferous forests (median 2113.2 μg kg^{-1}) was significantly higher than in soils under deciduous (median 580.8 μg kg^{-1}) or mixed (median 655.6 μg kg^{-1}) stands. Moreover, in medium humic mineral soil (2 – ≤ 5 mass-% humus) at >5–10 cm depth, the PAH concentration under deciduous forest (median

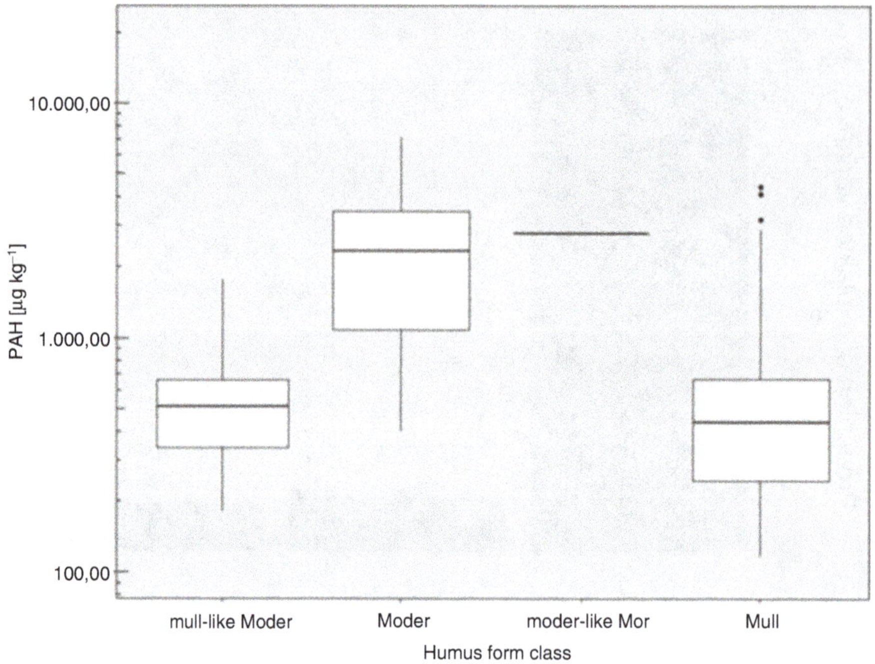

Fig. 8.3 PAH concentrations in the organic layer grouped by humus form classes under deciduous forest

183.1 µg kg^{-1}) was significantly higher than under coniferous forest (median 108.0 µg kg^{-1}).

In deciduous forest, the PAH concentration was lower in the organic layer than in extremely strong humic mineral soil (15 – ≤30 mass-% humus) at the first depth increment from 0 to 5 cm. Figure 8.3 presents the PAH concentrations in the organic layer plotted against the humus form class. The humus form class mull showed the lowest PAH concentration. It becomes obvious from Fig. 8.4 that the biological highly active humus form mull with more than 60% had the greatest share on the observed humus form classes.

The Federal Soil Protection and Contaminated Sites Ordinance (BBodSchV 1999) prescribes precaution values for PAH in fine soil depending on its humus concentration. The precaution value for soils with a humus concentration equal or lower than 8% is 3000 µg PAH kg^{-1} dry weight; the value for soils higher than 8% is 10,000 µg PAH kg^{-1} dry weight. The precaution values do not apply to organic layers. At only two of the investigated sampling sites, the PAH concentration exceeded the precaution value in the mineral soil in the first depth increment from 0 to 5 cm.

It was expected that PAH concentrations were universally decreasing with depth, due to the (historic) deposition. In deciduous forest, the PAH concentration was lower in the organic layer than in the first layer of extremely strong humic mineral soil. Mull was the dominant humus form class under deciduous forest (Fig. 8.4). It is characterized by high biological activity, and the litter layer is not present all year. On the one hand, the very active microorganisms supported the PAH degradation.

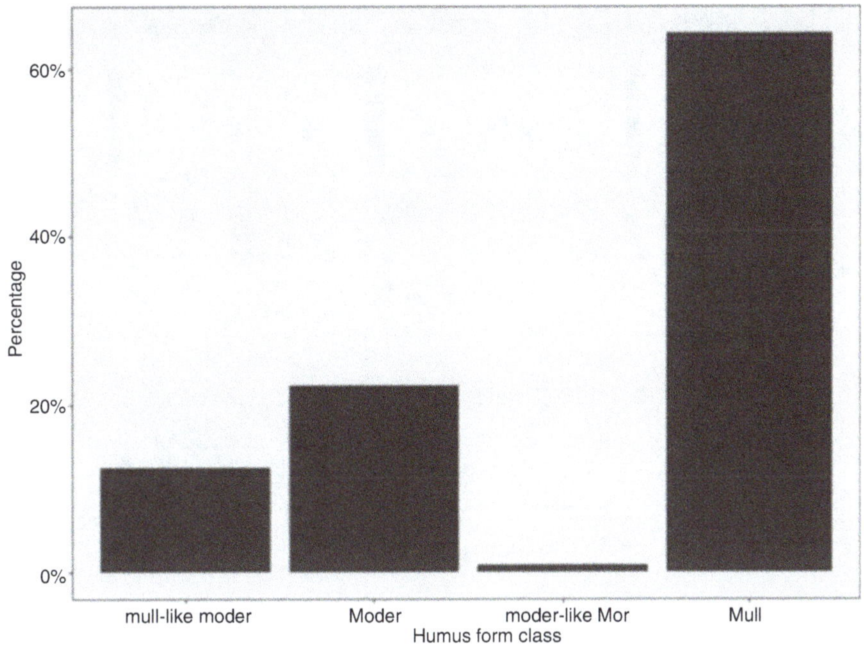

Fig. 8.4 Percentage of humus form classes under deciduous forest

On the other hand, biological activity supported an intensive mixing with the mineral soil. These processes presumably led to a higher PAH concentration in the mineral top soil compared to the organic layer. The same trend was observed for the other investigated substances.

It becomes obvious that degradation and relocation might both have been responsible factors for the PAH distribution within the organic layer and mineral soil. Furthermore, these factors seemed to be driven by additional site characteristics, e.g. humus form or TOC. High PAH concentrations in the O horizons or in the organic layers reflected the current airborne deposition of PAH, whereas high concentrations in mineral soils might stem from older deposition incorporated from organic layers into the mineral soil through biological activity.

PCB

The humus level was the factor explaining most of the PCB concentration in soils as revealed by SVM. It explained 32% of the total variance, followed by the forest type with 7% and soil depth with 5%. This made 44% of explained variance in total. A linear regression model explained only 33% of the total variance.

At some sampling sites under all three forest types, PCB was below quantification limit. This was especially obvious in weak humic soils (1 – ≤ 2 mass-% humus) under coniferous forest, where the median concentration amounted to 0.0 μg kg^{-1}. Maximum PCB concentration was 106.3 μg kg^{-1} under mixed forest in the organic layer (Fig. 8.5).

The PCB concentration in the organic layer of coniferous forest was with a median of 16.9 μg kg^{-1} significantly higher than that of deciduous forest with 8.6 μg kg^{-1}. In strong humic mineral soil (5 – ≤ 10 mass-% humus) at 0–5 cm

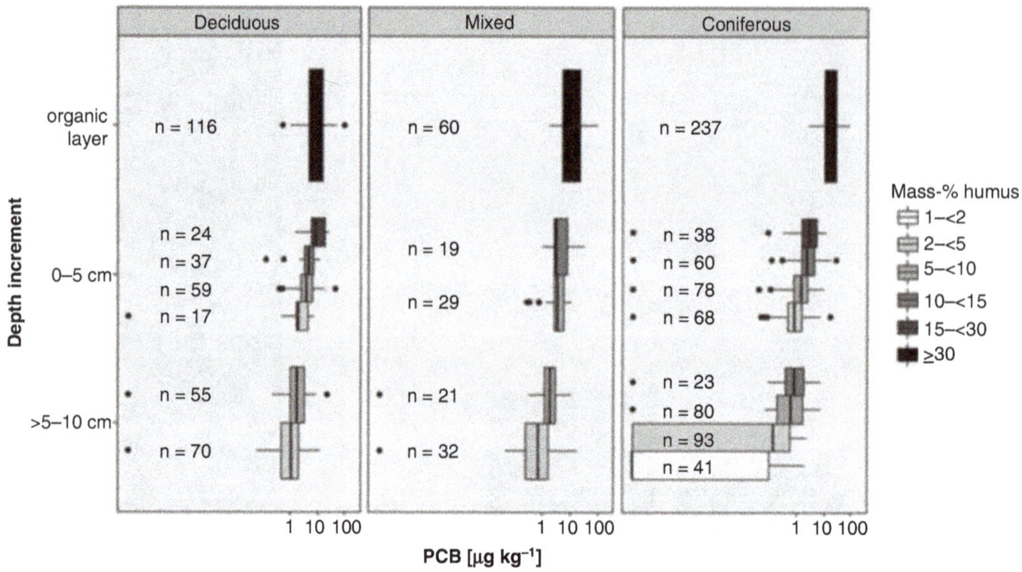

Fig. 8.5 PCB concentrations in the organic layer and the first two depth increments of the mineral soil grouped by forest type and humus level

depth, the PCB concentration was significantly lower in coniferous (median 1.4 µg kg^{-1}) compared to deciduous forests (median 3.8 µg kg^{-1}). The same was true for extremely strong humic soil (15 – ≤ 30 mass-% humus), where the median under deciduous forest amounted to 8.6 and to 3.0 µg kg^{-1} under coniferous forest.

As mentioned for PAH, the Federal Soil Protection and Contaminated Sites Ordinance (BBodSchV 1999) also prescribes precaution values for PCB6 in fine soil depending on its humus concentration, amounting to 50 µg kg^{-1} dry weight (≤8% humus) and 100 µg kg^{-1} dry weight (>8% humus concentration). At none of the investigated sampling sites, the PCB concentration exceeded the precaution values in the mineral soil, showing that the PCB concentration in German forest soils was universally low.

DDT and Its Metabolites

In the knowledge of the nationwide distribution of DDx, the distinctions between western and eastern federal states were applied in the SVM. In contrast to PAH and PCB, soil depth with 28% is the factor explaining most of the total variance of the DDx concentrations. Oppositely, the humus level contributes 1% to the explanation of the total variance. In total 43% of the total variance were explained by the given factors (39% with a linear model).

Figure 8.6 provides an overview of DDx concentrations grouped by depth, forest type, federal state, and humus level. Many humus levels are excluded because they contain too less sampling sites (<15), especially under the deciduous and mixed stands in the eastern part of Germany. For reasons of better visibility, the zero values (DDx below quantification limit) are not shown in Fig. 8.6.

As for PCB, at 14 sampling sites under all 3 forest types, DDx could not be detected and was set to 0.0 µg kg^{-1} due to the lower bound approach. Maximum

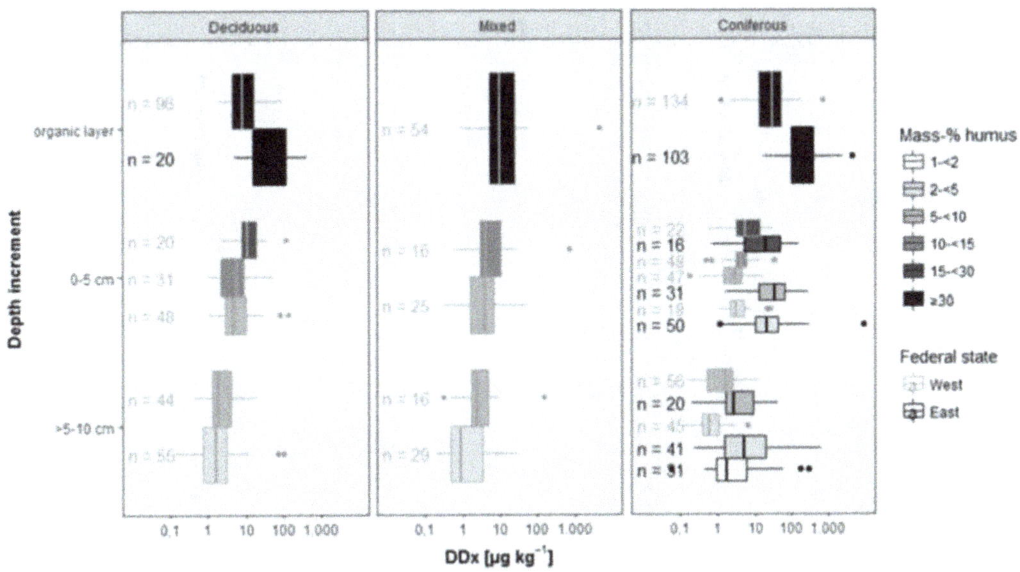

Fig. 8.6 DDx concentrations in the organic layer and the first two depth increments of the mineral soil grouped by eastern and western federal states, forest type, and humus level

concentration reached 8630.0 μg kg^{-1} in 0–5 cm depth of medium humic mineral soil (2 – ≤5 mass-% humus) under coniferous forest. Significant differences between DDx concentrations under coniferous forests in eastern compared to western soils were observed in the organic layer, in medium humic mineral soil (2 – ≤5 mass-% humus), and in 0–5 cm depth in strong humic mineral soil (5 – ≤10 mass-% humus). The DDx concentrations were lower under deciduous than under coniferous forest.

In the western federal states, the same pattern of explaining factors on the DDx concentrations was revealed, but only 31% (compared to 26% with linear regression) of the variance was explained by the model. In detail, soil depth had the highest proportion with 26%, forest type showed 4%, and humus level amounted to 1%. Due the lack of data in deciduous and mixed forests and because of the dominance of coniferous forests, SVM was not calculated separately for the eastern federal states.

HCB

Similar to DDx, soil depth with 27% was the factor explaining most of the total variance of the HCB concentrations. The forest type explained 6% and the humus level 2% of the total variance. In total 35% of the total variance was explained by these factors. A linear regression model gave 29% explanation.

As for the other substances, except PAH, also HCB was below the quantification limit at some sampling sites and was consequently set to 0.0 μg kg^{-1}. Maximum concentration amounted to 24.4 μg kg^{-1} in >5–10 cm depth in strong humic soil (5 – ≤10 mass-% humus) under mixed forest. Significant differences between the HCB concentrations in soils were observed in the organic layer. The concentration under coniferous forest (median 4.3 μg kg^{-1}) was higher than that under mixed forest (median 1.9 μg kg^{-1}) and under deciduous forest (median 1.2 μg kg^{-1}) (Fig. 8.7).

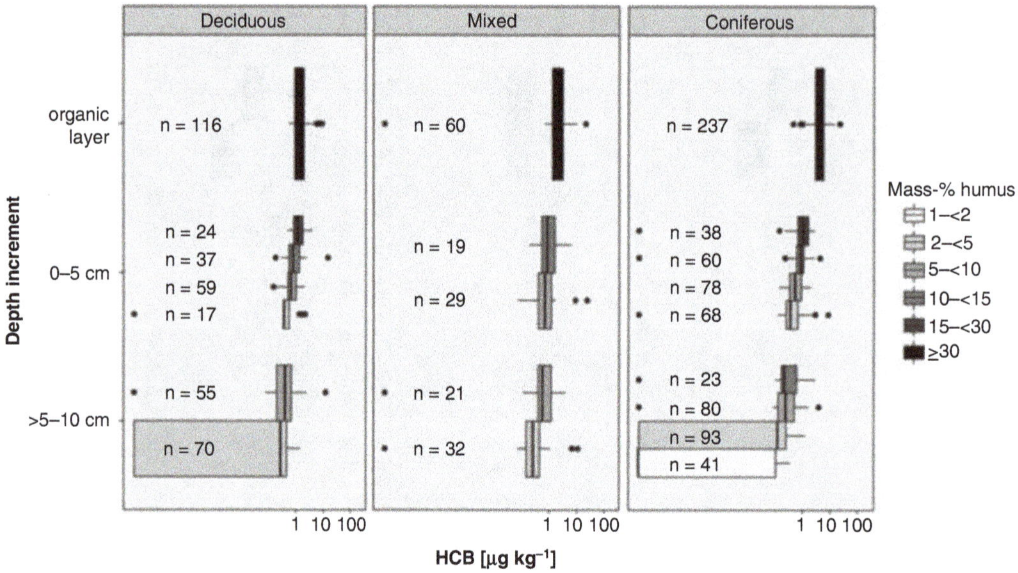

Fig. 8.7 HCB concentrations in the organic layer and the first two depth increments of the mineral soil grouped by forest type and humus level

Aichner et al. (2013) assessed the concentrations of HCB, like the other investigated organic substances, in German forest soils as low. Three main sources of HCB are given in the relevant literature. The application of HCB as a fungicide was discontinued in both parts of Germany between 1981 and 1984 (Heinisch et al. 2006b; UBA 2012). As a by-product, traces of HCB are included in various chlorine-containing chemicals. HCB can also arise from combustion processes of chlorine-containing compounds (Aichner et al. 2013).

dl-PCB

Data were available of 86 sites, which were only sampled in the organic layer. Only the 83 locations with ≥30 mass-% humus were evaluated. The dl-PCB concentration reached from 0.7 ng TEQ05 kg^{-1} in the organic layer under deciduous forest to 34.1 ng TEQ05 kg^{-1} in the organic layer of a mixed forest. The median concentration of dl-PCB under coniferous forest is significantly higher than that under deciduous forest (Fig. 8.8).

PCDD/F

As for dl-PCB, 83 sampling sites were evaluated. The PCDD/F concentration in the organic layer under coniferous forest was with 19.2 ng TEQ05 kg^{-1} significantly higher than under deciduous forest with 11.0 ng TEQ05 kg^{-1}. Minimum concentration was found in the organic layer under deciduous forest. Maximum concentration under coniferous forest amounted to 2.2 TEQ05 kg^{-1} and 105.5 TEQ05 kg^{-1}, respectively (Fig. 8.9).

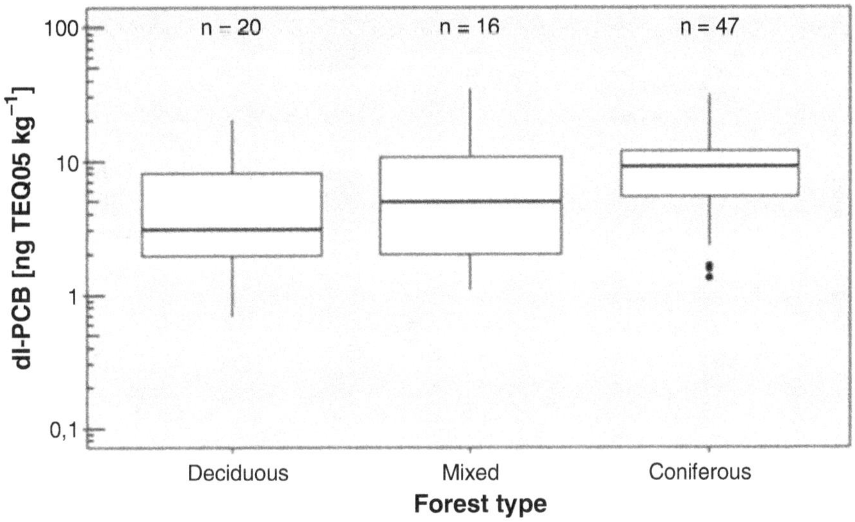

Fig. 8.8 dl-PCB concentration in the organic layer grouped by forest type

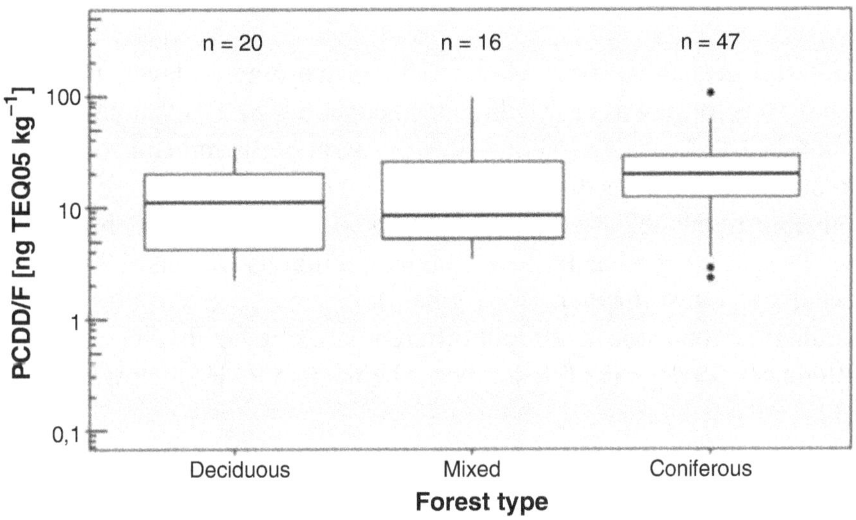

Fig. 8.9 PCDD/F concentration in the organic layer grouped by forest type

8.2.2 *Stocks of Organic Substances in German Forest Soils*

The stocks of all organic substances were differentiated according to depth increment and forest type, except for dl-PCB and PCDD/F. The stocks are also shown as total amount by forest type. The stocks of the organic layer were referred to the TOC stocks, since the organic matter mainly consists of TOC. This is also the binding agent for the organic substances.

PAH

PAH stocks ranged from 0.3 g ha^{-1} in the organic layer under deciduous forest to 3758.6 g ha^{-1} in mineral soil from 0 to 5 cm under mixed forest. In the organic layer and in both depth increments, the PAH stocks were significantly different between

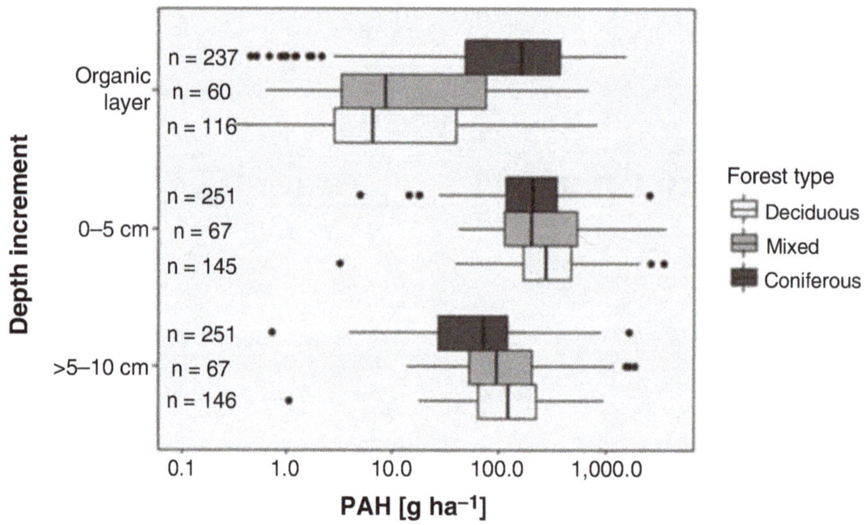

Fig. 8.10 PAH stocks in the organic layer and the first two depth increments of the mineral soil grouped by forest type

coniferous and deciduous forest (Fig. 8.10). In the organic layer, the PAH stock under coniferous forest was clearly higher (median 6.8 g ha^{-1}) than under deciduous forest (median 162.3 g ha^{-1}). In contrast, in 0–5 cm of the mineral soil, PAH stocks were significantly lower (median of 209.0 g ha^{-1}) under coniferous forest than under deciduous forest with 282.1 g ha^{-1}. The stocks in the soil depth increment from >5 to 10 cm were also significantly lower under coniferous (median 72.4 g ha^{-1}) than under deciduous forest (median 124.1 g ha^{-1}).

Regarding the total stocks without differentiating by depths, there are no significant differences between the forest types. The median PAH stock was 483.2 g ha^{-1} (Fig. 8.11).

PCB

PCB stocks showed the same pattern as PAH: stocks under coniferous forest were significantly higher than under deciduous forest (median 1.2 vs. 0.1 g ha^{-1}) and vice versa in both depth increment of mineral soil (median 0.7 vs. 1.9 g ha^{-1} and 0.1 vs. 0.7 g ha^{-1}, respectively) (Fig. 8.12).

Figure 8.13 shows significantly higher total PCB stocks under deciduous forest than under coniferous forest, amounting to 3.4 g ha^{-1} and 2.4 g ha^{-1}, respectively.

DDx

As for the DDx concentration, the stocks are shown separately for eastern and western federal states. Stocks were significantly higher in the organic layer under coniferous forest than under deciduous forest (median 17.1 vs. 0.5 g ha^{-1} in the eastern federal states, median 2.1 vs. 0.1 g ha^{-1} in the western federal states) and under mixed forest in the western federal states.

In the eastern federal states, no significant differences between the forest types in the mineral soil were detected. In the western federal states, the DDx stocks were significantly higher in 0–5 cm under deciduous forest than under coniferous forests

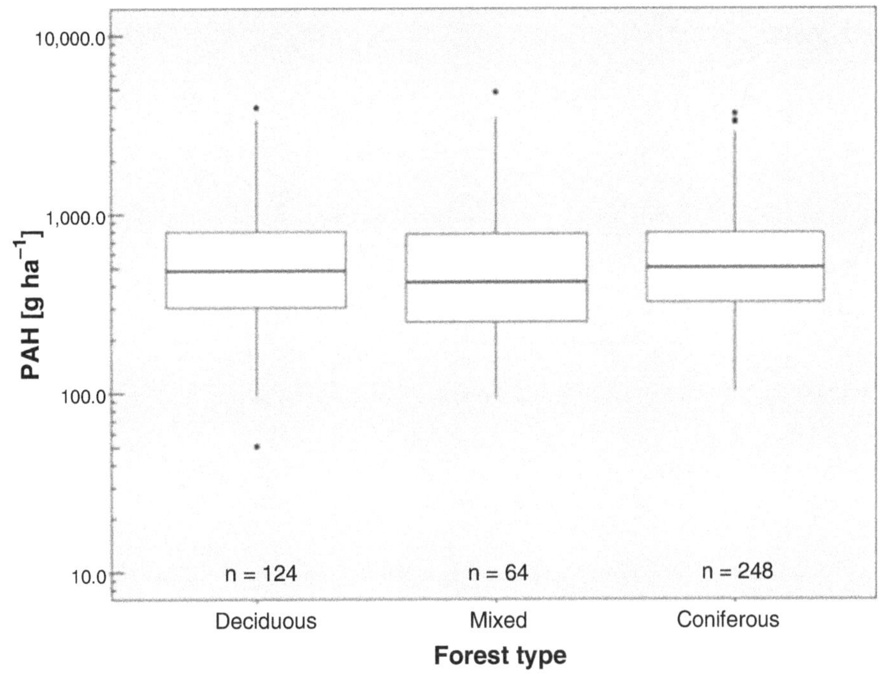

Fig. 8.11 Total PAH stocks grouped by forest type

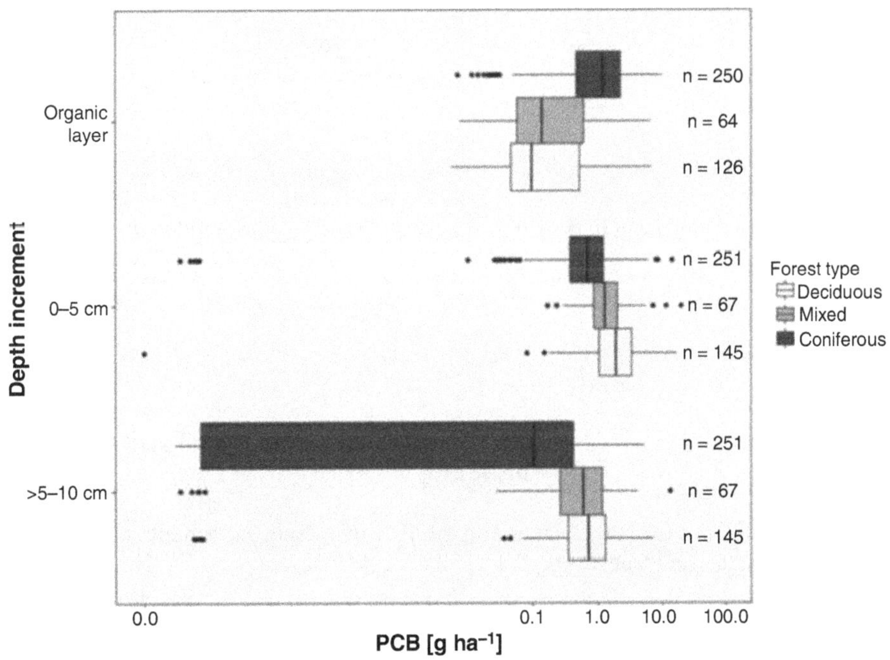

Fig. 8.12 PCB stocks in the organic layer and the first two depth increments of the mineral soil grouped by forest type

(median 2.7 g ha^{-1} vs. 1.5 g ha^{-1}). In >5–10 cm depth, DDx stocks under deciduous forest with a median of 0.8 g ha^{-1} and mixed forest with 0.7 g ha^{-1} were significantly higher than under coniferous forest with 0.4 g ha^{-1}.

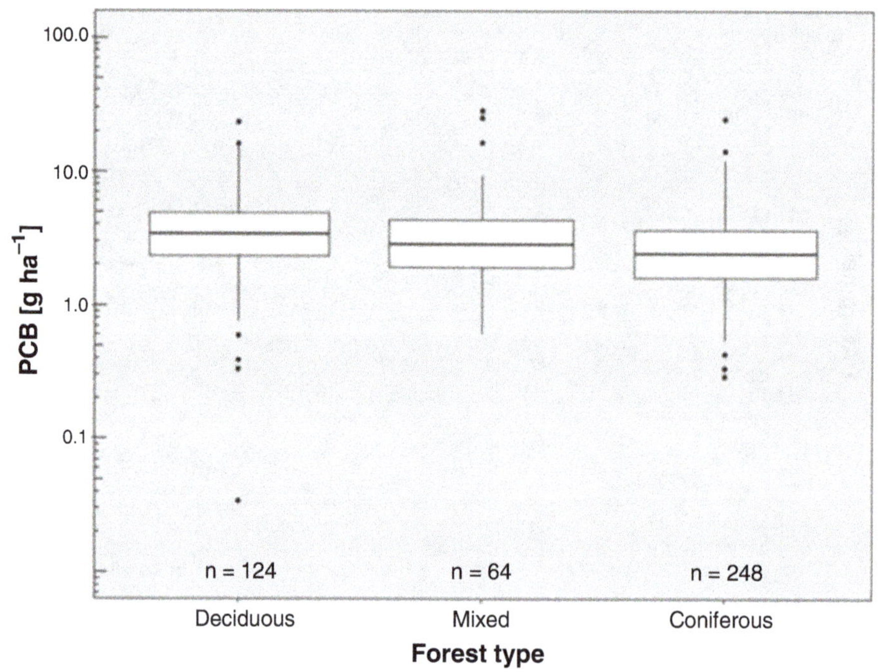

Fig. 8.13 Total PCB stocks grouped by forest type

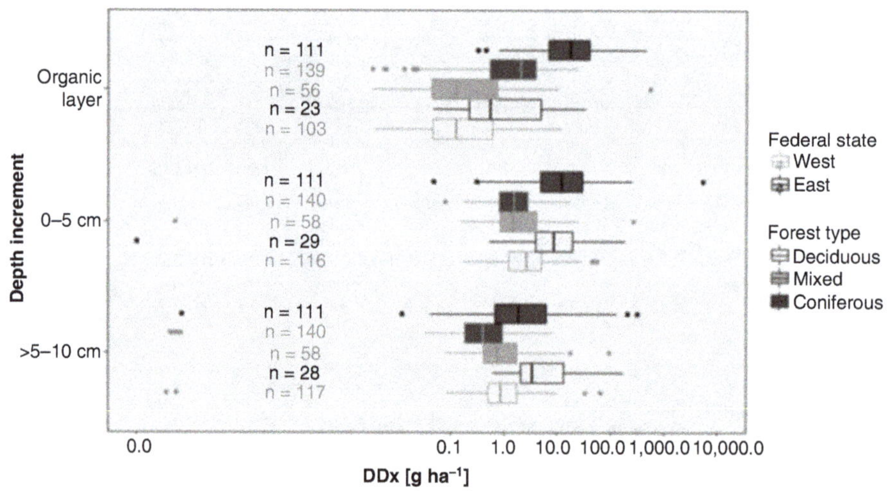

Fig. 8.14 DDx stocks in the organic layer and the first two depth increments of the mineral soil grouped by federal states and forest type

The stocks in the respective forest types were significantly higher in the eastern than in the western federal states (Fig. 8.14). Consequently, the total DDx stocks were significantly higher under deciduous and coniferous forests in the eastern than in the western federal states, as shown in Fig. 8.15. The differences within eastern and western federal states between the forest types were not significantly different and amounted to 38.3 g ha^{-1} in the eastern and to 4.4 g ha^{-1} in the western federal states.

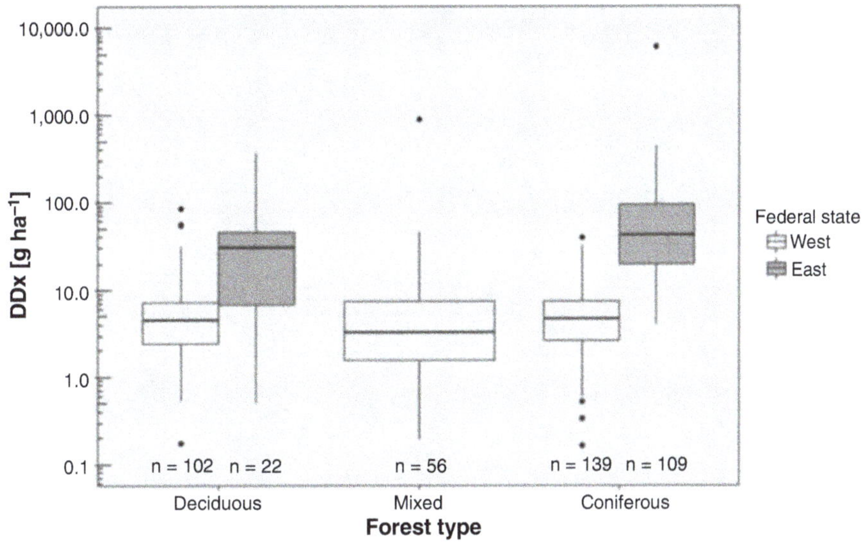

Fig. 8.15 Total DDx stocks grouped by federal states and forest type

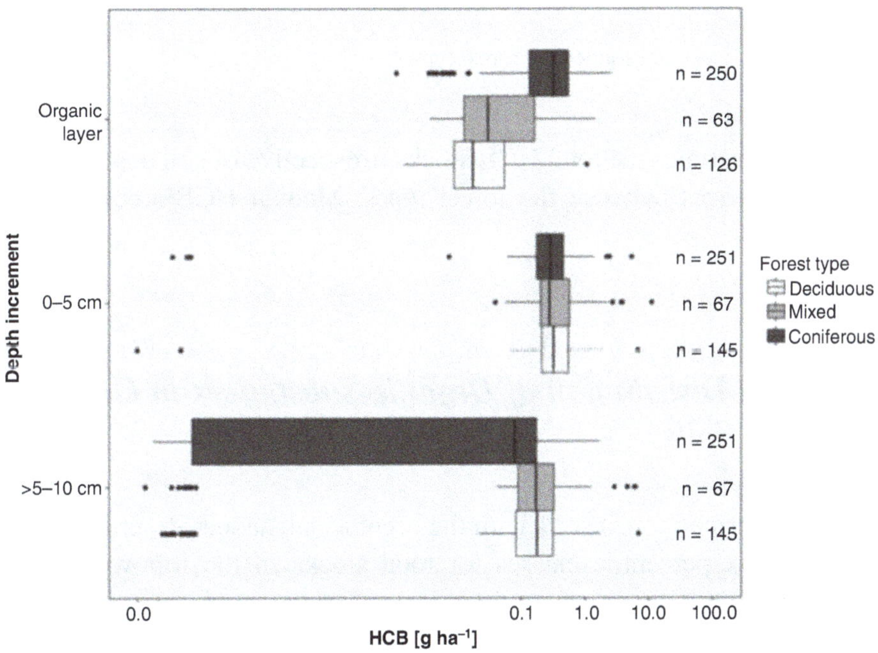

Fig. 8.16 HCB stocks in the organic layer and the first two depth increments of the mineral soil grouped by forest type

HCB

The HCB stocks were significantly higher in the organic layer under coniferous than under deciduous and mixed forest (median 0.3 g ha^{-1} vs. <0.1 g ha^{-1}). Oppositely, stocks in mineral soil from 5 to 10 cm were significantly lower under coniferous (median 0.1 g ha^{-1}) than under deciduous and mixed forest (median 0.2 g ha^{-1}). In the depth increment from 0 to 5 cm, no significant differences were detected (Fig. 8.16).

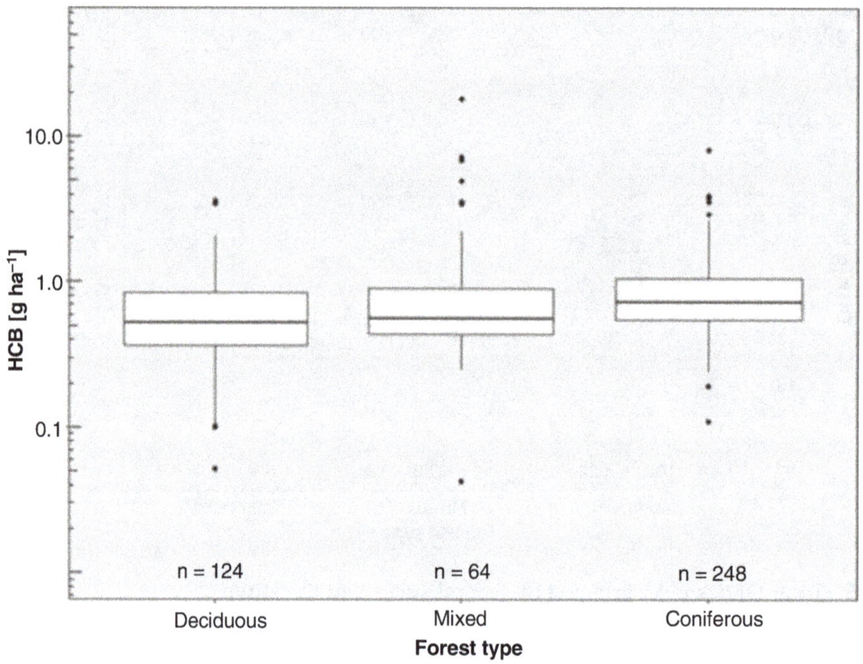

Fig. 8.17 Total HCB stocks grouped by forest type

Figure 8.17 illustrates the total HCB stocks irrespective of soil depth, showing no significant differences between the forest types. Median HCB stock amounted to 0.6 g ha^{-1}.

8.2.3 Spatial Distribution of Organic Substances in German Forest Soils

The burden of German forest soils with the organic substances described above was classified with six percentile classes for total stocks in the following maps. For dl-PCB and PCDD/F, the concentrations instead of the stocks are presented, since stocks were only available for the organic layer and not for the mineral soil.

PAH

Relatively low PAH stocks were found in the northern and southern part of Germany. High PAH stocks were present in the central German Uplands (Fig. 8.18).

The incomplete combustion of organic substances produces PAHs. Thus, their occurrence was due to various sources such as traffic or industrial exhaust fumes. Aichner et al. (2013, 2015) investigated in their studies with the same data as in the present study the PAH congeners. Based on a study of Khalili et al. (1995), they used the two- and three-ring PAHs as indicator for coal-borne sources. According to this, they assigned lignite opencast releasing coal by mining and (brown) coal-fired power plants as responsible sources for the relatively high low molecular weight PAH

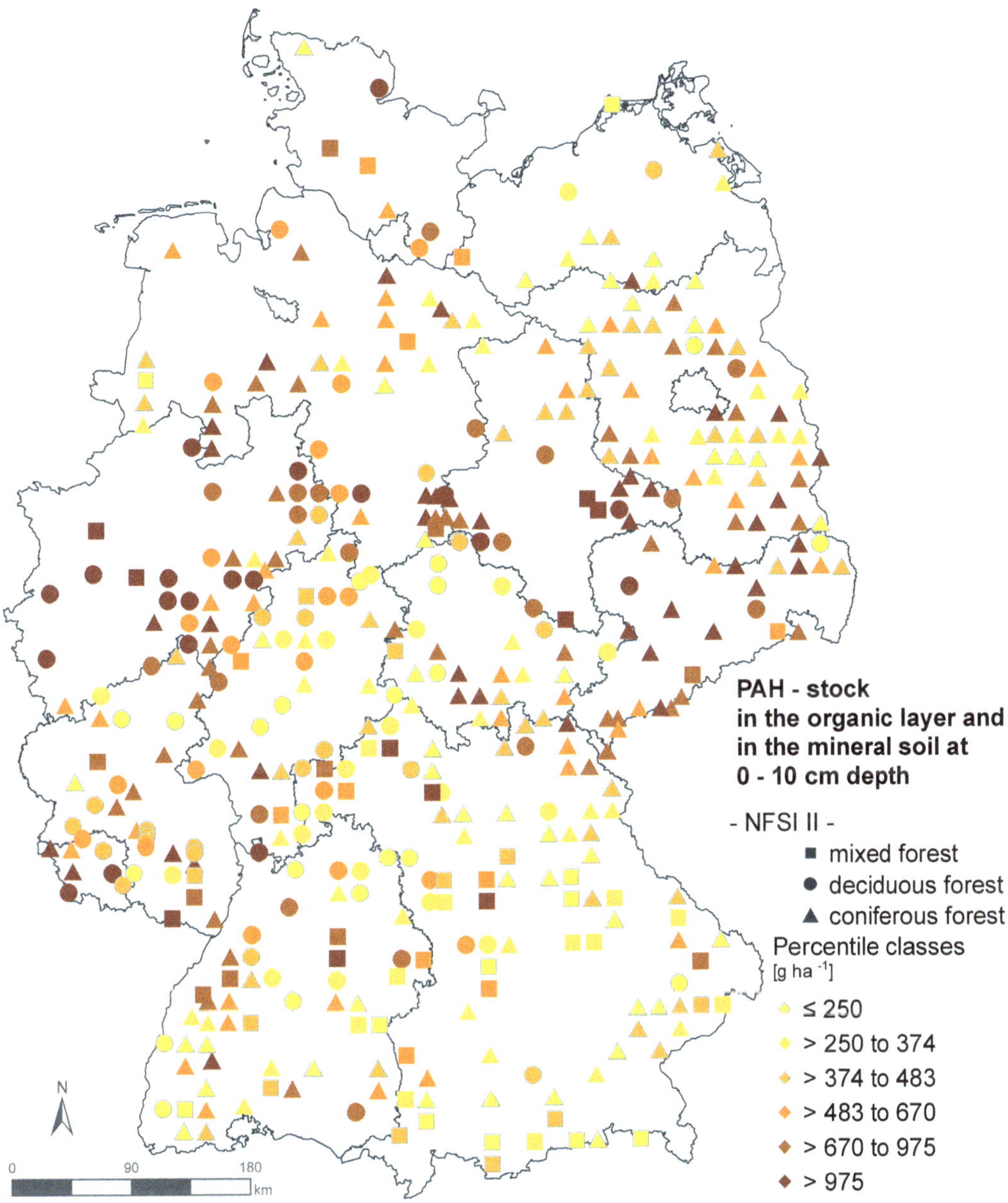

Fig. 8.18 Spatial distribution of total PAH stocks (organic layer and the first two depth increments of the mineral soil) classified with six percentile classes under different forest types

concentration in eastern Germany. Moreover, they hypothesized that the central German Uplands were a barrier for regional PAH emissions.

The topographic situation induced a distinct microclimate, i.e. high precipitation together with a change in temperature leading to a high PAH deposition rate. The two impact factors, namely, the vicinity to a main emitter and the topography inducing a distinct microclimate, were also reflected on the PAH stock distribution in German forest soils, as can be seen from the map in Fig. 8.18.

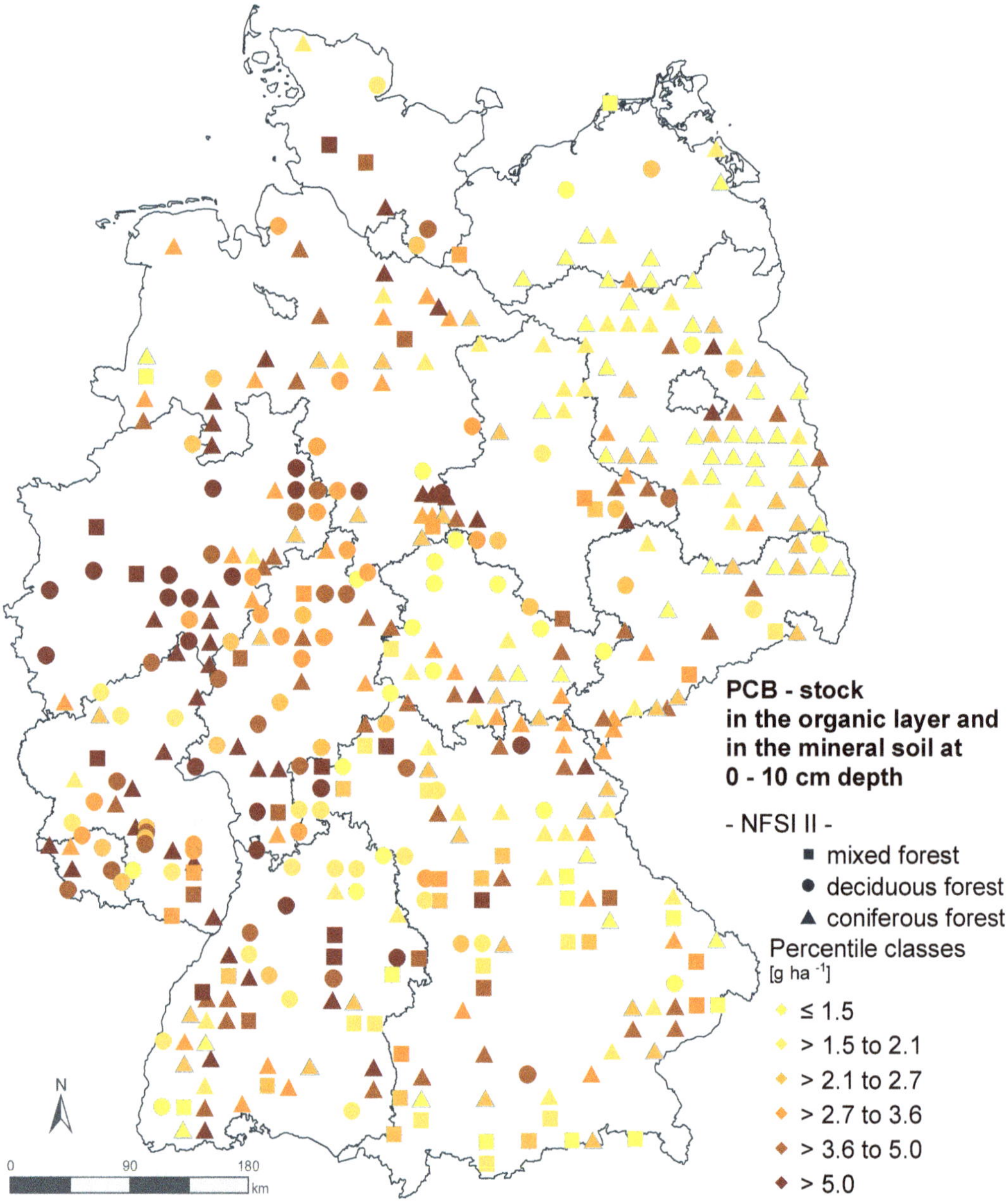

Fig. 8.19 Spatial distribution of total PCB stocks (organic layer and the first two depth increments of the mineral soil) classified with six percentile classes under different forest types

PCB, dl-PCB, and PCDD/F

PCB stocks show a high gradient in the spatial distribution between the (south-) western part of Germany, with relatively high values, and the north-eastern and eastern part, with lower values. Particularly high stocks were found in different regions in West Germany (Fig. 8.19).

Aichner et al. (2013) could not detect a specific distribution pattern from the congeners, regarding the same PCB data as in the present study. They explain the low PCB concentrations by re-evaporation and microbial degradation at least for tri- and tetrachlorinated PCB.

The PCB stock distribution seemed to be dependent upon regional emissions in former times. Aichner et al. (2013) mentioned studies of Breivik et al. (2002, 2007) who calculated a balance of PCB disposal from population density data. They estimated lower PCB emission in the eastern federal states. In combination with the tenfold higher PCB-containing waste disposal in the western federal states, this might explain the difference between higher stocks in the western compared to the north-eastern and eastern part of Germany. Additionally, it is under discussion that cities might contribute to recent PCB emissions (Jamshidi et al. 2007; Desborough and Harrad 2011).

Since stocks are not calculated for dl-PCB due to lacking data of mineral soil, only the concentration in the organic layer was regarded (not shown). The distribution of dl-PCB was like that of PCB: a gradient with relatively high values in the western part of Germany and lower values in the north-eastern part.

As for dl-PCB, stocks were not calculated for PCDD/F. There was a similar distribution pattern for PCDD/F concentrations in the organic layer as for dl-PCB, except for the south-western part of Germany, where low PCDD/F concentrations could be found (not shown).

The concentrations and consequently stocks of PCDD/F and dl-PCB were significantly higher in organic layers under coniferous soils compared to deciduous or mixed forests, respectively. The concentrations varied across Germany and depended most likely on the sources of emission, like discussed for the other organic substances.

The results for dl-PCB correlated well with these of PCB. Highest concentrations appeared mostly at sites of historical emissions from industrial applied PCB in the western federal states. Furthermore, sampling sites with higher PCB concentrations coincided with sites with elevated PCDD/F.

Note that stocks of dl-PCB and PCDD/F were only calculated for the organic layer. It might be misleading to conclude that stocks under coniferous forests were higher than under deciduous forest for the whole soil profile. It becomes obvious from the other investigated organic substances in the present study that total stocks (organic layer and mineral soil to 10 cm depth) were not significantly different in soils under a given forest type, except for PCB.

DDx

Opposite to PCB stocks, highest DDx stocks were found in the north-eastern part of Germany, while relatively lower stocks were present in western Germany. However, there were some hotspots mostly located in North Rhine-Westphalia (Fig. 8.20).

The spatial distribution of DDx in the investigated forest soils was generally attributed to the former use as insecticide and the different application times in eastern and western Germany. The application ended in the western federal states in 1972. In the eastern federal states, DDT was used until 1988 (Daly and Wania 2005; Heinisch et al. 2005, 2006a; Taniyasu et al. 2005; Breivik et al. 2007; Gasic et al. 2009). The high stocks in the eastern federal states, particularly in Brandenburg, were most likely due to the application of DDT between 1982 and 1984, in order to combat the dark beetle and the pine moth (Lymantria monacha; Heinisch et al.

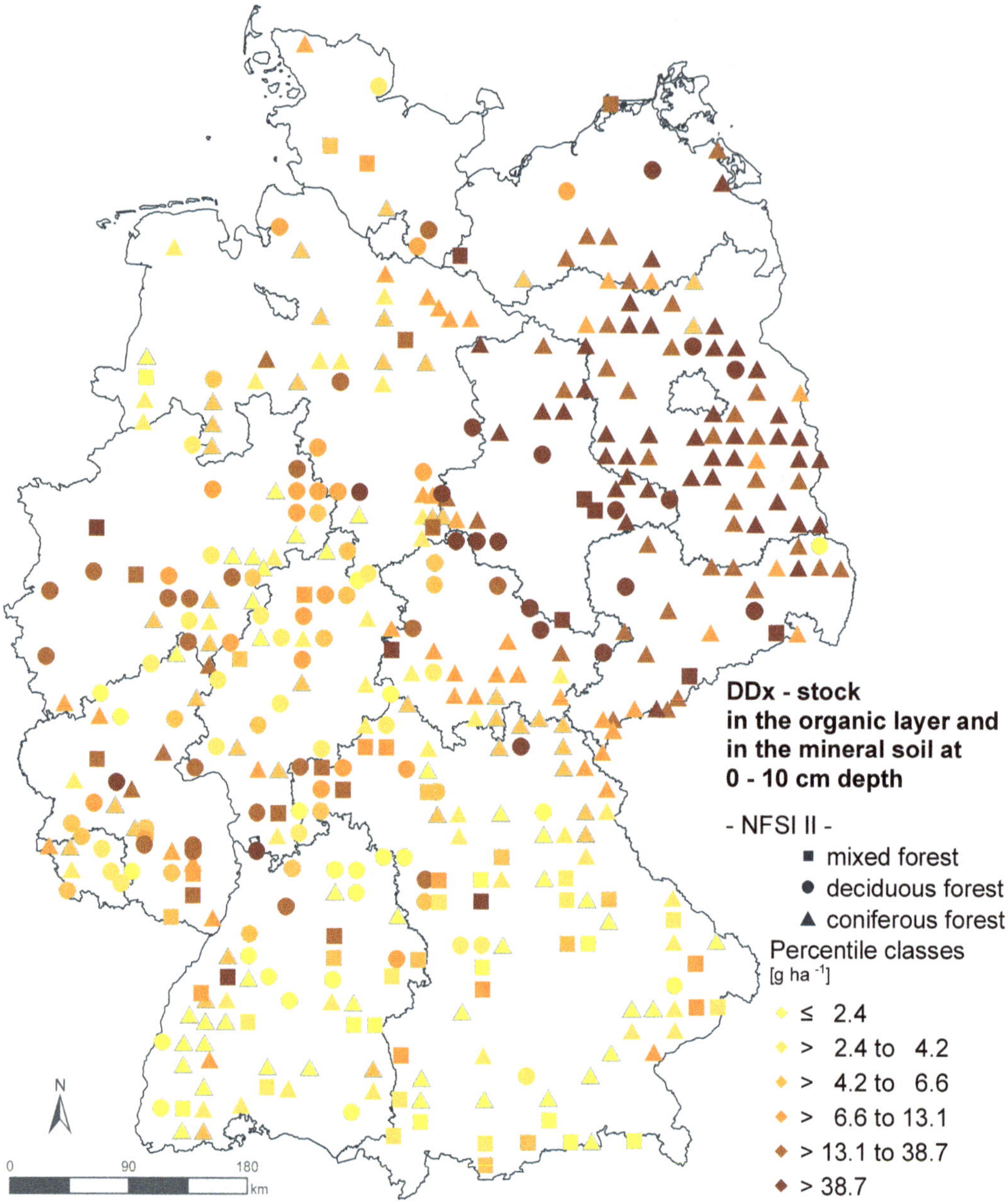

Fig. 8.20 Spatial distribution of total DDx stocks (organic layer and the first two depth increments of the mineral soil) classified with six percentile classes under different forest types

(1993, 2005)). Dimond and Owen (1996) estimated the half-life of DDT in soils to about 20–30 years in temperate climates, supporting the assumption for the DDx recovery in the mentioned regions today.

Aichner et al. (2013) differentiated with the same data as in the present study between technical (direct application) DDx, consisting mainly of 4,4′-DDT, and deposit-borne (degraded) DDx, consisting mainly of 4,4′-DDE. They used the specific proportion of DDT and DDE as a regionalized DDx fingerprint. Based on the results, they explained the exceptional high DDx concentration at some sampling sites in western Germany, particularly in North Rhine-Westphalia, with former emissions of the chemical industry. Moreover, in south-western part of Germany, the DDx fingerprint showed an advanced microbial breakdown of DDT. On the other hand, they found evidence from the DDx fingerprint for the application of technical DDT in the eastern part of Germany. These findings are very well reflected in the spatial DDx stock distribution in the present study.

HCB

The distribution of highest total HCB stocks was scattered across Germany. Hotspots were detected, e.g. in the south-eastern part of Bavaria, in the Rhein-Ruhr region of North Rhine-Westphalia, and, like PAH stocks, in north of the central German Uplands (Fig. 8.21).

Aichner et al. (2013) assessed the concentrations of HCB, like the other investigated organic substances, in German forest soils as low. Three main sources of HCB are given in the relevant literature. The application of HCB as a fungicide was discontinued in both parts of Germany between 1981 and 1984 (Heinisch et al. 2006b; UBA 2012). As a by-product, traces of HCB are included in various chlorine-containing chemicals. HCB can also arise from combustion processes of chlorine-containing compounds (Aichner et al. 2013). It is difficult to assign one specific main source to the detected HCB stock hotspots. For HCB concentrations, Aichner et al. (2013) named especially industrial combustion processes in the eastern federal states. In addition to this process, they suggested former HCB production as an additional source for the relatively high HCB concentrations in Bavaria and in the industrial regions along the river Rhine and in the Ruhr district. These suggestions also apply to the HCB stocks, as they were calculated from the concentration values.

8.2.4 Environmental Factors for the Distribution of Organic Substances

Aichner et al. (2013) pointed out the importance of local environment parameters for the distribution of organic substances. This is especially true if regional sources of organic substances are to be compared to atmospheric transport with subsequent deposition. The same authors performed a principal component analysis on the same data as in the present study with additional available environmental factors, e.g. climate, soil, and vegetation parameters. They found only weak statistical relationship between environmental parameters and the concentrations of organic substances. They concluded that concentrations of organic substances in German forest soils were mainly driven by former and current emission sources. Moreover, other factors as discussed above such as changing microclimate with altering topography can be decisive to explain pollution pattern on a regional scale.

In this context, it was expected that with decreasing temperatures because of increasing altitude, semivolatile organic compounds condense and precipitate. This effect suggests higher concentrations of organic substances depending on the topography as deposition barrier for organic substances. However, this is discussed

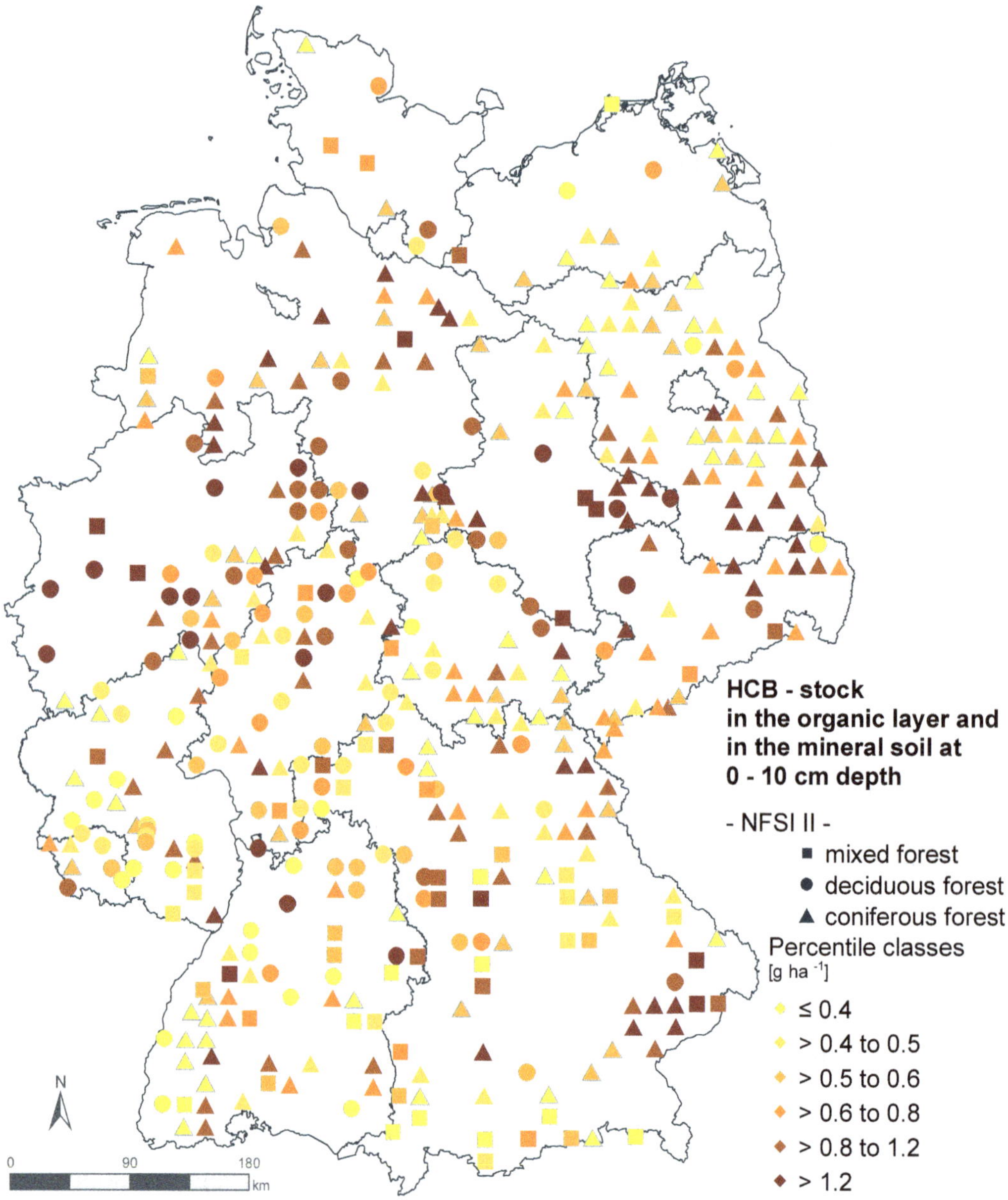

Fig. 8.21 Spatial distribution of total HCB stocks (organic layer and the first two depth increments of the mineral soil) classified with six percentile classes under different forest types

controversially and is not supported in general (Navarro-Ortega et al. 2012; Guzzella et al. 2011; Daly and Wania 2005; Kirchner et al. 2009; Hangen et al. 2011). Another temperature effect on the organic substances is the potential re-evaporation from soils, thereby reducing their concentrations (Brorström-Lundén and Löfgren 1998). Beside temperature, the amount of precipitation might influence the wet deposition and consequently the concentrations and stocks of organic substances.

In the present study, it is also concluded from the explained variance of the nonlinear regression analysis (SVM) that other factors than depth increment, TOC concentration, and forest type play a significant role in explaining the organic substances' concentrations and stocks as well as their distribution. More than half or half for PAH, respectively, of the variance remains unaccounted. It seems therefore reasonable to include regional characteristics into statistical models of future investigations in order to explain a greater share of the variance, particularly topography with changing climate parameters and distance to emitters.

8.3 Conclusion

To our knowledge, this is the first study to show stocks of organic substances in forest soils on a nationwide basis. As stock calculations took soil mass and soil volume by means of the bulk density, the depth of the investigated soil profiles, and the area into account, it was possible to compare the masses of organic substances per hectare in soil profiles to 10 cm depth under different forest types. The results showed that the organic substances stocks were not dependent on forest type for the investigated substances, except for PCB, showing higher stocks under coniferous than deciduous forest. This indicated that other factors than forest type explained the differences in total stock amounts across Germany, e.g. deposition or vicinity to emission sources.

The concentration of the investigated organic substances in German forest soils seemed to be generally low. This was underpinned by the values of PAH and PCB, which were lower than the precaution values prescribed in German soil protection legislation (BBodSchV). Nevertheless, soils store organic substances over long periods. Still, DDx applied as pesticide in the 1980s was detected in the organic layer, particularly in the eastern federal states. Additionally, deposition stemming from various sources cannot be ruled out and might become a problem in the future due to the persistence of organic substances. It is therefore recommended to find appropriate measures to reduce depositions, e.g. from fuel combustion, mineral, and metal industry. On the other hand, further soil samplings to study the development of organic substances concentrations under forest should be conducted.

The sampling campaign should include the search for impact factors responsible for the spatial distribution of organic substances (e.g. on a regional scale). Besides this, factors for the distribution within the soil profile can also be uncovered, i.e. the prevailing process of historic vs. current deposition. In this study, a greater share of the variance was explained with nonlinear regression compared to linear methods. It seems therefore reasonable to consider nonlinear trends in the evaluation of data from future studies. The outcome will improve models that forecast the distribution of organic substances in forest soils. Eventually, the results can contribute to further develop background, precaution, trigger, and action values in the German soil protection legislation.

Acknowledgements Financial support by the research fund of the Federal Environment Agency (UBA) is acknowledged (FKZ 3707 71 201). The data were generated in the BAM laboratories (Federal Institute for Materials Research and Testing, Berlin, Germany) and Helmholtz laboratories

(German Research Center for Environmental Health, Molecular EXposomics, Neuherberg, Germany). We gratefully thank Katja Kaminski, Sebastian Hein and Dino Berners (BAM), and Bernhard Henkelmann (Helmholtz) for processing the samples. We thankfully acknowledge Levke Godbersen and Friedrich Krone (BGR, Federal Institute for Geosciences and Natural Resources, Hannover, Germany) for measuring and providing TOC Data. We also thank Bernhard Aichner and Petra Lehnik-Habrink (BAM) and Marchela Pandelova and Karl-Werner Schramm (Helmholtz) for their very helpful comments on the draft manuscript.

References

Ad-HocAG_Boden (ed) (2005) Bodenkundliche Kartieranleitung (KA 5), vol 5. Schweizerbart'sche Verlagsbuchhandlung, Stuttgart, Germany

Aichner B, Bussian BM, Lehnik-Habrink P, Hein S (2013) Levels and spatial distribution of persistent organic pollutants in the environment: a case study of German forest soils. Environ Sci Technol 47(22):12703–12714. https://doi.org/10.1021/es4019833

Aichner B, Bussian BM, Lehnik-Habrink P, Hein S (2015) Regionalized concentrations and fingerprints of polycyclic aromatic hydrocarbons (PAHs) in German forest soils. Environ Pollut 203:31–39. https://doi.org/10.1016/j.envpol.2015.03.026

Aichner B, Glaser B, Zech W (2007) Polycyclic aromatic hydrocarbons and dioxin-like polychlorinated biphenyls in urban soils from Kathmandu, Nepal. Org Geochem 38 (4):700–715. https://doi.org/10.1016/j.orggeochem.2006.11.002

Aislabie JM, Richards NK, Boul HL (1997) Microbial degradation of DDT and its residues—a review. N Z J Agric Res 40(2):269–282. https://doi.org/10.1080/00288233.1997.9513247

Batool S, Ab Rashid S, Maah MJ, Sarfraz M, Ashraf MA (2016) Geographical distribution of persistent organic pollutants in the environment: a review. J Environ Biol 37(5):1125–1134

BBodSchG (1998) Bundes-Bodenschutzgesetz (Federal Soil Protection Act)

BBodSchV (1999) Bundes-Bodenschutz- und Altlastenverordnung (Federal Soil Protection and Contaminated Sites Ordinance)

Belis CA, Offenthaler I, Uhl M, Nurmi-Legat J, Bassan R, Jakobi G, Kirchner M, Knoth W, Krauchi N, Levy W, Magnani T, Moche W, Schramm KW, Simoncic P, Weiss P (2009) A comparison of Alpine emissions to forest soil and spruce needle loads for persistent organic pollutants (POPS). Environ Pollut 157(12):3185–3191. https://doi.org/10.1016/j.envpol.2009.05.035

BMELV (2006) Arbeitsanleitung für die zweite bundesweite Bodenzustandserhebung im Wald (BZE II). Bundesministerium für Ernährung, Landwirtschaft und Verbraucherschutz, Berlin, Germany

Boul HL (1995) DDT residues in the environment—a review with a New-Zealand perspective. N Z J Agric Res 38(2):257–277. https://doi.org/10.1080/00288233.1995.9513126

Breivik K, Sweetman A, Pacyna JM, Jones KC (2002) Towards a global historical emission inventory for selected PCB congeners—a mass balance approach—1. Global production and consumption. Sci Total Environ 290(1–3):181–198. https://doi.org/10.1016/s0048-9697(01)01075-0

Breivik K, Sweetman A, Pacyna JM, Jones KC (2007) Towards a global historical emission inventory for selected PCB congeners—a mass balance approach: 3. An update. Sci Total Environ 377(2–3):296–307. https://doi.org/10.1016/j.scitotenv.2007.02.026

Brorström-Lundén E, Löfgren C (1998) Atmospheric fluxes of persistent semivolatile organic pollutants to a forest ecological system at the Swedish west coast and accumulation in spruce needles. Environ Pollut 102(1):139–149. https://doi.org/10.1016/s0269-7491(98)00081-5

Daly GL, Wania F (2005) Organic contaminants in mountains. Environ Sci Technol 39(2):385–398

Desaules A, Ammann S, Blum F, Brandli RC, Bucheli TD, Keller A (2008) PAH and PCB in soils of Switzerland—status and critical review. J Environ Monit 10(11):1265–1277. https://doi.org/10.1039/b807206j

Desborough J, Harrad S (2011) Chiral signatures show volatilization from soil contributes to polychlorinated biphenyls in grass. Environ Sci Technol 45(17):7354–7357. https://doi.org/10.1021/es201895f

Dimond JB, Owen RB (1996) Long-term residue of DDT compounds in forest soils in Maine. Environ Pollut 92(2):227–230. https://doi.org/10.1016/0269-7491(95)00059-3

Dopico M, Gomez A (2015) Review of the current state and main sources of dioxins around the world. J Air Waste Manag Assoc 65(9):1033–1049. https://doi.org/10.1080/10962247.2015.1058869

Fiedler H, Hub M, Willner S, Hutzinger O (1995) Stoffbericht Hexachlorbenzol (HCB). Texte und Berichte zur Altlastenbearbeitung, vol 18/95. Landesanstalt für Umweltschutz Baden-Württemberg, Karlsruhe, Germany

Fortmann H, Meesenburg H (2007) Organische Schadstoffe in den Waldböden Niedersachsens—Bodendauerbeobachtung in Niedersachsen. GeoBerichte, vol 4. Landesamt für Bergbau, Energie und Geologie (LBEG), Hannover, Germany

Gasic B, Moeckel C, Macleod M, Brunner J, Scheringer M, Jones KC, Hungerbuhler K (2009) Measuring and modeling short-term variability of PCBs in air and characterization of urban source strength in Zurich, Switzerland. Environ Sci Technol 43(3):769–776. https://doi.org/10.1021/es8023435

Guzzella L, Poma G, De Paolis A, Roscioli C, Viviano G (2011) Organic persistent toxic substances in soils, waters and sediments along an altitudinal gradient at Mt. Sagarmatha, Himalayas, Nepal. Environ Pollut 159(10):2552–2564. https://doi.org/10.1016/j.envpol.2011.06.015

Hafner WD, Carlson DL, Hites RA (2005) Influence of local human population on atmospheric polycyclic aromatic hydrocarbon concentrations. Environ Sci Technol 39(19):7374–7379. https://doi.org/10.1021/es0508673

Hageman KJ, Hafner WD, Campbell DH, Jaffe DA, Landers DH, Simonich SL (2010) Variability in pesticide deposition and source contributions to snowpack in Western U.S. national parks. Environ Sci Technol 44(12):4452–4458. https://doi.org/10.1021/es100290q

Hangen E, Kirchner M, Kronawitter H, Mühlbacher T (2011) Organische Bodenschadstoffe (POPs) entlang eines Höhenprofils im Nationalpark Berchtesgaden. Paper presented at the Böden verstehen, Böden nutzen, Böden fit machen, Berlin and Potsdam, Germany

Heinisch E, Kettrup A, Wenzel-Klein S (1993) DDT/Lindan-Masseneinsätze in der DDR—Ökochemisch-ökotoxikologische Folgen. Umweltwissenschaften und Schadstoff-Forschung 5 (5):277–280. https://doi.org/10.1007/bf02937964

Heinisch E, Kettrup A, Bergheim W, Martens D, Wenzel S (2005) Persistent chlorinated hydrocarbons (PCHC), source-oriented monitoring in aquatic media—2. The insecticide DDT, constituents, metabolites. Fresenius Environ Bull 14(2):69–85

Heinisch E, Kettrup A, Bergheim W, Martens D, Wenzel S (2006a) Persistent chlorinated hydrocarbons (PCHC), source-oriented monitoring in aquatic media. 4. The chlorobenzenes. Fresenius Environ Bull 15(3):148–169

Heinisch E, Kettrup A, Bergheim W, Wenzel S (2006b) Persistent chlorinated hydrocarbons (PCHC), source-oriented monitoring in aquatic media—5. Polychlorinated biphenyls (PCBs). Fresenius Environ Bull 15(11):1344–1362

Horstmann M, McLachlan MS (1998) Atmospheric deposition of semivolatile organic compounds to two forest canopies. Atmos Environ 32(10):1799–1809. https://doi.org/10.1016/s1352-2310(97)00477-9

Jamshidi A, Hunter S, Hazrati S, Harrad S (2007) Concentrations and chiral signatures of polychlorinated biphenyls in outdoor and indoor air and soil in a major UK conurbation. Environ Sci Technol 41(7):2153–2158. https://doi.org/10.1021/es062218c

Jones KC, de Voogt P (1999) Persistent organic pollutants (POPs): state of the science. Environ Pollut 100(1–3):209–221. https://doi.org/10.1016/s0269-7491(99)00098-6

Kallenborn R (2007) Persistent organic pollutants (POPs) as environmental risk factors in remote high-altitude ecosystems. Ecotoxicol Environ Saf 63(1):100–107. https://doi.org/10.1016/j.ecoenv.2005.02.016

Khalili NR, Scheff PA, Holsen TM (1995) PAH source fingerprints for coke ovens, diesel and gasoline-engines, highway tunnels, and wood combustion emissions. Atmos Environ 29 (4):533–542. https://doi.org/10.1016/1352-2310(94)00275-p

Kirchner M, Faus-Kessler T, Jakobi G, Levy W, Henkelmann B, Bernhoft S, Kotalik J, Zsolnay A, Bassan R, Belis C, Krauchi N, Moche W, Simoncic P, Uhl M, Weiss P, Schramm KW (2009) Vertical distribution of organochlorine pesticides in humus along Alpine altitudinal profiles in relation to ambiental parameters. Environ Pollut 157(12):3238–3247. https://doi.org/10.1016/j.envpol.2009.06.011

Krauss M, Wilcke W, Zech W (2000) Polycyclic aromatic hydrocarbons and polychlorinated biphenyls in forest soils: depth distribution as indicator of different fate. Environ Pollut 110 (1):79–88. https://doi.org/10.1016/s0269-7491(99)00280-8

Lohmann R, Breivik K, Dachs J, Muir D (2007) Global fate of POPs: current and future research directions. Environ Pollut 150(1):150–165. https://doi.org/10.1016/j.envpol.2007.06.051

Maliszewska-Kordybach B, Smreczak B, Klimkowicz-Pawlas A, Terelak H (2008) Monitoring of the total content of polycyclic aromatic hydrocarbons (PAHs) in arable soils in Poland. Chemosphere 73(8):1284–1291. https://doi.org/10.1016/j.chemosphere.2008.07.009

Manz M, Wenzel KD, Dietze U, Schüürmann G (2001) Persistent organic pollutants in agricultural soils of central Germany. Sci Total Environ 277(1–3):187–198. https://doi.org/10.1016/s0048-9697(00)00877-9

Meijer SN, Ockenden WA, Sweetman A, Breivik K, Grimalt JO, Jones KC (2003) Global distribution and budget of PCBs and HCB in background surface soils: implications or sources and environmental processes. Environ Sci Technol 37(4):667–672. https://doi.org/10.1021/es0258091

Navarro-Ortega A, Ratola N, Hildebrandt A, Alves A, Lacorte S, Barceló D (2012) Environmental distribution of PAHs in pine needles, soils, and sediments. Environ Sci Pollut Res 19 (3):677–688. https://doi.org/10.1007/s11356-011-0610-5

Nizzetto L, Cassani C, Di Guardo A (2006) Deposition of PCBs in mountains: the forest filter effect of different forest ecosystem types. Ecotoxicol Environ Saf 63(1):75–83. https://doi.org/10.1016/j.ecoenv.2005.05.005

Offenthaler I, Bassan R, Belis C, Jakobi G, Kirchner M, Kräuchi N, Moche W, Schramm KW, Sedivy I, Simoncic P, Uhl M, Weiss P (2009) PCDD/F and PCB in spruce forests of the Alps. Environ Pollut 157(12):3280–3289. https://doi.org/10.1016/j.envpol.2009.05.052

Pandelova M, Henkelmann B, Bussian BM, Schramm KW (2018) Results of the second national forest soil inventory in Germany—interpretation of level and stock profiles for PCDD/F and PCB in terms of vegetation and humus type. Sci Total Environ 610:1–9. https://doi.org/10.1016/j.scitotenv.2017.07.246

Posado-Baquero R, Ortega-Calvo J-J (2011) Recalcitrance of polycyclic aromatic hydrocarbons in soil contributes to background pollution. Environ Pollut 159(12):3692–3699

R Core Team (2015) R: A language and environment for statistical computing. R Foundation for Statistical Computing, Vienna, Austria

Schmid P, Gujer E, Zennegg M, Bucheli TD, Desaules A (2005) Correlation of PCDD/F and PCB concentrations in soil samples from the Swiss soil monitoring network (NABO) to specific parameters of the observation sites. Chemosphere 58(3):227–234. https://doi.org/10.1016/j.chemosphere.2004.08.045

Schuster JK, Gioia R, Moeckel C, Agarwal T, Bucheli TD, Breivik K, Steinnes E, Jones KC (2011) Has the burden and distribution of PCBs and PBDEs changed in European background soils between 1998 and 2008? Implications for sources and processes. Environ Sci Technol 45 (17):7291–7297. https://doi.org/10.1021/es200961p

Sweetman AJ, Dalla Valle M, Prevedouros K, Jones KC (2005) The role of soil organic carbon in the global cycling of persistent organic pollutants (POPs): interpreting and modelling field data. Chemosphere 60(7):959–972. https://doi.org/10.1016/j.chemosphere.2004.12.074

Taniyasu S, Falandysz J, Swietojanska A, Flisak M, Horii Y, Hanari N, Yamashita N (2005) Clophen A60 composition and content of CBs, CNs, CDFs, and CDDs after 2D-HPLC, HRGC/LRMS, and HRGC/HRMS separation and quantification. J Environ Sci Health A Tox Hazard Subst Environ Eng 40(1):43–61. https://doi.org/10.1081/ese-200033521

UBA (2012) Umweltprobenbank des Bundes http://www.umweltprobenbank.de/en/documents/profiles/analytes/10052. Accessed 29.03.2019

UBA (2016) Polyzyklische Aromatische Kohlenwasserstoffe—Umweltschädlich! Giftig! Unvermeidbar? Hintergrundpapier. Umweltbundesamt, Dessau-Roßlau, Germany

UBA (2017) Dioxine und dioxinähnliche PCB in Umwelt und Nahrungsketten. Hintergrundpapier. Umweltbundesamt, Dessau-Roßlau, Germany

UNEP (2001) Final Act of the Plenipotentiaries on the Stockholm Convention on persistent organic pollutants. Geneva, Switzerland

van den Berg M, Birnbaum LS, Denison M, De Vito M, Farland W, Feeley M, Fiedler H, Hakansson H, Hanberg A, Haws L, Rose M, Safe S, Schrenk D, Tohyama C, Tritscher A, Tuomisto J, Tysklind M, Walker N, Peterson RE (2006) The 2005 World Health Organization reevaluation of human and mammalian toxic equivalency factors for dioxins and dioxin-like compounds. Toxicol Sci 93(2):223–241. https://doi.org/10.1093/toxsci/kfl055

Wang XP, Sun DC, Yao TD (2016) Climate change and global cycling of persistent organic pollutants: a critical review. Sci China Earth Sci 59(10):1899–1911. https://doi.org/10.1007/s11430-016-5073-0

Wania F, Mackay D (1996) Tracking the distribution of persistent organic pollutants. Environ Sci Technol 30(9):A390–A396. https://doi.org/10.1021/es962399q

Weiss P, Lorbeer G, Scharf S (2000) Regional aspects and statistical characterisation of the load with semivolatile organic compounds at remote Austrian forest sites. Chemosphere 40 (9–11):1159–1171. https://doi.org/10.1016/s0045-6535(99)00365-3

Wilcke W (2007) Global patterns of polycyclic aromatic hydrocarbons (PAHs) in soil. Geoderma 141(3–4):157–166. https://doi.org/10.1016/j.geoderma.2007.07.007

Chapter 9
Nutritional Status of Major Forest Tree Species in Germany

Ulrike Talkner, Winfried Riek, Inge Dammann, Martin Kohler, Axel Göttlein, Karl Heinz Mellert, and Karl Josef Meiwes

9.1 Introduction

Element contents and ratios in assimilation organs of trees are an essential component of a comprehensive forest condition diagnosis. They allow conclusions about the current nutritional status of trees (Sardans et al. 2016; Mellert et al. 2008; Fiedler and Höhne 1984; Wehrmann 1959). In addition, foliar chemistry can be used as an indicator for the exposure of forest ecosystems to atmospheric pollutants (Ewald 2005; Aber et al. 1989; Heinsdorf et al. 1988). While soil chemical properties indicate the nutrient supply, foliar element contents reflect the nutritional status of the tree itself. The nutritional status of trees is not only determined by the soil nutrient supply but also by the soil nutrient availability and the direct uptake of nutrients from the atmosphere as well as the direct nutrient loss from leaves and needles by leaching. A balanced nutrition of essential main and trace nutrients is a necessary condition for optimal plant growth (Aerts and Chapin 2000). In addition, optimally nourished forest trees are more resistant to external biotic and abiotic stress

U. Talkner (✉) · I. Dammann · K. J. Meiwes
Northwest German Forest Research Institute, Göttingen, Germany
e-mail: ulrike.talkner@nw-fva.de; inge.dammann@nw-fva.de; karl-josef.meiwes@nw-fva.de

W. Riek
University for Sustainable Development and Eberswalde Forestry State Center of Excellence, Eberswalde, Germany
e-mail: winfried.riek@hnee.de

M. Kohler
Chair of Silviculture, Albert-Ludwigs University, Freiburg, Germany
e-mail: martin.kohler@waldbau.uni-freiburg.de

A. Göttlein · K. H. Mellert
Chair of Forest Nutrition and Water Resources, Technical University Munich, Freising, Germany
e-mail: goettlein@forst.tu-muenchen.de; karl.mellert@tum.de

N. Wellbrock, A. Bolte (eds.), *Status and Dynamics of Forests in Germany*, Ecological Studies 237, https://doi.org/10.1007/978-3-030-15734-0_9

factors than poorly nourished trees (Flückiger and Braun 1999). Pronounced relationships between tree health and nutritional status can therefore be detected (Cape et al. 1990).

Chemical foliar analyses are a substantial part of the National Forest Soil Inventory (NFSI) in Germany. They reveal the current nutritional status of the forests. In addition, comparison of the nutritional data from NFSI I (1987–1992) and NFSI II (2007–2008) contributes significantly to the interpretation of forest ecosystem changes and the effectiveness of both air quality control and forestry stabilization measures (e.g. conversion to mixed-species forests, liming of acidified soils). Furthermore, the extensive data set of the NFSI offers additional opportunities for nutritional diagnoses of conifers by considering ratios between foliar element contents of older and current-year needles.

In this chapter, we evaluate the foliar chemistry of the following tree species: Norway spruce (*Picea abies* (L.) H. Karst.), Scots pine (*Pinus sylvestris* L.), European beech (*Fagus sylvatica* L.) and pedunculate (*Quercus robur* L.) and sessile oak (*Quercus petraea* (Matt.) Liebl.). In addition, we assess relationships between foliar chemistry and chemical parameters of the soil. The two oak species were pooled for all analyses. The nutritional diagnoses presented in this chapter are based on tree species-specific threshold values of an integrative assessment system, which was derived by Göttlein (2015). In this meta-analysis, the median threshold value for all assessment systems available in the literature for the major tree species is considered the most probable threshold value. The thresholds at the lower end of the range of normal nutrition are given in Table 9.1.

The evaluation of results depicted several noticeable phenomena for the four major tree species in Germany (Riek et al. 2016) (Fig. 9.1). Luxurious nutrition with nitrogen (N) was observed for all tree species. More than half of Scots pine and oak and more than a quarter of Norway spruce and European beech plots showed foliar N contents above the normal range. Only 8% of Norway spruce and 5% of Scots pine stands were within the range of (latent) N deficiency. Phosphorus (P) deficiencies were found for all tree species but were most pronounced on European beech plots, with almost two-thirds of plots in the (latent) deficiency range. For all tree species, at

Table 9.1 Threshold values for the four tree species Norway spruce (*Picea abies*), Scots pine (*Pinus sylvestris*), European beech (*Fagus sylvatica*) and oak (*Quercus petraea* and *Q. robur*) at the lower end of the range of normal nutrition according to Göttlein (2015)

Species	N mg g^{-1}	P mg g^{-1}	K mg g^{-1}	Ca mg g^{-1}	Mg mg g^{-1}	S mg g^{-1}	Fe µg g^{-1}	Mn µg g^{-1}	Cu µg g^{-1}	Zn µg g^{-1}
Picea abies	13.1	1.3	4.5	2.0	0.8	1.0	42	50	2	20
Pinus sylvestris	14.1	1.3	4.4	2.0	0.8	1.0	40	40	3	20
Fagus sylvatica	19.0	1.2	6.0	5.0	1.0	1.5	60	60	5	20
Quercus petraea +robur	20.0	1.4	6.1	5.0	1.2	1.2	70	66	6	15

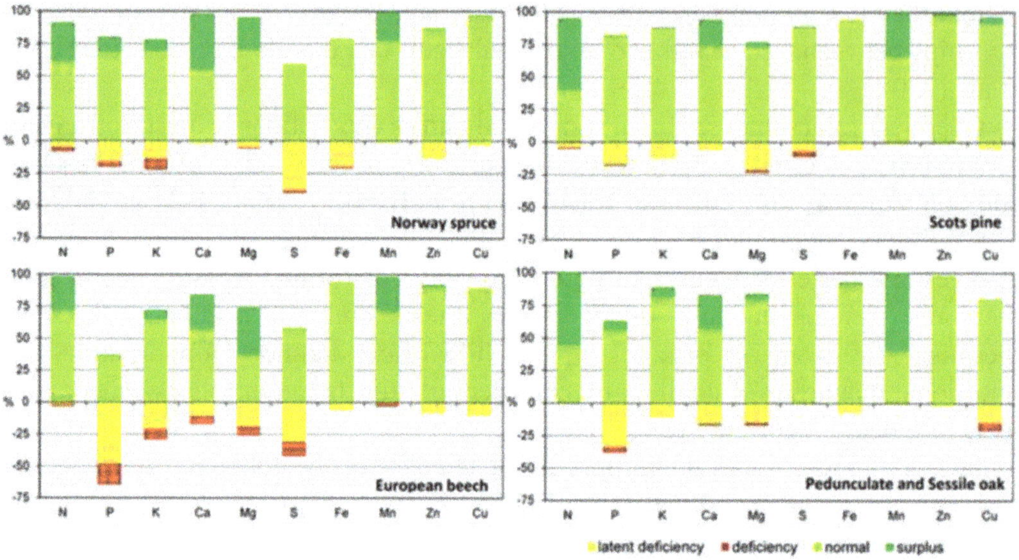

Fig. 9.1 Percentage of NFSI II plots in the four nutritional ranges according to the assessment system of Göttlein (2015) for Germany's major tree species, Norway spruce, Scots pine, European beech as well as pedunculate and sessile oak (pooled together)

least 72% of all plots showed normal or luxurious potassium (K), calcium (Ca) and magnesium (Mg) nutrition. The sulphur (S) nutritional status was evaluated as normal on most Scots pine and oak plots. However, in the case of Norway spruce and European beech, approximately 40% of the inventory plots were in the (latent) S deficiency range. Contents of the trace elements iron (Fe), zinc (Zn) and copper (Cu) were predominantly in the range of normal nutrition for all four tree species. For manganese (Mn), the high proportion of plots with trees in the luxurious range is striking. The results of NFSI II revealed that nutrient deficiencies are more widespread for European beech than for Norway spruce, oak and Scots pine.

Following from the noticeable phenomena found on the NFSI II plots (Fig. 9.1), the aim of this chapter is to test the following hypotheses:

1. Foliar N contents are related to atmospheric N deposition.
2. Foliar P contents of European beech are related to the acidification status of the soil and atmospheric N deposition.
3. Foliar S contents are related to the sulphate concentration in soil extracts as an indicator for former S deposition and for sulphate retention capacity of soils.
4. Liming of acidified soils increases foliar Ca and Mg contents but may have a negative impact on foliar K nutrition.
5. Air quality control measures lead to decreased foliar lead contents.
6. The nutritional diagnosis for conifers is improved by considering the ratios between foliar element contents of 2- (Scots pine) or 3- (Norway spruce) year-old needles and current-year needles.

9.2 Foliar Nitrogen Nutrition

Nitrogen is organically bound in plants in the form of proteins as well as amino and nucleic acids. It is an essential component of enzymes involved in metabolic processes, and N deficiency leads to metabolic disorders as well as reduced plant growth. In contrast, an N oversupply can either hinder the uptake of other nutrients or lead to nutrient losses from the foliage by canopy leaching, thereby inducing nutrient imbalances of, for example, K, Mg or P in relation to N (Schulze et al. 1989). In addition, an excess of N increases the vulnerability to pests and can affect a tree's resistance to frost.

Globally, N is considered to be the primary factor limiting growth in (near-natural) terrestrial ecosystems (Vitousek and Howarth 1991). Over centuries, biomass export—caused by intensive wood harvesting, litter raking and forest pasturing—has reduced N stocks of many Central European forest sites, thereby aggravating N shortage. However, during recent decades, this situation has fundamentally changed due to the atmospheric deposition of nitric oxides from industry and traffic as well as ammonium from agriculture (Gruber and Galloway 2008; Gauger et al. 2008). If N is no longer a limiting factor, other nutrients may reach a relative deficiency (Mellert and Göttlein 2013). Thus, growth stimulation by N input (Mellert et al. 2008; Spiecker et al. 2012) has the effect that other essential nutrients as well as water must be taken up in larger quantities (de Vries et al. 2006). In addition, direct uptake of N in the canopy may lead to nutrient losses from the foliage by canopy leaching of, for example, K (Klemm et al. 1989). The resulting higher nutrient and water requirements cannot always be met—particularly on sites with bedrock that has low levels of base cations—so that the N-based increase in growth may lead to nutritional imbalances involving N:Mg, N:K, N:Ca or N:P ratios and more frequent nutrient and water deficiency situations (Aber et al. 1989, 2003). Hence, one of the risk factors Central European forests currently face is the stress caused by high atmospheric N deposition (Glatzel et al. 1987; Matson et al. 2002; Meesenburg et al. 2016).

The foliar N contents of NFSI II plots stratified by soil substrate groups and current soil acidity levels showed only very weak differences between the various site-specific strata. Even the close relationship expected between humus form and N nutrition was nearly non-existent. This suggests that N compounds resulting from atmospheric deposition and available to plants supersede the influence of site-specific factors on N nutrition.

The fact that N deposition contributes to forest N nutrition was shown by the significant albeit weak linear relationship between N deposition and foliar N nutrition (Fig. 9.2 and Table 9.2).

In contrast to N deposition in Germany, the regional distribution of the tree species-specific N nutrition ranges in Germany showed only weak spatial patterns (Figs. 9.3 and 9.4). Most Scots pine and Norway spruce stands with normal N supply or even (latent) N deficiency were located in the north-eastern lowlands (Brandenburg, Saxony), the southern Black Forest and in Rhineland-Palatinate. In all other parts of Germany, foliar N nutrition was predominantly luxurious or normal.

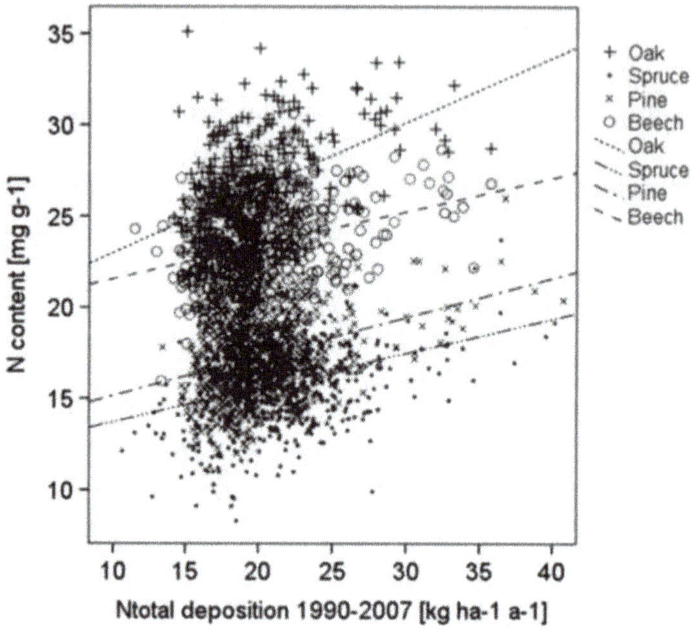

Fig. 9.2 Scatter plot of Norway spruce, Scots pine, European beech as well as pedunculate and sessile oak (pooled together) foliar N content in relation to total N deposition

Table 9.2 Parameters of the linear regression between foliar N content and total N deposition shown in Fig. 9.2: foliar N content = *a* * total N deposition + *b*

	a	*b*	R^2
Norway spruce	0.189	11.83	0.11**
Scots pine	0.215	13.00	0.12**
European beech	0.185	19.71	0.07**
Pedunculate and sessile oak	0.357	19.42	0.18**

* $p<0.05$; ** $p<0.01$; *** $p<0.001$

In addition to absolute foliar N contents, the relation of N to other nutrient elements can be used to evaluate nutrient imbalances (Flückiger and Braun 2003; Mellert and Göttlein 2012). In principle, a good nutritional state exists when the element contents are sufficient and present in a balanced relationship. Deviations from the reference values for "harmonic" element ratios can indicate possible imbalances in the nutrient supply.

With regard to a balanced nutrition, we found that high N to nutrient ratios were primarily associated with low values for each of the elements Mg, P, K and Ca (Fig. 9.5). In contrast, high foliar N contents did not automatically result in high ratios (Fig. 9.6). For stands with an N oversupply, very narrow as well as very high ratios occurred for all tree species. Only the highest individual values of the element ratios were generally associated with an exceptionally high level of N nutrition. Thus, no clear causal relationship between N surplus and disturbances to nutrient uptake could be derived.

Based on Fig. 9.5 and for the other main elements as well, it is clear that latent element deficiencies are associated with specific element ratios. In order to derive threshold ratios that lead to element deficiencies, the relationships between elements and element ratios were determined using smoothing functions in SPSS according to the LOESS method (Jacoby 2000). Subsequently, the individual intersection of the

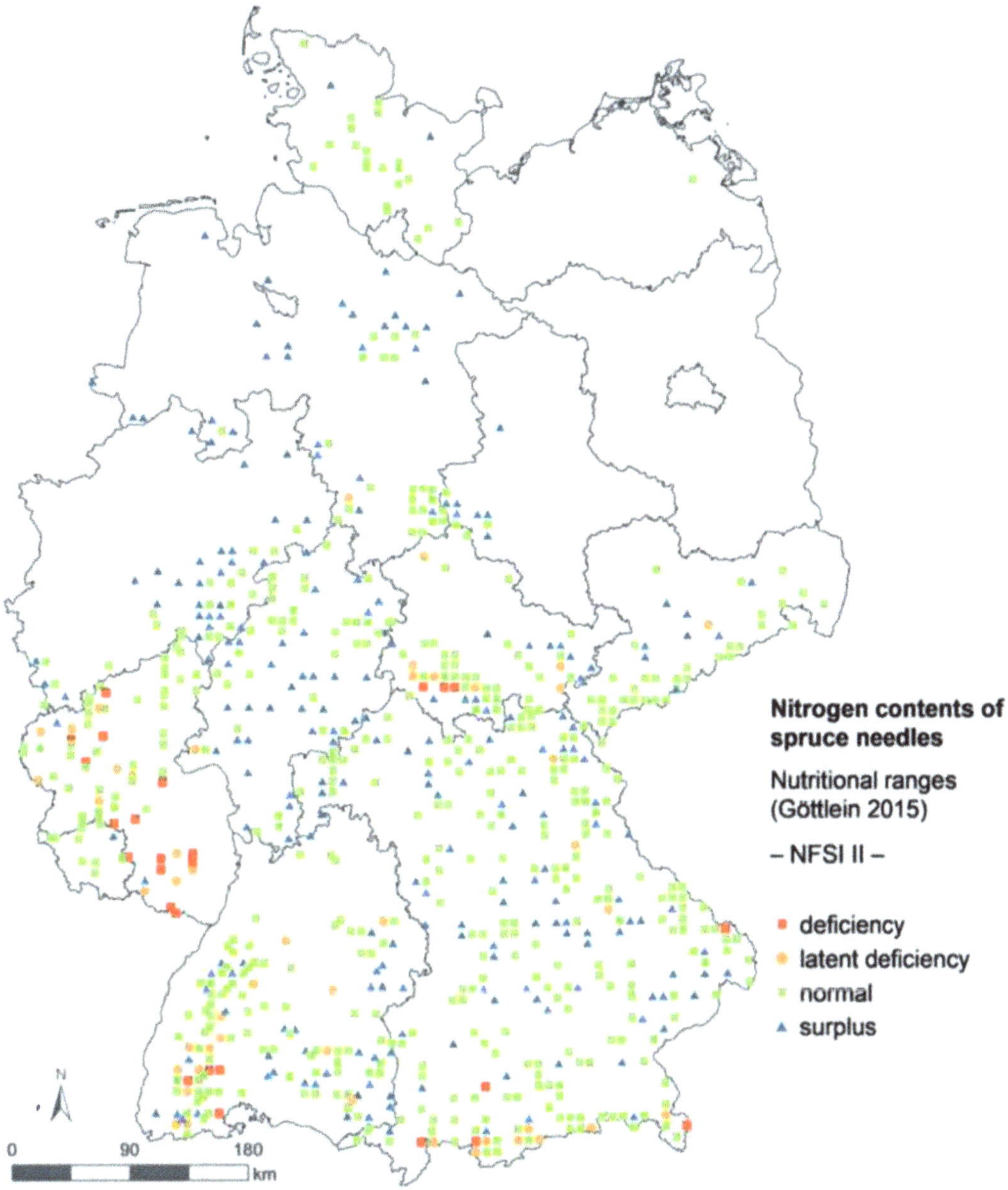

Fig. 9.3 Regional distribution of foliar nitrogen nutrition ranges for Norway spruce

fitted curve with the horizontal borderline between normal nutrition and latent deficiency was determined. The threshold values of N ratios calculated in this way are shown in Table 9.3. On a statistical average, exceeding these values resulted in latent deficiencies of the respective element.

The N:Mg, N:P and N:K ratios empirically derived from the NFSI data were almost identical to the values derived by van den Burg and reported by Mellert and Göttlein (2012). In general, the thresholds in Table 9.3 are slightly higher. However, the thresholds of Flückiger and Braun (2003) for Norway spruce and European beech are similar. Imbalanced N ratios can lead to an increased risk of possible infestation by harmful insects (e.g. Flückiger and Braun 2003). In order to avoid

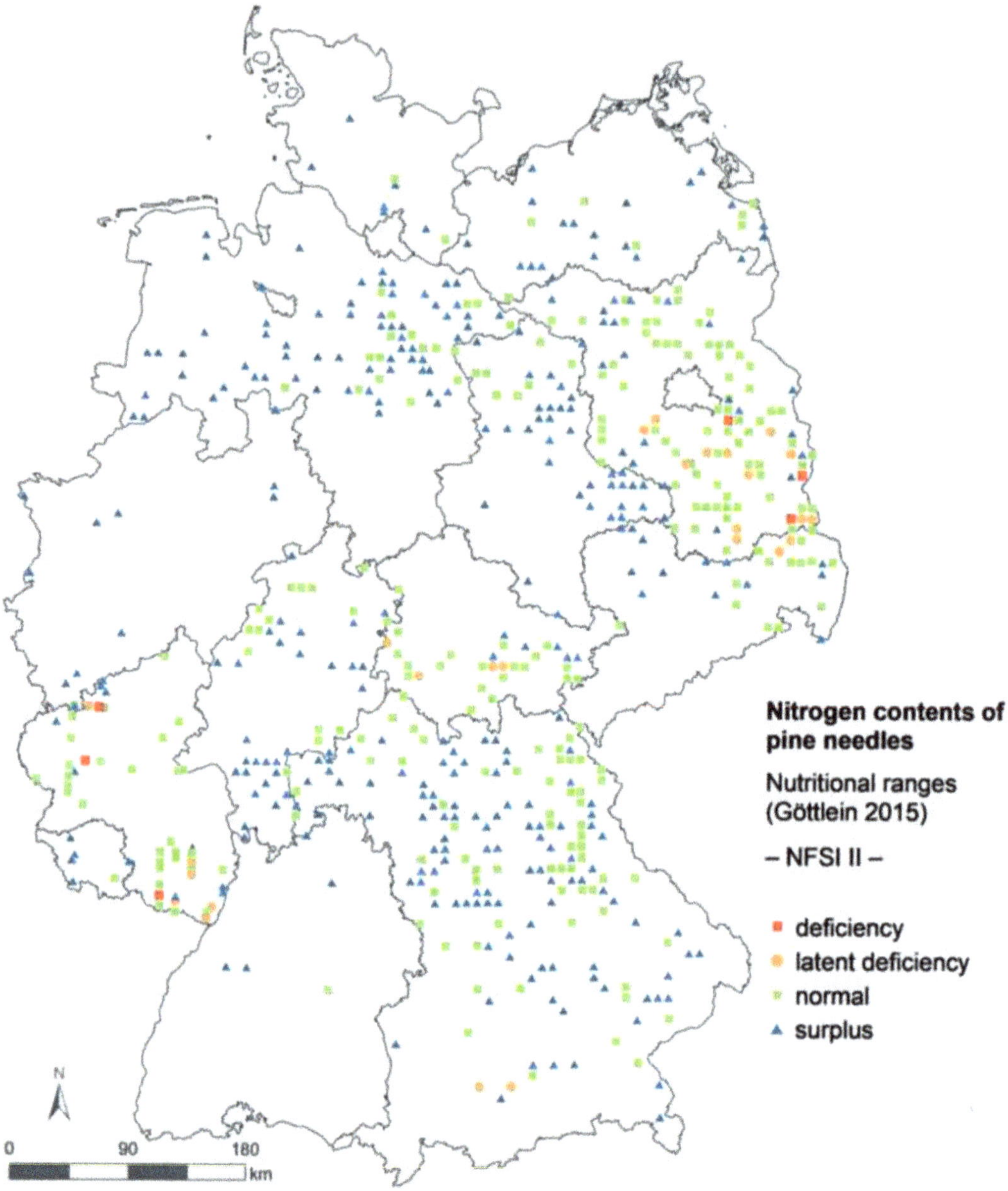

Fig. 9.4 Regional distribution of foliar nitrogen nutrition ranges for Scots pine

values clearly above the thresholds given in Table 9.3, more favourable element ratios may be achieved by soil protection liming, suitable fertilization measures or reduction of atmospheric N deposition.

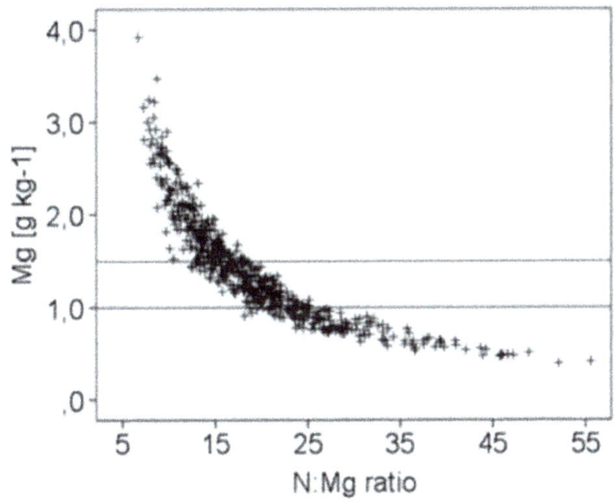

Fig. 9.5 Relationship between foliar Mg content and foliar N:Mg ratio in European beech; horizontal lines mark the normal range for Mg according to Göttlein (2015), i.e. values below the lower line indicate latent Mg deficiency

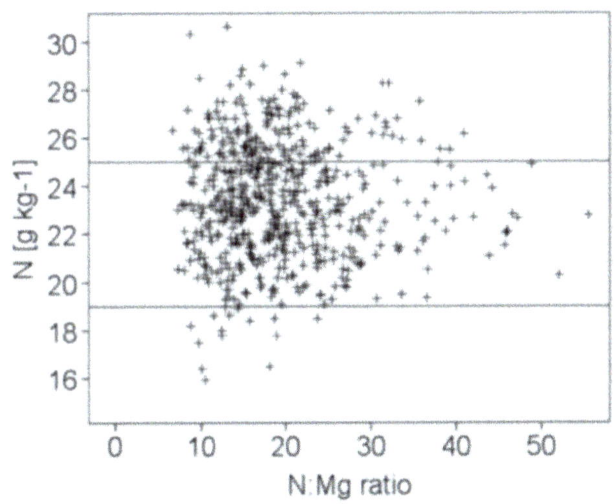

Fig. 9.6 Relationship between foliar N content and foliar N:Mg ratio in European beech; horizontal lines mark the normal range for N according to Göttlein (2015), i.e. values below the lower line indicate latent N deficiency

Table 9.3 Critical N ratios derived from the NFSI II data, which correspond to a latent element deficiency when exceeded

	Norway spruce	Scots pine	Pedunculate and sessile oak	European beech
N:Mg	22.0	22.8	23.5	23.5
N:P	12.2	14.2	19.8	19.7
N:K	3.6	4.3	4.5	4.0
N:Ca	7.8	8.9	5.6	4.7

9.3 Foliar Phosphorus Nutrition of European Beech

Foliar P nutrition of European beech is deficient across large areas of Europe (de Vries et al. 2000) and has decreased over the past 20 years (Talkner et al. 2015). Several factors may lead to P deficiency: (1) excessive N deposition, (2) climate change and (3) tree physiology. (1) N deposition influences foliar P nutrition primarily by negatively influencing mycorrhizal symbioses (Nilsson and Wallander 2003) but also through increased growth leading to unbalanced foliar nutrition (Nihlgård 1985; Aber et al. 1998). (2) Elevated atmospheric CO_2 concentrations as well as increased temperatures appear to increase the C:P ratio in plants (Sardans et al. 2012). In addition, drought events might decrease P contents in European beech trees (Peuke and Rennenberg 2004). (3) More frequent and more intense fructification events (Övergaard et al. 2007; Piovesan and Adams 2001) may deplete P reserves in trees when the mineralization of P from seeds and seed capsules is slow (Khanna et al. 2009) or seeds are eaten and displaced by mice and birds (Burschel et al. 1964). Foliar P contents represent a good indicator for plant availability of P at a site, since up to now, no single extraction method has been developed that can adequately quantify plant-available P in forest soils.

In Germany, the results of NFSI II have revealed that the foliar P content of 60% of European beech plots is either within the latent deficiency or the deficiency range (see Fig. 9.1). Foliar P contents tended to decrease between NFSI I and NFSI II. This finding is in accordance with results from other parts of Europe (de Vries et al. 2000; Talkner et al. 2015).

The foliar P content was not related to the soil pH value on NFSI II plots (Fig. 9.7). One would expect the highest P availability and therewith the highest foliar P contents on soils with intermediate pH values. In calcareous soils high in pH,

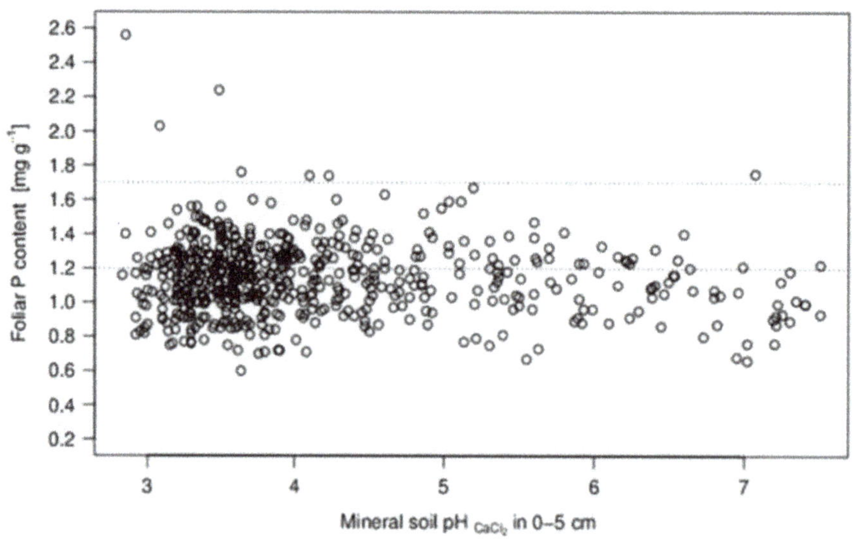

Fig. 9.7 Foliar P content versus pH in $CaCl_2$ in 0–5 cm soil depth at European beech plots; horizontal dashed lines mark the normal range for P according to Göttlein (2015), i.e. values below the lower line indicate latent P deficiency

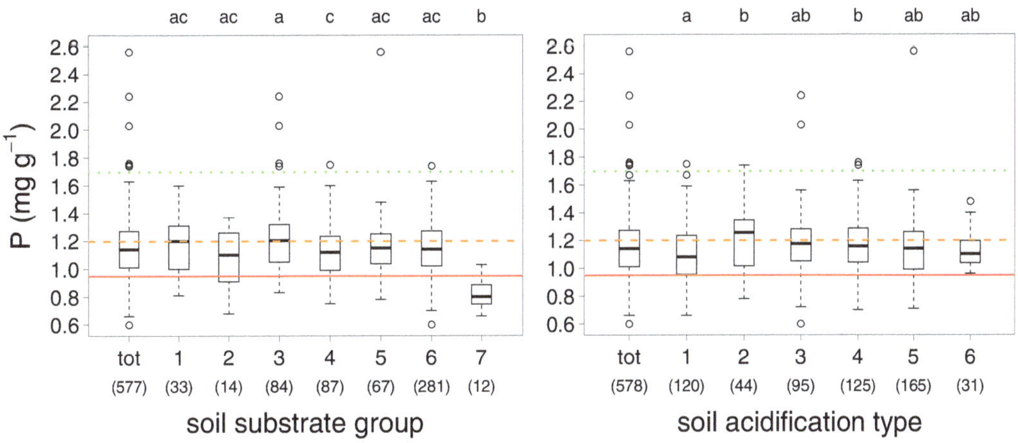

Fig. 9.8 Foliar P content of different soil substrate groups (left) and soil acidification types (right) at European beech NFSI II plots; horizontal dashed lines mark the nutritional ranges for P according to Göttlein (2015); tot: all groups/types taken together; soil substrate groups: [1]soils from base-poor unconsolidated sediment, [2]soils of alluvial plains, [3]loamy soils of the lowland, [4]soils from weathered carbonate bedrock, [5]soils from basic-intermediate bedrock, [6]soils from base-poor hard bedrock, [7]soils from the Alps; for definition of soil acidification types, see Chap. 2; different letters above the boxes depict significantly different foliar P contents

Ca phosphates of low solubility are formed; hence, P availability is low in such soils. However, these Ca phosphates are easily dissolved when soils slightly acidify, leading to higher P availability. In non-calcareous soils with pH values below 5, the solubility of P is known to be low since it is adsorbed to Fe and aluminium (Al) oxides (Blume et al. 2016; Jönsson Belyazid and Belyazid 2012).

Even though we did not find the expected relationship between foliar P contents and soil pH values, foliar P contents differed among soil acidification types and soil substrate groups in the expected way (Fig. 9.8). National Forest Soil Inventory II plots on soils with very high base saturation (>85%) in the mineral top- and subsoil showed the lowest foliar P contents. In contrast, European beech trees on plots with slightly reduced base saturation (50–85%) in the mineral topsoil showed the highest foliar P contents. Foliar P contents of European beech trees on soils with low base saturation have been in between these two groups. The group of NFSI plots with very high base saturation contains the plots in the calcareous Alps and plots on other pure calcareous soils. Indeed, on alpine soils, foliar P contents were lowest compared to all other soil parent material groups. In the calcareous Alps, P causes low vitality of European beech (Ewald 2000) and appears to be the most important nutrient limiting growth of Norway spruce (Mellert and Ewald 2014). The group of plots with slightly reduced base saturation in the mineral topsoil and highest foliar P contents contained shallowly acidified calcareous sites.

The foliar P content was weakly related to the N:P ratio of the forest floor and mineral soil: the larger the N:P ratio, the smaller the foliar P content (Fig. 9.9). The N:P ratio of the forest floor is influenced by site variables such as tree species composition and climate but also by N deposition. This would lead to the

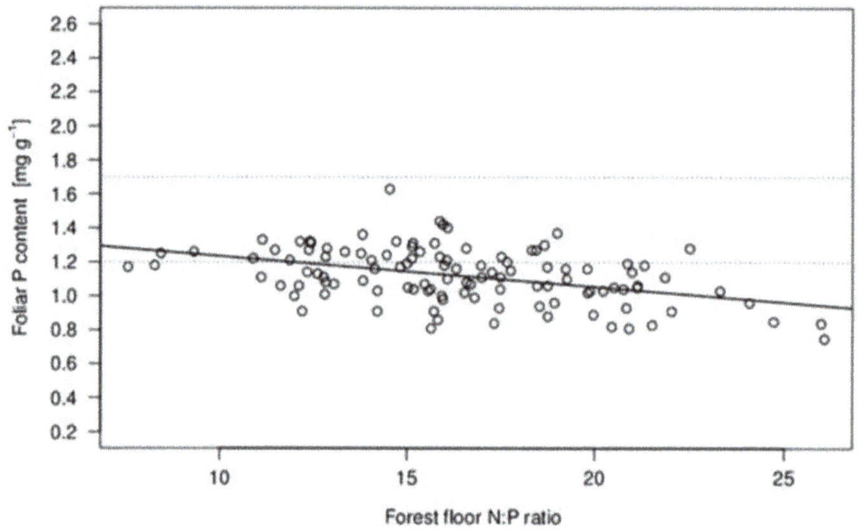

Fig. 9.9 Relationship between foliar P content and the N:P ratio in the forest floor (L + Of + Oh) at European beech plots ($R^2 = 0.18$, $p < 0.001$); horizontal dashed lines mark the normal range for P according to Göttlein (2015), i.e. values below the lower line indicate latent P deficiency

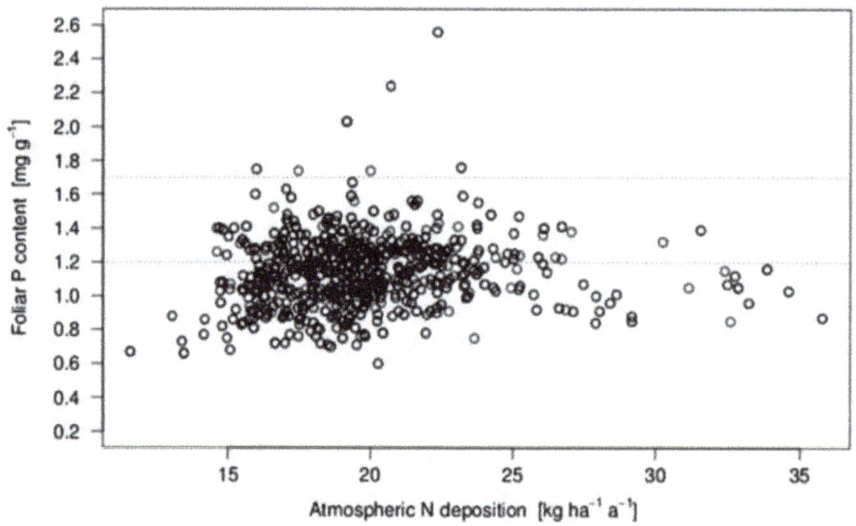

Fig. 9.10 Foliar P content versus total atmospheric N deposition at European beech plots; horizontal dashed lines mark the normal range for P according to Göttlein (2015), i.e. values below the lower line indicate latent P deficiency

expectation that foliar P contents are negatively related to N deposition (Braun et al. 2010). However, this could not be verified by the NFSI II dataset (Fig. 9.10).

The proportion of the variance in foliar P contents that was explained by the N:P ratio of all forest floor layers and mineral soil depths, i.e. the coefficient of determination, was much smaller than that found in the evaluation of the ICP Forests dataset, which contains time series of monitoring data (Talkner et al. 2015). The lower coefficient of determination may be due to the fact that foliar nutrients comprise high interannual variances, primarily due to meteorological differences among years (Evers 1972). Therefore, it is recommended to sample leaves in several (subsequent) years in order to be able to evaluate foliar nutritional status at a given site (Wehrmann 1959). Unfortunately, such a high sampling effort could not be realized in the NFSI. Hence, the data may represent the mean nutritional status throughout Germany and may even show coarse regional differences; however, the data may not be suitable for identifying correlations between foliar nutritional status and soil chemical parameters. This assumption is confirmed by the fact that all significant relationships between foliar P and soil chemical parameters found in the ICP Forests dataset when mean foliar P contents of several years were selected (Talkner et al. 2015) are no longer detectable when foliar ICP Forests data from just a single year were selected (data not published). In the latter case, close to the same coefficients of determination were found as with the NFSI II data set, which also comprised data from just 1 sampling year.

Since the determination of foliar P contents in several years is time- and resource-intensive, an indicator for foliar P status is urgently needed. Several soil extraction methods have been tested to assess the P status of a site (Kohlpaintner et al. 2017). For European beech, the best relationship was found between foliar P and citrate-extractable P in 0–10 cm soil depth. In addition, the P content of the forest floor seemed to be a rather good indicator, since, by litterfall, it directly represents the foliar P status of European beech (Talkner et al. 2015). Indeed, a significant positive relationship between foliar P content and the P content of the forest floor was found for European beech plots in the NFSI II (Fig. 9.11). However, the relationship was much weaker for the NFSI II compared to the ICP Forests dataset (Talkner et al. 2015), probably due to the constraints of the NFSI foliar data mentioned above.

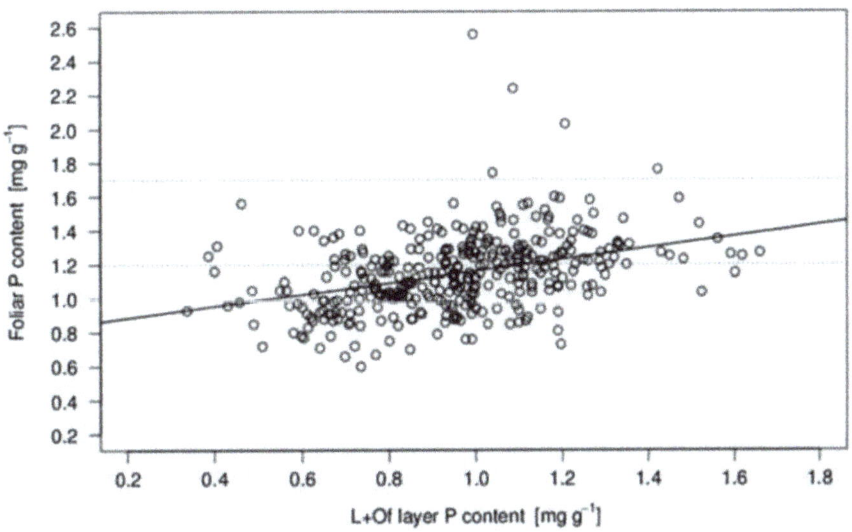

Fig. 9.11 Relationship between foliar P content and the P content in the L + Of layer of the forest floor at European beech plots ($R^2 = 0.16, p < 0.001$); horizontal dashed lines mark the normal range for P according to Göttlein (2015), i.e. values below the lower line indicate latent P deficiency

Proper separation of the forest floor and the mineral soil is very important when evaluating P. Therefore, only the L or the L + Of layer should be considered.

In summary, the foliar P nutrition of European beech was deficient across large areas of Germany and deteriorated over time in some regions. Foliar P contents seem to be lowest on calcareous soils that are high in base saturation, indicating that the geological substrate (Augusto et al. 2017) may have influenced foliar P nutrition. The influence of atmospheric N deposition on P nutrition (Braun et al. 2010) can only be proven indirectly by the relationship between foliar P content and N:P ratio of the soil. Foliar nutritional data should be determined over several years in order to be able to evaluate the foliar P status of a given site. In addition to the search for single soil extraction methods that are able to adequately quantify plant-available P in soils, effort is required to determine if the P content of the forest floor would be a suitable indicator of the foliar P status of European beech forest sites.

9.4 Foliar Sulphur Nutrition

Sulphur dioxide is produced mainly during the combustion of fuels containing S. Since 1990, emissions have been reduced by 92%, with the largest reductions occurring between 1990 and 1998 (UBA 2015), and according to this, also the S input to forest ecosystems decreased (Fig. 9.12 upper graphs). This was primarily due to the closure of power plants and other industrial plants in the former German Democratic Republic (GDR) and the implementation of exhaust gas cleaning technologies. Also the use of fuel with low or no S content had a decisive impact.

In 1987, when sampling for NFSI I began, SO_2 emissions were high, showing a heavily polluted area in the southern part of the former GDR (Fig. 9.12, upper left). Corresponding to this, excessive foliar S nutrition was observed in eastern Germany (Fig. 9.12, lower left). In the other parts of Germany, foliar S nutrition was mostly within normal range. In the southern part of Bavaria, however, at the edge of the Alps, S deficiency was detected at many NFSI plots. Twenty years later, atmospheric SO_2 concentrations were very low throughout Germany (Fig. 9.12, upper right), and only some single NFSI plots still had an excessive foliar S nutrition (Fig. 9.12, lower right). In contrast to NFSI I, many plots showed latent or clear deficiency, even in the eastern part of Germany. Regions with widespread S deficiency were again southern Bavaria with the Alps and additionally many regions with granitic bedrock, such as the Black Forest (in the southwest), the Bavarian Forest and the Ore Mountains (in the east at the border to the Czech Republic) and the Harz Mountains (in central Germany). Thus, forests experienced a dramatic change in S availability, from high and even too high concentrations in the atmosphere, leading to an above ground uptake path, to low concentrations in the atmosphere, causing sole availability of S from the soil.

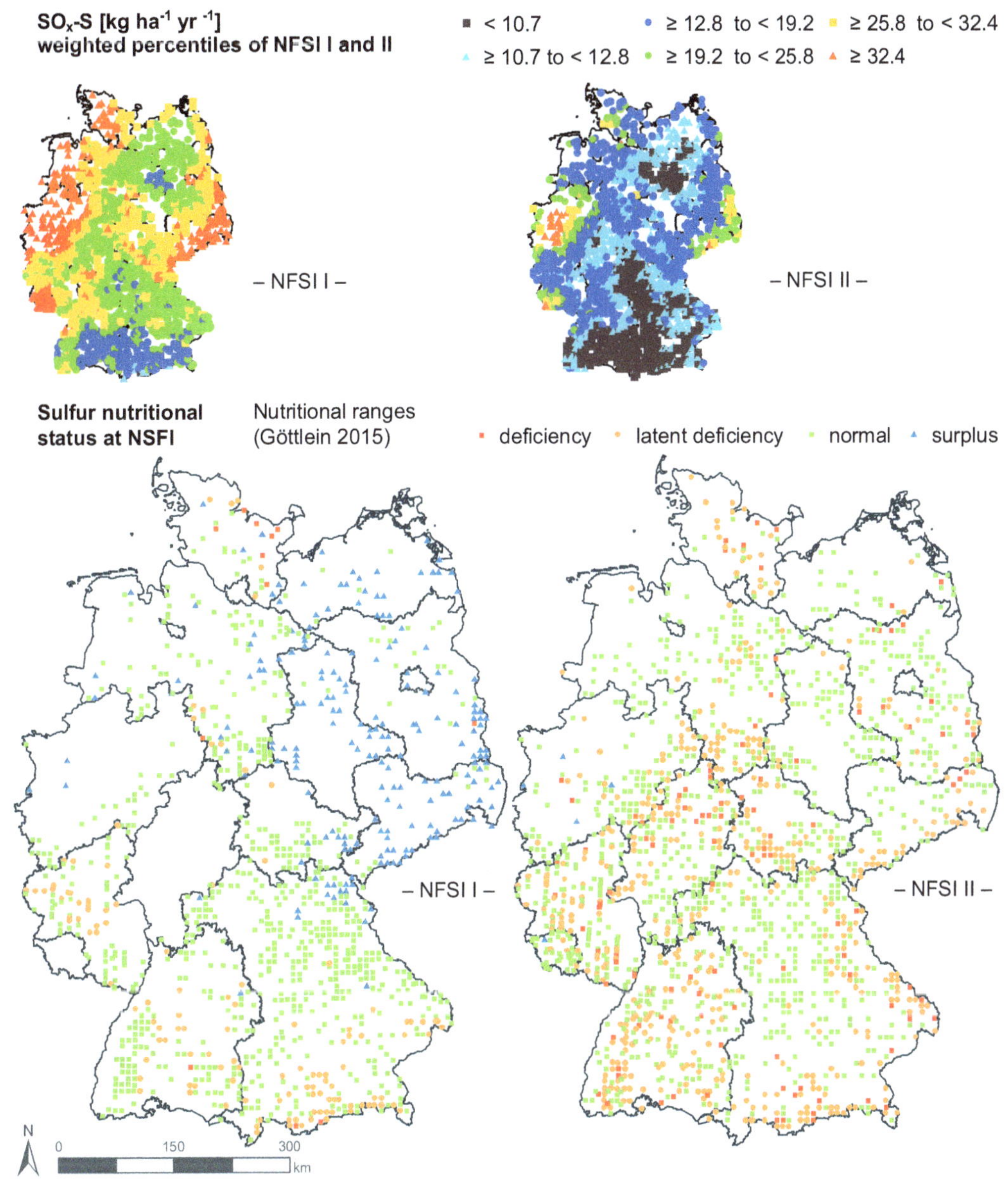

Fig. 9.12 Comparison of foliar S content for NFSI I and NFSI II; nutritional ranges are according to Göttlein (2015); the upper two graphs show the development of total S deposition from NFSI I to NFSI II

In the 1970s/1980s, sulphate accumulated in the mineral soil, depending on the atmospheric S input and sulphate retention capacity of soils (Meiwes et al. 1980). Sulphur that accumulated in soils in times of high S emissions has been slowly released after the reduction of S input (Alewell 2001). In 2007/2008, foliar S contents were still dependent on water-extractable sulphate in the deeper mineral soil, while water-extractable sulphate in the upper soil layers had no effect on foliar S (Fig. 9.13). This means that the foliar S content is still influenced by inorganic sulphate retained in the subsoil, although atmospheric S input has decreased to about

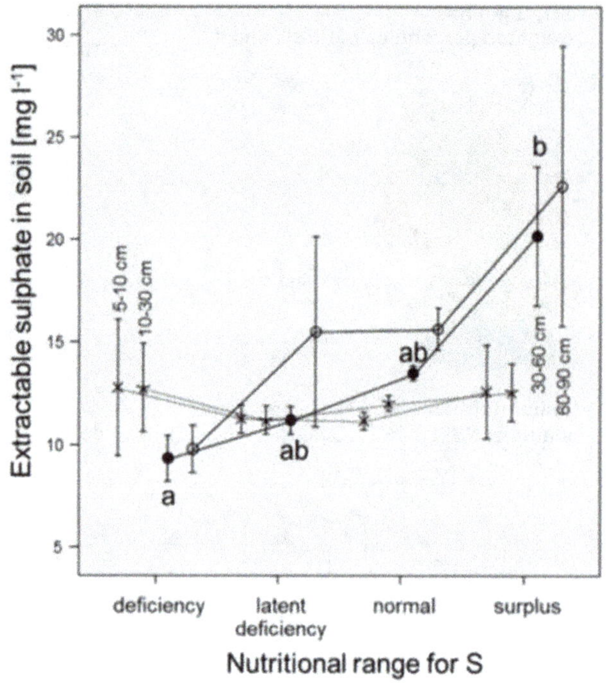

Fig. 9.13 Water-extractable sulphate (1:2 extract) of the nutritional ranges in different soil depths; average ± standard error; different letters differ significantly with $p < 0.05$; because data for water-extractable sulphate were available only in some federal states, this graph represents only a subset of NFSI II

10% compared to 1980. Under these conditions, forests growing on parent material low in S in particular appear to run to S deficiency according to the expert-based thresholds for S of Göttlein (2015) (Table 9.1). The nearly constant sulphate concentrations in the topsoil, independent from the nutritional status (Fig. 9.13), indicate that regulation processes of the ecosystem lead to a minimal amount of S cycling in the system.

Looking at the S content of the parent material (Table 9.4), acidic magmatic rocks, metamorphic rocks (with the exception of rich paragneiss) and sandstones have a very low S content. Although the geological parent material is often covered

Table 9.4 Sulphur content of various types of bedrock according to analyses by Ricke (1960), also including some other citations given by this author; values given as 5th-percentile, median and 95th-percentile

Type	Bedrock	*n*	S [%] 5th perc.–median–95th perc.
Magmatic rock	Acid (e.g. granite)	31	0.006–**0.014**–0.047
	Intermediate (e.g. diorite)	17	0.009–**0.030**–0.199
	Basic (e.g. basalt)	18	0.013–**0.033**–0.138
Metamorphic rock	Orthogneiss, quartzite, mica schist, etc.	6	0.013–**0.019**–0.068
	Paragneiss	1	0.203
Sedimentary rock	Sandstones	9	0.006–**0.023**–0.036
	Greywacke	6	0.128–**0.154**–0.238
	Clay, shale	21	0.014–**0.155**–0.500
	Limestone, dolomite	16	0.012–**0.081**–0.211

Table 9.5 Threshold values for the range of normal S nutrition according to Göttlein et al. (2011); the age-separated values for European beech were calculated from the data set used in this publication

Range of normal foliar S nutrition [mg g^{-1}]			
Type	Norway spruce	Scots pine	European beech
Young	0.9–1.6	1.0–1.4	1.2–2.7
Old	0.8–1.4	1.0–1.6	1.2–3.0

by aeolic (loess) or other layers, regions with bedrock poor in S correspond well with regions with insufficient S nutrition in NFSI II (Fig. 9.12).

Sulphur nutrition deteriorated from NFSI I to NFSI II (Fig. 9.12). This finding is supported by results from the ICP Forests Level II network (Jonard et al. 2015). The authors from that study found indications that mineral nutrition of trees progressively deteriorated in Europe with respect to P, S and base cation nutrition. However, it has been questioned whether the threshold values used were reasonable, indicating that a large number of NFSI II plots were within the range of latent or clear S deficiency. Particularly, this question arises because many threshold values were established in times of high S emissions when a real S deficiency hardly occurred. In the extensive collection of nutritional data of van den Burg (1985, 1990), it is possible to distinguish between young and old European beech, Norway spruce and Scots pine trees. Young and old trees are not separated by a given age but by how the respective nutritional values were elaborated. Pot trials, experiments with nutrient solutions or studies using sand cultures can be done only using small (young) plants. Such experiments can be run under well-controlled conditions and thus should not be influenced by the SO_2 emission regime. As shown in Table 9.5, the lower threshold values for normal S nutrition derived for young trees, under well-controlled conditions, and for old trees, under less-controlled conditions, are very close. This indicates that the previous higher SO_2 emission had little impact on the threshold for S deficiency. Thus, the results of NFSI II must be interpreted in a way that on parent materials poor in S, forests tend towards S shortage, which might induce a growth limitation in the long term. Sulphur is an element that should be increasingly considered for forestry practices that aim to be sustainable in terms of nutrients.

9.5 Effects of Liming

Forest ecosystems have been exposed to elevated atmospheric deposition of S and N for several decades (Waldner et al. 2014). Deposition of these compounds can contribute to acidification, fertilization and eutrophication (Ulrich 1981; Aber et al. 1989; Galloway et al. 2003), especially in nutrient-poor forest soils. Until the 1990s, S deposition has been the main factor causing soil acidification. With the strong reduction of S emissions, the relative contribution of N to soil acidification became more important (Schöpp et al. 2003; Meesenburg et al. 2016). Among other consequences, soil acidification causes the decline of soil cation exchange capacity and the loss of nutrient cations (de Vries et al. 2014). To mitigate the negative consequences of acidification, liming proved to be useful in improving soil chemical properties, foliar nutrition and tree vitality (Jonard et al. 2010). The aim of liming in German forests is to compensate for anthropogenic acidifying deposition and hence to ensure good nutritional status of trees (LAF 2000; von Wilpert et al. 2013). The decision criteria for liming include the impact of acidifying deposition, soil cation exchange characteristics of soils and foliar nutrition. Although atmospheric input of acids is decreasing, the critical loads of acidity are still exceeded in parts of Germany and Europe (Nagel et al. 2014; Waldner et al. 2014, 2015; Meesenburg et al. 2016). In addition, the prediction of time scales of reversibility of acidification is dependent on soil properties (Prechtel et al. 2001) and the development of deposition rates; a minimum of a few decades would be expected (Meesenburg et al. 2016). Therefore leaching of cationic nutrients with seepage water is an ongoing process, and forest liming will remain a common practice in some federal states of Germany. During the period from 1980 to 2012, approximately 3.2 million hectares of forest area were limed in Germany, in some cases twice (Jacob and Andreae 2012). Usually, dolomitic limestone containing Mg was applied. In some regions P was added.

To assess the effect of liming on foliar nutrition of the major tree species in Germany, we compared limed and unlimed NFSI II plots of those German states that limed on a large scale until the NFSI II (Rhineland-Palatinate, Lower Saxony, North Rhine-Westphalia, Baden-Wurttemberg, Hesse, Saxony, Thuringia). The selection of NFSI II plots that are considered to be sensitive to soil acidification was done according to the liming concepts of the respective federal states. The group of unlimed plots comprised only NFSI II plots that are sensitive to soil acidification but which have not yet been limed.

The foliar N content of the four major tree species did not differ on limed and unlimed NFSI II plots (Figs. 9.14, 9.15, 9.16, and 9.17). This result is in accordance with studies by Huber et al. (2006), Kulhavý et al. (2009), Jonard et al. (2010) and Greve et al. (2016), who determined no or slight liming effects in the foliar N content. The most likely explanation is the current high N deposition rates in Germany.

For Norway spruce, the foliar P content was slightly lower on limed NFSI II plots than on unlimed plots. The N:P ratio was not affected. However, the P status in the foliage of the other tree species showed no treatment effect. This result was unexpected, as in some regions, liming was combined with P fertilization. However, the applied P dosages stayed mostly undocumented, so their influence on foliar P content at NFSI II plots cannot be estimated. Results of other studies (von Wilpert 2003; Guckland et al. 2011) have found no reaction of the foliar P content after liming. However, Kulhavý et al. (2009) detected an increase of P in current-year needles of Norway spruce after liming, and Huber et al. (2006) reported on lower P contents in 1- and 2-year-old needles of Norway spruce 20 years after liming.

The foliar Ca content of European beech and oak was higher on limed than on unlimed plots. However, the Ca content for both tree species was mainly within normal range for both limed and unlimed plots. For Norway spruce and Scots pine, no liming effect on foliar Ca contents was observed. This does not correspond with results of field experiments. An increase in foliar Ca content after liming is known

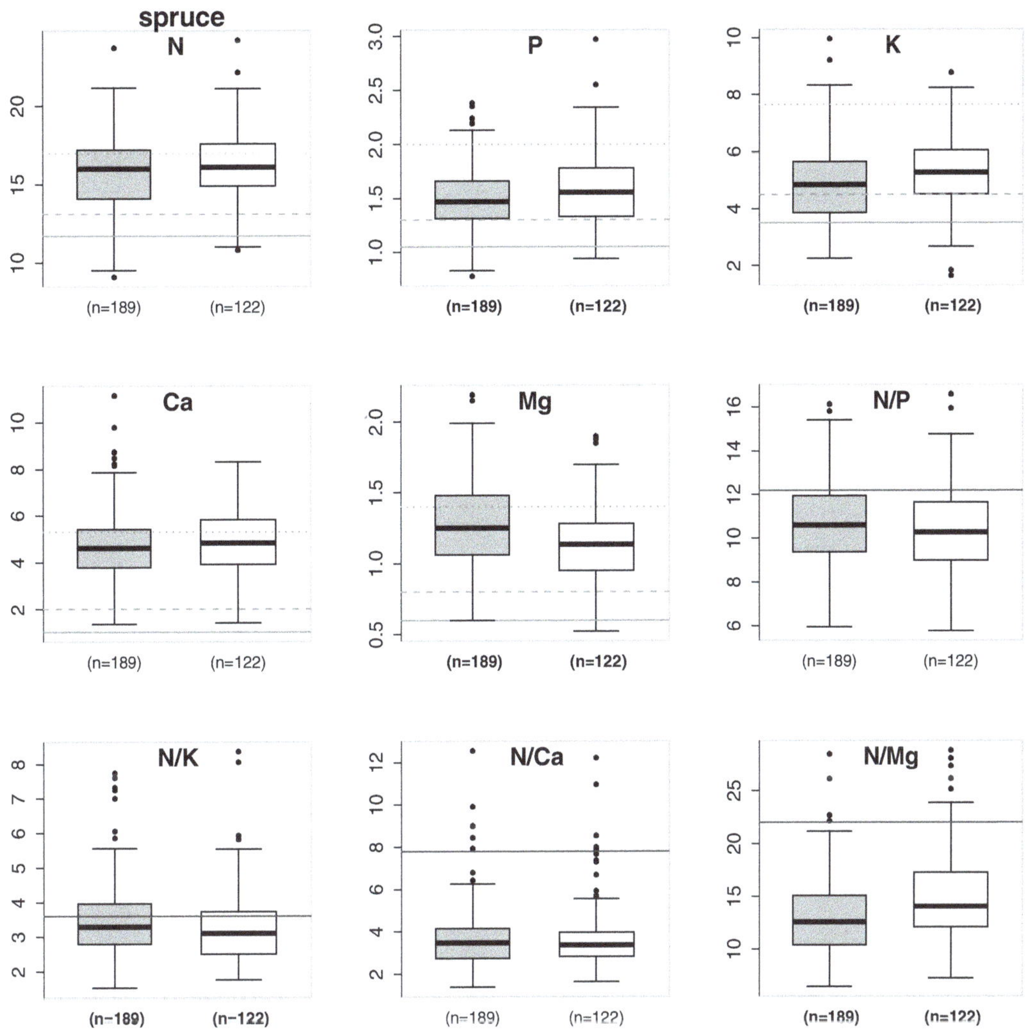

Fig. 9.14 Foliar N, P, K, Ca and Mg contents (mg g^{-1}) and ratios in current-year needles on limed (grey box) and unlimed (white box) NFSI II plots for Norway spruce. . . . .. upper threshold of normal range, --- lower threshold of normal range, __threshold of symptom range (Göttlein 2015) and critical N ratio, respectively (Table 9.3). In brackets, number of NFSI II plots; bold, significant difference ($p < 0.05$)

from field experiments that compared limed and unlimed study plots under comparable site conditions (von Wilpert 2003; Kulhavý et al. 2009; Guckland et al. 2011; Greve 2014; Jonard et al. 2010). This discrepancy may be due to preferential liming of the most acidic soils, while the acidity of the reference group of unlimed, but still acidified soils, may be less.

The foliar Mg contents of limed Norway spruce, Scots pine, European beech and oak plots were higher than those of unlimed plots. In addition, they were within the normal range on limed plots, while they were in the (latent) deficiency range at up to 45% of unlimed European beech plots. Furthermore, the N:Mg ratio was significantly smaller on limed than on unlimed plots. These positive effects of liming are consistent with results from field experiments (Evers et al. 2008; Kulhavý et al. 2009; Jonard et al. 2010). At the NFSI II plots, foliar Mg content of Norway spruce and European beech increased with the frequency of liming (not shown). A lime

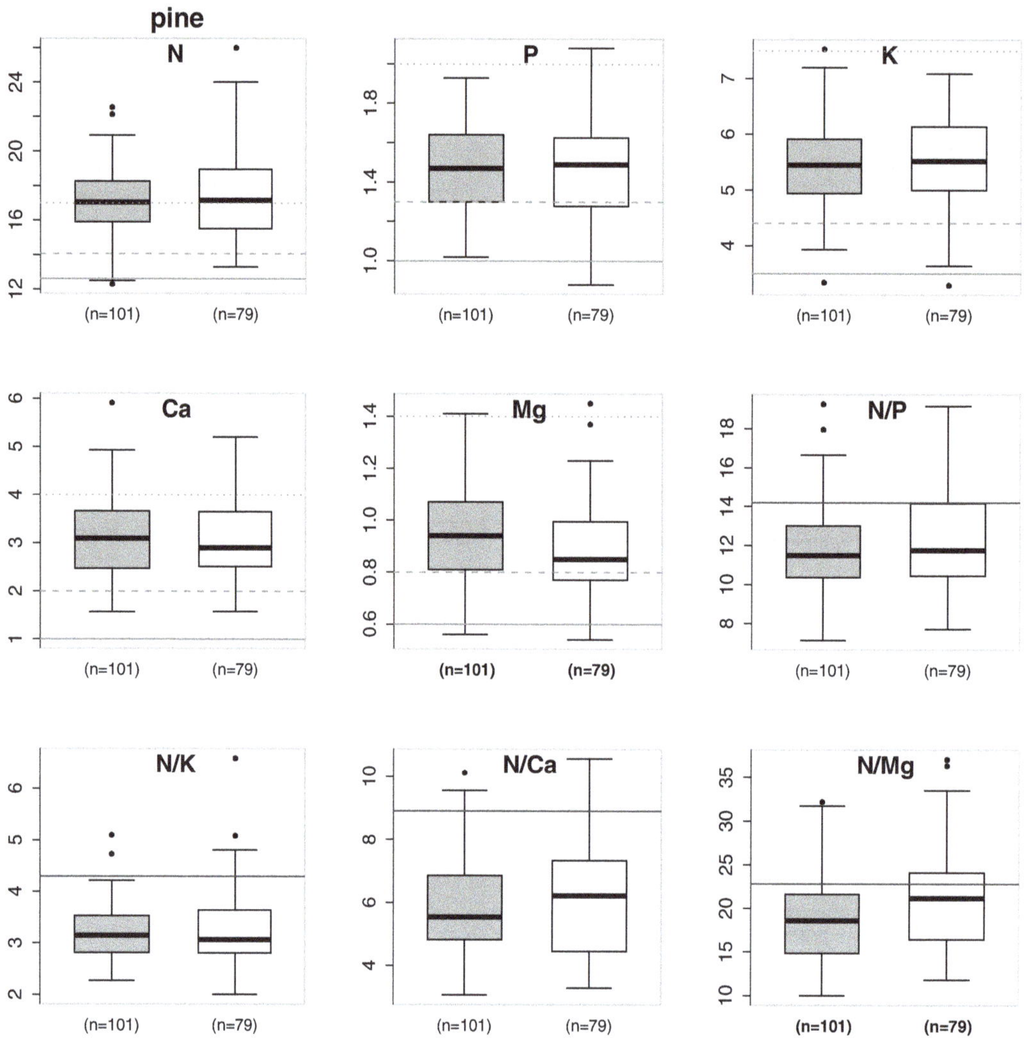

Fig. 9.15 Foliar N, P, K, Ca and Mg contents (mg g^{-1}) and ratios in current-year needles on limed (grey box) and unlimed (white box) NFSI II plots for Scots pine. . . .. upper threshold of normal range, --- lower threshold of normal range, __threshold of symptom range (Göttlein 2015) and critical N ratio, respectively (Table 9.3). In brackets, number of NFSI II plots; bold, significant difference ($p < 0.05$)

dosage of 3 t ha^{-1} already improved the foliar Mg content; additional application provoked further significant increase.

In contrast to Ca and Mg, foliar K contents of Norway spruce and European beech were lower on limed than on unlimed plots. There is inconsistent evidence regarding reduction of K in the foliage after liming in literature. In some studies, a decrease of K has been described (Evers et al. 2008; Weis et al. 2009; Ouimet et al. 2013; von Wilpert et al. 2013), whereas other studies have found no effect of liming (Huber et al. 2006; Kulhavý et al. 2009; Jonard et al. 2010). Guckland et al. (2011) assumed that the reduction of foliar K contents of Norway spruce and European beech is a matter of ion competition during nutrient uptake. Foliar K contents indicating (latent) deficiency were observed on about 40% and unbalanced N:K ratios on 39% of limed Norway spruce and European beech NFSI II plots. For both tree

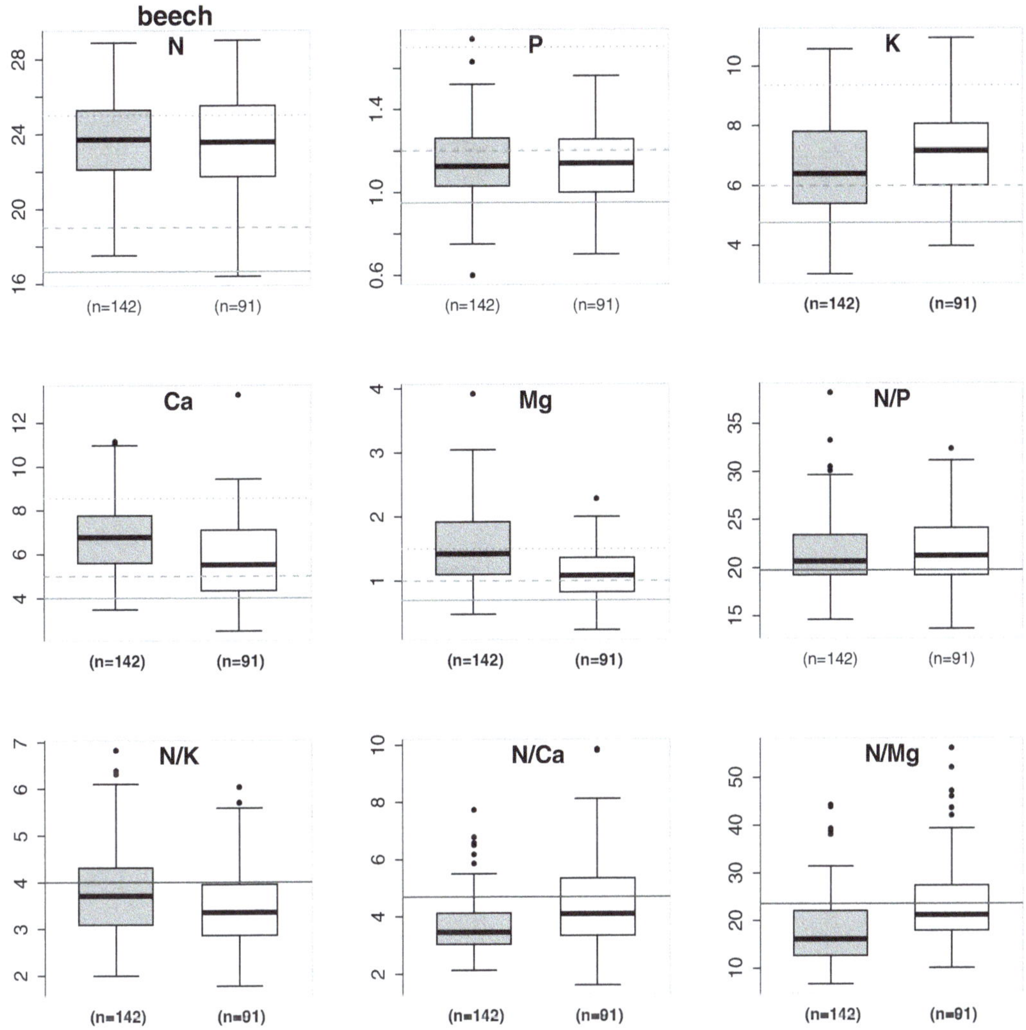

Fig. 9.16 Foliar N, P, K, Ca and Mg contents (mg g^{-1}) and ratios in current-year needles on limed (grey box) and unlimed (white box) NFSI II plots for European beech. . . .. upper threshold of normal range, --- lower threshold of normal range, __threshold of symptom range (Göttlein 2015) and critical N ratio, respectively (Table 9.3). In brackets, number of NFSI II plots; bold, significant difference ($p < 0.05$)

species, the percentage of foliar K deficiency was about 25% on the unlimed NFSI II plots. On plots that were limed once with 3 t ha^{-1}, foliar K contents were similar to those on unlimed plots, while on plots that were limed twice (6 t ha^{-1}) or more, K contents were significantly lower than on unlimed plots. This indicates that after an application of at least 6 t ha^{-1} lime, a lowering effect on the foliar K content must be considered.

The findings at the NFSI II plots showed that liming of sites that are sensitive to soil acidification can contribute to sufficient and balanced foliar Mg content of Norway spruce, Scots pine, European beech as well as pedunculate and sessile oak. Regarding Ca, the application of lime improved the nutrition of European beech and oak. However, a decrease of foliar K content of Norway spruce and

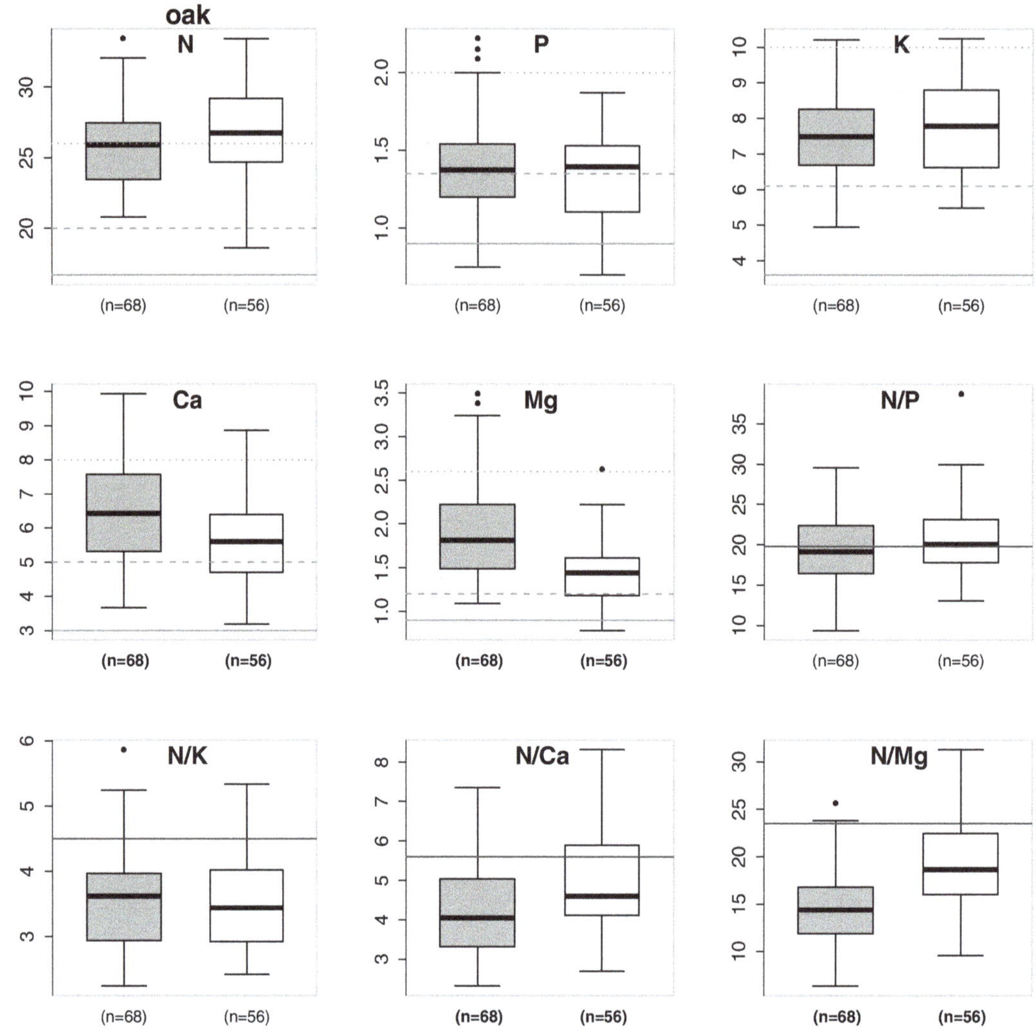

Fig. 9.17 Foliar N, P, K, Ca and Mg contents (mg g^{-1}) and ratios in current-year needles on limed (grey box) and unlimed (white box) NFSI II plots for pedunculate and sessile oak (pooled together). . . .. upper threshold of normal range, --- lower threshold of normal range, ___threshold of symptom range (Göttlein 2015) and critical N ratio, respectively (Table 9.3). In brackets, number of NFSI II plots; bold, significant difference ($p < 0.05$)

European beech following lime dosages greater than 3 t ha^{-1} should be considered in the planning of liming.

9.6 Effectiveness of Air Quality Control Measures with Respect to Lead

It is well known that the canopy can act as a filter for air pollutants. For example, approximately 50–80% of the lead (Pb) in needles of conifers may be adsorbed or taken up directly from the atmosphere (Shahid et al. 2017). This was demonstrated

Table 9.6 Lead content in current-year needles of Norway spruce ($n = 139$) and Scots pine ($n = 170$) in NFSI I (1987–1992) and NFSI II (2007–2008)

	Foliar Pb content (mg kg^{-1} DW)			
	Norway spruce		Scots pine	
	NFSI I	NFSI II	NFSI I	NFSI II
Median	1.00	0.34	2.80	0.45
Mean	1.16	0.46	3.18	0.56
Std. dev.	0.84	0.32	2.35	0.34

by washing needles with water and analysing needles before and after washing (Al-Alawi and Mandiwana 2007), by removing and analysing the wax layer (Gandois and Probst 2012) or by analysing the dust particles separated from the wax (Staszewski et al. 2012). Klaminder et al. (2005) assessed foliar Pb uptake relating ^{206}Pb:^{207}Pb ratios in tree compartments to the corresponding isotopic ratios in different soil layers. Foliar Pb that is taken up directly from the atmosphere enters the leaf or needle mainly through stomata, cuticular cracks, lenticels, ectodesmata and aqueous pores (Shahid et al. 2017). Because of the Pb adsorption to the cuticle and the high rate of direct foliar Pb uptake, Pb content of unwashed foliage can be used as an indicator of Pb pollution in the atmosphere.

Mean Pb contents in unwashed current-year needles of Norway spruce (*Picea abies*) and Scots pine (*Pinus sylvestris*) decreased from NFSI I to NFSI II by 60% and 82%, respectively (Table 9.6). This reflects the decreasing trend of Pb emissions in Germany, with a reduction from 2130 t a^{-1} in 1990 to 366 t a^{-1} in 2007 (UBA 2017). Lead in fuel for vehicles was the main Pb emission source. Since 1980, the use of unleaded fuel has increased, and since 2000, leaded fuel is no longer allowed within the European Union.

Mean Pb contents in current-year needles of Norway spruce and Scots pine measured on NFSI II plots are higher than Pb values measured in needles of lodgepole pine (*Pinus contorta*) and Sitka spruce (*Picea sitchensis*) in Western Ireland (0.15–0.17 mg Pb kg^{-1} DW) (Asam et al. 2014) and in needles of silver fir (*Abies alba*) in the French Pyrenees (median ± SD, 0.2 ± 0.13 mg Pb kg^{-1} DW) and Vosges Mountains (mean ± SD, 0.33 ± 0.11 mg Pb kg^{-1} DW) (Gandois and Probst 2012; Gandois et al. 2010). Since the two investigated sites in Western Ireland are located about 30 km from the Atlantic coast, with prevailing west winds, the low Pb values in conifer needles may be considered as background values.

9.7 Ratios of Nutrient Contents from Needles of Different Ages (Norway Spruce and Scots Pine)

In addition to nutrient contents of current-year needles, element contents of 2- (Scots pine) and 3-year-old needles (Norway spruce) were analysed in the National Forest Soil Inventory II (NFSI II) in Germany. This analysis facilitates a more comprehensive diagnosis of the nutritional state of these two coniferous species. Furthermore, the analysis of element content in needles of different ages facilitates inferences about plant physiological processes, such as the translocation of nutrients from older to younger needles. For this purpose, the NFSI II is an outstanding database with 610 Scots pine and 409 Norway spruce sites. However, contents in needles of the NFSI dataset do not represent a real time series, as needles were sampled in the same year. Thus, it must be assumed that older needles had element contents in the year of formation that are comparable to contents of the current needle year. However, element contents in foliage can vary from year to year, especially during or after climatically extreme years (Göttlein et al. 2009; Jonard et al. 2009; Schleppi et al. 2000). This may be a problem for the evaluation of the dataset if there were a systematic deviation of element contents in the respective needle-formation year at all NFSI sites. However, this can be ruled out, since needle sampling was not simultaneous at all federal states but was spread over a 3-year period (2006–2008). Therefore, the mean values of the whole data set should allow the importance and tendencies of element ratios in Norway spruce and Scots pine needles of different ages to be assessed, assuming that changes in nutrient concentrations are not due to changes in the organic fraction of the needle (e.g. waxes, phenolics).

In adult trees, a large proportion of nutrients is stored in the foliar biomass (Beets and Pollock 1987; Fife et al. 2008; Alriksson and Eriksson 1998; Scarascia-Mugnozza et al. 2000). Therefore, it stands to reason that this nutrient pool is used for internal reallocation processes. For example, Miller (1986) has shown that for the annual growth of 40-year-old European black pine, 40–50% of N and more than 50% of P were recruited by retranslocation processes running mainly within the foliar biomass. According to Miller (1986), retranslocation of nutrients might play an increasing role with advancing tree age and is particularly relevant in stands that have reached canopy closure and maximum crown sizes.

Retranslocation processes are more likely for nutrients deemed to be relatively mobile in plants such as N, P, K, Mg or S (Nambiar and Fife 1991). Especially in senescent needles, retranslocation of nutrients must be considered (Helmisaari 1992). However, nutrient retranslocation has been observed already in 6-month-old needles of several coniferous species (Fife et al. 2008).

There is no consistent pattern in literature regarding the question as to whether the nutritional state itself is controlling the extent of nutrient retranslocation. Miller et al. (1979) found increased N retranslocation within needles of European black pine in stands with poor N nutrition. At the same time, increased element retranslocation has been observed with increasing plant availability of the respective element. For instance, in an N-fertilized *Pinus radiata* stand, Fife and Nambiar (1997) detected a growth increase of 45% and simultaneously an N retranslocation that rose by a factor of 4.5.

All of the studies mentioned above were case studies involving only one or a small number of sites. We tested the extent to which the findings of these studies of nutrient retranslocation in the foliar biomass are reflected by the extensive data set of the NFSI II using Norway spruce and Scots pine. We examined whether nutritional diagnosis can be improved for these two coniferous species by considering ratios between foliar element contents of current-year needles and of 2- (Scots pine) or 3- (Norway spruce) year-old needles. In addition, we assigned these ratios to one of four nutritional ranges based on the assessment system of Göttlein (2015) in order to evaluate nutrient ratios (and potential retranslocation processes) dependent on the nutritional state of the respective element in the current-year needles.

As shown in Fig. 9.18, ratios of N contents differ for the two tree species. For Norway spruce, there is a slight decrease (10%) in relative N contents in 3-year-old needles at sites with N deficiency. With improving N nutrition, the relative N contents in older needles decrease significantly, reaching 83% of the N contents in the current needle year at sites classified as N surplus sites. N is deemed to be a mobile nutrient in plants. Therefore, it is likely that the observed decrease in N contents in 3-year-old needles is caused by retranslocation processes. However, a weight increase among older needles may also contribute to the N content decrease in needles of this age. Consistent with the results of the study by Fife and Nambiar (1997), our findings suggest that the reduction of relative N contents in older needles with increasing N nutrition is likely caused by additional biomass growth. Thus, at many Norway spruce sites with normal or surplus N nutrition, N stimulates additional biomass growth. In contrast, this appears not to be the case for Scots pine. For this species, N contents of older needles are slightly higher or similar compared to contents of current-year needles, and there is no significant pattern between sites of different N nutritional states. The growth of Scots pine at sites with normal and surplus N nutrition seems to be limited by other factors. This is likely also true for Scots pine at N-deficient sites, because otherwise a difference of N contents between younger and older needles would be expected, as reported in Miller et al. (1979).

The ratios of P show similar patterns for Norway spruce and Scots pine. With increasing P nutrition, relative P contents in older needles decrease significantly, reaching only 72% of current-year P content at sites with surplus P nutrition for Norway spruce and 87% for Scots pine, respectively. P is also considered a readily mobile nutrient. Thus, retranslocation of P from older to current-year needles is likely, even if—as mentioned above—a weight increase in older needles contributes to the decrease of P contents in needles of this age. For Norway spruce, lowered P contents in older needles were detected already at sites with P deficiency, whereas at Scots pine sites with P deficiency, P contents did not differ between current-year and 2-year-old needles. This suggests that P retranslocation between different aged Scots pine needles is less effective for Scots pine than Norway spruce. However, at $N = 7$, the sample size for this nutritional range is very low compared to the other ranges observed in this study.

Even for K, relative contents in the older needles of Norway spruce and Scots pine decrease significantly with increasing K nutrition range in current-year needles. For trees with surplus K nutrition, the relative K contents in older needles drop to less than 80% of contents measured in current-year needles. This was observed for both Norway spruce and Scots pine. Again, K is considered highly mobile in plants. Therefore, increasing retranslocation of K with rising K nutrition is highly likely and supports the hypothesis discussed in Nambiar and Fife (1991) that the extent of retranslocation of K, P and N mostly shows a positive relationship to the absolute quantity of the respective element stored in the foliar biomass. However, a weight

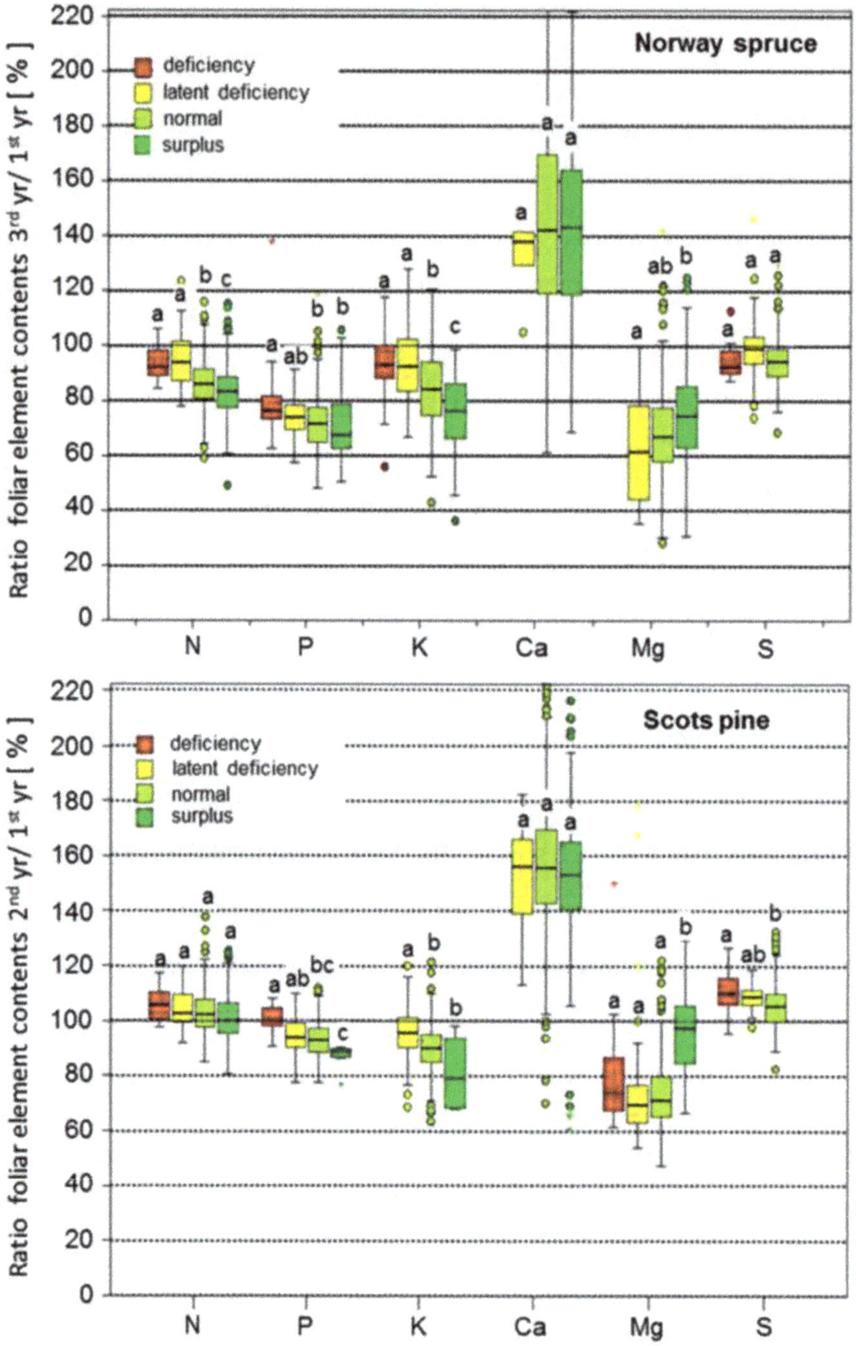

Fig. 9.18 Ratios between foliar element contents of 2- (Scots pine) or 3-year-old needles (Norway spruce) and current-year needles, grouped by nutritional ranges based on the assessment system of Göttlein (2015). Letters above boxplots indicate statistically significant differences between nutritional ranges of the respective element (Scheffé test, $\alpha < 0.05$)

increase among older needles may also contribute to the K content decrease in previous year needles.

For Ca, distinctly higher contents in older needles were observed at Norway spruce sites (+40%) as well as at Scots pine sites (+55%). This relative increase of Ca contents in older needles was visible in all nutritional ranges; i.e. it is independent of Ca contents in current-year needles. This is highly plausible as Ca in plant tissues is largely organically bound and therefore not translocated through the phloem. Furthermore, Ca is constantly supplemented by transpiration flow (Kadereit et al. 2014), leading to an enrichment of Ca contents in older needles which has been already observed in previous studies for Norway spruce and Scots pine (e.g. Bauer et al. 2000). It is notable that Ca ratios show the greatest variability compared to the other main nutrients shown in Fig. 9.18. This is consistent with the results of previous case studies that have observed the highest annual fluctuation of Ca contents in needles of Norway spruce (Evers 1972) and Scots pine (Hippeli and Branse 1992) compared to other main nutrients. It is quite likely that this is also reflected in the large-scale data set of the NFSI II.

Compared to the other nutrients, Mg shows the greatest reduction in relative contents in older needles for both tree species (Fig. 9.18). This reduction is highest in trees with latent Mg deficiency, where contents in previous year needles drop to 60% for Norway spruce and 70% for Scots pine. In contrast, relative Mg contents in older needles amount to 80% (Norway spruce) and nearly 100% (Scots pine) for trees assigned to the surplus Mg nutrition range. These findings are consistent with the results of other studies, where an increase in Mg retranslocation has been detected when Mg nutrition declined (Le Goaster et al. 1990; Reemtsma 1986).

Although S appears to be organically bound in plants, in the literature it is considered a rather mobile nutrient (Nambiar and Fife 1991; Rennenberg 1999; Köstner et al. 1998). However, only very small or no S content gradients between current-year and older needles were detected for Norway spruce. In contrast, a slight enrichment of S in older needles of Scots pine was observed in all nutritional ranges. While this enrichment decreases slightly with increasing S nutrition in current-year needles of Scots pine, we found no statistically significant differences between S nutrition ranges for Norway spruce (Fig. 9.18). As S can also be taken up directly by the foliage, S enrichment in older needles cannot be excluded, although airborne deposition of S has reduced considerably in Germany in recent years.

In summary, the ratios of contents of the main nutrients between current-year and older needles at Norway spruce and Scots pine sites of the NFSI II provide a means for drawing plausible conclusions. For both tree species, the immobile element Ca shows an enrichment in older needles, whereas for nutrients considered relatively mobile in plants (N, P, K, Mg), retranslocation from older into current-year needles is likely. The extent of element translocation and enrichment, respectively, corresponds to the general nutritional state for these two tree species (Fig. 9.1). For Scots pine, a species with relatively low nutrient requirements, the NFSI data set reveals a generally better nutritional situation and at the same time a higher enrichment of the immobile Ca compared to Norway spruce. The retranslocation of mobile nutrients shows similar patterns for both species, but the extent of foliar element retranslocation appears to be lower in most cases for Scots pine.

Above, we discussed whether nutritional state itself controls the extent of nutrient retranslocation; even the large-scale dataset of the NFSI II cannot provide a clear answer. In the case of Mg, retranslocation from older to current-year needles increases with declining Mg nutrition in current-year needles for both tree species. In contrast, our findings suggest that P and K are increasingly translocated with improving nutritional range of the respective element in current-year needles at Norway spruce and Scots pine sites. For N, this effect applies only to Norway spruce, whereas for Scots pine, no significant differences of N contents between current-year and older needles were observed over all N nutritional ranges, suggesting that at many Scots pine sites, growth seems to be limited by other factors.

Precise determination of foliar element translocation within needle years requires that the nutritional data should originate from real time series, and furthermore, the weight of foliar biomass should be assessed for each needle year in order to calculate retranslocation on the level of annual changes of element quantities. As these data are not available, we were not able to exclude the possibility that a weight increase among older needles contributed to foliar element content changes that are not related to retranslocation processes.

9.8 Conclusions

Nutrient contents in the leaves and needles of the four major tree species on NFSI plots provided valuable information about the current nutritional status in Germany and revealed effects of increased N deposition, forest soil liming and air quality control measures on nutritional status. The hypotheses tested were largely confirmed by the NFSI dataset:

1. Foliar N contents are related to atmospheric N deposition.
2. Foliar P contents of European beech are deficient through large areas of Germany. Both geological substrate and atmospheric N deposition may influence foliar P nutrition.
3. Foliar S contents are related to sulphate concentration in water extracts of subsoil samples. Due to a marked decrease in S deposition, stands at sites low in geogenic S tend to have an S shortage.
4. Liming of acidified soils leads to increased foliar Ca and Mg contents, while foliar K contents decrease.
5. Air quality control measures lead to decreased foliar lead contents.
6. The nutritional diagnosis for conifers can be improved by considering the ratios between foliar element contents of 2- (Scots pine) or 3- (Norway spruce) year-old needles and current-year needles.

We showed that foliar element contents are important for evaluating the current nutritional status of forest ecosystems and facilitate conclusions about the effectiveness of forest stabilization and air quality control measures.

However, field and laboratory experiments are needed to reveal the causes for decreasing foliar P and K contents. The presence of deficient foliar P, S and K contents and high N to nutrient ratios indicates a critical nutrient status and should be considered in forest management planning, especially forest biomass harvesting. In general, foliar element contents should be included in decision support systems for sustainable forestry.

When working with NFSI data, it should always be considered that leaves and needles were collected in a single year only and that the sampling year varied among the federal states of Germany. Hence, annual fluctuations in foliar nutrient contents cannot be accounted for and may affect the differences in foliar nutritional status among NFSI plots and relationships to other parameters.

References

Aber JD, Nadelhoffer KJ, Steudler P, Melillo JM (1989) Nitrogen saturation in northern forest ecosystems. Bioscience 39(6):378–386

Aber JD, McDowell W, Nadelhoffer K, Magill A, Berntson G, Kamakea M, McNulty S, Currie W, Rustad L, Fernandez I (1998) Nitrogen saturation in temperate forest ecosystems—hypotheses revisited. Bioscience 48(11):921–934. https://doi.org/10.2307/1313296

Aber JD, Goodale CL, Ollinger SV, Smith M-L, Magill AH, Martin ME, Hallett RA, Stoddard JL (2003) Is nitrogen deposition altering the nitrogen status of northeastern forests? AIBS Bull 53 (4):375–389

Aerts R, Chapin FS (2000) The mineral nutrition of wild plants revisited: a re-evaluation of processes and patterns. Adv Ecol Res 30:1–67

Al-Alawi MM, Mandiwana KL (2007) The use of Aleppo pine needles as a bio-monitor of heavy metals in the atmosphere. J Hazard Mater 148(1–2):43–46. https://doi.org/10.1016/j.jhazmat.2007.02.001

Alewell C (2001) Predicting reversibility of acidification: the European sulfur story. Water Air Soil Pollut 130(1–4):1271–1276. https://doi.org/10.1023/a:1013989419580

Alriksson A, Eriksson HM (1998) Variations in mineral nutrient and C distribution in the soil and vegetation compartments of five temperate tree species in NE Sweden. For Ecol Manag 108 (3):261–273. https://doi.org/10.1016/s0378-1127(98)00230-8

Asam Z, Nieminen M, Kaila A, Laiho R, Sarkkola S, O'Connor M, O'Driscoll C, Sana A, Rodgers M, Zhan XM, Xiao LW (2014) Nutrient and heavy metals in decaying harvest residue needles on drained blanket peat forests. Eur J For Res 133(6):969–982. https://doi.org/10.1007/s10342-014-0815-5

Augusto L, Achat DL, Jonard M, Vidal D, Ringeval B (2017) Soil parent material—a major driver of plant nutrient limitations in terrestrial ecosystems. Glob Chang Biol 23(9):3808–3824. https://doi.org/10.1111/gcb.13691

Bauer G, Persson H, Persson T, Mund M, Hein M, Kummetz E, Matteucci G, Van Oene H, Scarascia-Mugnozza G, Schulze E-D (2000) Linking plant nutrition and ecosystem processes. In: Schulze ED (ed) Carbon and nitrogen cycling in European forest ecosystems. Springer, Berlin, pp 63–98

Beets P, Pollock D (1987) Accumulation and partitioning of dry matter in Pinus radiata as related to stand age and thinning. N Z J For Sci 17(2):246–271

Blume H-P, Brümmer GW, Fleige H, Horn R, Kögel-Knabner I, Kandeler E, Kretzschmar R, Stahr K, Wilke B-M (2016) Scheffer/Schachtschabel Soil Science. Springer, Berlin

Braun S, Thomas VF, Quiring R, Flückiger W (2010) Does nitrogen deposition increase forest production? The role of phosphorus. Environ Pollut 158(6):2043–2052

Burschel P, Huss J, Kalbhenn R (1964) Die natürliche Verjüngung der Buche: mit 63 Tabellen. Sauerländer

Cape JN, Freer-Smith PH, Paterson IS, Parkinson JA, Wolfenden J (1990) The nutritional status of Picea abies (L) Karst. across Europe, and implications for "forest decline". Trees 4(4):211–224 de Vries W, Reinds GJ, Gundersen P, Sterba H (2006) The impact of nitrogen deposition on carbon sequestration in European forests and forest soils. Glob Chang Biol 12:1151–1173

de Vries W, Reinds GJ, van Kerkvoorde MS, Hendriks CMA, Leeters EEJM, Gross CP, Voogd JCH, Vel EM (2000) Intensive monitoring of forest ecosystems in Europe. EC, UN/ECE, Brussels, Geneva

de Vries W, Dobbertin MH, Solberg S, Van Dobben HF, Schaub M (2014) Impacts of acid deposition, ozone exposure and weather conditions on forest ecosystems in Europe: an overview. Plant Soil 380(1–2):1–45

Evers FH (1972) Die jahrweisen Fluktuationen der Nährelementkonzentrationen in Fichtennadeln und ihre Bedeutung für die Interpretation nadelanalytischer Befunde. Allgemeine Forst- und Jagdzeitung 143:68–74

Evers J, Dammann I, Noltensmeier A, Nagel R-V (2008) Auswirkungen von Bodenschutzkalkungen auf Buchenwälder (*Fagus sylvatica L.*). In: Ergebnisse angewandter Forschung zur Buche. Beiträge aus der NW-FVA, vol 3. Nordwestdeutsche Forstliche Versuchsanstalt (NW-FWA), Göttingen, pp 21–50

Ewald J (2000) Does phosphorus deficiency cause low vitality in European beech (*Fagus sylvatica* L.) in the Bavarian Alps? Forstwissenschaftliches Centralblatt 119(5):276–296

Ewald J (2005) Ecological background of crown condition, growth and nutritional status of *Picea abies* (L.) Karst. in the Bavarian Alps. Eur J For Res 124(1):9–18

Fiedler HJ, Höhne H (1984) Das NPK-Verhältnis in Kiefernnadeln als Erscheinung und Mittel zur Ernährungsdiagnose. Beitr Forstwirtsch 18:128–132

Fife DN, Nambiar EKS (1997) Changes in the canopy and growth of *Pinus radiata* in response to nitrogen supply. For Ecol Manag 93(1):137–152

Fife DN, Nambiar EKS, Saur E (2008) Retranslocation of foliar nutrients in evergreen tree species planted in a Mediterranean environment. Tree Physiol 28(2):187–196

Flückiger W, Braun S (1999) Nitrogen and its effect on growth, nutrient status and parasite attacks in beech and Norway spruce. Water Air Soil Pollut 116(1-2):99–110

Flückiger W, Braun S (2003) Critical limits for nutrient concentrations and ratios for forest trees—a comment. Empirical critical loads for nitrogen. Swiss Agency for the Environment, Forests and Landscape (SAEFL), Berne

Galloway JN, Aber JD, Erisman JW, Seitzinger SP, Howarth RW, Cowling EB, Cosby BJ (2003) The nitrogen cascade. AIBS Bull 53(4):341–356

Gandois L, Probst A (2012) Localisation and mobility of trace metal in silver fir needles. Chemosphere 87(2):204–210

Gandois L, Nicolas M, Vanderheijden G, Probst A (2010) The importance of biomass net uptake for a trace metal budget in a forest stand in north-eastern France. Sci Total Environ 408 (23):5870–5877

Gauger T, Haenel H-D, Rösemann C, Dämmgen U, Bleeker A, Erisman JW, Vermeulen AT, Schaap M, Timmermanns RMA, Builtjes PJH (2008) National implementation of the UNECE convention on long-range transboundary air pollution (effects). Part 1: Deposition loads: methods, modelling and mapping results, trends. UBA-Texte vol 08/38. Umweltbundesamt, Dessau-Roßlau

Glatzel G, Kazda M, Grill D, Halbwachs G, Katzensteiner K (1987) Ernährungsstörungen bei Fichte als Komplexwirkung von Nadelschäden und erhöhter Stickstoffdeposition – ein Wirkungsmechanismus des Waldsterbens. Allg Forst Jagdztg 158(5/6):91–97

Göttlein A (2015) Grenzwertbereiche für die ernährungsdiagnostische Einwertung der Hauptbaumarten Fichte, Kiefer, Eiche, Buche. Allg Forst Jagdztg 186(5/6):110–116

Göttlein A, Rodenkirchen H, Häberle K, Matyssek R (2009) Nutritional effects triggered by the extreme summer 2003 in the free air ozone fumigation experiment at the Kranzberger Forst. Eur J For Res 128(2):129–134

Göttlein A, Baier R, Mellert KH (2011) Neue Ernährungskennwerte für die forstlichen Hauptbaumarten in Mitteleuropa - Eine statistische Herleitung aus van den Burg's Literaturzusammenstellung. Allg Forst Jagdztg 182(9/10):173–186

Greve M (2014) Langfristige Auswirkungen der Waldkalkung auf Bodenzustand, Sickerwasser und Nadelspiegelwerte von drei Versuchsanlagen in Rheinland-Pfalz. Forstarchiv 85(2):35–46
Greve M, Block J, Schröck H-W, Schutze J, Werner W, Wies K (2016) Nährstoffversorgung Rheinland-Pfälzischer Wälder. Mitteilungen aus der FAWF, vol 76. Forschungsanstalt für Waldökologie und Forstwirtschaft Rheinland-Pfalz, Trippstadt
Gruber N, Galloway JN (2008) An Earth-system perspective of the global nitrogen cycle. Nature 451(7176):293–296
Guckland A, Paar U, Dammann I, Evers J, Meiwes KJ, Mindrup M (2011) Einfluss der Kalkung auf die Bestandesernährung. AFZ Wald 6:23–25
Heinsdorf D, Krauss H, Hippeli P (1988) Ernährungs-und bodenkundliche Unterschungen in Fichtenbeständen des mittleren Thüringer Waldes unter Berücksichtigung der in den letzten Jahren aufgetretenen Umweltbelastungen. Beitr Forstwirtsch 22:160–167
Helmisaari H-S (1992) Nutrient retranslocation in three *Pinus sylvestris* stands. For Ecol Manag 51 (4):347–367
Hippeli P, Branse C (1992) Veränderungen der Nährelementkonzentrationen in den Nadeln mittelalter Kiefernbestände auf pleistozänen Sandstandorten Brandenburgs in den Jahren 1964 bis 1988. Forstwiss Centralbl 111(1):44–60
Huber C, Weis W, Göttlein A (2006) Tree nutrition of Norway spruce as modified by liming and experimental acidification at the Höglwald site, Germany, from 1982 to 2004. Ann For Sci 63 (8):861–869
Jacob F, Andreae H (2012) Medizin für den Wald. Forstliche Bodenschutzkalkungen – Ein bundesdeutscher Überblick. Paper presented at the 1. Deutsches Kalkungssymposium, Dresden
Jacoby WG (2000) Loess: a nonparametric, graphical tool for depicting relationships between variables. Elect Stud 19(4):577–613
Jonard M, André F, Dambrine E, Ponette Q, Ulrich E (2009) Temporal trends in the foliar nutritional status of the French, Walloon and Luxembourg broad-leaved plots of forest monitoring. Ann For Sci 66(4):1–10
Jonard M, André F, Giot P, Weissen F, Perre R, Ponette Q (2010) Thirteen-year monitoring of liming and PK fertilization effects on tree vitality in Norway spruce and European beech stands. Eur J For Res 129(6):1203–1211. https://doi.org/10.1007/s10342-010-0410-3
Jonard M, Furst A, Verstraeten A, Thimonier A, Timmermann V, Potocic N, Waldner P, Benham S, Hansen K, Merila P, Ponette Q, De La Cruz AC, Roskams P, Nicolas M, Croise L, Ingerslev M, Matteucci G, Decinti B, Bascietto M, Rautio P (2015) Tree mineral nutrition is deteriorating in Europe. Glob Chang Biol 21(1):418–430. https://doi.org/10.1111/gcb.12657
Jönsson Belyazid U, Belyazid S (2012) Phosphorus cycling in boreal and temperate forest ecosystems—a review of current knowledge and the construction of a simple phosphorus model. Belyazid Consulting and Communication AB, Malmö
Kadereit JW, Körner C, Kost B, Sonnewald U (2014) Strasburger – Lehrbuch der Pflanzenwissenschaften, 37th edn. Springer Spektrum, Berlin
Khanna PK, Fortmann H, Meesenburg H, Eichhorn J, Meiwes KJ (2009) Biomass and element content of foliage and aboveground litterfall on the three long-term experimental beech sites: Dynamics and significance. In: Brumme R, Khanna PK (eds) Functioning and management of European beech ecosystems. Springer, Berlin, pp 183–205
Klaminder J, Bindler R, Emteryd O, Renberg I (2005) Uptake and recycling of lead by boreal forest plants: quantitative estimates from a site in northern Sweden. Geochim Cosmochim Acta 69 (10):2485–2496. https://doi.org/10.1016/j.gca.2004.11.013
Klemm O, Kuhn U, Beck E, Katz C, Oren R, Schulze E-D, Steudle E, Mitterhuber E, Pfanz H, Kaiser W (1989) Leaching and uptake of ions through above-ground Norway spruce tree parts. In: Schulze E-D, Lange OL, Oren R (eds) Forest decline and air pollution. Springer, Berlin, pp 210–237
Kohlpaintner M, Fäth J, Mellert KH, Blum U, Göttlein A (2017) Können einfache Extraktionsmethoden einen Beitrag zur Abschätzung des pflanzenverfügbaren Phosphors in

Waldböden leisten?. Paper presented at the Tagung der Sektion Waldernährung im Deutschen Verband Forstlicher Forschungsanstalten (DVFFA), Gotha

Köstner B, Schupp R, Schulze E-D, Rennenberg H (1998) Organic and inorganic sulfur transport in the xylem sap and the sulfur budget of Picea abies trees. Tree Physiol 18(1):1–9

Kulhavý J, Marková I, Drápelová I, Truparová S (2009) The effect of liming on the mineral nutrition of the mountain Norway spruce (*Picea abies* L.) forest. J For Sci 55(1):1–8

LAF (2000) Leitfaden Forstliche Bodenschutzkalkung in Sachsen. Schriftenreihe der Sächsischen Landesanstalten für Forsten (LAF), vol 21/2000

Le Goaster S, Dambrine E, Ranger J (1990) Mineral supply of healthy and declining trees of a young spruce stand. Water Air Soil Pollut 54(1):269–280

Matson P, Lohse KA, Hall SJ (2002) The globalization of nitrogen deposition: consequences for terrestrial ecosystems. Ambio 31(2):113–119. https://doi.org/10.1639/0044-7447(2002)031[0113:tgondc]2.0.co;2

Meesenburg H, Ahrends B, Fleck S, Wagner M, Fortmann H, Scheler B, Klinck U, Dammann I, Eichhorn J, Mindrup M (2016) Long-term changes of ecosystem services at Solling, Germany: recovery from acidification, but increasing nitrogen saturation? Ecol Indic 65:103–112

Meiwes KJ, Khanna PK, Ulrich B (1980) Retention of sulphate by an acid brown earth and its relationship with the atmospheric input of sulphur to forest vegetation. J Plant Nutr Soil Sci 143 (4):402–411

Mellert KH, Ewald J (2014) Nutrient limitation and site-related growth potential of Norway spruce (Picea abies L. Karst) in the Bavarian Alps. Eur J For Res 133(3):433–451. https://doi.org/10.1007/s10342-013-0775-1

Mellert KH, Göttlein A (2012) Comparison of new foliar nutrient thresholds derived from van den Burg's literature compilation with established central European references. Eur J For Res 131 (5):1461–1472. https://doi.org/10.1007/s10342-012-0615-8

Mellert K, Göttlein J (2013) Identification of nutrient thresholds and limiting nutrient factors of Norway spruce by applying new critical foliar nutrient concentrations and modern regression. Allg Forst Jagdztg 184:197–203

Mellert KH, Prietzel J, Straussberger R, Rehfuess KE, Kahle HP, Perez P, Spiecker H (2008) Relationships between long-term trends of air temperature, precipitation, nitrogen nutrition and growth of coniferous stands in Central Europe and Finland. Eur J For Res 127(6):507–524. https://doi.org/10.1007/s10342-008-0233-7

Miller HG (1986) Carbon × nutrient interactions—the limitations to productivity. Tree Physiol 2 (1–2–3):373–385

Miller HG, Cooper JM, Miller JD, Pauline OJL (1979) Nutrient cycles in pine and their adaptation to poor soils. Can J For Res 9(1):19–26

Nagel HD, Schlutow A, Scheuschner T (2014) Modellierung und Kartierung atmosphärischer Stoffeinträge und kritischer Belastungsschwellen zur kontinuierlichen Bewertung der ökosystemspezifischen Gefährdung der Biodiversität in Deutschland-PINETI (Pollutant INput and EcosysTEM Impact). Teilbericht 4 Critical Load, Exceedance und Belastungsbewertung. UBA-Texte, vol 63/2014. Umweltbundesamt, Dessau-Roßlau

Nambiar EKS, Fife DN (1991) Nutrient retranslocation in temperate conifers. Tree Physiol 9 (1–2):185–207

Nihlgård B (1985) The ammonium hypothesis: an additional explanation to the forest dieback in Europe. Ambio 14(1):2–8

Nilsson LO, Wallander H (2003) Production of external mycelium by ectomycorrhizal fungi in a Norway spruce forest was reduced in response to nitrogen fertilization. New Phytol 158 (2):409–416. https://doi.org/10.1046/j.1469-8137.2003.00728.x

Ouimet R, Moore JD, Duchesne L, Camire C (2013) Etiology of a recent white spruce decline: role of potassium deficiency, past disturbances, and climate change. Can J For Res 43(1):66–77. https://doi.org/10.1139/cjfr-2012-0344

Övergaard R, Gemmel P, Karlsson M (2007) Effects of weather conditions on mast year frequency in beech (*Fagus sylvatica* L.) in Sweden. Forestry 80(5):553–563. https://doi.org/10.1093/forestry/cpm020

Peuke A, Rennenberg H (2004) Carbon, nitrogen, phosphorus, and sulphur concentration and partitioning in beech ecotypes (*Fagus sylvatica* L.): phosphorus most affected by drought. Trees 18(6):639–648. https://doi.org/10.1007/s00468-004-0335-x

Piovesan G, Adams JM (2001) Masting behaviour in beech: Linking reproduction and climatic variation. Can J Bot 79(9):1039–1047

Prechtel A, Alewell C, Armbruster M, Bittersohl J, Cullen JM, Evans CD, Helliwell R, Kopácek J, Marchetto A, Matzner E (2001) Response of sulphur dynamics in European catchments to decreasing sulphate deposition. Hydrol Earth Syst Sci 5(3):311–326

Reemtsma J (1986) Der Magnesium-Gehalt von Nadeln niedersächsischer Fichtenbestände und seine Beurteilung. Allg Forst Jagdztg 157:196–200

Rennenberg H (1999) The significance of ectomycorrhizal fungi for sulfur nutrition of trees. Plant Soil 215(2):115–122

Ricke W (1960) Ein Beitrag zur Geochemie des Schwefels. Geochim Cosmochim Acta 21 (1–2):35–80

Riek W, Russ A, Hannemann J, Kallweit R (2016) Bodenzustand und Baumernährung: Kennwerte aus BZE und Level II-Programm. In: 30 Jahre forstliches Umweltmonitoring in Brandenburg. Eberswalder Forstliche Schriftenreihe, vol 63. Ministerium für Ländliche Entwicklung Umwelt und Landwirtschaft des Landes Brandenburg, Eberswalde, pp 40–60

Sardans J, Rivas-Ubach A, Penuelas J (2012) The C:N:P stoichiometry of organisms and ecosystems in a changing world: a review and perspectives. Perspect Plant Ecol Evol Syst 14(1):33–47. https://doi.org/10.1016/j.ppees.2011.08.002

Sardans J, Alonso R, Janssens IA, Carnicer J, Vereseglou S, Rillig MC, Fernandez-Martinez M, Sanders TGM, Penuelas J (2016) Foliar and soil concentrations and stoichiometry of nitrogen and phosphorous across European *Pinus sylvestris* forests: Relationships with climate, N deposition and tree growth. Funct Ecol 30(5):676–689. https://doi.org/10.1111/1365-2435.12541

Scarascia-Mugnozza G, Bauer G, Persson H, Matteucci G, Masci A (2000) Tree biomass, growth and nutrient pools. In: Schulze E-D (ed) Carbon and nitrogen cycling in European forest ecosystems. Springer, Berlin, pp 49–62

Schleppi P, Tobler L, Bucher JB, Wyttenbach A (2000) Multivariate interpretation of the foliar chemical composition of Norway spruce (*Picea abies*). Plant Soil 219(1):251–262

Schöpp W, Posch M, Mylona S, Johansson M (2003) Long-term development of acid deposition (1880–2030) in sensitive freshwater regions in Europe. Hydrol Earth Syst Sci 7(4):436–446

Schulze ED, Lange OL, Oren R (1989) Forest decline and air pollution: a study of spruce (Picea abies) on acid soils. Springer, Berlin

Shahid M, Dumat C, Khalid S, Schreck E, Xiong TT, Niazi NK (2017) Foliar heavy metal uptake, toxicity and detoxification in plants: a comparison of foliar and root metal uptake. J Hazard Mater 325:36–58. https://doi.org/10.1016/j.jhazmat.2016.11.063

Spiecker H, Mielikäinen K, Köhl M, Skovsgaard JP (2012) Growth trends in European forests: studies from 12 countries. Springer, Berlin

Staszewski T, Lukasik W, Kubiesa P (2012) Contamination of Polish national parks with heavy metals. Environ Monit Assess 184(7):4597–4608. https://doi.org/10.1007/s10661-011-2288-z

Talkner U, Meiwes KJ, Potočić N, Seletković I, Cools N, De Vos B, Rautio P (2015) Phosphorus nutrition of beech (*Fagus sylvatica* L.) is decreasing in Europe. Ann For Sci 72(7):919–928. https://doi.org/10.1007/s13595-015-0459-8

UBA (2015) Daten zur Umwelt. Umwelttrends in Deutschland. Umweltbundesamt, Dessau-Roßlau

UBA (2017) Schwermetall-Emissionen. Umweltbundesamt. https://www.umweltbundesamt.de/daten/luft/luftschadstoff-emissionen-in-deutschland/schwermetall-emissionen#textpart-1. Accessed 07/05/2017

Ulrich B (1981) Destabilisierung von Waldökosystemen durch Akkumulation von Luftverunreinigungen. Forst Holzwirt 36(21):525–532

van den Burg J (1985) Foliar analysis for determination of tree nutrient status: a compilation of literature data. Rijksinstituut voor Onderzoek in de Bos- en Landschapsbouw "De Dorschkamp", Wageningen

van den Burg J (1990) Foliar analysis for determination of tree nutrient status: a compilation of literature data: 2: Literature 1985–1989. "De Dorschkamp" Institute for Forestry and Urban Ecology, Wageningen

Vitousek PM, Howarth RW (1991) Nitrogen limitation on land and in the sea: how can it occur? Biogeochemistry 13(2):87–115

von Wilpert K (2003) Drift des Stoffhaushalts im Fichten-Düngeversuch Pfalzgrafenweiler. Allg Forst Jagdztg 174(2/3):21–30

von Wilpert K, Hartmann P, Schäffer J (2013) Regenerationsorientierte Bodenschutzkalkung. Merkblatt 54/2013. Forstliche Versuchs-und Forschungsanstalt Baden-Württemberg, Freiburg

Waldner P, Marchetto A, Thimonier A, Schmitt M, Rogora M, Granke O, Mues V, Hansen K, Karlsson GP, Zlindra D, Clarke N, Verstraeten A, Lazdins A, Schimming C, Iacoban C, Lindroos AJ, Vanguelova E, Benham S, Meesenburg H, Nicolas M, Kowalska A, Apuhtin V, Napa U, Lachmanova Z, Kristoefel F, Bleeker A, Ingerslev M, Vesterdal L, Molina J, Fischer U, Seidling W, Jonard M, O'Dea P, Johnson J, Fischer R, Lorenz M (2014) Detection of temporal trends in atmospheric deposition of inorganic nitrogen and sulphate to forests in Europe. Atmos Environ 95:363–374. https://doi.org/10.1016/j.atmosenv.2014.06.054

Waldner P, Thimonier A, Pannatier EG, Etzold S, Schmitt M, Marchetto A, Rautio P, Derome K, Nieminen TM, Nevalainen S, Lindroos AJ, Merila P, Kindermann G, Neumann M, Cools N, de Vos B, Roskams P, Verstraeten A, Hansen K, Karlsson GP, Dietrich HP, Raspe S, Fischer R, Lorenz M, Iost S, Granke O, Sanders TGM, Michel A, Nagel HD, Scheuschner T, Simoncic P, von Wilpert K, Meesenburg H, Fleck S, Benham S, Vanguelova E, Clarke N, Ingerslev M, Vesterdal L, Gundersen P, Stupak I, Jonard M, Potocic N, Minaya M (2015) Exceedance of critical loads and of critical limits impacts tree nutrition across Europe. Ann For Sci 72 (7):929–939. https://doi.org/10.1007/s13595-015-0489-2

Wehrmann J (1959) Methodische Untersuchungen zur Durchführung von Nadelanalysen in Kiefernbeständen. Forstwiss Centralbl 78(3):77–97

Weis W, Gruber A, Huber C, Gottlein A (2009) Element concentrations and storage in the aboveground biomass of limed and unlimed Norway spruce trees at Hoglwald. Eur J For Res 128(5):437–445. https://doi.org/10.1007/s10342-009-0291-5

Chapter 10
Plants as Indicators of Soil Chemical Properties

Hagen S. Fischer, Barbara Michler, Daniel Ziche, and Anton Fischer

10.1 Introduction

It has long been known that the distribution of plant species in space depends on site factors. Humboldt and Bonpland (1807) described the dependence of plant species on altitude in tropical countries, i.e. across a temperature gradient. The year 2013 was the centenary anniversary of the publication of Braun-Blanquet's first paper in the field of vegetation science (Braun-Blanquet 1913), in which he asserted that chemical properties of soils, specifically acidity and content of nutrients, are the prime factors that determine the distribution patterns of plant species. Based on these findings, Ellenberg (1950) developed a system to indicate soil conditions on agricultural fields based on the weed species growing between the crops. He assigned ordinal values ranging from 1 (little) to 5 (much) for weed species for five main site factor groups: soil acidity (pH), nitrogen (N) availability, water supply, temperature and continentality. The arithmetic mean of the indicator values for all weed species at a given place was used as an indication of site condition.

Later, Ellenberg extended the concept to all Central European vegetation types and plant species, respectively, and based the system on a 9-point scale (Ellenberg 1974). This concept of indicator values has been improved in recent years (Ellenberg et al. 2001, 2003) and has been adapted for other countries, such as Switzerland

H. S. Fischer (✉)
ifanos-Natur&Landschaft, Roettenbach, Germany
e-mail: hagensfischer@t-online.de

B. Michler · A. Fischer
Technical University Munich, Freising, Germany
e-mail: b.michler@wzw.tum.de; hfischer@wzw.tum.de

D. Ziche
Thünen Institute of Forest Ecosystems, Eberswalde, Germany
e-mail: daniel.ziche@thuenen.de

N. Wellbrock, A. Bolte (eds.), *Status and Dynamics of Forests in Germany*, Ecological Studies 237, https://doi.org/10.1007/978-3-030-15734-0_10

(Landolt 1977; Landolt et al. 2010), Great Britain (Hill et al. 1999), Hungary (Zólyomi et al. 1967; Soó 1980; Borhidi 1995) and Poland (Zarzycki et al. 2002).

Today, indicator values remain a very popular tool for the integrated assessment of site conditions. For example, Persson (1981) used indictor values to support the interpretation of ordination diagrams.

Without detracting the credit these works deserve, it should be noted that the underlying concept suffers from several deficits:

1. The assignment of indicator values to plant species is primarily based on the personal experience of the investigators. They "reflect the opinion of a (few) experts" (Wildi 2016) rather than being a result of the analysis of measured habitat conditions. Direct measurements have been used only marginally to define the indicator values, as systematic measurements of site conditions in phytosociological relevés were very rare. The comparison of mean indicator values with measurements of certain factors has been applied a posteriori only by way of example (Ellenberg et al. 2001).
2. The calculation of arithmetic means of indicator values is problematic, as ordinal variables should not be averaged. Möller (1992) suggests the use of medians instead of arithmetic means. Averaging the indicator values of the species occurring on a given plot is permissible only under the (implicit) auxiliary hypothesis that Ellenberg's indicator value classes have (at least approximately) equal class widths.
3. Furthermore, there is debate as to whether either averages weighted or unweighted by cover should be used (Rodenkirchen 1982). The idea is that a species is to be expected to have a higher cover value when growing close to its ecological optimum. But cover also has a strong species-specific component: some species—orchids, for example—never occur at high cover, while others often do. The difference between averages weighted and unweighted by cover, however, depends on the evenness of the vegetation of the plot: for plots with high evenness, the difference is small because all species have more or less the same weight. Relevant differences are expected only in plots with a pronounced dominance structure. However, using cover as a weight for a weighted average is problematic in any case: measurement of the specific response of a certain species to the site factor under discussion would be necessary to determine the width of the ecological amplitude (Peppler-Lisbach 2008). Species with very narrow ecological amplitude should receive a higher weight than species that occur over a wider range of ecological conditions. But such a measure is completely lacking in all indicator systems to date. The cover-weighted average is only a substitute for this shortcoming.
4. Mean indicator values have also been used for environmental monitoring. Ewald et al. (2013) showed an increase in mean N values in forests in Central Europe over a period of nearly one century. Ewald and Ziche (2017), however, showed that Ellenberg's mean N values are positively correlated with several nutrients but not with N concentration in topsoil. So the question arises: which measurable environment factor caused the increase of the N values?

Several attempts have already made to overcome these shortcomings: a posteriori calibration tries to justify the use of indicator values (Ellenberg et al. 2001). In most cases, the expected correlation is significant even though there is a low coefficient of determination. In some cases, however, the sign of the regression coefficient is even opposite to expected (Wamelink et al. 2002).

Hill et al. (1999), for example, recalibrated Ellenberg's indicator values based on the national database of British vegetation samples. Ewald and Ziche (2017) analysed which physical units correspond to the mean indicator value for nutrients.

Nevertheless, none of these approaches addresses the major concern: the indicator system should indicate *measurable* properties of the soil (and other habitat variables) directly. Fischer (1986) used ordination methods to predict soil properties (pH) directly from vegetation relevés. However, prior to the second National Forest Soil Inventory (NFSI II) in Germany, data required to follow this approach was limited to small datasets covering only small spatial ranges and few vegetation types.

The goal of this paper is to create a species-based indicator system to assess soil properties. The indicator system is based on intensive soil analyses and vegetation surveys that yielded the German NFSI II data.

10.2 Climate, Soil, and Vegetation Data

The NFSI II recorded soil data as well as vegetation relevés on a regular grid of 8 km grid size at 1838 sampling points (BMELV 2006; Wellbrock et al. 2016). Sampling was conducted between 2006 and 2008. The vegetation was sampled on an area of 400 m^2 within a 30 m circle around the central soil pits of the plots. On the NFSI plots, 819 vascular plant species were recorded overall. 425 plant species occurred on more than 4 plots. Nomenclature of species follows Jansen and Dengler (2008).

Soil variables that were accounted for in our work included pH (H_2O), pH(KCl), pH($CaCl_2$), cation exchange capacity and the stocks of total nitrogen, total phosphorous (aqua regia extracts), organic and inorganic carbon and the cations Ca, Mg, K, Na, Al, Fe, Mn and H. Furthermore, the available field capacity was estimated based on texture analyses. Details on analytical methods can be found in HFA (GAFA 2014). Initial studies have shown that correlation between vegetation and soil is closest for the top 10 cm of soil (Ziche et al. 2016). Consequently, the first two depths (0–5 and 5–10 cm) were used for further analyses. Additionally, climate information was attributed to the NFSI plots using geostatistical methods such as ordinary kriging (precipitation) or regression kriging (temperature, Ziche and Seidling 2010). Complete datasets with all recognized variables were available for 1619 sampling plots.

Within the framework of the intensive (Level II) monitoring of forest ecosystems in Germany,[1] another independent dataset was compiled that comprises soil data as

[1] http://blumwald.thuenen.de/level-ii/allgemeine-informationen/

well as vegetation relevés (Seidling 2005). This second dataset with 48 sampling points was used to evaluate our indicator system. All soil analyses were performed after 2005 and at depth intervals of 0–5, 5–10, 10–20, 20–40 and 40–80 cm.

Correspondence analysis (CA) using the programme CANOCO was employed to identify outliers (ter Braak and Šmilauer 2002). Species' cover values were transformed according to Pudlatz (1975) to compensate the extremely skewed distribution of cover values. Rare species with a frequency of less than 5% were omitted from the analysis because they do not contribute to the general floristic structure of the vegetation. Only 20% of the omitted species have their main occurrence in forests (Schmidt et al. 2011). Forty-five percent are regarded as species of forests as well as open land. They are less typical for forests. Thirty-four percent of the species are not mentioned at all in the list of Schmidt et al. (2011). These are either typical species of open habitats that were geminated in the forest only by chance or taxa that could not be determined to species level (e.g. *Quercus* sp.). The CA indicates that no plots with outlying species composition were present. After omitting rare species, one relevé contained no other species and hence was omitted.

10.3 Environmental Impact on Species Composition

With the remaining 1618 relevés, forward selection of environmental variables using a Monte Carlo significance test with a canonical correspondence analysis (CCA) was applied to select soil variables that most affect the species composition of the vegetation. Forward selection of environmental variables in CCA is a stepwise regression approach: a variable is selected only if it contributes additionally to explaining species pattern compared to the variables already selected. The variable that explained the most and hence was selected first is base saturation (BS). C/N ratio and pH also show high predictive power. All variables appeared to significantly affect vegetation structure. This may be due to the large sample size. Moreover, significant relationships may not be relevant. For the purpose of indicators, the focus should be on the variables with the highest F value (Table 10.1).

Organic carbon is very highly correlated with total N. It was excluded from the analysis to avoid problems with collinearity.

The new indicator system is described below using base saturation as an example. Nevertheless, the model was developed for all variables listed in Table 10.1. However, the best results are expected for the first few variables in the list.

Finally, CCA was performed to identify the major gradients in vegetation. A biplot of the CCA with relevés and soil factors as environmental variables (Fig. 10.1) revealed a soil acidity gradient along the first axis with base saturation (BS) and pH increasing to the left and C/N ratio (CN) to the right. The second axis displays a climatic gradient with a positive correlation with altitude above sea level (Alti) and precipitation (Prec) and a negative correlation with temperature (Temp). The floristic structure of Germany's forests is primarily determined by a soil acidity gradient and secondarily by a temperature gradient. Hence, soil properties associated with acidity

Table 10.1 Ranking of environmental variables according to forward selection in CANOCO

	Df	AIC	F	Pr(>F)	
Base saturation (BS)	1	7030.8	54.0537	0.005	**
C/N ratio (CN)	1	7041.1	43.5222	0.005	**
pH	1	7043.9	40.6465	0.005	**
Altitude (Alti)	1	7051.1	33.2755	0.005	**
Total nitrogen (Nges)	1	7053.5	30.8410	0.005	**
Exchangeable Ca	1	7054.7	29.6279	0.005	**
Cation exchange capacity (KAK)	1	7056.1	28.1512	0.005	**
Exchangeable K	1	7057.5	26.8032	0.005	**
Precipitation (Prec)	1	7059.3	24.9029	0.005	**
Mean annual temperature (Temp)	1	7059.7	24.4844	0.005	**
Exchangeable Mn	1	7068.2	15.8808	0.005	**
Exchangeable Na	1	7068.5	15.6035	0.005	**
Total P	1	7069.0	15.0751	0.005	**
Exchangeable Mg	1	7070.7	13.3911	0.005	**
Inorganic carbon (Cco3)	1	7073.5	10.6246	0.005	**
Exchangeable Al	1	7076.5	7.5715	0.005	**
Usable field capacity (nFK)	1	7077.3	6.7302	0.005	**
C/P ratio (CP)	1	7078.6	5.4352	0.005	**
Exchangeable H^+	1	7079.7	4.4004	0.005	**
Exchangeable Fe	1	7079.9	4.2086	0.005	**

** Pr(>F) < 0.01

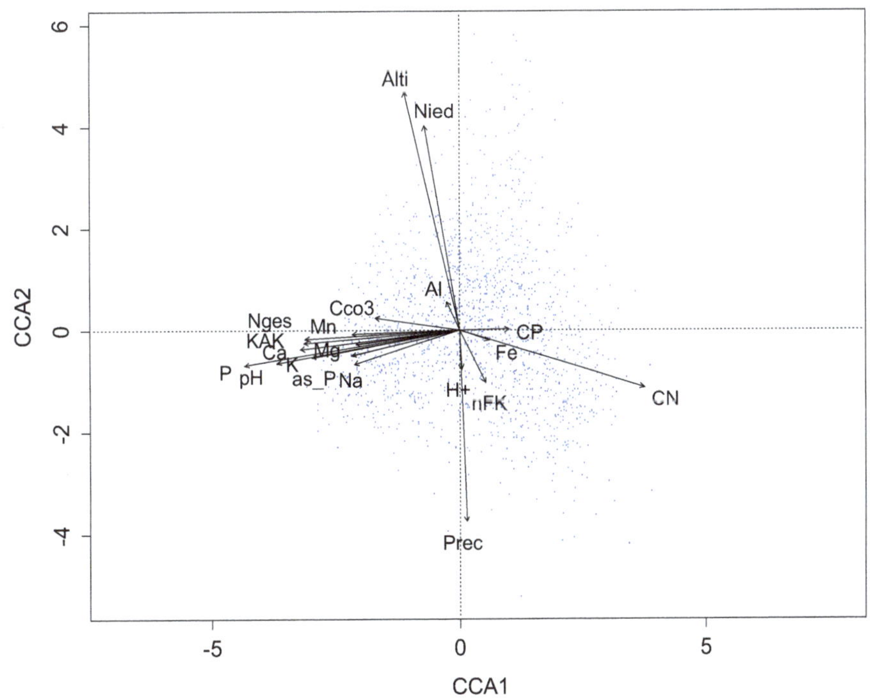

Fig. 10.1 Biplot of the CCA with relevés (blue dots) and site factors

(base saturation, pH, C/N ratio) can be expected to be good indicated in a species-based indicator system. The significant effect of temperature on forest vegetation demonstrates the high sensitivity of forests to the expected climate changes.

10.4 Modelling Species Response to Soil Properties

We fitted generalized linear models (GLM), applying second-order polynomial logistic regression (McCullagh and Nelder 1989). Soil variables serve as independent variables to the presence/absence data of the species, the dependent variable. These models describe the conditional probability of the occurrence of the species depending on the soil conditions:

$$\log\left(\frac{p(V|x)}{1-p(V|x)}\right) = b_0 + b_1x + b_2x^2 \tag{10.1}$$

where $p(V|x)$ is the conditional probability for finding species V at soil condition x and b_0, b_1 and b_2 are the coefficients of the regression equation. This approach is equivalent to fitting a symmetric, unimodal response curve to the occurrences of the species (Jongman et al. 1995). This is a parametric approach. Parametric methods have the advantage that the species response curve can be described using terms such as optimum and tolerance (Jongman et al. 1995). These values for base saturation are given in the appendix.

Unimodal response curves are the underlying principle of the commonly used CCA (ter Braak 1987; ter Braak and Šmilauer 2002). However, it remains unclear whether the unimodal model is appropriate to describe species response curves (Austin and Meyers 1996). While some authors state that it is "unlikely ecologically" that species have a multimodal response to a single variable (Roberts n.d.), others favour bimodal and skewed response curves (Zelený and Tichý n.d.). Therefore, we also applied generalized additive models (GAM), which allow any kind of response curve, to test for the occurrence of species with bimodal or skewed response curves. For this, we compared 95% confidence bands of the GLM with the GAM curves. If the GAM curve lies within the confidence band of the GLM, the latter is obviously an adequate description of the species response, and there is no reason to consider the more complicated GAM. Fitting of GLM and GAM was performed using the software package R (R Core Team 2016).

In our dataset, most species fit very well to the unimodal model. This was expected as the unimodal model is the basis for the widely used correspondence analysis. If a majority of species violates the basic assumption of this method, it would not perform as well as it does. Frequent and successful application of this method also suggests that the unimodal model is appropriate for most species. We were able to show that bimodal response curves are the exception rather than the rule.

Figure 10.2 shows some examples: *Circaea lutetiana* is a species with a clear, bell-shaped distribution within the base saturation gradient with an ecological optimum at 0.7, whereas beech (*Fagus sylvatica*) is an example of a species with wide ecological amplitude with respect to base saturation. *Deschampsia flexuosa* and *Mercurialis perennis* are species clearly preferring the lower or upper end of the gradient, respectively, with an optimum at the lower or upper end of the range, respectively.

Nevertheless, some species cannot be described adequately with a unimodal response curve. One example is *Acer pseudoplatanus* (Fig. 10.2e), which has one optimum around 30% base saturation and a second one at 100%. The GAM of this species lies outside the 95% confidence bands of the GLM. Such species cannot serve as indicator species in any indicator system as they do not have a clear optimum. Therefore, bimodal species were excluded from our indicator system. Fortunately, these cases are very rare, and most species in the dataset fit to the unimodal model.

In a few cases, the response curve was U-shaped instead of bell-shaped. This is the case if the coefficient b_2 in the logistic regression is significantly greater than 0. In fact, this is a special case of a bimodal response curve. Such species also do not qualify as indicator species and were excluded from the indicator system. An example is *Picea abies* (Fig. 10.2f).

10.5 Predicting Soil Properties by Species Composition

In the next step, we needed to derive the conditional probabilities of different values of the soil factor to be indicated $p(x|\vec{V})$ depending on the species combination $\vec{V}$ found in the forest stand. This is performed by means of the Bayes' formula (Fischer 1990, 1994):

$$p(x|\vec{V}) = \frac{p(\vec{V}|x) \cdot p(x)}{p(\vec{V})} \tag{10.2}$$

The multivariate conditional probability of finding the species combination $\vec{V}$ at soil condition x is estimated from the univariate response curves assuming independence of the effects:

$$p(\vec{V}|x) = p(V_1|x) \cdot p(V_2|x) \cdot \ldots \cdot p(V_n|x) \tag{10.3}$$

The prior probabilities of the soil conditions $p(x)$ correspond to the frequency distribution of x in the area of investigation. Kernel density estimation with *R* function "density" was used to determine $p(x)$. The prior probability of the species

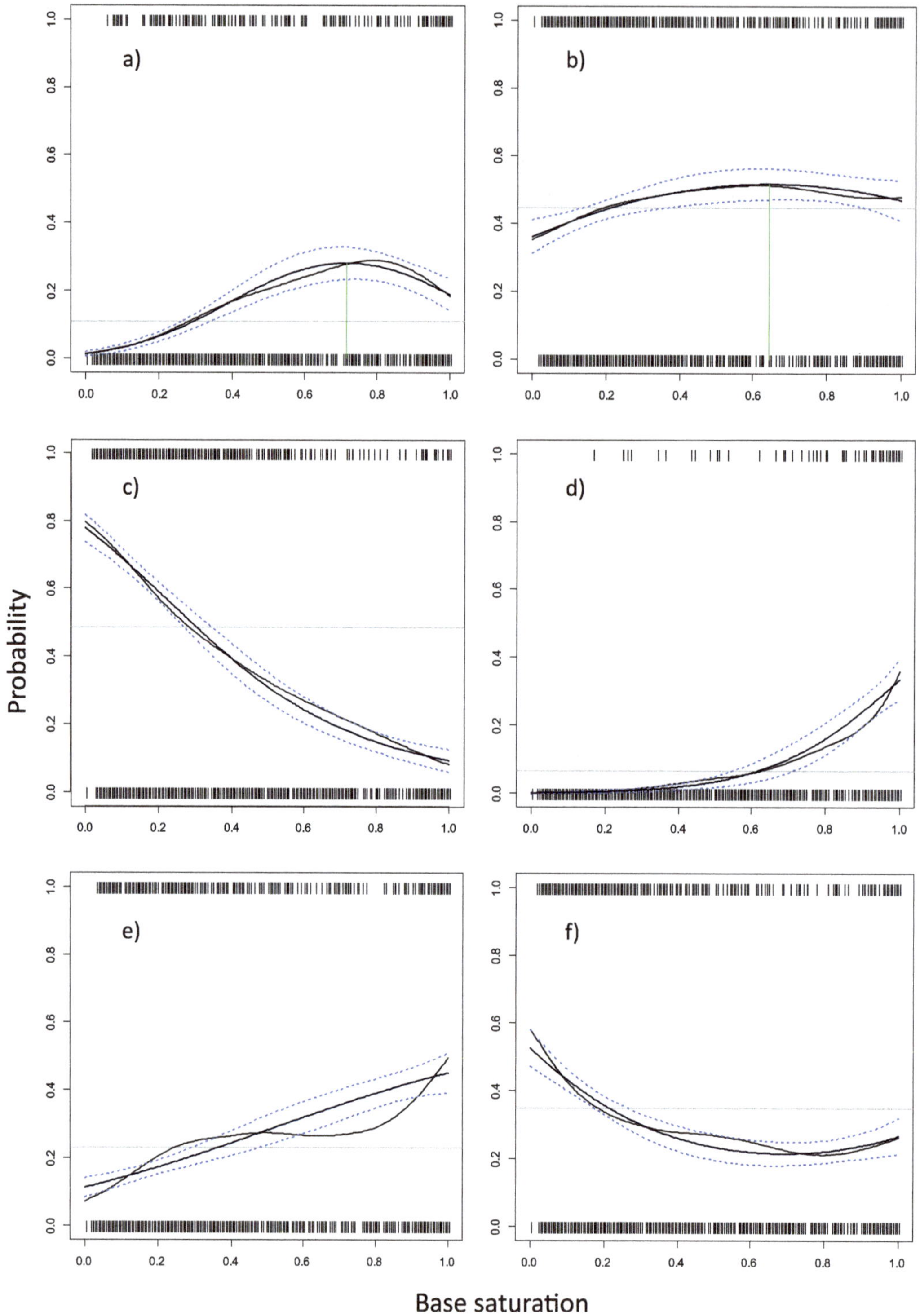

Fig. 10.2 Examples of species response curves with respect to base saturation (BS). The tics at probability 0 and 1 (bottom, top) show the location of relevés with and without the particular species, respectively. *Circaea lutetiana* has an optimum around 0.7 (70%), *Fagus sylvatica* is an indifferent species, *Deschampsia flexuosa* has a clear preference for low base saturations and *Mercurialis perennis* has a preference for high base saturation. *Acer pseudoplatanus* is a species with a bimodal response curve with respect to base saturation: there is one local maximum around 0.3 and another one 1.0. *Picea abies* is an example for a species with U-shaped response curve

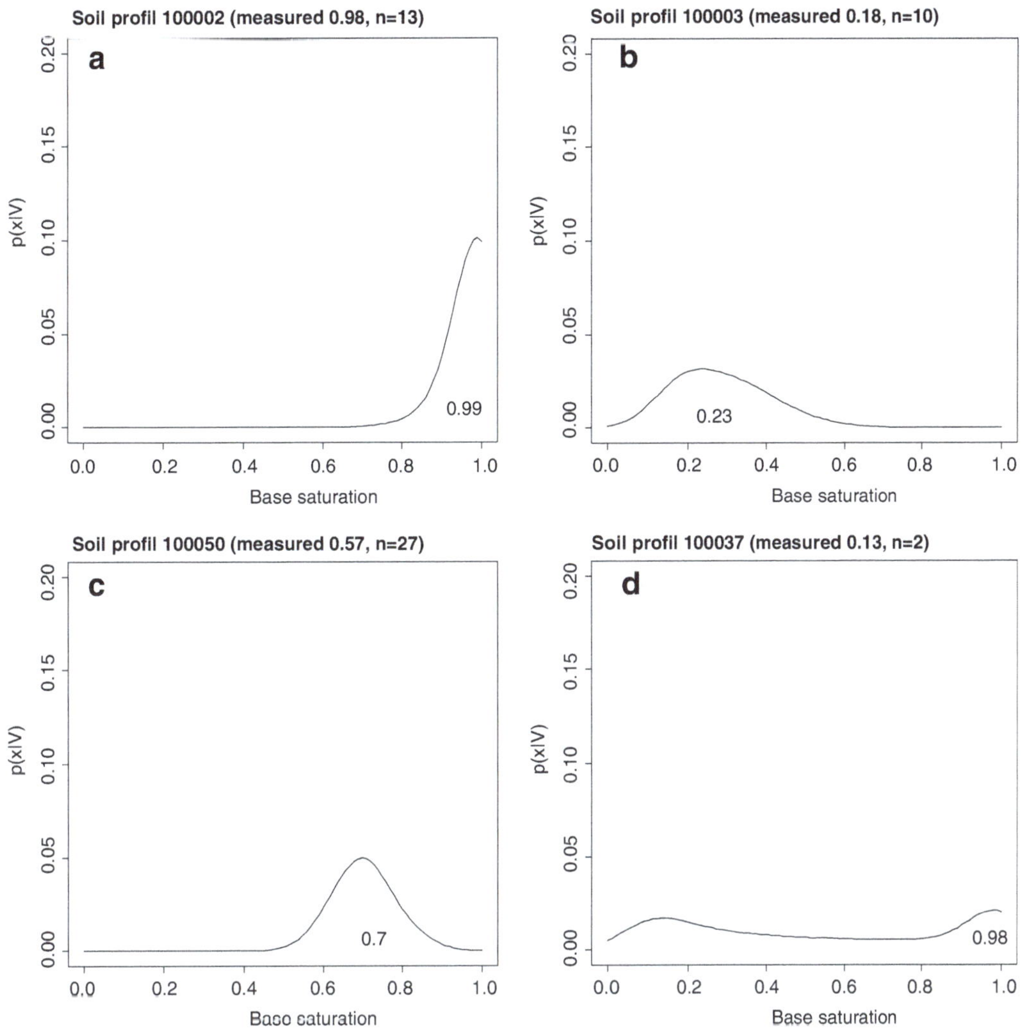

Fig. 10.3 Examples of probability profiles for relevés. The maxima of the curves correspond to the indicated base saturation (*n*, number of indicator species). Relevé 100037 is an example of a flat profile with a bimodal distribution indicating uncertain indication

composition $p(\vec{V})$ is a constant at a given point in the landscape and hence does not need to be considered. The value of the soil condition x with the highest conditional probability $p(x|\vec{V})$ is then taken as the result of the indication.

The results of the application of the indicator model are curves describing the probability of different values of the soil factor on certain relevés based on the species growing at those points. Four examples from different regions in Germany are shown in Fig. 10.3 for the variable base saturation. In most cases, these profiles display a clear peak. The base saturation with the highest probability is the indicated base saturation value. Only in very few cases is the profile a flat curve indicating uncertain indication. An example is given in Fig. 10.3d.

Comparison of the indicated and the measured base saturation in the independent reference dataset yields a quite high coefficient of determination of $R^2 = 0.6$

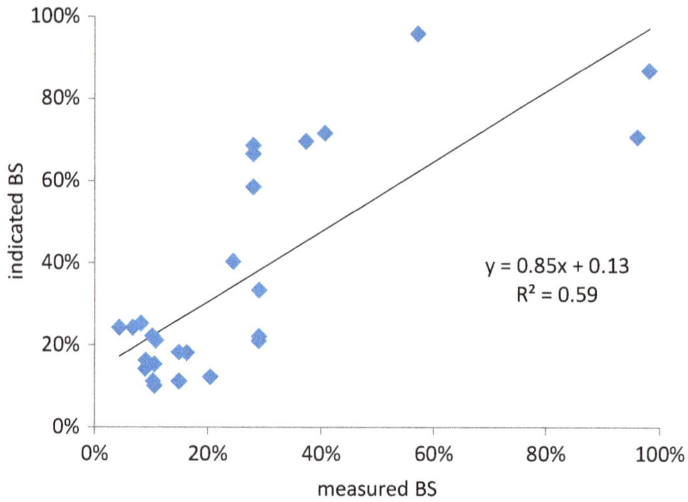

Fig. 10.4 Regression between measured and indicated base saturation in the reference dataset

(Fig. 10.4). The second important variable according to forward selection of environmental variables (Table 10.1) is the ratio of organic carbon and total N (C/N ratio). With an R^2 of 0.38, nearly 40% of the variability of the C/N ratio can be predicted with the indicator system. With pH, the third variable in the ranking of the forward selection of environmental variables (Table 10.1), the coefficient of determination (R^2) was 0.40.

10.6 The WeiWIS Indicator System

We implemented the algorithm described above in a MS-ACCESS application called "*Wei*henstephan *W*ood *I*ndicator *S*ystem" or WeiWIS. WeiWIS combines vegetation relevés with a species' calibration data and generates indicated values for habitat factors of all relevés and habitat factor profiles for selected plots (Fig. 10.5).

With WeiWIS as a new indicator tool, many unresolved problems with existing indicator systems become obsolete. The question as to whether the arithmetic mean or the median should be used to characterize a site does not apply to our approach, because the response curves themselves are considered, not only the maxima. This is also the reason why the question as to whether weighted or unweighted means leads to better result is not relevant here. In fact, species are weighted implicitly with the width of the ecological amplitude: species with a narrow and high peak have a higher weight in the system compared to species with a wider, flatter curve.

Furthermore, the resulting probability profile for each relevé gives an indication of the reliability of the result. A wide and flat profile suggests a less reliable result, probably due to a smaller number of species or a species combination that includes contradictory species.

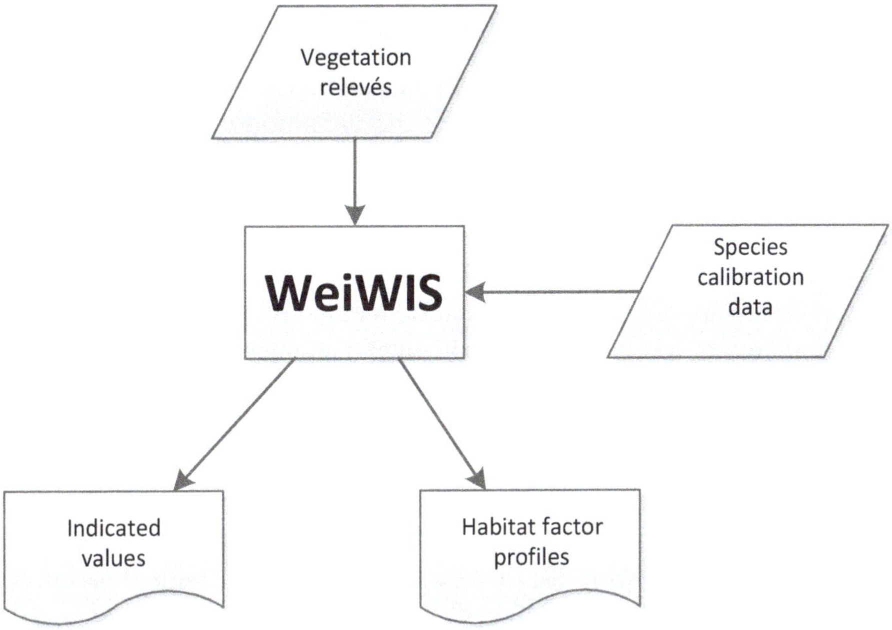

Fig. 10.5 Structure of the WeiWIS tool. Users can input their own vegetation relevés. They are analysed with the species calibration data, and indicated values for all relevés and habitat factor profiles are generated

10.7 Discussion

Although the application of indicator values is very popular, comparisons with measured values are quite rare in the literature. Michler (1994) found significant correlations between measured soil variables and mean indicator values with an R^2 of up to 0.2 (pH with mean R value, K^+, Zn^{2+} and phosphate with mean N value). Total N was not significantly correlated with the N value. Ewald and Ziche (2017) found base saturation was the best variable to predict Ellenberg's indicator value for nutrients (mN). They found a correlation of −0.55 between C/N ratio and mN, corresponding to an R^2 of 0.30.

Our indicator system, however, shows a significant correlation between measured and indicated base saturation, with $R^2 = 0.6$. This is a remarkable improvement compared to Ewald and Ziche (2017), especially since they used the same BZE II dataset as our study.

A major advantage of our indicator system is that it indicates soil properties (such as base saturation or C/N ratio) directly and not via a dimensionless value (mean indicator value) that needs to be correlated a posteriori with the desired measurable parameter.

Several attempts have already been made to define indicator values for plant species on the basis of measurements. Wamelink et al. (2005) used the mean of pH values for all relevés where a species is present as an indicator value. However, this concept is biased by the extent of the range of the observed gradient. If the calibration dataset does not cover the entire range of habitat conditions that a species can occupy, the mean cannot correctly reflect the species' optimum. Furthermore, the mean is influenced by the frequency distribution of the habitat factor. If some habitat conditions occur more frequently in the calibration dataset than others, these conditions will receive a higher weight for the definition of the indicator values. Gégout et al. (2003) overcame this problem by using GLM to define the ecological optimum. The value of the habitat factor with the highest probability is taken as the indicator value for the species. However, both approaches have their indicator based on the arithmetic mean of the indicator values of the species present in a plot.

A very different approach is the imputation method of Tichý et al. (2010). Here, the vegetation relevé for which environmental variables should be indicated is compared with the relevés of a reference dataset for which environmental measurements are available. The mean of the values of the measured environmental variables of floristically similar relevés in the reference dataset is used for indication. This approach requires a huge reference dataset where all vegetation types are represented with statically representative sample sizes. Otherwise, indicator values cannot be identified for all relevés. In contrast to our approach, Tichý's method produces only one indicator value but provides no measure of the reliability of the value. With his recommended parameter (maximum distance = 0.5), only 22% of our relevés could be assigned a value, showing that our reference dataset with 1619 relevés is still too small for Tichý's approach. With a maximum distance of 1, all relevés were assigned. But the variance of the estimated values was significantly smaller ($p = 0.09$) than the measured ones. Consequently, the imputed values are clearly biased.

Peppler-Lisbach (2008) based his indicator concept on niche overlap: the indicated value is the centre of the overlap of the niches of all species in a plot. Obviously, this concept can only be applied if *all* species in a plot have overlapping niches.

10.8 Conclusions

To date, all species-based systems to indicate site conditions, such as the famous Ellenberg indicator values, reflect "expert opinions only" (Wildi 2016), rather than an outcome of sound statistical analysis of measured data. With WeiWIS, we have developed an indicator system for soil properties that is based on a large, statistically representative sample of measurements. We have undertaken the step away from systems developed during the last half century that are primarily subjective towards a quantitative system that is based on a systematic sampling design and which follows a sound statistical approach.

Species-based indicator systems that indicate complex factors, like Ellenberg's system, should not be a substitute for ecological measurements in environmental monitoring (Michler 1994), because changes, for example, in the mean N value, cannot be attributed to a concrete, measurable variable. However, this step is necessary to set up effective, sustainable planning in the management of habitats and to avoid misinterpretations. Our system avoids these problems because it directly indicates measurable values. The method can be applied to all environmental variables that effect species distribution.

Moreover, our indicator systems can help to estimate the conditions in the past, for which no measurements are available, and then compare them to the current situation within the same system.

Acknowledgements The National Forest Soil Inventory is a joint project of the federal government and the German states. We thank the individual federal states for providing the datasets, as well as the Federal Ministry of Food and Agriculture (BMEL) for facilitating the data analysis. We acknowledge in particular all site investigators and the persons in charge of the individual federal research stations (http://www.blumwald.de/bze/ansprechpartner-bze) for their assistance and support. This work was supported by the Thünen Institute, the German Federal Research Institute for Rural Areas, Forestry and Fisheries. A previous version that was limited to Bavaria was supported by project F49 of the board of trustees of the Bavarian Agency for Forest and Forestry and the Bavarian State Ministry for Agriculture and Forestry.

References

Austin MP, Meyers JA (1996) Current approaches to modelling the environmental niche of eucalypts: implication for management of forest biodiversity. For Ecol Manag 85(1–3):95–106. https://doi.org/10.1016/s0378-1127(96)03753-x

BMELV (2006) Arbeitsanleitung für die zweite bundesweite Bodenzustandserhebung im Wald (BZE II). Bundesministerium für Ernährung, Landwirtschaft und Verbraucherschutz, Berlin

Borhidi A (1995) Social behaviour types, the naturalness and relative ecological indicator values of the higher plants in the Hungarian flora. Acta Bot Hungar 39(1–2):97–181

Braun-Blanquet J (1913) Die Vegetationsverhältnisse der Schneestufe in den Rätisch-Lepontischen Alpen: Ein Bild des Pflanzenlebens an seinen äussersten Grenzen. Schweizerische Naturforschende Gesellschaft, Zurich

Ellenberg H (1950) Unkrautgemeinschaften als Zeiger für Klima und Boden. Eugen Ulmer, Stuttgart

Ellenberg H (1974) Zeigerwerte der Gefässpflanzen Mitteleuropas. Scripta Geobotanica IX. Erich Goltze, Göttingen

Ellenberg H, Weber HE, Düll R, Wirth V, Werner W (2001) Zeigerwerte von Pflanzen in Mitteleuropa. Scripta Geobotanica XVIII, 3rd edn. Erich Goltze, Göttingen

Ellenberg H, Weber HE, Düll R, Wirth V, Werner W (2003) Zeigerwerte von Pflanzen in Mitteleuropa. Scripta Geobotanica XVIII, Datenbank. Erich Goltze, Göttingen

Ewald J, Hennekens S, Conrad S, Wohlgemuth T, Jansen F, Jenssen M, Cornelis J, Michiels HG, Kayser J, Chytry M, Gegout JC, Breuer M, Abs C, Walentowski H, Starlinger F, Godefroid S (2013) Spatial and temporal patterns of Ellenberg nutrient values in forests of Germany and adjacent regions—a survey based on phytosociological databases. Tuexenia 33:93–109

Ewald J, Ziche D (2017) Giving meaning to Ellenberg nutrient values: National Forest Soil Inventory yields frequency-based scaling. Appl Veg Sci 20(1):115–123. https://doi.org/10.1111/avsc.12278

Fischer H (1986) Zur Vorhersage ökologischer Parameter aufgrund der floristischen Struktur der Vegetation. Tuexenia 6:405–414

Fischer HS (1990) Simulating the distribution of plant communities in an alpine landscape. Coenoses 5:37–43

Fischer HS (1994) Simulation der räumlichen Verteilung von Pflanzengesellschaften auf der Basis von Standortskarten. Dargestellt am Beispiel des MaB-Testgebiets Davos = Simulation of the spatial distribution of plant communities based on the maps of site factors. Veröffentlichungen des Geobotanischen Instituts der Eidgenössischen Technischen Hochschule, Stiftung Rübel, vol 122. ETH Zürich, Zurich

GAFA (ed) (2014) Handbuch Forstliche Analytik (HFA). Grundwerk und 1. – 5. Ergänzung des Gutachterausschuss Forstliche Analytik (GAFA). Federal Ministry of Food, Agriculture and Consumer Protection, Northwest German Forest Research Institute, Bonn

Gégout J-C, Hervé J-C, Houllier F, Pierrat J-C (2003) Prediction of forest soil nutrient status using vegetation. J Veg Sci 14(1):55–62

Hill MO, Mountford J, Roy D, Bunce RGH (1999) Ellenberg's indicator values for British plants. Ecofact Research Report Series 2b. Institute of Terrestrial Ecology, Huntingdon

Humboldt A, Bonpland A (1807) Ideen zu einer Geographie der Pflanzen nebst einem Naturgemälde der Tropenländer. Cotta, Tübingen

Jansen F, Dengler J (2008) GermanSL -Eine universelle taxonomische Referenzliste für Vegetationsdatenbanken in Deutschland. Tuexenia 28:239–253

Jongman RHG, ter Braak CJF, van Tongeren OFR (1995) Data analysis in community and landscape ecology. Cambridge University Press, Cambridge

Landolt E (1977) Ökologische Zeigerwerte zur Schweizer Flora. Veröffentlichungen des Geobotanischen Instituts der Eidgenössischen Technischen Hochschule, Stiftung Rübel, vol 64. ETH Zürich, Zurich

Landolt E, Bäumler B, Erhardt A, Hegg O, Klötzli F, Lämmler W, Nobis M, Rudmann-Maurer K, Schweingruber F, Theurillat J (2010) Flora indicativa: Ökologische Zeigerwerte und biologische Kennzeichen zur Flora der Schweiz und Alpen, 2nd edn. Haupt Verlag, Bern

McCullagh P, Nelder JA (1989) Generalized linear models. Monographs on statistics and applied probablilty, vol 37, 2nd edn. Chapman and Hall/CRC, London, New York

Michler B (1994) Vegetationskundliche, bodenkundliche, phänologische und morphologische Untersuchungen zur Variabilität sekundärer Pflanzeninhaltsstoffe am Beispiel der Pyrrolizidinalkaloide in Symphytum officinale agg. Friedrich-Alexander-University Erlangen-Nürnberg, Erlangen

Möller H (1992) Zur Verwendung des Medians bei Zeigerwertberechnungen nach Ellenberg. Tuexenia 12:25–28

Peppler-Lisbach C (2008) Using species-environmental amplitudes to predict pH values from vegetation. J Veg Sci 19(4):437–U434. https://doi.org/10.3170/2008-8-18394

Persson S (1981) Ecological indicator values as an aid in the interpretation of ordination diagrams. J Ecol 69(1):71–84

Pudlatz H (1975) Zur Transformation der Variablen bei mangelnder Normalverteilung. Gießener Geogr Schriften 32:29–33

R Core Team (2016) R: A language and environment for statistical computing. R Foundation for Statistical Computing, Vienna

Roberts D (n.d.) R labs for vegetation ecologists. LAB 5—Modeling species/environment relations with generalized additive models. http://ecology.msu.montana.edu/labdsv/R/labs/lab5/lab5.html

Rodenkirchen H (1982) Wirkungen von Meliorationsmaßnahmen auf die Bodenvegetation eines ehemals streugenutzten Kiefernstandortes in der Oberpfalz. Forstliche Forschungsberichte München, vol 53. Chair for soil sience at University Munich, Munich

Schmidt M, Kriebitzsch W-U, Ewald J (2011) Waldartenlisten der Farn-und Blütenpflanzen, Moose und Flechten Deutschlands. BfN, Bonn

Seidling W (2005) Ground floor vegetation assessment within the intensive (Level II) monitoring of forest ecosystems in Germany: chances and challenges. Eur J For Res 124(4):301–312. https://doi.org/10.1007/s10342-005-0087-1

Soó R (1980) Synopsis systematico-geobotanica florae vegetationisque Hungariae VI. Akadémiai Kiadó, Budapest ter Braak CJF (1987) Unimodal models to relate species to environment. University Wageningen, Wageningen

ter Braak CJF, Šmilauer P (2002) CANOCO reference manual and CanoDraw for Windows user's guide: software for canonical community ordination (version 4.5). Biometris, Wageningen

Tichý L, Hájek M, Zelený D (2010) Imputation of environmental variables for vegetation plots based on compositional similarity. J Veg Sci 21(1):88–95. https://doi.org/10.1111/j.1654-1103.2009.01126.x

Wamelink GWW, Goedhart PW, van Dobben HF, Berendse F (2005) Plant species as predictors of soil pH: replacing expert judgement with measurements. J Veg Sci 16(4):461–470. https://doi.org/10.1111/j.1654-1103.2005.tb02386.x

Wamelink GWW, Joosten V, van Dobben HF, Berendse F (2002) Validity of Ellenberg indicator values judged from physico-chemical field measurements. J Veg Sci 13(2):269–278. https://doi.org/10.1658/1100-9233(2002)013[0269:voeivj]2.0.co;2

Wellbrock N, Bolte A, Flessa H (eds) (2016) Dynamik und räumliche Muster forstlicher Standorte in Deutschland: Ergebnisse der Bodenzustandserhebung im Wald 2006 bis 2008. Thünen Report, vol 43. Johann Heinrich von Thünen Institute, Federal Research Institute for Rural Areas, Forestry and Fisheries, Braunschweig

Wildi O (2016) Why mean indicator values are not biased. J Veg Sci 27(1):40–49

Zarzycki K, Trzcińska-Tacik H, Różański W, Szeląg Z, Wołek J, Korzeniak U (2002) Ecological indicator values of vascular plants of Poland. Biodiversity of Poland, vol 2. W. Szafer Institute of Botany, Polish Academy of Science, Krakow

Zelený D, Tichý L (n.d.) Species response curves in JUICE. http://www.sci.muni.cz/botany/zeleny/hof.php

Ziche D, Michler B, Fischer HS, Kompa T, Höhle J, Hilbrig L, Ewald J (2016) Boden als Grundlage biologischer Vielfalt. In: Wellbrock N, Bolte A, Flessa H (eds) Dynamik und räumliche Muster forstlicher Standorte in Deutschland. Ergebnisse der Bodenzustandserhebung im Wald 2006 bis 2008. Thünen Report, vol 43. Johann Heinrich von Thünen Institute, Federal Research Institute for Rural Areas, Forestry and Fisheries, Braunschweig, pp 292–342

Ziche D, Seidling W (2010) Homogenisation of climate time series from ICP Forests Level II monitoring sites in Germany based on interpolated climate data. Ann For Sci 67(8):804. https://doi.org/10.1051/forest/2010051

Zólyomi B, Baráth Z, Fekete G, Jakucs P, Kárpáti I, Kárpáti V, Kovács M, Máté I (1967) Einreihung von 1400 Arten der ungarischen Flora in ökologische Gruppen nach TWR-Zahlen. Fragm Bot Mus Hist Nat Hung 4:101–142

Chapter 11
Spatial Response Patterns in Biotic Reactions of Forest Trees and Their Associations with Environmental Variables in Germany

Nadine Eickenscheidt, Heike Puhlmann, Winfried Riek, Paul Schmidt-Walter, Nicole Augustin, and Nicole Wellbrock

11.1 Introduction

Forest soils show diverse conditions and are subject to natural and anthropogenic changes, as was demonstrated in the previous chapters that evaluated the results of the first National Forest Soil Inventory (NFSI I) and second National Forest Soil Inventory (NFSI II) of Germany. With regard to soil acidification, a slow recovery has been observed since the NFSI I in the 1990s. However, constant high loads of atmospheric nitrogen (N) from industrialization, burning of fossil fuels, traffic and

N. Eickenscheidt (✉)
State Agency for Nature, Environment and Consumer Protection of North Rhine-Westphalia, Recklinghausen, Germany
e-mail: nadine.eickenscheidt@lanuv.nrw.de

H. Puhlmann
Forest Research Institute Baden-Württemberg, Freiburg, Germany
e-mail: heike.puhlmann@forst.bwl.de

W. Riek
University for Sustainable Development and Eberswalde forestry State Center of Excellence, Eberswalde, Germany
e-mail: winfried.riek@hnee.de

P. Schmidt-Walter
Northwest German Forest Research Institute, Göttingen, Germany
e-mail: paul.schmidt-walter@nw-fva.de

N. Augustin
Department of Mathematical Sciences, University of Bath, Bath, UK
e-mail: n.h.augustin@bath.ac.uk

N. Wellbrock
Thünen Institute of Forest Ecosystems, Eberswalde, Germany
e-mail: nicole.wellbrock@thuenen.de

N. Wellbrock, A. Bolte (eds.), *Status and Dynamics of Forests in Germany*, Ecological Studies 237, https://doi.org/10.1007/978-3-030-15734-0_11

intensive agriculture (Berge et al. 1999; Galloway et al. 2008), which are associated with eutrophication and acidification (Aber et al. 1998; de Vries et al. 2014), are still worrying. The nutrition of forest trees and soil vegetation indicated an oversupply of N. Due to the strong increase in carbon (C), C/N ratios in the organic layer and upper mineral soil layers significantly increased. Heavy metals showed a translocation from the organic layer to the upper mineral soil. Furthermore, the modelled time series of drought stress indices and stored soil water available to plants indicated an increase in the intensity of water deficiency since 1990 and a decrease in the number of years characterized by good water supply. Based on these findings, the question of greatest relevance is how forest trees respond to the conditions and changes in forest soils.

In the present chapter, the biotic reactions of forest trees to conditions and changes in forest soils and to environmental conditions in general were examined. The focus was the four main tree species of Germany: Norway spruce (*Picea abies* (L.) Karst), Scots pine (*Pinus sylvestris* L.), European beech (*Fagus sylvatica* L.) and pedunculate and sessile oak (*Quercus robur* L. and *Q. petraea* (Matt.) Liebl., considered together), as well as the European silver fir (*Abies alba* Mill.). Tree defoliation, tree growth and tree nutrition were chosen as biotic indicators of tree vitality. Tree defoliation denotes the loss of needles or leaves in the crown of a tree compared to a local or absolute reference tree with full foliage. Defoliation is assessed as part of the Forest Condition Survey. The Forest Condition Survey represents a basic part of the Forest Monitoring in addition to the NFSI and the Intensive Forest Monitoring. In Germany, the condition of forest trees was recorded first in 1984, and the survey has been conducted annually throughout Germany since 1990 (see Chap. 1). Tree growth is not a part of the Forest Condition Survey and NFSI, but growth rings were evaluated on NFSI plots in some federal states in Germany. Tree nutrition was recorded as an obligatory parameter during both NFSIs (see Chaps. 1 and 9). The following sections deal with (1) the secondary growth response to drought, (2) the definition of defoliation development types and reasons for differences among these types, (3) the definition of forest nutrition types and reasons for differences among these types, as well as (4) the joint evaluation of defoliation development types and forest nutrition types in Germany. The overall aim was to identify regions at risk of tree vitality loss and risk factors. The findings could contribute to choosing appropriate political and management measures to sustain and improve tree vitality.

11.2 The Secondary Growth Response to Drought

This section discusses the extent to which the secondary growth of trees is associated with the availability of soil water. For the analysis, drill cores were extracted within the federal states of Baden-Wuerttemberg, Hesse, Lower Saxony, Bremen and Saxony-Anhalt. 197 drill cores were available for spruce, 193 for beech, 30 for fir, 174 for pine and 98 for oak (common and sessile oak). The vast majority of these

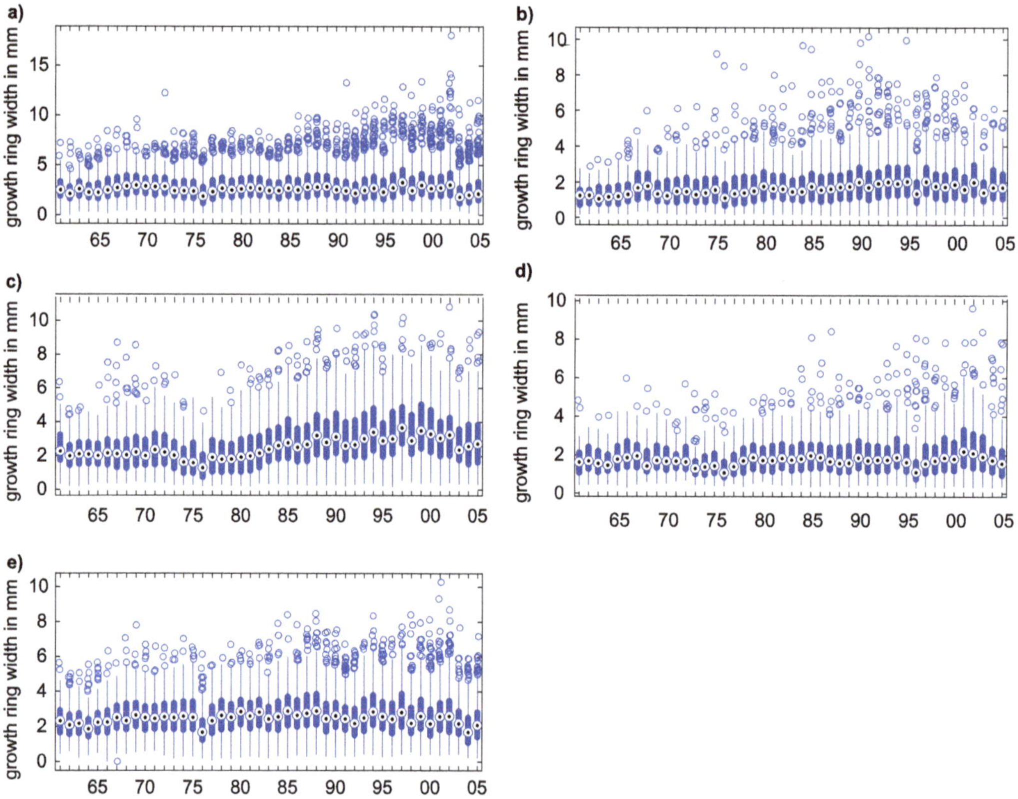

Fig. 11.1 Boxplots of the annual growth ring widths for spruce (**a**), pine (**b**), fir (**c**), oak (**d**) and beech (**e**) from 1961 to 2005. Note that years 1960 to 1990 include measurements only from Baden-Wuerttemberg

drill cores were taken at plots located on brown soils. Soils that were affected by groundwater and stagnant water were not included in further analysis. In Baden-Wuerttemberg, the widths of annual growth rings were measured for the entire core up to the centre of the stem, whereas in other federal states only the last 15 annual growth rings prior to sampling (2006–2008) were available (Thormann 2014). All further analysis refers to the period from 1961 (beginning of the LWF-Brook90 modelling) to 2005. In total, this led to growth ring measurements from 299 plots with on average 2–3 trees per plot; Fig. 11.1 gives an overview of the time series used. In addition to the time series of the absolute annual growth ring widths, various adjusted growth trends and standardized time series were examined for their correlations with climate and water balance variables. Correlations with absolute annual growth ring widths were greatest; hence, only these are discussed below.

The annual growth ring data were linked to the results of the soil water balance simulations with LWF-Brook90 and other climatic variables (see Chap. 10). A total of 134 different climate and soil water variables were assigned to each annual growth ring for the corresponding NFSI plot and year. Table 11.1 gives an overview of the correlations between the annual growth ring widths and some of the climate and water regime variables. Many of these correlations were statistically significant due

Table 11.1 Correlations (Pearson coefficient) between annual growth ring widths and characteristic values for climate and soil water availability at the NFSI plots

			Spruce	Pine	Silver fir	Oak	Beech
Air temperature	Minimum temperature in May [°C]	tmin_may	0.054	0.079	0.137	0.178	0.112
	Mean annual temperature [°C]	tmean_y	0.054	0.088	0.148	0.178	0.113
	Temperature sum in veg. period [°C]	tsum_vp	n.s.	0.063	0.139	0.179	0.138
	Number of days > 5 °C [d]	gdd5_y	0.074	0.105	0.165	0.166	0.107
	Day degree sum > 5 °C [°C]	gdd5_tsum_y	0.054	0.075	0.158	0.194	0.130
	Begin of vegetation period [Julian day]	vp_start	−0.060	−0.111	−0.145	−0.122	−0.108
	Duration of vegetation period [d]	vp_dauer	0.056	0.103	0.130	0.121	0.114
Water balance	Annual precipitation sum [mm]	prec_y	n.s.	0.139	n.s.	n.s.	0.049
	Precipitation sum in veg. period [mm]	prec_vp	0.039	0.144	n.s.	n.s.	0.167
	Precipitation sum May to July [mm]	prec_mayjul	0.036	0.118	n.s.	0.090	0.177
	Grass reference evapotranspiration [mm]	et0_vp	0.029	n.s.	0.151	n.s.	0.125
	Climatic water balance [mm]	kwb_vp	0.034	0.089	n.s.	n.s.	0.099
	Deep percolation [mm]	vrfln_vp	n.s.	n.s.	−0.056	0.263	0.104
Soil water	Plant available soil water 0-100 cm [mm]	$S_{p0,100}$_y	n.s.	n.s.	n.s.	0.289	0.127
	Plant available soil water 0-100 cm [mm]	$S_{p0,100}$_vp	n.s.	n.s.	n.s.	0.323	0.140
	Plant available soil water 0-10 cm [mm]	$S_{p0,10}$_y	−0.076	0.120	0.045	n.s.	0.106
	Plant available soil water 0-10 cm [mm]	$S_{p0,10}$_vp	−0.071	0.101	n.s.	n.s.	0.164
	Plant available soil water 10-30 cm [mm]	$S_{p,1030}$_y	−0.071	0.070	n.s.	n.s.	0.089
	Plant available soil water 10-30 cm [mm]	$S_{p,1030}$_vp	−0.058	n.s.	n.s.	0.185	0.126
	Plant available soil water 30-60 cm [mm]	$S_{p,3060}$_y	n.s.	0.038	n.s.	0.259	0.124
	Plant available soil water 30-60 cm [mm]	$S_{p,3060}$_vp	n.s.	n.s.	n.s.	0.280	0.121

(continued)

Table 11.1 (continued)

			Spruce	Pine	Silver fir	Oak	Beech
Soil water	Plant available soil water 60-90 cm [mm]	$S_{p,6090}$_y	0.039	n.s.	n.s.	0.314	0.097
	Plant available soil water 60-90 cm [mm]	$S_{p,6090}$_vp	0.033	n.s.	n.s.	0.285	0.098
	Total soil water 0-100 cm [mm]	$S_{t,0100}$_y	n.s.	0.051	0.071	0.203	0.119
	Total soil water 0-100 cm [mm]	$S_{t,0100}$_vp	n.s.	0.042	0.067	0.228	0.126
	Total soil water 0-10 cm [mm]	$S_{t,010}$_y	−0.089	0.099	0.123	−0.131	0.157
	Total soil water 0-10 cm [mm]	$S_{t,010}$_vp	−0.088	0.089	0.094	n.s.	0.192
	Total soil water 10-30 cm [mm]	$S_{t,1030}$_y	−0.078	0.105	0.071	−0.153	0.086
	Total soil water 10-30 cm [mm]	$S_{t,1030}$_vp	−0.070	0.074	0.056	n.s.	0.108
	Total soil water 30-60 cm [mm]	$S_{t,3060}$_y	n.s.	n.s.	0.051	0.113	0.077
	Total soil water 30-60 cm [mm]	$S_{t,3060}$_vp	n.s.	n.s.	0.054	0.127	0.073
	Total soil water 60-90 cm [mm]	$S_{t,6090}$_y	0.043	n.s.	0.068	0.360	0.112
	Total soil water 60-90 cm [mm]	$S_{t,6090}$_vp	0.039	n.s.	0.066	0.348	0.112
Water shortage	Number of days with $\psi w < -1200$ hPa [d]	d_Ψw_{1200}_vp	−0.043	−0.107	n.s.	−0.171	−0.157
	Integral sum of matric potential below $\psi w < -1200$ hPa [MPa]	v_Ψw_{1200}_vp	0.078	0.119	0.066	n.s.	0.188
	Number of days with REW < 0.4 [d]	d_REW40_vp	−0.032	−0.083	n.s.	−0.151	−0.159
	Integral sum of relative plant available soil water below REW < 0.4 [mm]	v_REW40_vp	−0.069	−0.111	−0.062	n.s.	−0.180
	Transpiration ratio [mm mm^{-1}]	Tratio_vp	0.053	0.010	n.s.	n.s.	0.143
	Tratio sum below Tratio < 0.5 [mm mm^{-1}]	v_Tratio50_vp	−0.066	−0.101	n.s.	n.s.	−0.142
	Transpiration difference [mm]	Tdiff_vp	−0.068	−0.106	n.s.	n.s.	−0.155

Significance levels of the correlations: dark grey = $p < 0.001$, light grey = $p < 0.01$, white = not significant; white numbers: covariable in boosted regression trees. Variable ending _y: means/totals over the whole year; _vp: dynamic vegetation period from LWF-Brook90 to the large sample size, but the correlations between annual growth rings and climate and water regime variables were generally weak; i.e. they statistically differed from zero only by small amounts. In Table 11.1, variables showing very weak correlations for all tree species were omitted for a better overview. There were clear differences between the tree species considered: soil water availability in particular was associated with secondary growth in beech trees. Both the absolute soil water storage (S_t, S_p) and the derived water shortage indices were significantly correlated with annual growth ring widths of beech. Oak also showed an association with soil water capacity. Compared to beech, however, lower depths (30–60, 60–90 cm) play a greater role. Correlations with water scarcity indices were, in most cases, not significant for oak. However, oak appeared to benefit from more frequent excess of water, as the comparatively strong correlation to seepage values suggested. Among the conifer species, spruce had the strongest correlation between annual growth ring width and soil water retention. Pine showed a stronger correlation with precipitation totals, while the secondary growth of fir was mostly correlated with temperature variables.

The mean values of effective root depth compared to mean values for NFSI or standard depth (0–100 cm) were not more strongly correlated with annual growth ring widths. Water scarcity indices derived from the modelled soil water contents and matrix potentials were also only partly more strongly correlated than the absolute soil water storage values. Of the water scarcity indices considered, the shortfall of a critical matrix potential in the root space ($v\_\Psi w_{1200}\_vp$) showed the closest relation to secondary growth. $v\_\Psi w_{1200}\_vp$ was highly significantly correlated with annual growth ring widths for all tree species with the exception of oak.

Based on a preselection of possible explanatory variables (Table 11.1), boosted regression trees (BRTs) were used to estimate the annual growth ring widths of a given tree species as a function of climate and soil water variables (R version 3.3.1, package "dismo"; Elith et al. 2008). Only variables that were statistically significantly correlated with width of annual growth rings and whose functional correlation depicted in the BRTs was useful and justifiable in soil science and plant physiology were permitted as covariables in the BRTs. The BRTs explained between 19.3% (spruce) and 61.6% (oak) of the variance in the measured annual growth ring widths. The explanatory grades of the BRTs for beech (35.1%), pine (37.1%) and fir (26.0%) were similar. Figure 11.2 gives an overview of the covariables considered in the BRTs and their relative influence on the explained variance.

Data on soil water capacity and the resulting water shortage indices were included as covariables in all BRT models. As expected, lower soil water availability or more pronounced dry periods led to a decline of annual growth ring widths. This relationship was particularly evident in spruce and beech. Similar reactions were observed by Alavi (2002) and Scharnweber et al. (2011). Together, covariables describing water availability were responsible for 48% (oak) to 100% (spruce) of the variance indicated by the BRTs. With the exception for the spruce model, air temperature was another important covariable: an increase in the width of annual growth rings with rising temperatures was observed in the lower temperature ranges, while at higher temperature ranges, a decline or plateau of the annual growth ring widths was

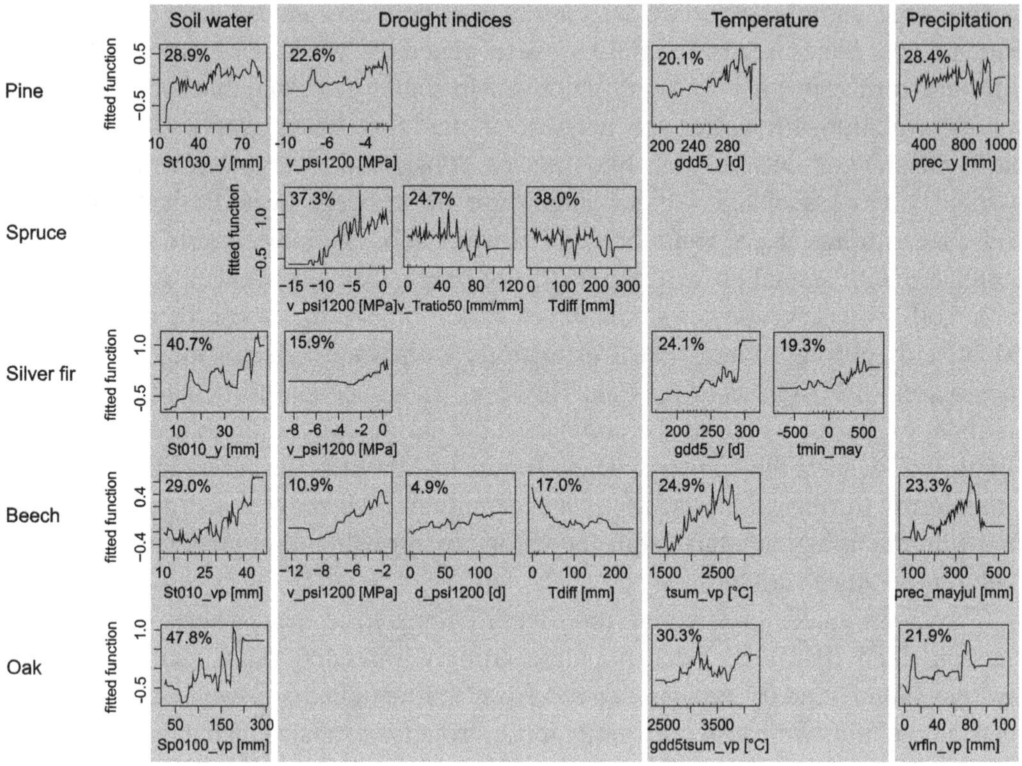

Fig. 11.2 Relationship between prediction of the model (y-axis) and BRT covariables (x-axis) for the various tree species. Percentage: share of the covariable in the variance of annual growth ring widths declared by the BRT

evident for all tree species. Precipitation (for pine and beech) or rather seepage (for oak) was another covariable of the BRTs.

11.3 Defoliation Development Types and Associated Risk Factors

The Forest Condition Survey is mandatory across Europe and has been conducted annually on a 16 × 16 km grid throughout Germany since 1990. Between 2006 and 2008, the Forest Condition Survey took place on the denser grid of the NFSI II [mainly 8 × 8 km; with exception of the federal states Rhineland-Palatinate (4 × 12 km + 16 × 16 km), Saarland (2 × 4 km) and Schleswig-Holstein (8 × 4 km)]. Data from the corresponding denser grids were also available for Baden-Wuerttemberg, Hesse, Lower Saxony and Saxony-Anhalt from 2005 to 2015, for Mecklenburg-Western Pomerania from 1991 to 2015, for Rhineland-Palatinate additionally from 2009 to 2010 and for Saarland from 2009 to 2015. A partial dataset based on the denser grid was also provided by Bavaria from 2009 to 2015. In addition, changes of the grid over time occurred, such as shifts of the initial grid to coincide with the grid of the National Forest Inventory in Bavaria in 2006 and in

Brandenburg in 2009. Tree defoliation represents the main parameter assessed during the Forest Condition Survey. The estimation of defoliation takes place visually using binoculars, and defoliation is recorded in 5% classes from 0% (no defoliation) to 100% (dead tree). In addition to defoliation, several other parameters (e.g. insect infestation and fructification) are investigated. A detailed description of the survey and quality assurance can be found in Wellbrock et al. (2016) for Germany and in Eichhorn et al. (2016) for Europe. The objectives of this section include (1) identifying regions of similar level and temporal development of defoliation (age-independent defoliation development types) for the four main tree species of Germany and (2) determining reasons for differences in defoliation among defoliation development types. Particular focus was placed on regions at risk of high levels of tree defoliation and risk factors. In Sect. 11.3.1 defoliation development types are defined, in Sect. 11.3.2 variables associated with defoliation are determined, and in Sect. 11.3.3 an integrated analysis of defoliation development types and associated variables is conducted.

11.3.1 Defining Age-Independent Defoliation Development Types

In the first step, defoliation development types were defined; these types characterize regions of similar level and temporal development of defoliation. The definition of defoliation development types posed two problems. First, complete time series of defoliation of a plot were necessary. Complete time series could only be available for the 16 × 16 km basic grid, if at all. However, changes of this grid have resulted in a loss of many plots, including all of Bavaria and Brandenburg. Hence, the original defoliation data were not very useful for defining defoliation development types for all of Germany. Second, if nevertheless the few complete time series were used for defining defoliation development types, the resulting defoliation development types mainly reflected tree age due to the strong species-specific dependence of defoliation on age (Eickenscheidt et al. 2016, 2019). The aim, however, was to determine age-independent defoliation development types, since regions at risk of high levels of defoliation and risk factors which could be attributed to factors other than age, e.g. human activities, were the primary focus. Thus, age adjustment was necessary to make an age-independent statement about regions at risk. In order to achieve both goals (complete time series and age adjustment) at once, we draw on our spatio-temporal models for defoliation, which were also developed as part of the evaluations of the NFSI II data. A detailed description and results can be found in Eickenscheidt et al. (2016, 2019). Spatio-temporal modelling was conducted by species using generalized additive mixed models (GAMMs; Augustin et al. 2009; Lin and Zhang 1999; Wood 2006a, b):

$$\log_{it} E(y_{it}) = \log_{it}(\mu_{it}) = f_1(\text{stand age}_{it}) + f_2(\text{easting}_i, \text{northing}_i, \text{year}_t) \quad (11.1)$$

where y_{it} is the mean defoliation of one of spruce, pine, beech or oak for sample plot $i = 1, \ldots, n$ and for year $t = 1, \ldots, 26$, averaged over all trees of the respective species at sample plot i. Before averaging, the defoliation class of a single tree was converted into a continuous variable by using the midpoint of the class. The logit link was used since defoliation represents an estimated percentage ranging between 0 and 100; hence defoliation was divided by 100, and the logit link ensured that fitted values were bounded in (0,1). A one-dimensional smooth function of stand age was applied using a penalized cubic regression spline basis for smoothing. A three-dimensional smooth function of year and of coordinates (easting and northing of the Gauß-Krüger coordinate system, GK4) was used; this is a tensor product smoother constructed from a two-dimensional marginal smooth for space and a marginal smooth for time (Augustin et al. 2009). The marginal bases were a two-dimensional thin-plate regression spline basis for easting and northing and a cubic regression spline basis for year. The tensor product of the two marginal smooths was chosen so that different penalties for space (metre) and time (years) were used (Wood 2006a, b). A normal distribution was assumed for the error term. The number of trees per plot was considered as weights. The temporal correlation was modelled by a first-order autoregressive-moving average process (ARMA(1,1)) (Pinheiro and Bates 2000). All evaluations were performed using R 3.4.1 (R Core Team 2017). The R package mgcv (Version 1.8-18; Wood 2017) was utilized for spatio-temporal modelling of defoliation. Although the models were adequate, the model approach has inherent uncertainties. Models generally have model errors, although here a substantial part of the total variance was explained [adjusted R^2 for spruce, 0.54 ($n = 10182$); pine, 0.41 ($n = 9252$); beech, 0.47 ($n = 9283$); and oak, 0.47 ($n = 6098$)]. To take this uncertainty into account, defoliation time series for each plot and tree species were repeatedly simulated (40 times) from the predictive distribution of defoliation, and cluster analysis was then carried out for each of the simulated time series. The predictive distribution was obtained by sampling from the multivariate normal posterior distribution of the model parameters, which itself was obtained by using Bayes' theorem (Augustin et al. 2009; Silverman 1985; Wahba 1983; Wood 2006a, c). Time series were simulated for each plot of the densified grid of 2008 (mainly 8 × 8 km). Stand age was assumed to be 70 years for spruce and pine and 90 years for beech and oak, which roughly corresponded to the weighted median age of the species of the basic grid in 2008. Based on the simulated time series, 40 model-based cluster analyses (R package mclust, version 5.3; Fraley et al. 2017) were conducted for each species. The number of clusters ranged between seven and nine. The assigned 40 clusters per plot and species were summarized in a string and the pairwise string comparison resulted in a string-distance matrix (restricted Damerau-Levenshtein distance; R package stringdist, version 0.9.6.4; van der Loo et al. 2017). Subsequently, a hierarchical cluster analysis was performed using the string-distance matrix. Plots having 80% agreement in the initial 40 cluster analyses were assigned to one cluster using the hierarchical cluster analysis. For the resulting clusters, time trends (median) and Bayesian credible intervals (2.5 and 97.5 percent quantiles) were estimated based on the 40 simulations. Clusters with similar levels of defoliation, identical characteristic peaks and similar time trends were further combined to one cluster. Plots that had not been assigned to a cluster in the first step were then also assigned to clusters in this way. Assignment of these plots was clear, with the exception of four pine and oak plots, which were assigned to the spatially adjacent cluster. Our approach thus led to nine clusters (defoliation development types) for spruce, beech and oak and ten clusters for pine (Fig. 11.3). Finally, these original defoliation development types could be summarised to five broad defoliation development types for each tree species (Fig. 11.3). The summary was based on relatively similar levels of defoliation, identical characteristic peaks and relatively similar time trends. In Fig. 11.3 it can be seen, which of the original clusters were summarized to a broad cluster. The median time trends and Bayesian credible intervals (2.5 and 97.5 percent quantiles) were again estimated based on the 40 simulations, assuming a stand age of 70 years for spruce and pine and 90 years for beech and oak.

Large-scale rather than small-scale spatial defoliation development types were detected for the four main tree species (Fig. 11.3). Clear north-south and east-west differences were found in the defoliation development (Figs. 11.4, 11.5, 11.6, 11.7). For all tree species beginning in 2004, the highest defoliation and strongest increase, respectively, were observed in the defoliation development types that included the south-western part of Germany (e.g. Black Forest). A similar but slightly weaker trend was found in the adjacent north-western regions of Germany (e.g. Rhineland-Palatinate, Saarland, Forest of Odes, Spessart). An opposite trend was shown for the defoliation development types that included the north-eastern part of Germany: at the beginning of the 1990s, high defoliation was reported from the eastern part of the North German Lowlands, but defoliation sharply decreased until the mid-1990s and since then has remained relatively constant and on a low level. A trend unlike the trend in south-western Germany was also observed for the defoliation development types including south-eastern Germany. In south-eastern Germany, defoliation was generally high at the beginning of the 1990s but decreased over time. In north-western Germany, defoliation of all tree species, with the exception of oak, was comparably low and showed only few temporal dynamics. Defoliation development types of pine and oak showed similar spatial distribution. Distribution of defoliation development types of spruce and beech was also similar but slightly different from the types found for pine and oak. Differences mainly occurred in the North German Lowlands (where only one cluster including the north and east of the Lowlands was observed) and southern Germany (where a smaller cluster in south-western Germany but a larger cluster in south-eastern Germany ranging to Saxony was observed).

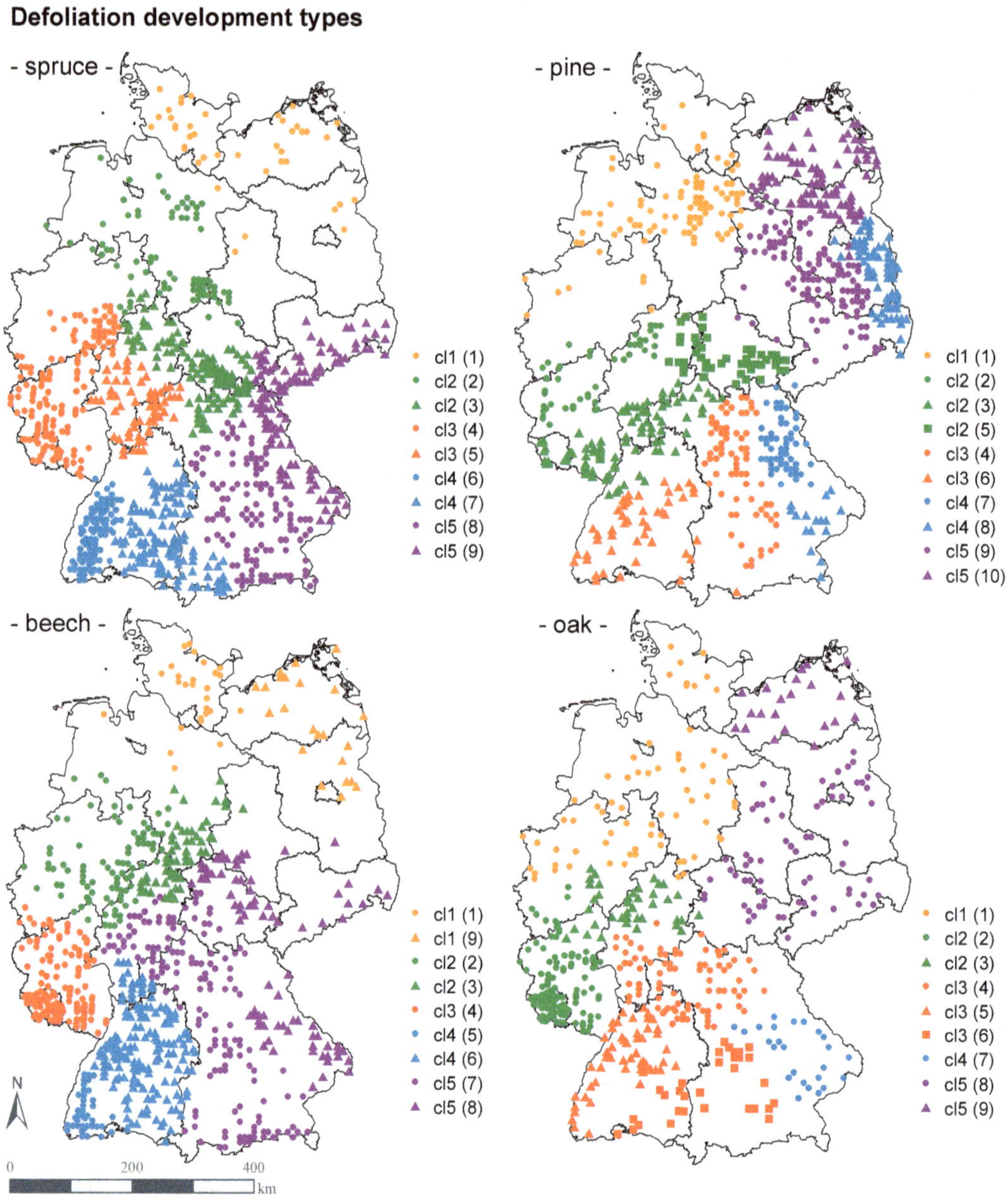

Fig. 11.3 Regional distribution of age-independent defoliation development types (cluster) of spruce (top, left), pine (top right), beech (bottom, left) and oak (bottom, right). The colours indicate the five broad defoliation development types (cl1 to cl5). The combinations of colour and symbol indicate the nine and ten (number in brackets) original defoliation development types, respectively

11.3.2 *Variables Associated with Defoliation*

In a second step, variables associated with defoliation were investigated on a species by species basis. Unlike Sect. 11.4, where discriminant analysis was used to investigate which site-specific and environmental variables were decisive for assignment to a specific nutrition type, variables for defoliation were initially examined

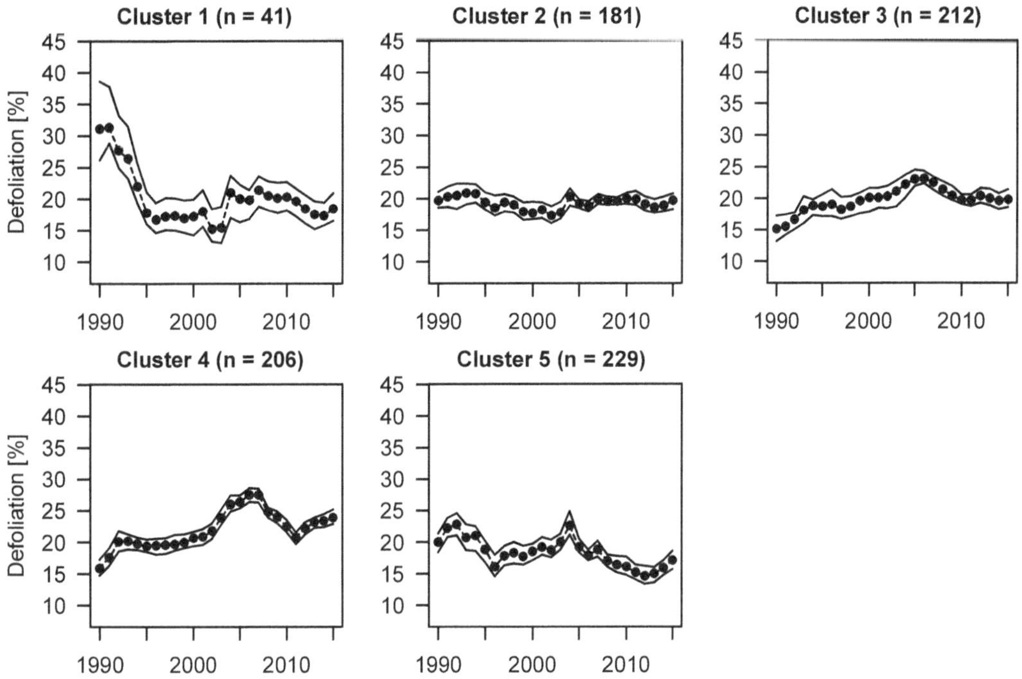

Fig. 11.4 Mean time trends of defoliation and credible intervals of the five broad defoliation development types of spruce. The time trends were estimated based on 40 simulations and assuming a stand age of 70 years. The sample size is given in parenthesis

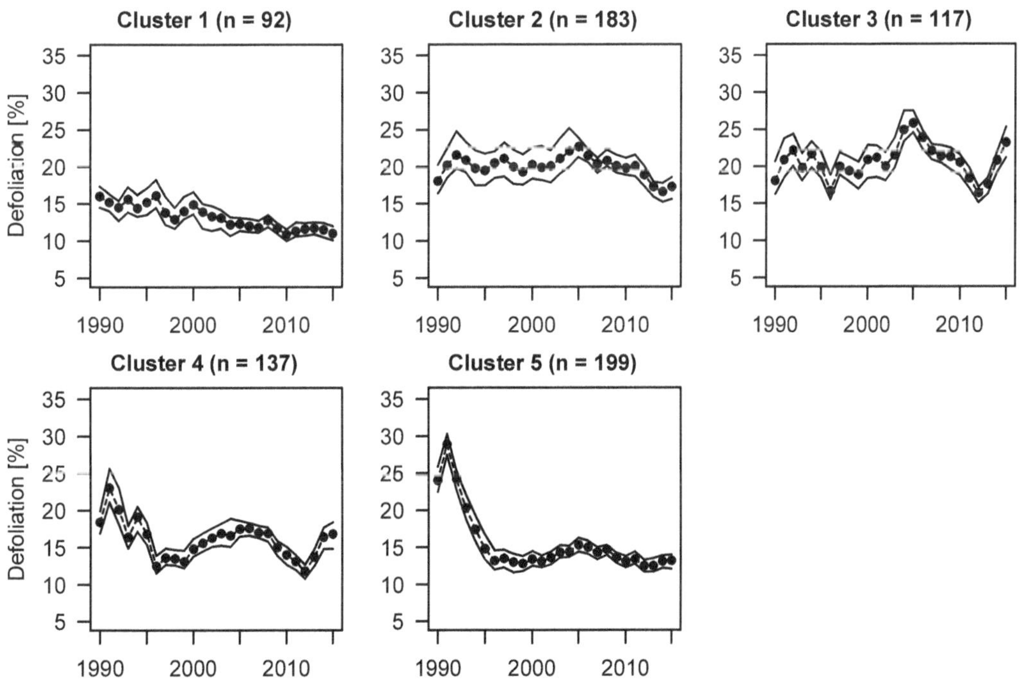

Fig. 11.5 Mean time trends of defoliation and credible intervals of the five broad defoliation development types of pine. The time trends were estimated based on 40 simulations and assuming a stand age of 70 years. The sample size is given in parenthesis

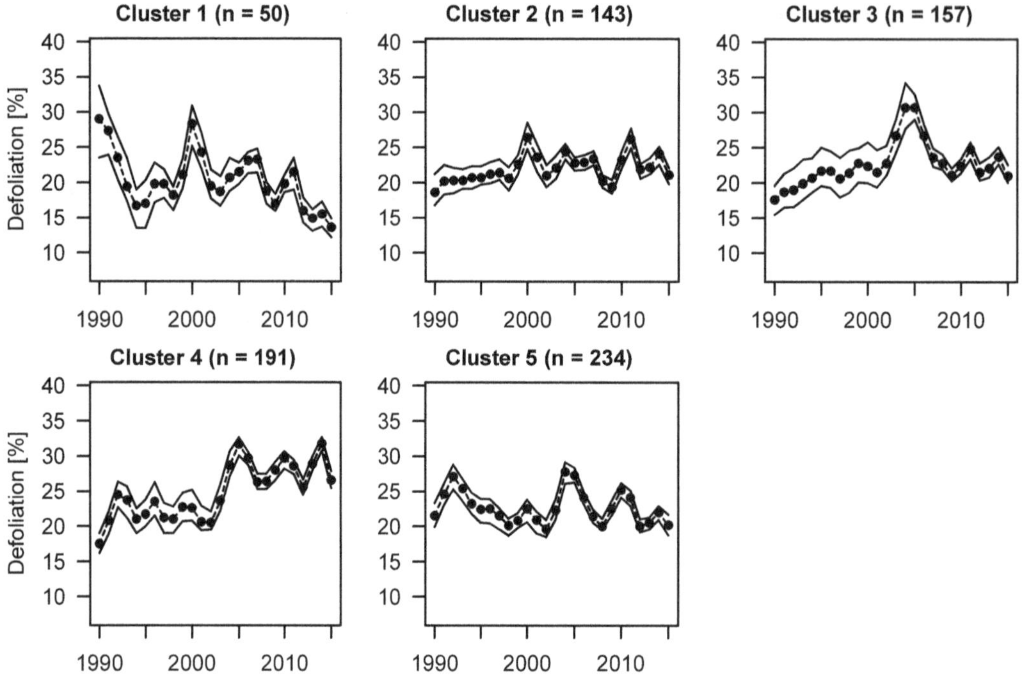

Fig. 11.6 Mean time trends of defoliation and credible intervals of the five broad defoliation development types of beech. The time trends were estimated based on 40 simulations and assuming a stand age of 90 years. The sample size is given in parenthesis

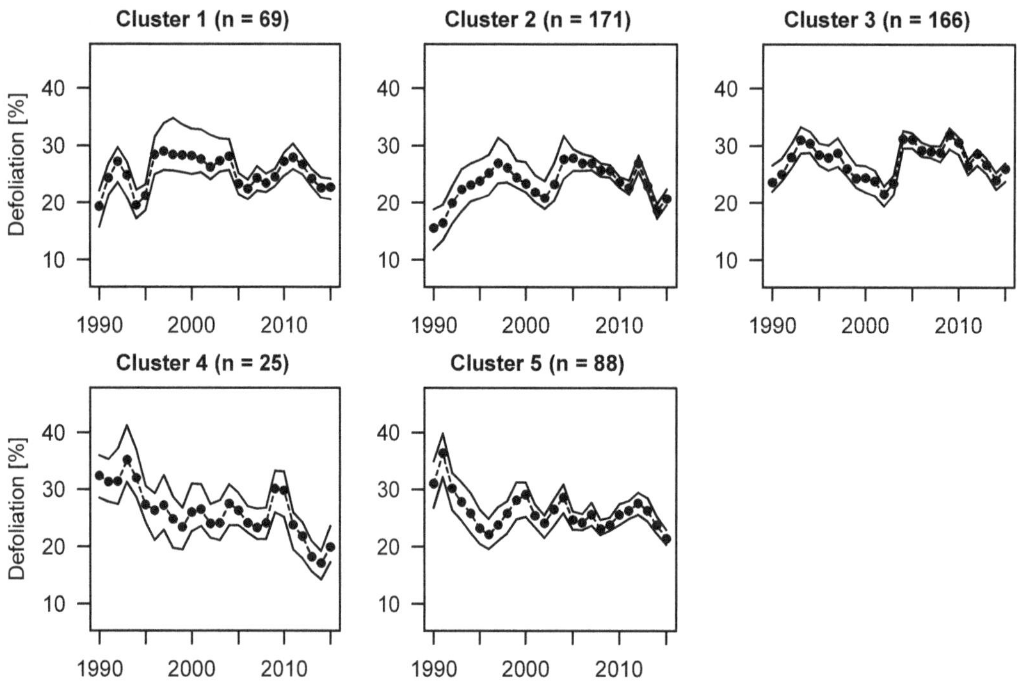

Fig. 11.7 Mean time trends of defoliation and credible intervals of the five broad defoliation development types of oak. The time trends were estimated based on 40 simulations and assuming a stand age of 90 years. The sample size is given in parenthesis

independently from the defoliation development types since annually varying time series of defoliation formed the basis of these types. As the data are observational, we could not make any inference on causality. Here we have explored which of the available explanatory variables had a statistical effect (rather than a causal effect) on defoliation via model selection. This exploratory analysis resulted in a list of variables that were strongly associated with defoliation and hence might be important in explaining the process leading to defoliation. Associated variables were determined for (1) 1991–2010 (referred to as time series) and (2) the period of the NFSI II (referred to as NFSI period). The two approaches were chosen due to differences in the spatial and temporal resolution and due to differences in the number of available variables. In the first case (time series), the annual defoliation values of the plots of the 16 × 16 km basic grid were considered. In the second case (NFSI period), the denser grid of the NFSI II was used, and a mean defoliation value for 2006 to 2008 was calculated for each plot and species. Species-specific defoliation of the plots remained relatively constant between 2006 and 2008, with the exception of beech, where higher defoliation was observed at several plots in 2006, probably due to pronounced fructification. For the time series, fewer variables and plots per year were available, but information was provided annually. Potential influencing variables which were available included stand age, fructification (only from 1999), insect infestation, deposition and weather conditions including deviations from the long-term mean (1961–1990). Lag effects were also considered by using the previous year's values. The soil water balance values regarded for tree growth (see Sect. 11.2) were ultimately not used for defoliation because values were not available for all plots due to shifts in grids and because weather conditions proved equally suitable. For the NFSI period, a large number of variables were available, primarily originating from the NFSI II. In addition to the variables considered for the time series listed above, which were averaged for the time period, additional variables included parameters of soil condition (e.g. C/N ratio, base saturation, stocks of total and exchangeable nutrients, heavy metal stocks), forest nutrition and accompanying information (e.g. liming). GAMMs were used for analyses of association with defoliation. Thus, it was possible to include categorical factors as well as continuous variables and to detect linear as well as non-linear effects. Weights and temporal autocorrelations could also be considered. Examples and a detailed description of the model selection process can be found in Eickenscheidt et al. (2016). In brief, co-linearity among variables was considered, and forward selection and the Bayesian information criterion (BIC; Schwarz 1978) were used for model selection. Model residuals were checked. Stand age represented the most important variable associated with defoliation by far; thus, stand age was included in every model from the beginning.

Defoliation increased species-specific with stand age. For spruce and in particular for beech, defoliation increased nearly linearly, whereas for oak and in particular for pine, defoliation clearly increased until a stand age of approximately 60 years and 40 years, respectively, and only little association of defoliation with age occurred for older trees (Eickenscheidt et al. 2016, 2019).

Table 11.2 Results of the final models for the four main tree species of Germany (1) for the time series (1991–2010) and (2) for the NFSI period (mean 2006–2008)

	Spruce	Pine	Beech	Oak
Time series:				
Stand age [year]	***	***	***	***
Fructification [classes]			***[a]	
Insect infestation [yes/no]		***		***
NH_x deposition [kg N ha^{-1} $year^{-1}$]		***		
temp [°C]		***		
prec_py [mm $year^{-1}$]			***	
te(temp_dev, prec_dev) [°C] [mm $year^{-1}$]		***	***	
te(temp_py_dev, prec_py_dev) [°C] [mm $year^{-1}$]	***		***	
te(temp, temp_dev) [°C] [°C]	***			***
R^2	0.49	0.22	0.37/0.44[a]	0.33
n	3910	3729	3121/1932[a]	1748
NFSI period:				
Stand age [year]	***	***	***	***
Fructification [yes/no]			**	
Insect infestation [yes/no]				***
N deposition [kg N ha^{-1} $year^{-1}$]	***			
temp [°C]	***			
temp_dev [°C]	*			
prec [mm $year^{-1}$]	**			
te(temp, prec) [°C] [mm $year^{-1}$]		***		
et [mm $year^{-1}$]		***	**	***
et_dev [mm $year^{-1}$]	**		*	***
C stock of organic layer [t C ha^{-1}]	**			
C/N ratio in 0–5 cm soil depth [–]	**	***		
C/P ratio in 0–5 cm soil depth [–]				***
N concentration of needles/leaves [g kg^{-1}]			***	
R^2	0.54	0.37	0.46	0.44
n	756	666	463	282

The significance of variables associated with defoliation, the coefficient of determination (R^2) of the final model and the sample size (n) are indicated

temp temperature, *prec* precipitation, *et* evapotranspiration, *_dev* deviation from the long-term mean 1961–1990, *_py* previous year's value, *te()* tensor product()

*$p < 0.05$; **$p < 0.01$; ***$p < 0.001$; [a]Model only for 1999–2010

11.3.2.1 Time Series

For the time series, weather conditions had a direct association with annual defoliation of all four tree species (Table 11.2). Deviations of annual mean temperature and precipitation sums from the long-term mean played a major role. Seidling (2007) also reported relationships between defoliation and deviations from long-term means of temperature and precipitation for all four tree species of the German 16 × 16 km

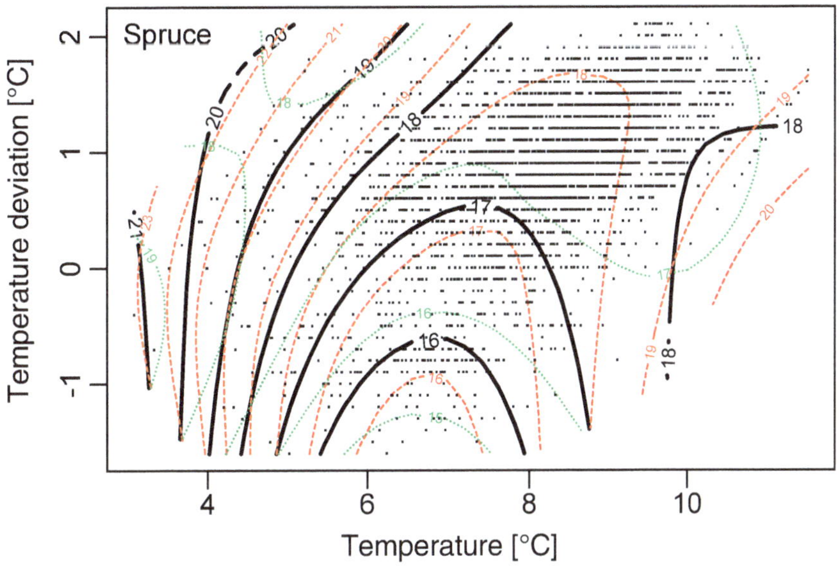

Fig. 11.8 Relationships of temperature and deviations of temperature and their interaction to defoliation of spruce. Black contour lines indicate defoliation in percent. Red and green lines indicate the corresponding standard error of defoliation. Black points reflect the sample distribution

grid. Lagged effects, especially drought in the previous year, and cumulated drought in several preceding years were shown to be closely linked to defoliation in the following year (Ferretti et al. 2014; Klap et al. 2000; Seidling 2007; Zierl 2004); this result was corroborated by our findings.

In the following the results are described for the four tree species, and some figures are presented as examples.

For Norway spruce, mean temperature of the recent year and the deviation of this temperature and the interaction were associated with defoliation of spruce. Higher defoliation in general was found for plots with lower mean temperatures than for plots with higher mean temperatures (Fig. 11.8). Defoliation was highest when positive temperature deviations also occurred on these plots with lower mean temperatures. Lowest defoliation occurred at negative temperature deviations at annual mean temperatures between 6 and 8 °C. Furthermore, previous year's deviations of temperature and precipitation and their interaction had an association with defoliation of the recent year. Years that were cooler and drier compared to the long-term mean resulted in lower defoliation, but warmer and drier years resulted in increased defoliation (compare to beech and Fig. 11.9). Warmer and wetter years were associated with almost no changes in defoliation of spruce. Seidling (2007) also reported higher defoliation of spruce after high temperature (and low rainfall) in the previous and also in the current year, in particular following 2003. For the conifers, higher defoliation especially in the summer 1 year after drought might be attributed to higher needle fall in autumn of the drought year (Solberg 2004). In general, needle loss is still visible years after the event because conifers keep several needle sets.

For Scots pine, recent year's deviations of temperature and precipitation and their interaction showed an association with defoliation. Defoliation of pine was highest

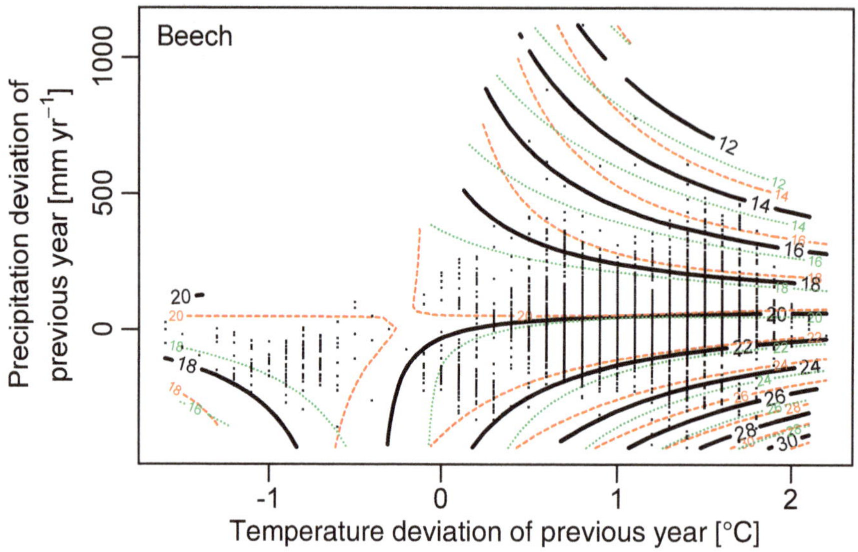

Fig. 11.9 Relationship of previous year's deviations of temperature and precipitation and their interaction with defoliation of beech of the recent year. Black contour lines indicate defoliation in percent. Red and green lines indicate the corresponding standard error of defoliation. Black points reflect the sample distribution

when the temperature deviation was approximately 2 °C (which corresponded to the highest temperature deviations observed), and precipitation did not deviate. In cooler and wetter years, decreased defoliation of pine was observed. Defoliation further decreased nearly linearly with increasing annual mean temperature. Pine trees in Germany are commonly known to be relatively drought-tolerant (Ellenberg 1996), for example, due to a deep taproot system and early and rapid stomata closure (Seidling 2007 and references therein). However, notable temperature surplus accompanied by precipitation deficits, as observed in 2003, was associated with visible drought stress even in pine trees, which was also reported by Seidling (2007). Besides direct effects of weather, indirect effects were also associated with defoliation. In pine, insect infestation was associated with increased defoliation, another result also documented by Seidling (2001) and Seidling and Mues (2005). Severe infestation is most likely a result of exceptional climatic situations, which favour the development of insects and simultaneously make trees susceptible to infestation. Furthermore, in our study, a negative linear relationship of defoliation and NH_x deposition was observed.

For European beech, previous year's deviations of temperature and precipitation and their interaction had an association with defoliation of the recent year (Fig. 11.9) similar as found for spruce. However, warmer and wetter years were associated with decreased defoliation of beech which was different from spruce. Furthermore, recent year's deviations of temperature and precipitation and their interaction also showed an association with defoliation of beech but which was different from the association observed for pine. Similar to the effect of deviations of temperature and precipitation of the previous year, defoliation of beech increased in the event of positive temperature deviations and drier conditions, whereas cooler and drier years were associated with a decrease in defoliation. Cooler and wetter years also resulted in increased defoliation of beech. Additionally for beech, very low precipitation (<500 mm $year^{-1}$) and in particular very high precipitation (>1500 mm $year^{-1}$) of the previous year came along with increased defoliation. The latter was rarely found and was linked to high elevation (e.g. Alps, Black Forest), which was most probably an effect of low temperature and short vegetation period or oxygen deficiency within the soil. Sensitivity of beech to drought is well known, although drought resistance varies among beech populations (Bolte et al. 2016). Beech usually develops from natural rejuvenation and thus is adapted to site conditions. Increased defoliation of beech at low previous year summer or annual precipitation was also reported by Seidling (2006) for Germany and by de Marco et al. (2014) for Europe. Furthermore, above-average previous summer temperature was shown to have a negative association with defoliation of beech in Germany (Seidling 2007; Seidling et al. 2012). The present study additionally revealed that an increase in fructification resulted in increased defoliation of beech (see also Eickenscheidt et al. 2016). Both low precipitation and high temperature of the previous year might be attributed to fructification in the current year, as well as directly to drought stress in the previous year. Weather conditions in the previous early summer determine the production of flower buds and leaf buds, respectively. Hence, fructification is directly linked to higher defoliation but also indirectly because of deterioration of the branch structure as well as development of small leaves. Eichhorn et al. (2005) and Seidling (2007) also reported enhanced defoliation in mast years.

For pedunculate and sessile oak similar as for spruce, the mean temperature of the recent year and the deviation of this temperature and the interaction were associated with defoliation (Fig. 11.10). Lowest defoliation was primarily observed at similar

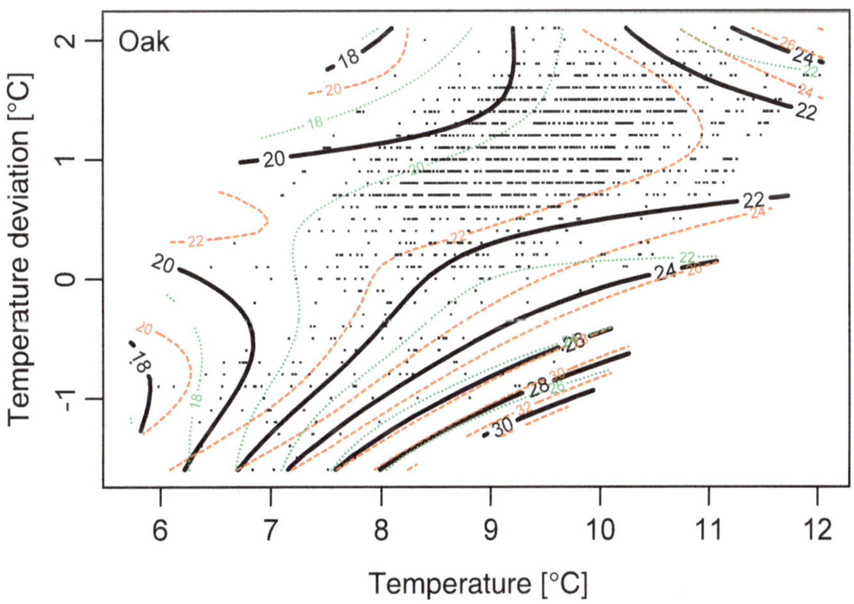

Fig. 11.10 Relationships of temperature and deviations of temperature and their interaction to defoliation of oak. Black contour lines indicate defoliation in percent. Red and green lines indicate the corresponding standard error of defoliation. Black points reflect the sample distribution

conditions as for spruce. However, mean temperatures were in general higher (6–12 °C) than for spruce (3.5–12 °C). Higher defoliation of oak was found at higher mean temperatures. On plots with higher mean temperatures, positive temperature deviations resulted in increased defoliation, although on plots with average or moderately high mean temperatures, negative deviations of temperature (−1 °C) were associated with the highest defoliations observed for oak. Oak tolerates a wide range of climatic conditions and soil water availability. It can be found on soils with stagnant soil water, but it is also known to be drought-tolerant due to its taproot system and fast stomatic response. However, similar to our findings, several studies have demonstrated a negative impact of high summer temperatures on defoliation of oak in Europe (de Marco et al. 2014; de Vries et al. 2014). At the same time, the observed strong reaction of oak to negative temperature deviations might indicate sensitivity to damage by late frost. Our study further indicated a strong association between defoliation and insect infestation of oak, which was corroborated for Germany by Eichhorn et al. (2005) and on the European scale by Seidling and Mues (2005).

11.3.2.2 NFSI Period

Analyses regarding the NFSI period further revealed soil and nutrient parameters as relevant variables aside from stand age and direct and indirect weather conditions (Table 11.2).

For spruce, a linear decrease in defoliation with increasing stocks of organic C in the organic layer was observed (not shown). The organic C in the organic layer reflects the total mass of the organic layer and thus the humus type. A large mass of organic matter might be less susceptible to drying up and protects trees against drought stress, which was also hypothesized by Seidling (2007). Furthermore, defoliation of spruce showed a negative linear relationship to N deposition (not shown). N deposition and N nutrition were generally closely linked and thus, N deposition might indicate the N nutrition status. In addition, an increase in defoliation with increasing C/N ratios in the 0–5 cm soil depth beginning at a C/N ratio of 30 occurred (Fig. 11.11). Ratios larger than 30 can usually be found in soils having low turnover of organic matter and low N and other nutrient supply.

Defoliation of pine also exhibited a relationship to the C/N ratio. However, the relationship was in contrast to the relationship detected for spruce: the highest defoliation was found at small C/N ratios (Fig. 11.11). Soil pH and exchangeable calcium (Ca) were highly correlated with the C/N ratio. Hence, both variables could be used in the GAMM of pine instead of the C/N ratio but had less explanatory power than the ratio. Soils with high pH values, high Ca stocks and high turnover rates (low C/N ratios) were accompanied by high defoliation, which might indicate antagonisms between Ca and other nutrients, especially potassium (K) [Ca-K antagonism; Evers and Hüttl (1992); Zech (1970)]. This antagonistic effect may be enhanced since calcareous soils are often shallow and prone to drought, which exacerbates the K uptake under drought stress (Evers and Hüttl 1992). Riek and

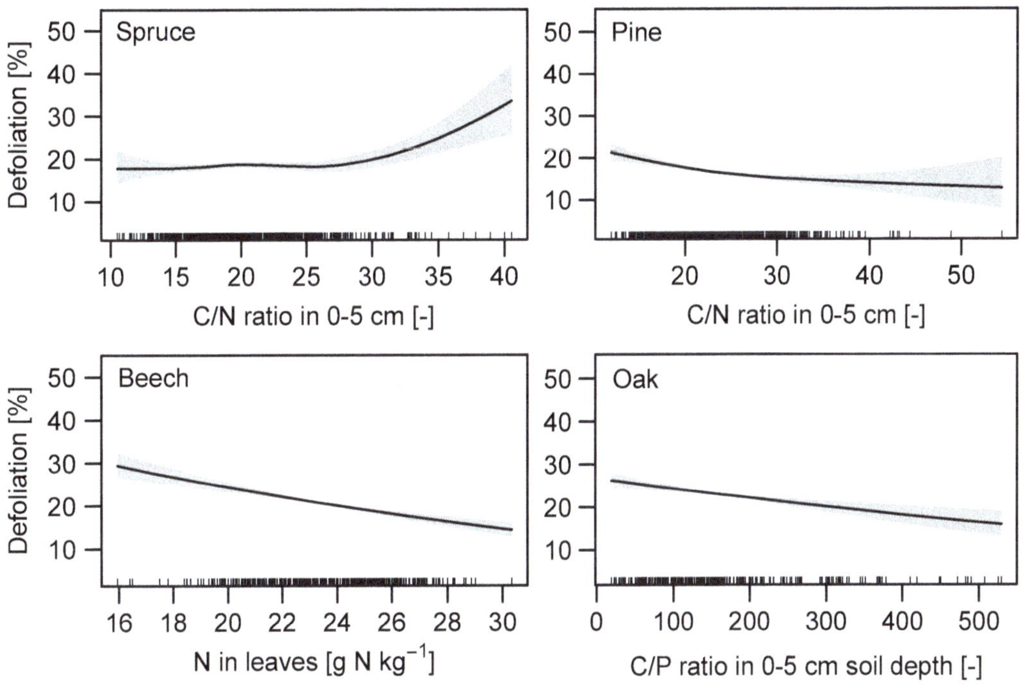

Fig. 11.11 Relationship of C/N ratio in the 0–5 cm soil depth to defoliation of spruce (top, left) and pine (top, right), of N concentration of leaves to defoliation of beech (bottom, left) and of C/P ratio in 0–5 cm soil depth to defoliation of oak (bottom, right). The lines at the *x*-axis reflect the observed values. The grey-shaded area indicates the 95% credible interval

Wolff (1999) also reported a negative correlation between defoliation of pine trees of the Level I sites and soil Ca stocks and needle concentrations of Ca, respectively, during the NFSI I. However, these soil conditions also might be an indicator of low mass of organic layer and thus little protection of the soil against drying up.

A negative linear relationship between defoliation and the N concentration of the leaves existed for beech (Fig. 11.11). Findings by Seidling (2004) for the German Intensive Forest Monitoring plots corroborate a negative correlation between foliar N supply and defoliation. Low N nutrition below the normal nutrition range according to Göttlein (2015; 19.0–25.0 g N kg^{-1}) probably resulted in N deficiency, which might limit tree growth. Interestingly, negative effects of N surplus on defoliation were not indicated.

For oak, a negative linear relationship between defoliation and the ratio of C to phosphorus (P) was observed (Fig. 11.11). The pH value and the organic C stock in the organic layer could also be used in the GAMM of oak instead of the C/P ratio since the variables were highly correlated. High pH values, low C stocks and low C/P ratios were accompanied by high defoliation, which again suggested that the relationship presumably was an indicator of the mass of the organic layer and thereby protection against drying up.

11.3.3 *Integrated Analysis of Defoliation Development Types and Associated Variables*

In the final step, the variables identified in the previous section to be associated with defoliation were further investigated and regarded at the level of the defoliation development types. Results of the previous section highlighted weather conditions and in particular their deviation from the long-term mean as important variables associated with defoliation. Thus, model-based cluster analyses were conducted separately for the time series of relative deviation from the long-term mean of annual mean temperature and annual precipitation sum for Germany. This was done to see whether the resulting clusters coincide with the defoliation development types from Sect. 11.3.1. The results of the model-based cluster analyses were combined by concatenating the cluster indices of the separate analyses to one a string. Subsequently, a hierarchical cluster analysis was performed using the string-distance matrix (see Sect. 11.3.1). The result revealed 11 different weather deviation clusters (Fig. 11.12). Although these weather deviation clusters were more differentiated, they showed similarities to the landscape regions of Germany (Fig. 11.12), which are derived from geomorphological, geological, hydrological, biogeographical and pedological characteristics.

The landscapes and their weather conditions are presented in brief. The North German Lowlands are subdivided into the north-western and the north-eastern Lowlands based on differences in climatic conditions especially regarding precipitation. The Central Upland Range is bordering in the south. The range is also divided into a western part (e.g. Rhenish Slate Mountains, Harz) and an eastern part (e.g. Thuringian Forest, Ore Mountains, Bavarian Forest). To the south(-western) of these low mountain ranges are the Southwest German Scarplands (e.g. Spessart, Franconian and Swabian Albs, Black Forest). In the south the Alpine Foreland and finally the Bavarian Alps follow. The North German Lowlands had on average the highest mean temperature of the landscape regions. The lowest mean temperatures could be found at high altitudes in southern Germany (e.g. Black Forest, Bavarian Forest) and especially in the Alps (data not shown). These regions were also characterized by high precipitation. The north-eastern Lowlands had the lowest mean precipitation, whereas the north-western Lowlands showed a maritime influence. All weather deviation clusters were characterized by a relative mean increase in temperature between 1990 and 2010 compared to the long-term mean (Table 11.3). The mean increase was lowest in the Lowlands (in particular in the eastern part) and highest in the south of Germany, especially in the Alps. Simultaneously, no relative changes in mean precipitation compared to the long-term mean were observed in south-western Germany, whereas precipitation slightly increased in south-eastern Germany and the most in the Lowlands (in particular in the eastern part) (Table 11.4).

The spatial patterns of the defoliation development types of oak (Fig. 11.3) matched well with the landscape regions (Fig. 11.12) and with the weather deviation clusters (Fig. 11.12). Thus, the results of oak are discussed in detail as an example.

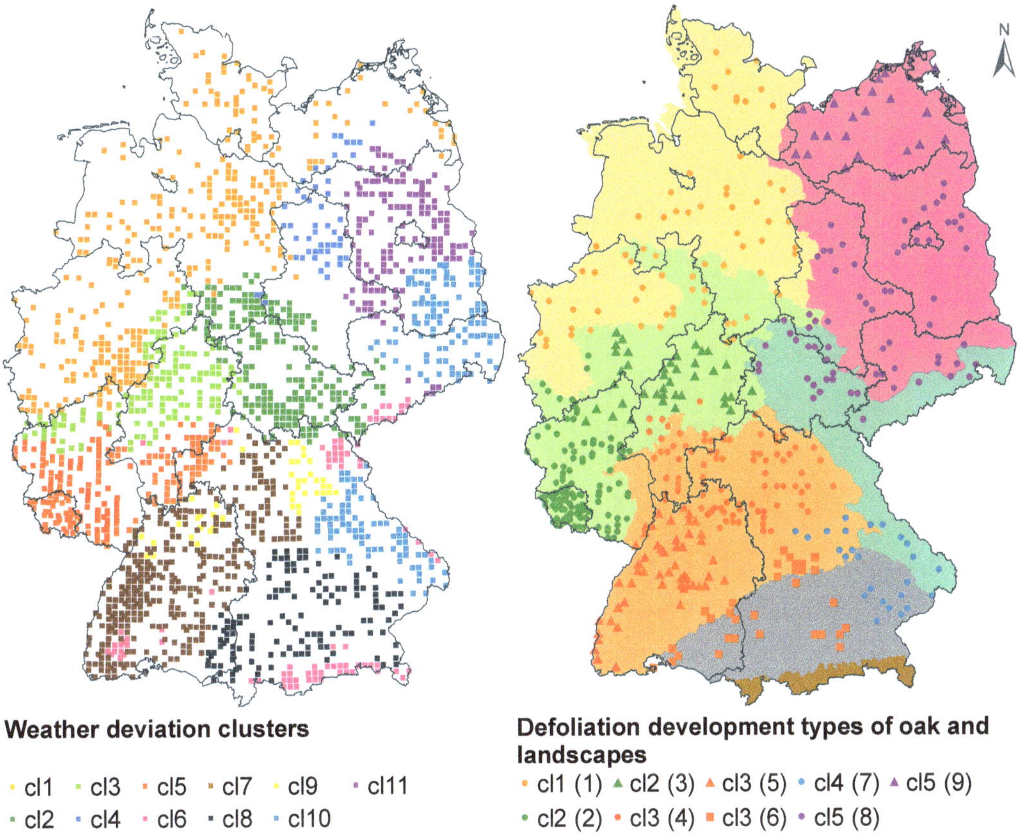

Fig. 11.12 Regional distribution of the 11 weather deviation clusters (left) and of age-independent defoliation development types of oak (Fig. 11.3) plotted together with the regional distribution of the landscape regions of Germany (right). The weather deviation clusters are based on the relative deviation of annual mean temperature and precipitation sum from the long-term mean (1961–1990). Defoliation development types: The point colours indicate the five broad defoliation development types (cl1–cl5) and the combinations of point colour and symbol indicate the nine original defoliation development types (number in brackets), respectively. Landscapes (background colour): north-western Lowlands (yellow), north-eastern Lowlands (pink), western Central Upland Range (green), eastern Central Upland Range (blue), Southwest German Scarplands (orange), Alpine Foreland (light grey), Alps (brown). (Data source of landscapes: Bundesamt für Naturschutz (BfN), supplied in July 2018)

Oaks grow from the northern Lowlands to the low mountain ranges but rarely in pronounced mountainous or cooler regions. In Sect. 11.3.2 it was shown that a further temperature increase (which was observed on average for all of Germany) resulted in an increase in defoliation in regions having high mean temperatures (Fig. 11.10). Defoliation in the warm Lowlands was on a relatively constant and high level since the mid-1990s (defoliation development type 1 and 5; Fig. 11.7). At the beginning of the time series (especially in 1991), defoliation of type 5 was notably higher than that of type 1. In 1991, the north-eastern Lowlands were notably drier on a relative basis than the north-western Lowlands. However, methodological differences in defoliation assessment after the introduction of the Forest Condition Survey in former eastern Germany, especially in the federal state Mecklenburg-Western

Table 11.3 Relative deviation [%] of mean annual temperature from the long-term mean (1961–1990) for the 11 weather deviation clusters

	1	2	3	4	5	6	7	8	9	10	11
1990	17	17	16	18	13	18	14	13	14	15	17
1991	3	2	1	3	3	1	2	0	2	1	3
1992	14	15	13	16	12	22	14	15	14	16	14
1993	1	2	4	2	6	9	7	7	7	5	2
1994	15	21	20	15	19	34	24	27	25	23	15
1995	8	8	10	7	11	9	9	9	9	9	6
1996	−14	−17	−13	−16	−9	−12	−9	−10	−10	−13	−16
1997	8	8	10	7	9	16	10	10	9	8	6
1998	9	11	10	10	8	14	12	13	13	13	9
1999	18	18	18	18	14	15	14	12	16	15	17
2000	19	23	20	20	19	29	22	22	23	23	20
2001	9	12	12	8	11	14	13	13	14	11	6
2002	15	17	17	13	17	29	19	21	19	18	12
2003	13	17	17	12	18	25	19	18	19	16	11
2004	9	9	8	9	7	10	9	9	10	9	8
2005	12	11	13	11	11	7	9	6	10	9	9
2006	18	19	18	19	16	21	17	14	17	16	17
2007	20	23	20	22	17	29	20	21	22	24	21
2008	15	18	14	19	11	21	15	17	17	19	18
2009	12	13	13	13	12	19	13	14	13	15	10
2010	−6	−7	−4	−5	−2	−3	−1	0	−2	−2	−6

Pomerania, cannot be ruled out as a reason for particularly high defoliation at the beginning of the 1990s (Riek and Wolff 1999). Since the mid-1990s defoliation was slightly lower in the eastern part of the Lowlands (defoliation development type 5), which might be attributed to a lesser relative increase in mean temperature and at the same time higher relative increase in precipitation compared to the western part (type 1). In addition, the western part was notably more frequently affected by insect infestation (data not shown). Defoliation development type 2 (western Central Upland Range) and type 3 [Southwest German Scarplands to Alpine Foreland (original defoliation development type 6)] showed similarities in particular regarding the strong increase in defoliation following 2003. In 2003, an exceptional drought and heat occurred in Germany that was most pronounced in south and south-western Germany. Defoliation development type 2 particularly was characterized by a relative increase in drought between 2003 and 2010, whereas type 3 was characterized by a relative increase in temperature. For the latter, highest mean defoliation was found in the last years starting in 2004. Already in 1994, high defoliation at level comparable to that in 2003 was observed for type 3. In 1994 highest relative temperature increase occurred in south Germany. For defoliation development type 4, highest defoliation was found in this year. Since 1994 defoliation on average decreased and was the lowest defoliation observed nowadays. This type could not be

Table 11.4 Relative deviation [%] of annual precipitation sums from the long-term mean (1961–1990) for the 11 weather deviation clusters

	1	2	3	4	5	6	7	8	9	10	11
1990	5	−4	−1	0	−2	−3	−1	3	−5	−8	3
1991	−14	−24	−24	−21	−27	−15	−25	−8	−25	−20	−21
1992	2	3	5	6	1	−2	1	1	−3	−4	−1
1993	21	13	10	20	1	7	1	8	3	14	17
1994	21	20	12	26	3	3	7	2	3	10	25
1995	1	17	6	−1	19	21	20	14	27	20	9
1996	−20	−8	−17	−17	−20	−6	−7	−6	−13	−11	−15
1997	−8	−5	−15	−7	−9	−8	−11	−12	−11	−8	−8
1998	31	23	17	24	8	9	6	4	17	13	16
1999	3	5	1	−2	6	16	15	17	6	−2	−7
2000	1	2	5	0	22	9	1	12	−3	−1	−4
2001	19	16	11	20	20	19	21	19	37	20	16
2002	32	33	24	39	19	26	28	31	32	39	31
2003	−17	−22	−22	−23	−27	−22	−29	−25	−28	−30	−25
2004	10	8	1	7	−6	0	−3	−5	−2	3	4
2005	1	0	−5	5	−16	5	−5	5	−6	4	10
2006	−5	−3	−7	−10	−3	2	0	−6	3	−5	−14
2007	33	40	22	48	8	15	14	10	17	19	45
2008	3	−2	−10	11	−5	−4	−2	−8	−2	0	11
2009	0	10	1	9	0	2	1	0	5	8	9
2010	7	17	−2	31	9	6	9	8	15	24	34

assigned to one landscape but was located in a region of comparably low temperatures and high precipitation. The development of defoliation may indicate that the temperature increase might in general be a benefit for this oak cluster.

Defoliation development types of pine showed spatial distributions and time series of defoliation that were similar to those of oak (Fig. 11.3). Other than for oak, the spatial pattern of defoliation development types of pine matched even better with the weather deviation clusters than with the landscape regions. The division of defoliation development type 4 in two spatially independent areas (Fig. 11.3) also occurred for the weather deviation cluster 10. In addition, the three weather deviation clusters 2, 3 and 4 matched well with the original defoliation development types 5, 2 and 3, which were summarized to the broad defoliation development type 2.

Spatial distributions of defoliation development types of spruce and beech were similar but deviated from the spatial distributions observed for pine and oak (Fig. 11.3). Defoliation of both species showed a similar trend for the five defoliation development types, although defoliation of beech in general showed more pronounced fluctuations than defoliation of spruce. Moreover, weather conditions were similarly associated with defoliation of both species (see Sect. 11.3.2).

As an example the results for beech will be discussed in the following. Similarities between the spatial patterns of defoliation development types and landscapes were not as immediately apparent as for oak but also existed. First of all, slight differences between the western and eastern Lowlands were indicated (original defoliation types 1 and 9). However, beech predominantly grows in moist and cooler regions of the low mountain ranges and the Alpine Foreland and is rarely found in the Lowlands (Fig. 11.3). The eastern Central Upland Range was covered well by the original defoliation development type 8. The western Central Upland Range was also covered well but by two different broad defoliation development types (types 2 and 3). The spatial distribution of type 3, which reflects the federal states Rhineland-Palatinate and Saarland, also roughly existed for the weather deviation clusters (cluster 5; Fig. 11.12). Regarding south Germany, again a prominent change of defoliation development types occurred along the border of two federal states (Baden-Wuerttemberg and Bavaria). Again, this spatial distribution was also partly present in the weather deviation clusters (clusters 6 and 8).

Differences among the five broad defoliation development types of beech regarding time series of precipitation and temperature deviations from 1990 to 2010 are exemplarily presented (Fig. 11.13). Time series of all clusters had notable positive temperature deviations from 1997 to 2009, whereas no systematic deviations occurred for precipitation. The year 1996 was exceptionally cold and dry, and the year 2003 was exceptionally warm and dry throughout Germany. Both events resulted in increased defoliation. However, the intensities differed within clusters.

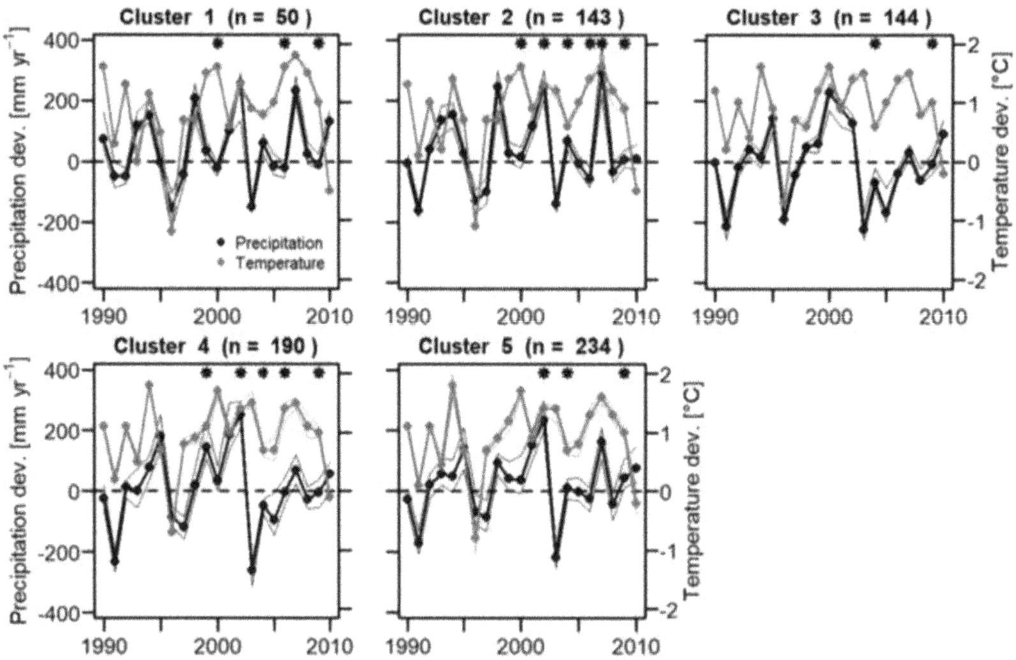

Fig. 11.13 Deviations of precipitation and temperature from the long-term mean 1961–1990 for the five defoliation development types of beech. The median and the intervals representing the 25 and 75 percent quantiles are shown. Fructification of beech in 50% of plots or more for one cluster is indicated by a star beginning in 1999. The dashed line represents the line of no deviation. The sample size is given in parenthesis

At the beginning of the 1990s, highest defoliation of beech was observed for defoliation development type 1. For type 1, the year 1990 was very warm compared to the long-term mean and compared to the other clusters. The conditions at the end of the 1980s and beginning of the 1990s might be the reason for the observed high defoliation, but methodological differences presumably played a role as well. Deviations of temperature and precipitation for defoliation development types 1 and 2 followed a similar pattern, but positive temperature deviations were slightly lower and positive precipitation deviations were slightly higher for type 2. Defoliation development type 2 showed relatively constant mean defoliation between 1990 and 2015, whereas type 1 showed more fluctuation. Both types were characterized by the defoliation peak in 2000 that followed the notable positive temperature deviations at average precipitation in the consecutive years 1999 and 2000. These weather conditions were only observed in the regions of types 1 and 2. In 2000, pronounced fructification additionally occurred for beech of both types. The year 2003 featured positive temperature deviations of on average 1 °C and a notable negative deviation of precipitation of more than 100 mm per year for these types. These extreme conditions were accompanied by fructification and a defoliation peak in 2004, which however was lower than the peak in 2000. In contrast, defoliation development types 3 to 5 were characterized by pronounced fructification and high defoliation in 2004, most likely because the deviations of temperature increase and precipitation decrease in 2003 were notably extremer in south and south-western Germany. The highest negative deviation of precipitation accompanied by a high positive deviation of temperature occurred for type 4. In 2004, the greatest increase in defoliation was also observed for this type. Similar to type 3, extreme weather conditions were observed again in the subsequent years. However, beech of type 4 had additionally recurring pronounced fructification, and defoliation remained on the highest level observed for all defoliation development types. In summary, exceptional high temperatures were primarily associated with high defoliation of beech in the recent year, whereas the combination of high temperature and drought was mainly associated with high defoliation in the next year. However, the direct impact of weather was most likely overimposed by the impact of fructification (indirect weather effect).

Significant differences ($p < 0.0001$, ANOVA) among defoliation development types were further observed regarding the key soil parameters identified to be associated with defoliation (see Sect. 11.3.2). The defoliation development types representing south-western Germany had the lowest C stock and thus mass of the organic layer (spruce, type 4), the lowest C/N ratio (pine, type 3), the lowest N concentration of leaves of beech during the NFSI II (beech, type 4) and also the lowest C/P ratio (oak, type 3) (Fig. 11.14).

In conclusion, weather conditions and in particular deviations of temperature and precipitation from the long-term mean explained a large proportion of the differences among defoliation development types of the four tree species. An adaption of forest stands to the long-term average of the local water balance is generally assumed (Zierl 2004 and literature therein); hence, it is obvious that deviations might affect tree vitality. However, in part, defoliation of trees reacted differently to weather

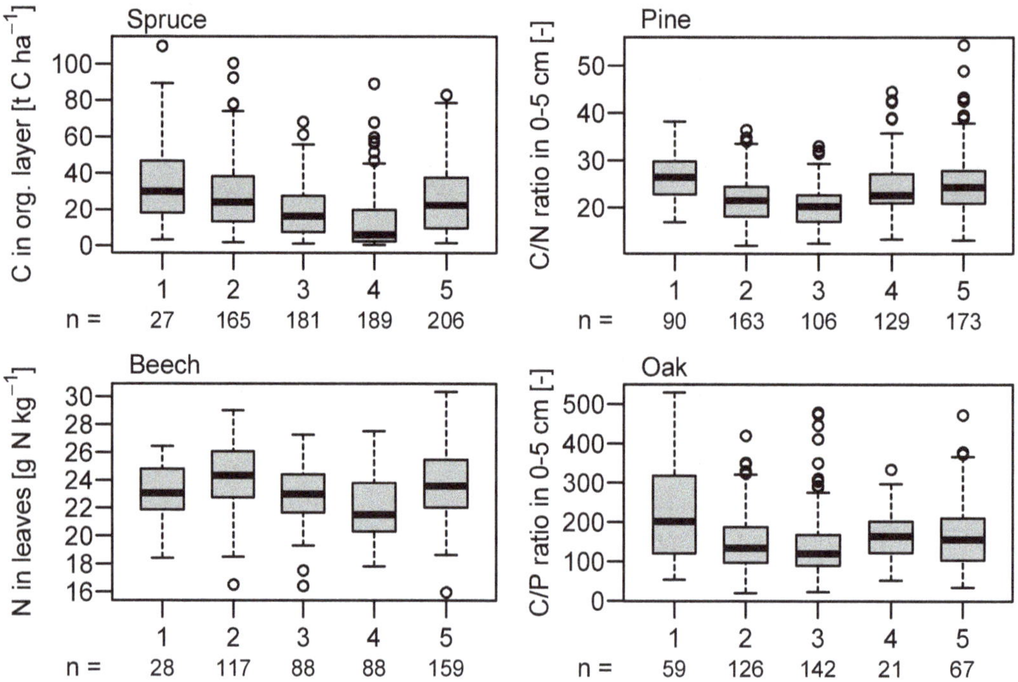

Fig. 11.14 Boxplots of C stock in the organic layer for spruce (top, left), C/N ratio in the 0–5 cm soil depth for pine (top, right), N concentration of leaves of beech (bottom, left) and C/P ratio in the 0–5 cm soil depth for oak (bottom, right) for the five defoliation development types of each species

conditions, possibly as consequence of tree species, provenance and differences in other stand conditions. Extraordinarily high temperature deviations of approximately 1.5 °C and more and in particular when accompanied by extraordinarily high negative precipitation deviations were associated with increased defoliation of all species most likely due to drought stress. These conditions have cumulatively occurred since 2003 especially in south-western Germany. Anders et al. (2004) also corroborated this result, although all of Germany was affected by drought stress in 2003: the highest water deficit calculated as the difference between the climatic water budget of 2003 and the long-term water budget of the reference period 1961–1990 occurred in south-western Germany. Hilly and mountainous areas as, e.g. found in large areas of south-western Germany, are further often characterized by shallow and stony soils, which have low water storage capacity and hamper deep rooting. Additionally, in south-western Germany, low mass of the organic layer was frequently found, which might increase drying up of the soil. In contrast, generally deeper soils and thicker organic layers could be found in the North German Lowlands. Drought stress was detected as the most important risk factor associated with high defoliation. In addition to variables indicating drought stress, fructification of beech, insect infestation (pine, oak) as well as deficiencies in N nutrition proved to be other stress factors associated with high defoliation. South-western Germany emerged as the area with greatest risk of defoliation, most likely due to drought stress in recent years. However, areas being prone to drought stress due to pedological conditions generally represent areas with risk of high defoliation in particular with regard to climate change.

11.4 Defining Forest Nutrition Types

The nutritional status of trees can be interpreted as an integrative overall expression of the site-specific water and nutrient supply and the influence of climate and deposition. Below, we assume that at a given site with given environmental conditions, there are identifiable nutritional categories specific to tree species. According to Göttlein (2015), classes can be assigned on the basis of levels of element concentrations in needles and leaves, and, in particular, trees can be assessed for deficiency or surplus of specific elements (see Chap. 9). Typification of the nutritional situation goes beyond classification of the levels of individual elements and collates specific combinations of element supplies. With this approach, we provide definitions for nutrition types for the tree species Norway spruce (*Picea abies* (L.) H. Karst.), Scots pine (*Pinus sylvestris* L.), European beech (*Fagus sylvatica* L.) and pedunculate (*Quercus robur* L.) and sessile oak (*Quercus petraea* (Matt.) Liebl.) by analysing the level of five key elemental nutrients (Ca, Mg, K, P and N) according to Göttlein (2015). For operational reasons, the categories "latent deficiency" and "deficiency", as well as "normal" and "surplus", are pooled. The distinction between normal and surplus nutrition was maintained only for the element nitrogen due to the special considerations for nitrogen (see Chap. 9). In contrast to element-specific classification of the nutrition data, this approach is thus based on simultaneous analysis of multiple key nutrients. Defined nutrition types and their frequencies in the NFSI II sampling data are presented in Table 11.5. Combinations of classifications with a sample size of $n < 10$ were not considered and are not listed in the table. Thus, there are 9 different nutrition types for spruce, 10 for Scots pine, 13 for beech and 6 for oak. Assignment to a nutrition type varied by tree species: 91% (spruce), 86% (pine, beech) and 76% (oak) of all plots that were studied could be assigned to one of the nutrition types defined in Table 11.5. These are arranged in the table by species and according to decreasing frequency. The codes presented in Table 11.5 are also used in the descriptions of the identified types in the text that follows.

The most frequent nutrition type for spruce was characterized by an adequate supply of all the key nutrients (Ca0/K0/Mg0/P0/N0). The second most common type featured a surplus of N but otherwise a balanced nutrient supply (Ca0/K0/Mg0/P0/N+). According to frequency of occurrence, there followed sites with deficiencies of K (Ca0/K−/Mg0/P0/N0) and P (Ca0/K0/Mg0/P−/N0). Combined with a deficiency of K and surplus of N (Ca0/K−/Mg0/P0/N+), deficiencies of both P and N (Ca0/K0/Mg0/P−/N−), P and K (Ca0/K−/Mg0/P−/N0) and P, K and N (Ca0/K−/Mg0/P−/N−) were less common.

For the other tree species, the most common nutrition type was the type featuring N surplus (pine, oak) and the type featuring P deficiencies (beech). The nutrition type with an adequate supply of all key nutrients ranked in second place for pine, oak and

Table 11.5 Nutrition types based on the combined classification of the levels of Ca, K, Mg, P and N according to Göttlein (2015)

	Ca	K	Mg	P	N	Code	n
Spruce (n=797)	grey	grey	grey	grey	grey	Ca0/K0/Mg0/P0/N0	271
	grey	grey	grey	grey	black	Ca0/K0/Mg0/P0/N+	178
	grey	white	grey	grey	grey	Ca0/K-/Mg0/P0/N0	93
	grey	grey	grey	white	grey	Ca0/K0/Mg0/P-/N0	67
	grey	white	grey	grey	black	Ca0/K-/Mg0/P0/N+	28
	grey	grey	grey	white	white	Ca0/K0/Mg0/P-/N-	26
	grey	white	grey	white	grey	Ca0/K-/Mg0/P-/N0	25
	grey	grey	grey	grey	white	Ca0/K0/Mg0/P0/N-	21
	grey	white	grey	white	white	Ca0/K-/Mg0/P-/N-	17
Beech (n=575)	grey	grey	grey	white	grey	Ca0/K0/Mg0/P-/N0	104
	grey	grey	grey	grey	grey	Ca0/K0/Mg0/P0/N0	71
	grey	white	grey	white	grey	Ca0/K-/Mg0/P-/N0	53
	grey	grey	grey	grey	black	Ca0/K0/Mg0/P0/N+	51
	grey	grey	white	white	grey	Ca0/K0/Mg-/P-/N0	50
	grey	white	grey	grey	grey	Ca0/K-/Mg0/P0/N0	32
	grey	grey	grey	white	black	Ca0/K0/Mg0/P-/N-	30
	white	grey	white	white	grey	Ca-/K0/Mg-/P-/N0	27
	grey	white	grey	grey	black	Ca0/K-/Mg0/P0/N+	24
	grey	white	white	white	grey	Ca0/K-/Mg-/P-/N0	14
	grey	grey	white	grey	grey	Ca0/K0/Mg-/P0/N0	13
	white	grey	white	grey	grey	Ca-/K0/Mg-/P0/N0	12
	grey	white	grey	white	black	Ca0/K-/Mg0/P-/N+	12
Pine (n=610)	grey	grey	grey	grey	black	Ca0/K0/Mg0/P0/N+	198
	grey	grey	grey	grey	grey	Ca0/K0/Mg0/P0/N0	126
	grey	grey	white	grey	black	Ca0/K0/Mg-/P0/N+	57
	grey	grey	white	grey	grey	Ca0/K0/Mg-/P0/N0	32
	grey	white	grey	grey	grey	Ca0/K-/Mg0/P0/N0	25
	grey	grey	grey	white	grey	Ca0/K0/Mg0/P-/N0	23
Pine (n=610)	grey	grey	grey	white	black	Ca0/K0/Mg0/P-/N+	22
	grey	grey	white	white	black	Ca0/K0/Mg-/P-/N+	18
	white	grey	white	grey	grey	Ca-/K0/Mg-/P0/N0	11
	grey	white	grey	white	black	Ca0/K-/Mg0/P-/N+	11
Oak (n= 318)	grey	grey	grey	grey	black	Ca0/K0/Mg0/P0/N+	96
	grey	grey	grey	grey	grey	Ca0/K0/Mg0/P0/N0	61
	grey	grey	grey	white	grey	Ca0/K0/Mg0/P-/N0	42
	grey	grey	grey	white	black	Ca0/K0/Mg0/P-/N+	23
	white	grey	grey	grey	black	Ca-/K0/Mg0/P0/N+	11
	grey	grey	white	grey	black	Ca0/K0/Mg-/P0/N+	10

The colours in the element columns and the symbols after the elements symbolize adequate supply (grey, 0), (latent) deficiency (white, –) and nitrogen surplus (black, N+)

beech. For pine, Mg deficiency in combination with N surplus was frequently found (Ca0/K0/Mg−/P0/N+), whereas for beech, P deficiency combined with K deficiency (Ca0/K−/Mg0/P−/N0) and Mg deficiency (Ca0/K0/Mg−/P−/N0), respectively, occurred frequently. Furthermore, surplus of N without concurrent deficiency of other nutrients (Ca0/K0/Mg0/P0/N+) was commonly observed for all tree species.

There do not appear to be any spatial patterns in the regional distribution of nutrition types. Using the spruce by way of example, the cartogram in Fig. 11.15 shows that a deficiency in K with otherwise balanced nutrient supply (Ca0/K−/Mg0/P0/N0) or even in combination with a surplus of N (Ca0/K−/Mg0/P0/N+) occurred

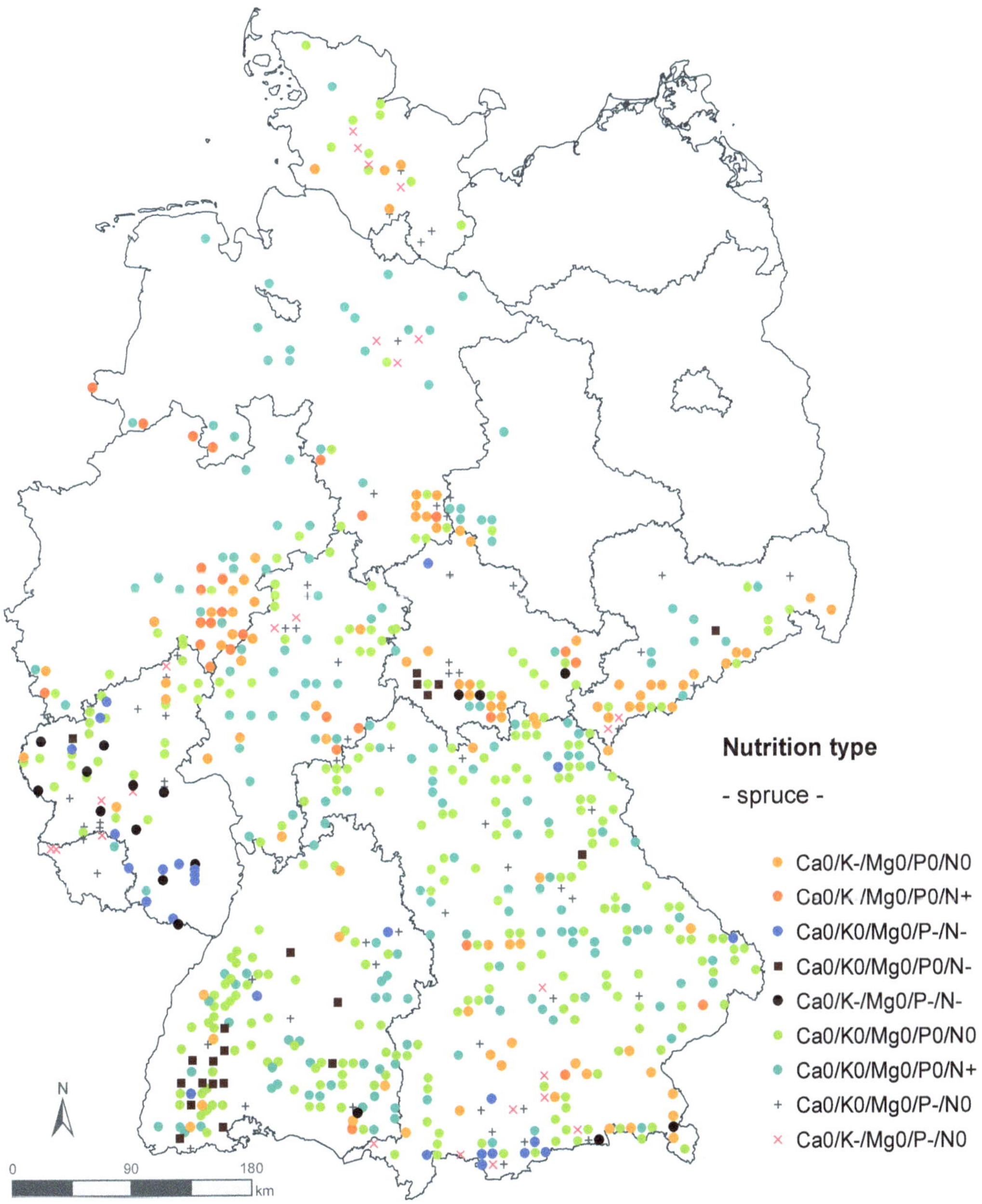

Fig. 11.15 Regional distribution of nutrition types using spruce as the example

notably in the regions of the Erzgebirge, the Thuringian Forest, the Harz and the Sauerland and Siegerland. The parent material is most likely not the reason for the K deficiency in these regions. However, the spruce stands in these regions were limed several times, so that K deficiency is probably due to cation antagonism with Ca. Types with N deficiencies occurred regionally: in the Limestone Alps region in combination with a P deficiency (Ca0/K0/Mg0/P−/N−), in Baden-Wuerttemberg (especially in the Black Forest) and in the northern regions of the Thuringian Forest with otherwise balanced nutrient supply (Ca0/K0/Mg0/P0/N−) and in the Rhineland-Palatinate area in combination with deficiencies of both K and P (Ca0/K−/Mg0/P−/N−). All other nutrition types were distributed evenly throughout Germany with no particular regional hotspots.

We used discriminant analysis to investigate which site-specific and environmental factors were decisive for assignment to a specific nutrition type. In advance, the potential influencing parameters from the NFSI database were subjected to a principal component analysis. This procedure ensured that the linear correlations between the variables were eliminated. The number of variables to be extracted, or principal components, was determined based on the Kaiser criterion. A varimax rotation was used to simplify the interpretation of the results. Principal component values were calculated by regression using the statistical programme SPSS Statistics Release 22.0.0.0. Individual parameters for which the frequency distribution was not suitable for principal component analysis were log-transformed a priori.

Overall, 42 potential influencing factors were identified from the NFSI database and subjected to principal component analysis. These included parameters for soil chemistry, climate, modelled variables of soil water budget (see Chap. 3) and regionalized deposition data and critical load exceedances for nitrogen. Applying the Kaiser criterion, the principal component analysis extracted ten principal components that describe 79.2% of all the total variance for all parameters considered. The first five components alone accounted for more than half the variance. Assignment of individual variables to principal components was evident using the rotated component matrix.

Based on the component matrix, input variables were selected for the discriminant analysis. The following variables were chosen:

- Precipitation [mm year^{-1}]
- Evapotranspiration [mm year^{-1}]
- Number of limings [–]
- Ntot deposition 1990–2007 [kg ha^{-1} year^{-1}]
- K concentration (30–60 cm) [% of cation exchange capacity]
- C stock (organic layer to 90 cm) [kg ha^{-1}]
- C/P ratio (0–5 cm) [–]
- pH(H_2O) value in 30–60 cm [–]
- C/N ratio (organic layer) [–]
- Available water capacity (root zone) [mm]
- Relative water storage (root zone) [%]
- N stock (organic layer) [kg ha^{-1}]

With these variables, a total of 38 stepwise discriminant analyses were performed according to the number of nutrition types, and the percentage classification probabilities for each nutrition type were calculated using linear combination of the influencing parameters. For two nutrition types for beech (Ca0/K−/Mg−/P−/N0 and Ca0/K0/Mg−/P0/N0) and one type for pine (Ca0/K0/Mg−/P−/N+), no significant discriminant function could be derived based on the variables listed. The results of the discriminant function analysis presented in Table 11.6 first indicate the numbers of each correctly classified site. In addition, for each variable, the table lists the correlation coefficient with the calculated percent classification probability for all sites for each tree species. Correlation coefficients for variables that were also included in the model based on stepwise discriminant function analysis are shown in bold.

Although the number of correctly classified sites was somewhat low at times, some variables showed strong and plausible correlations with the classification probabilities. For example, the probability of classifying nutrition types with surpluses of N is positively correlated with N deposition. However, in cases of both deficiency and normal levels of N, there was a predominantly negative correlation. Considering P deficiency, there was often a strong correlation with the C/P ratio at 0–5 cm. This is the case especially for spruce (Ca0/K0/Mg0/P−/N0 and Ca0/K−/Mg0/P−/N0 types) and pine (Ca0/K0/Mg0/P−/N+).

One notable result was the relationship of deficiency in K (sometimes in combination with a surplus of N and a deficiency of P) and the number of limings. This result was consistent with the findings of Chap. 9. For example, this relationship could be seen in scatter plots for the spruce nutrition types Ca0/K−/Mg0/P0/N0 (Fig. 11.16) and Ca0/K−/Mg0/P0/N+ (Fig. 11.17). For the K deficiency type with no N surplus, there was a relatively strong correlation of classification probability with the plant available K in the soil. In addition, there was a high classification probability with narrow C/N ratios, which suggested possible competition between K and both NH_4^+ and Ca on alkaline-rich sites. There was also a positive correlation with N deposition. The classification probability increased considerably with the number of limings; this result might be explained by Ca-K antagonism. In examining the probability of classification to the N surplus nutrition-type Ca0/K−/Mg0/P0/N+, it is clear that there was no longer a relationship to available K in the soil; instead, in this case, the N deposition appeared to induce the K deficiency. For this nutrition type as well, an increased number of limings was associated with a higher classification probability. Moreover, the probability of a K deficiency is higher with lower quantities of soil water available to plants.

Thus, overall the results indicated that the supply of K in the soil (as applicable, in combination with liming events) is extremely important for the Ca0/K−/Mg0/P0/N0 nutrition type, while the N deposition is more likely responsible for the K deficiency in the Ca0/K−/Mg0/P0/N+ nutrition type.

The following logistical relationship can be calculated for the relation between classification probability and N deposition illustrated in Fig. 11.16:

Table 11.6 Correlations of site-specific and environmental variables with the classification probabilities to nutrition type for the tree species spruce, beech, pine and oak

	Nutrition type	Correctly classified [%]	Precipitation (year) [mm]	Evapotranspiration (year) [mm]	Number of liming	Nt deposition [kg/ha/a]	K (30–60 cm) [%]	C (humus-90 cm) [t/ha]	C/P (0–5 cm)	pH (H_2O) (30–60 cm)	C/N (humus)	AWC root-zone [mm]	Relative water storage [mm]	N (humus) [kg/ha]
Spruce ($n = 797$)	Ca0/K0/Mg0/P0/N0	54.6		0.341	**−0.582**	**−0.67**		−0.417	**−0.631**	0.31				−0.421
	Ca0/K0/Mg0/P0/N+	60.6	**−0.545**		−0.317	**0.524**					**−0.336**			**−0.335**
	Ca0/K−/Mg0/P0/N0	74.7		−0.332	**0.683**	**0.51**	**−0.353**				**−0.552**			
	Ca0/K0/Mg0/P−/N0	71.1					**0.455**	0.319	**0.811**	−0.274				0.289
	Ca0/K−/Mg0/P0/N+	80.5		−0.39	**0.581**	**0.801**						**−0.357**		
	Ca0/K0/Mg0/P/N−	68.8				**−0.807**					**0.659**			−0.316
	Ca0/K−/Mg0/P−/N0	79.1						0.52	**0.981**					0.342
	Ca0/K0/Mg0/P0/N−	74.1	**0.823**								**0.592**		0.291	
	Ca0/K−/Mg0/P−/N−	71.5			**0.589**	**−0.648**								

(continued)

Table 11.6 (continued)

	Nutrition type	Correctly classified [%]	Precipitation (year) [mm]	Evapotranspiration (year) [mm]	Number of liming	Nt deposition [kg/ha/a]	K (30–60 cm) [%]	C (humus-90 cm) [t/ha]	C/P (0–5 cm)	pH (H_2O) (30–60 cm)	C/N (humus)	AWC root-zone [mm]	Relative water storage [mm]	N (humus) [kg/ha]
Beech (n = 575)	Ca0/K0/Mg0/P–/N0	68.9					**0.251**	**−0.309**		**0.759**				−0.264
	Ca0/K0/Mg0/P0/N0	67.5	−0.342	0.264					**−0.462**			**0.531**	**−0.642**	
	Ca0/K–/Mg0/P–/N0	68.9		−0.281	**0.953**				**0.236**					
	Ca0/K0/Mg0/P0/N+	54.3	**−0.704**					−0.322	**−0.673**				−0.318	−0.351
	Ca0/K0/Mg–/P–/N0	63.7				**−0.57**					**0.679**		**0.403**	−0.354
	Ca0/K–/Mg0/P0/N0	68.9	**0.829**						**−0.559**					−0.301
Beech (n = 575)	Ca0/K0/Mg0/P–/N+	70.1	−0.27						**0.511**	**0.356**		**0.426**	**−0.504**	
	Ca–/K0/Mg–/P–/N0	70.8			**−0.503**				**0.644**	**−0.696**				0.468
	Ca0/K–/Mg0/P0/N+	76.2		−0.343	**0.655**						**−0.542**	**−0.287**	**0.412**	0.288
	Ca–/K0/Mg–/P0/N0	79.1			**−0.325**						−0.345	**0.504**		**0.618**
	Ca0/K–/Mg0/P–/N+	83.7		−0.285	**0.509**	**0.648**			**0.455**					

Pine ($n = 610$)	Ca0/K0/ Mg0/P0/ N+	63.0				**0.632**					**−0.722**			**−0.288**
	Ca0/K0/ Mg0/P0/ N0	57.0	−0.319			**−0.779**					**0.74**		−0.248	
	Ca0/K0/ Mg−/P0/ N+	66.4	**−0.366**					**0.377**			**−0.701**			
	Ca0/K0/ Mg−/P0/ N0	68.4			**0.425**		**0.637**				**0.516**			
	Ca0/K−/ Mg0/P0/ N0	74.4	−0.321	0.31		−0.279					**0.859**		**−0.549**	
	Ca0/K0/ Mg0/P−/ N0	78.4	**0.44**	**0.225**	**0.28**	**−0.237**			**0.506**		**0.303**	**−0.312**		
	Ca0/K0/ Mg0/P−/ N+	79.5						0.564	**0.912**					**0.312**
	Ca−/K0/ Mg−/P0/ N0	63.0		**−0.97**									**0.427**	
	Ca0/K−/ Mg0/P−/ N+	87.5					−0.25	**0.46**		**0.794**				
Oak ($n = 318$)	Ca0/K0/ Mg0/P0/ N+	60.1	0.256			**0.471**			**−0.731**					−0.353
	Ca0/K0/ Mg0/P0/ N0	59.4				**−0.845**		**−0.729**	−0.382				−0.289	
	Ca0/K0/ Mg0/P−/ N0	72.3			**0.728**	**−0.671**								

(continued)

Table 11.6 (continued)

	Nutrition type	Correctly classified [%]	Precipitation (year) [mm]	Evapotranspiration (year) [mm]	Number of liming	Nt deposition [kg/ha/a]	K (30–60 cm) [%]	C (humus-90 cm) [t/ha]	C/P (0–5 cm)	pH (H_2O) (30–60 cm)	C/N (humus)	AWC root-zone [mm]	Relative water storage [mm]	N (humus) [kg/ha]
	Ca0/K0/Mg0/P–/N+	78.0				0.25		0.396	**0.823**	**0.283**				0.284
	Ca–/K0/Mg0/P0/N+	78.3				0.288		0.313	0.453	−0.367	−0.314		**0.594**	**0.784**
	Ca0/K0/Mg–/P0/N+	73.9	**0.66**											**−0.753**

Bold correlation coefficients indicate variables that were also included in the model based on stepwise discriminant function analysis

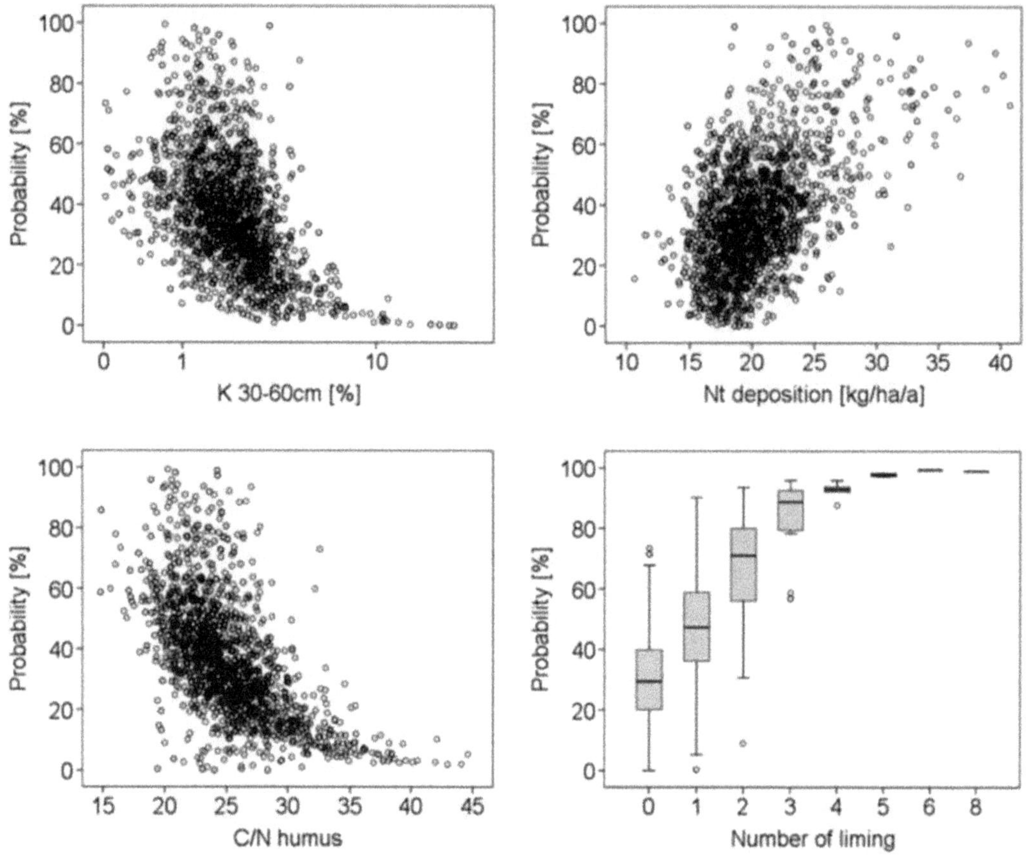

Fig. 11.16 Relationships between the classification probability to spruce nutrition-type Ca0/K−/Mg0/P0/N0 and exchangeable K at a depth of 30–60 cm, N deposition, C/N ratio in the organic layer and the number of liming events

$$\text{Probability} = 1/(1 + 3184.64^{*}\ 0.7212\ \text{Nt});\quad R^2 = 0.71 \tag{11.2}$$

This equation shows that a K deficiency was increasingly likely (probability $> 50\%$) with N input when input rates are greater than 24.7 kg ha^{-1} year^{-1} and at input rates above 33.7 kg ha^{-1} year^{-1}, there was an exceptionally high risk of K deficiency (probability $> 95\%$). Under these conditions, the growth stimulated by the addition of N means that nutrients and water must be taken up at greater quantities and antagonism between NH_4^+ and other cationic nutrients becomes increasingly important. Beyond the thresholds, the demand for K can apparently no longer be met and a K deficiency arises, largely independent of the K supply at the site.

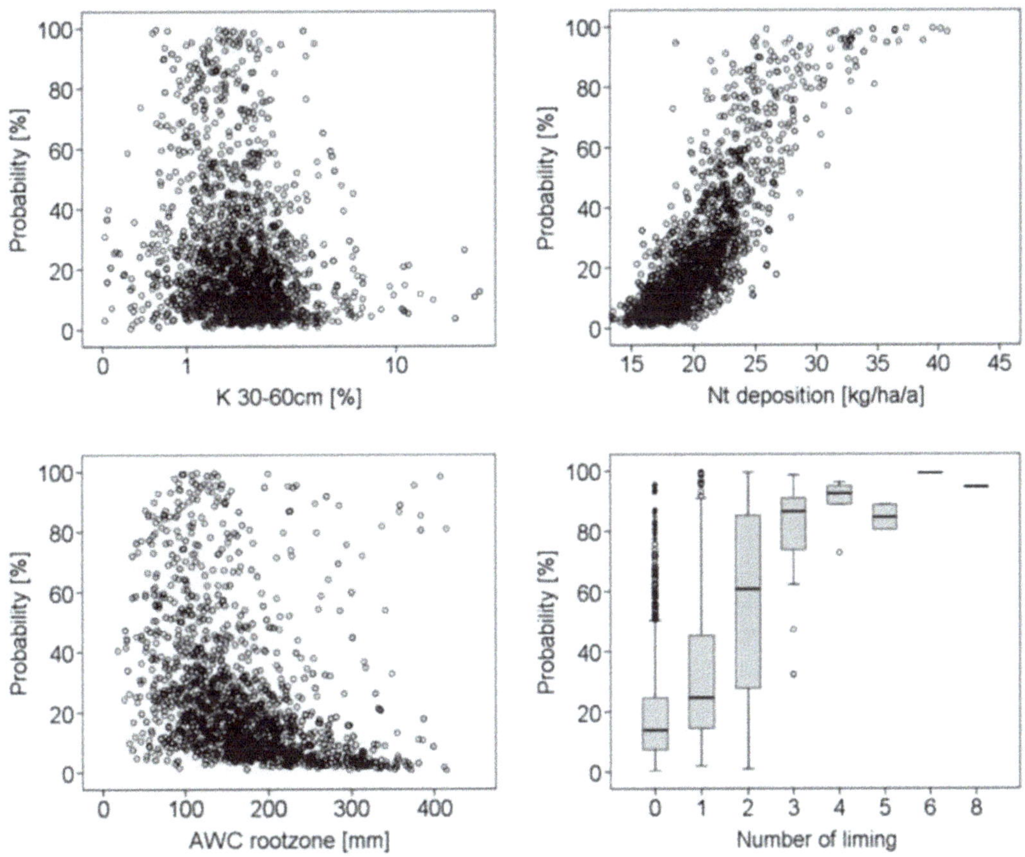

Fig. 11.17 Relationships between the classification probability to spruce nutrition-type Ca0/K–/Mg0/P0/N+ and exchangeable K at a depth of 30–60 cm, N deposition, water capacity in the root zone available to plants and the number of liming events

11.5 Combined Defoliation Development Types and Nutrition Types

Defoliation development types and nutrition types were cross-tabled. There were few clear spatial patterns in the regional distribution of combined defoliation development types and nutrition types, which is most likely due to different factors that were decisive for assignment to a specific defoliation development type (see Sect. 11.3) and nutrition type (see Sect. 11.4), respectively. For spruce, however, various nutrient deficiencies (P, K, N) could be observed for defoliation development type 3 in particular (Table 11.7), which also was characterized by comparably high defoliation in the last decade. Defoliation development type 4 had the highest defoliation in recent years, and N deficiency was predominantly found for this type. Defoliation of beech showed a strong negative association to N nutrition (see Sect. 11.3.2). Defoliation development types 2 and 5 of beech frequently belonged to the N surplus nutrition type and their defoliation time series showed no trends, which was in

Table 11.7 Example of a cross table of defoliation development and nutrition types for spruce

		Defoliation development type of spruce					
Nutrition type		1	2	3	4	5	Row_{tot}
Ca0/K−/Mg0/ P−/N−	Count	0	2	11	1	3	17
	Expected count	0.55	3.90	3.79	3.63	5.14	
	Std. residuals	−0.74	−0.96	3.70	−1.38	−0.94	
Ca0/K−/Mg0/ P−/N0	Count	3	5	5	2	7	22
	Expected count	0.71	5.05	4.91	4.69	6.65	
	Std. residuals	2.72	−0.02	0.04	−1.24	0.14	
Ca0/K−/Mg0/ P0/N+	Count	0	8	12	1	4	25
	Expected count	0.81	5.74	5.57	5.33	7.55	
	Std. residuals	−0.90	0.95	2.72	−1.88	−1.29	
Ca0/K−/Mg0/ P0/N0	Count	3	21	17	8	29	78
	Expected count	2.52	17.90	17.39	16.63	23.56	
	Std. residuals	0.30	0.73	−0.09	−2.12	1.12	
Ca0/K0/Mg0/ P−/N−	Count	0	0	8	7	3	18
	Expected count	0.58	4.13	4.01	3.84	5.44	
	Std. residuals	−0.76	−2.03	2.00	1.61	−1.05	
Ca0/K0/Mg0/ P−/N0	Count	3	15	16	10	17	61
	Expected count	1.97	13.99	13.60	13.01	18.43	
	Std. residuals	0.73	0.27	0.65	−0.83	−0.33	
Ca0/K0/Mg0/ P0/N−	Count	0	3	1	11	2	17
	Expected count	0.55	3.90	3.80	3.63	5.14	
	Std. residuals	−0.74	−0.46	−1.43	3.87	−1.38	
Ca0/K0/Mg0/ P0/N+	Count	3	42	26	30	44	145
	Expected count	4.69	33.26	32.33	30.92	43.81	
	Std. residuals	−0.78	1.52	−1.11	−0.17	0.03	
Ca0/K0/Mg0/ P0/N0	Count	6	40	39	55	73	213
	Expected count	6.88	48.86	47.49	45.42	64.35	
	Std. residuals	−0.34	−1.27	−1.23	1.42	1.08	
Col_{tot}		20	142	138	132	187	619

Observed count, the expected count from the chi-square test and the standardized residuals are presented

contrast to the other types. Defoliation development type 1 was the only type where adequate nutrient supply was primarily observed and defoliation was lowest in recent years. Hence, nutrient deficiencies may enhance the sensitivity of trees to drought stress.

11.6 Conclusion

This chapter focused on spatial patterns in the biotic reactions of forest trees and their association with conditions and changes in forest soils and other environmental variables. Secondary tree growth, tree defoliation and tree nutrition, primarily of the four main tree species of Germany, were considered as biotic reactions indicating tree vitality. Associations of growth ring widths and defoliation with indices of drought stress were found. Growth ring widths, in particular of spruce and beech, decreased with low soil water availability and with pronounced dry periods. In addition, at higher temperature ranges, a decline or plateau of the annual growth ring widths was evident. High annual mean temperature deviations of approximately 1.5 °C and more from the long-term mean (1961–1990) and particularly when accompanied by high negative precipitation deviations were also associated with increased defoliation of all tree species. Defoliation development types that showed clear large-scale spatial distribution patterns were identified. The defoliation development types were in good accordance with the landscape regions of Germany. Weather conditions and in particular relative deviations of temperature and precipitation from the long-term mean explained a large proportion of differences among defoliation development types. The defoliation development types found in the south-western part of Germany (especially the area of Baden-Wuerttemberg but also adjacent areas in southern Hesse and areas of Rhineland-Palatinate and Saarland) have been characterized by the highest defoliation of all species in the last years of observation, most likely due foremost to the strongest and frequent positive deviations of temperature with negative deviations of precipitation beginning in 2003. Hence, south-western Germany is an area with high risk for high defoliation, at least in recent years. Large parts of this area also had a low mass of organic layer, which might protect the soil against drying up. Although there were few clear spatial patterns in the regional distribution of nutrition types as well as of combined defoliation development types and nutrition types, different nutrition deficiencies (K, P, N) were cumulatively observed in parts of south-western Germany. Nutrient deficiencies may enhance the sensitivity of trees to drought stress. The results of the investigation of nutrition types revealed that nutrition of trees generally was adequate or characterized by N surplus with the exception of beech, where P deficiency was most common. The situation of N nutrition was mainly associated with N deposition. In regions with high N deposition, deficiency of K was also related to N deposition, presumably because of antagonistic effects between K and NH_4^+. However, in regions with low or normal N deposition, K deficiency was mainly a result of low K stocks in the soil. Independent of N deposition, liming was also associated with K deficiency, again presumably because of antagonistic effects (Ca-K antagonism), which also appeared to play a role in tree defoliation, especially in pine.

In considering the results of the NFSI II presented in the previous chapters of this book, changes in weather conditions and in particular in available soil water (see Chap. 3) certainly had the strongest impact on biotic tree reactions. Changes in C and total stocks in the organic layer (see Chap. 6) were also important in this context, since the organic layer most likely reduces drying up and thus drought stress. Hence, drought stress was the main risk factor detected. In central Europe, a further increase in the frequency of summer drought (high temperature combined with low precipitation) is expected as consequence of climate change (Lindner et al. 2010). Environmental policy should primarily aim at mitigating climate change. Additionally, silviculture measures are crucial: for example, consideration should be given to increased diversity of tree species adapted to the site conditions. Positive and negative aspects of liming in terms of nutrition supply need to be evaluated on a plot-by-plot basis. Liming might cause further drying of soils, because it alters the decomposition and distribution of organic matter in the organic layer and soil; these effects should be considered and discussed. Attention should also be paid to the effect of chronic high N depositions and possible interactions with the effects of climate change. Measures should aim to reduce stress factors originating from air pollution and nutrient deficiencies in order to facilitate the regenerative capability of forest trees and adaption to climate change.

Acknowledgments We would like to thank the Northwest German Forest Research Institute for providing their drill core data. We further thank the responsible institutions of the federal states for providing the data of the Forest Condition Survey of their denser grids. In addition, we acknowledge the discussion with Inge Dammann (Northwest German Forest Research Institute).

References

Aber JD, McDowell W, Nadelhoffer K, Magill A, Berntson G, Kamakea M, McNulty S, Currie W, Rustad L, Fernandez I (1998) Nitrogen saturation in temperate forest ecosystems—hypotheses revisited. Bioscience 48(11):921–934. https://doi.org/10.2307/1313296

Alavi G (2002) The impact of soil moisture on stem growth of spruce forest during a 22-year period. For Ecol Manag 166(1–3):17–33

Anders S, Beck W, Lux W, Müller J, Fischer R, König A, Küppers JG, Thoroe C, Kätzel R, Löffler S, Heydeck P, Möller K (2004) Auswirkung der Trockenheit 2003 auf Waldzustand und Waldbau. Interim Report, vol BMVEL 533-7120/1. Federal Research Institute for Forest and Timber Industries, Eberswalde

Augustin NH, Musio M, von Wilpert K, Kublin E, Wood SN, Schumacher M (2009) Modeling spatiotemporal forest health monitoring data. J Am Stat Assoc 104(487):899–911. https://doi.org/10.1198/jasa.2009.ap07058

Berge E, Bartnicki J, Olendrzynski K, Tsyro SG (1999) Long-term trends in emissions and transboundary transport of acidifying air pollution in Europe. J Environ Manag 57(1):31–50. https://doi.org/10.1006/jema.1999.0275

Bolte A, Czajkowski T, Cocozza C, Tognetti R, de Miguel M, Pšidová E, Ditmarova L, Dinca L, Delzon S, Cochard H, Ræbild A, de Luis M, Cvjetkovic B, Heiri C, Müller J (2016) Desiccation and mortality dynamics in seedlings of different European beech (Fagus sylvatica l.) populations under extreme drought conditions. Front Plant Sci 7:1–12. https://doi.org/10.3389/fpls.2016.00751

de Marco A, Proietti C, Cionni I, Fischer R, Screpanti A, Vitale M (2014) Future impacts of nitrogen deposition and climate change scenarios on forest crown defoliation. Environ Pollut 194:171–180. https://doi.org/10.1016/j.envpol.2014.07.027

de Vries W, Dobbertin MH, Solberg S, Van Dobben HF, Schaub M (2014) Impacts of acid deposition, ozone exposure and weather conditions on forest ecosystems in Europe: an overview. Plant Soil 380(1–2):1–45

Eichhorn J, Icke R, Isenberg A, Paar U, Schönfelder E (2005) Temporal development of crown condition of beech and oak as a response variable for integrated evaluations. Eur J For Res 124 (4):335–347. https://doi.org/10.1007/s10342-005-0097-z

Eichhorn J, Roskams P, Potočić N, Timmermann V, Ferretti M, Mues V, Szepesi A, Durrant D, Seletković I, Schroeck H-W, Nevalainen S, Bussotti F, Paloma G, Wulff S (2016) Part IV Visual assessment of crown condition and damaging agents. In: UNECE ICP Forests Programme Co-ordinating Centre (ed) Manual on methods and criteria for harmonized sampling, assessment, monitoring and analysis of the effects of air pollution on forests. Thünen Institute of Forest Ecosystems, Eberswalde, 54 pp

Eickenscheidt N, Augustin NH, Wellbrock N, Dühnelt P-E, Hilbrig L (2016) Kronenzustand – Steuergrößen und Raum-Zeit-Entwicklung von 1989–2015. In: Wellbrock N, Bolte A, Flessa H (eds) Dynamik und räumliche Muster forstlicher Standorte in Deutschland – Ergebnisse der Bodenzustandserhebung im Wald 2006 bis 2008. Thünen Report 43. Johann Heinrich von Thünen Institute, Federal Research Institute for Rural Areas, Forestry and Fisheries, Braunschweig, pp 387–456

Eickenscheidt N, Augustin NH, Wellbrock N (2019) Spatio-temporal trend estimation of tree defoliation in Germany from 1989 to 2015 and grid examination using generalized additive mixed models. iForest-Biogeosciences and Forestry. Accepted for publication

Elith J, Leathwick JR, Hastie T (2008) A working guide to boosted regression trees. J Anim Ecol 77 (4):802–813. https://doi.org/10.1111/j.1365-2656.2008.01390.x

Ellenberg H (1996) Vegetation Mitteleuropas mit den Alpen in ökologischer Sicht. Ulmer, Stuttgart

Evers FH, Hüttl RF (1992) Magnesium-, Calcium-und Kaliummangel bei Waldbäumen – Ursachen, Symptome, Behebung. Merkblätter der Forstlichen Versuchs- und Forschungsanstalt Baden-Württemberg, vol 42. Forest Research Institute Baden-Württemberg, Kassel

Ferretti M, Nicolas M, Bacaro G, Brunialti G, Calderisi M, Croisé L, Frati L, Lanier M, Macchcrini S, Santi E, Ulrich E (2014) Plot-scale modelling to detect size, extent, and correlates of changes in tree defoliation in French high forests. For Ecol Manag 311:56–69. https://doi.org/10.1016/j.foreco.2013.05.009

Fraley C, Raftery AE, Scrucca L, Murphy TB, Fop M (2017) Mclust: Gaussian mixture modelling for model-based clustering, classification, and density estimation, R package Version 5.3. https://cran.rproject.org/web/packages/mclust/index.html

Galloway JN, Townsend AR, Erisman JW, Bekunda M, Cai ZC, Freney JR, Martinelli LA, Seitzinger SP, Sutton MA (2008) Transformation of the nitrogen cycle: recent trends, questions, and potential solutions. Science 320(5878):889–892. https://doi.org/10.1126/science.1136674

Göttlein A (2015) Grenzwertbereiche für die ernährungsdiagnostische Einwertung der Hauptbaumarten Fichte, Kiefer, Eiche, Buche. Allg Forst Jagdztg 186(5/6):110–116

Klap JM, Voshaar JHO, De Vries W, Erisman JW (2000) Effects of environmental stress on forest crown condition in Europe. Part IV: statistical analysis of relationships. Water Air Soil Pollut 119(1–4):387–420. https://doi.org/10.1023/a:1005157208701

Lin XH, Zhang DW (1999) Inference in generalized additive mixed models by using smoothing splines. J R Stat Soc Ser B Stat Methodol 61:381–400. https://doi.org/10.1111/1467-9868.00183

Lindner M, Maroschek M, Netherer S, Kremer A, Barbati A, Garcia-Gonzalo J, Seidl R, Delzon S, Corona P, Kolstrom M, Lexer MJ, Marchetti M (2010) Climate change impacts, adaptive capacity, and vulnerability of European forest ecosystems. For Ecol Manag 259(4):698–709. https://doi.org/10.1016/j.foreco.2009.09.023

Pinheiro JC, Bates DM (2000) Mixed-effects models in S and S-PLUS Springer. Springer, New York

R Core Team (2017) R: A language and environment for statistical computing. R Foundation for Statistical Computing, Vienna, Austria

Riek W, Wolff B (1999) Integrierende Auswertung bundesweiter Waldzustandsdaten. Final Report on the project "Integrierende Auswertung bundesweiter Waldschadens-, Bodenzustands-, Klima- und Immissionsdaten mittels multivariat-statistischer Modellbildung zur Interpretation neuartiger Waldschäden und Ableitung von Maßnahmenempfehlungen". Federal Research Institute for Forestry and Timber Industries, Eberswalde

Scharnweber T, Manthey M, Criegee C, Bauwe A, Schröder C, Wilmking M (2011) Drought matters—declining precipitation influences growth of Fagus sylvatica L. and Quercus robur L. in north-eastern Germany. For Ecol Manag 262(6):947–961

Schwarz G (1978) Estimating the dimension of a model. Ann Stat 6(2):461–464

Seidling W (2001) Integrative studies on forest ecosystem conditions: multivariate evaluation on tree crown condition for two areas with distinct deposition gradients. Work report of the Institute for World Forestry. Federal Research Centre for Forestry and Forest Products, Hamburg

Seidling W (2004) Crown condition within integrated evaluations of Level II monitoring data at the German level. Eur J For Res 123(1):63–74

Seidling W (2006) Auswirkungen von klimatischem Trockenstress auf den Waldzustand. Arbeitsbericht (Working Report). Federal Research Institute for Forest and Timber Industries (Centre for Forest Ecology and Inventory) & Eberswalde University for Applied Sciences, Eberswalde

Seidling W (2007) Signals of summer drought in crown condition data from the German Level I network. Eur J For Res 126(4):529–544. https://doi.org/10.1007/s10342-007-0174-6

Seidling W, Mues V (2005) Statistical and geostatistical modelling of preliminarily adjusted defoliation on an European scale. Environ Monit Assess 101(1–3):223–247

Seidling W, Ziche D, Beck W (2012) Climate responses and interrelations of stem increment and crown transparency in Norway spruce, Scots pine, and common beech. For Ecol Manag 284:196–204. https://doi.org/10.1016/j.foreco.2012.07.015

Silverman BW (1985) Some aspects of the spline smoothing approach to non-parametric regression curve fitting. J R Stat Soc Ser B Methodol 47(1):1–52

Solberg S (2004) Summer drought: a driver for crown condition and mortality of Norway spruce in Norway. For Pathol 34(2):93–104. https://doi.org/10.1111/j.1439-0329.2004.00351.x

Thormann B (2014) Zuwachsverhalten von Buche und Fichte im Hinblick auf die Trockenjahre 1976 und 2003: Eine Analyse von dendrochronologischen Zuwachsdaten der Bodenzustandserhebung in Nord-West Deutschland. Master theses. Georg August University, Göttingen

van der Loo M, van der Laan J, R Core Team, Logan N (2017) stringdist: approximate string matching and string distance functions, R Package Version 0.9.6.4. https://cran.r-project.org/web/packages/stringdist/index.html

Wahba G (1983) Bayesian "confidence intervals" for the cross-validated smoothing spline. J R Stat Soc Ser B Methodol 45(1):133–150

Wellbrock N, Bolte A, Flessa H (eds) (2016) Dynamik und räumliche Muster forstlicher Standorte in Deutschland: Ergebnisse der Bodenzustandserhebung im Wald 2006 bis 2008. Thünen Report 43. Johann Heinrich von Thünen Institute, Federal Research Institute for Rural Areas, Forestry and Fisheries, Braunschweig

Wood SN (2006a) Generalized additive models: an introduction with R. Chapman & Hall/CRC Press, Boca Raton

Wood SN (2006b) Low-rank scale-invariant tensor product smooths for generalized additive mixed models. Biometrics 62(4):1025–1036. https://doi.org/10.1111/j.1541-0420.2006.00574.x

Wood SN (2006c) On confidence intervals for generalized additive models based on penalized regression splines. Aust N Z J Stat 48(4):445–464. https://doi.org/10.1111/j.1467-842X.2006.00450.x

Wood SN (2017) mgcv: mixed GAM computation vehicle with automatic smoothness estimations, R package version 1.8-18. https://cran.rproject.org/web/packages/mgcv/index.html

Zech W (1970) Nadelanalytische Untersuchungen über die Kalkchlorose der Waldkiefer (Pinus silvestris). J Plant Nutr Soil Sci (Z Pflanzenernähr Bodenkd) 125(1):1–16

Zierl B (2004) A simulation study to analyse the relations between crown condition and drought in Switzerland. For Ecol Manag 188(1–3):25–38. https://doi.org/10.1016/j.foreco.2003.07.019

Chapter 12
Sustainable Use and Development of Forests and Forest Soils: A Resume

Andreas Bolte, Joachim Block, Johannes Eichhorn, Tanja G. M. Sanders, and Nicole Wellbrock

12.1 Introduction

Environmental conditions determine forests' species composition, structure, and growth. Only with profound knowledge of these factors we can sustainably preserve and utilize forest ecosystems and their services. In order to gain this knowledge, we need to assess and monitor these ecosystems systematically. This knowledge was gained through two National Forest Soil Inventories (NFSI I and II) reflecting climate and weather conditions, as well as the deposition regime of nutrients and pollutants to German forests. Additionally, plots of the German Intensive Forest Monitoring (Level II), part of the International Co-operative Programme on Forests (ICP Forests), are included. Human activities resulting in greenhouse gases and air pollutant emissions are changing the physical and chemical growth conditions of individual trees and whole forests. The same applies to forest management due to, e.g., changes of the tree species composition, timber use, and liming. The forest soil condition bears witness of all these different natural and anthropogenic impacts. Therefore, care and conservation of forest soils are fundamental elements of a sustainable use of forest ecosystems besides sustainable and environmentally responsible forest management and timber utilization (BMEL 2015).

The first NFSI (NFSI I) studied the morphological, physical, and chemical status of German forest soils within the assessment period 1989–1992. A repetition after 15 years allows an insight in the temporal dynamics of forest soil conditions and

A. Bolte (✉) · T. G. M. Sanders · N. Wellbrock
Thünen Institute of Forest Ecosystems, Eberswalde, Germany
e-mail: andreas.bolte@thuenen.de; tanja.sanders@thuenen.de; nicole.wellbrock@thuenen.de

J. Block
Research Institute for Forest Ecology and Forestry Rheinland-Pfalz, Trippstadt, Germany

J. Eichhorn
Northwest German Forest Research Institute, Göttingen, Germany
e-mail: johannes.eichhorn@nw-fva.de

N. Wellbrock, A. Bolte (eds.), *Status and Dynamics of Forests in Germany*, Ecological Studies 237, https://doi.org/10.1007/978-3-030-15734-0_12

their spatial patterns due to comparable sampling and analysis methods (see Chap. 1). Integration of the assessment of crown condition, forest nutrition, stand structure, and forest vegetation at the same samples plots provided a valuable and unique opportunity to evaluate the options and limitations of human utilization of forest ecosystems at a national scale.

In the following chapters, main results of the NFSIs, complemented by Level II, and their implications for sustainable forest management as well as forest and environmental policies in Germany are discussed and evaluated.

12.2 Clean Air Policies and Forest Liming Take Effect Against Soil Acidification

After the acknowledgment of lake and stream acidification found in large areas of Scandinavia in the mid-1970s (Almer et al. 1974), and the observation of *forest decline* at the end of this decade, the effects of acidifying sulfur and nitrogen deposition on forests, forest soils, and their chemical status, as well as on forest nutrition, moved in the focus of politics and scientists. A deep-reaching acidification and base cation depletion, the mobilization of root-toxic aluminum ion in the soil solution, and a deficiency of magnesium, calcium, and phosphorus nutrition were identified as serious impacts on forest soils and ecosystems (e.g., Ulrich 1986b; Hüttl and Schaaf 1997; see also Chap. 3; Carreira et al. 1997). The establishment and development of clean air policies on both the international scale [Convention of Long-Range Transboundary Air Pollution CLRTAP, UNECE (1979)] and the national scale [Large Combustion Plant Directive in 1983, 13. BImSchV, BR (1983)] have significantly reduced emissions, in particular of acidifying sulfur oxide (SO_x, Fig. 12.1).

Successful measures starting in the early 1980s included the removal of sulfur from waste gases of coal power plants and the introduction of catalyzers for motor vehicles in the Federal Republic of Germany (FRG). Contrasting policies of the German Democratic Republic (GDR) regime and neighbor countries (CSSR, Poland) during the 1960s to 1980s led to regionally high deposition of sulfur oxide and base-rich fly ash emitted by lignite combustion in particular of coal power plants. Whereas fly ash deposition buffered acid deposition in less and mid-distant areas to power plants (Hofmann et al. 1990), remote and elevated regions like the Ore Mountains with considerably lower ash deposition received extremely high and unbuffered sulfur oxide loads resulting in extensive forest damages (Zimmermann et al. 2002), comparable to regions with high sulfur oxide emissions in West Germany like the Harz and the Fichtel Mountains (Matzner et al. 2004). After the reunification in 1990, consistent clean air policies were implemented throughout Germany. As a result sulfur and total acid deposition strongly decreased throughout Germany (Fig. 12.1). However, considerably less reduction is visible for the emission and deposition of acidifying nitrogen

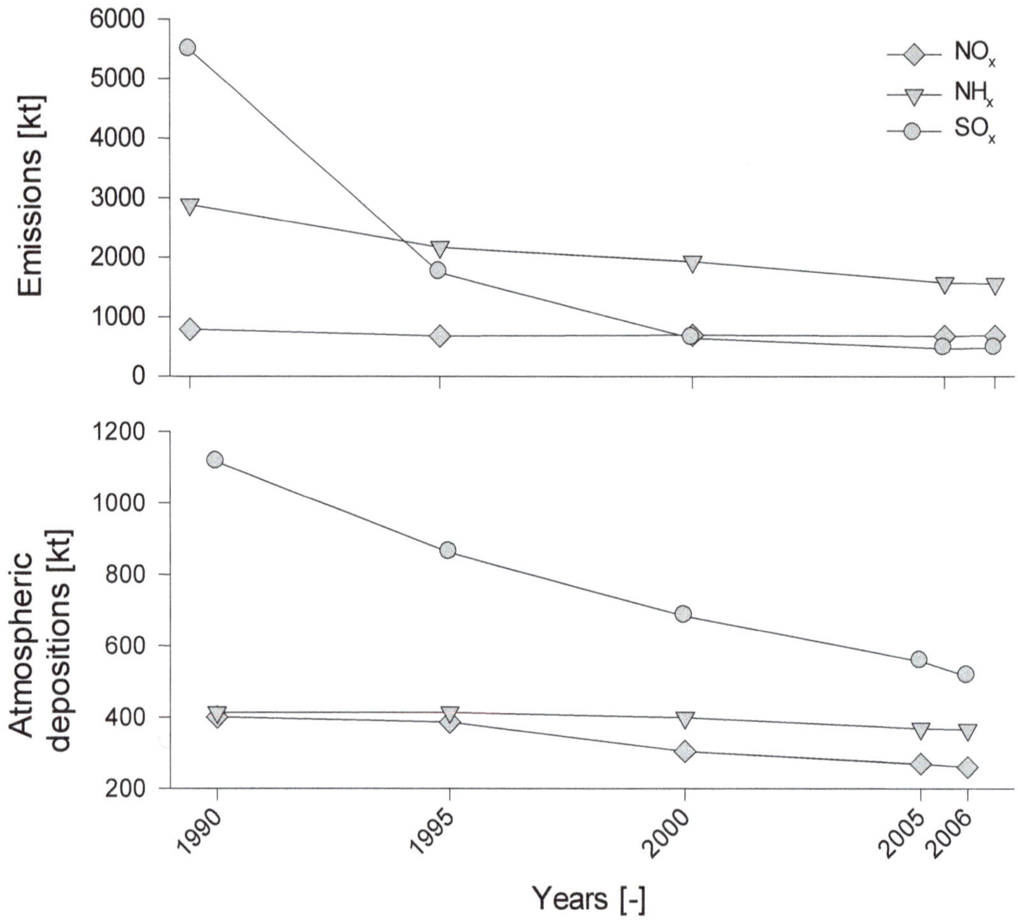

Fig. 12.1 Temporal trend (1990/NFSI I to 2006/NFSI II) of air pollutant (NH_x, NO_x SO_x) emissions (above, data source: UBA 2018) and total deposition in forests (below, modelled values; see Chap. 2)

compounds [NO_x, NH_x (Fig. 12.1)] (UBA 2013), leading to the initiation of the National Emission Ceilings Directive (NECD, 2016/2284/EU) in order to monitor the impact of acidifying and eutrophying compounds. In 2015 acidifying nitrogen contributed far more to the exceedance of critical loads for acidity (CLF_{aci}) than sulfur (see Chap. 2).

At a local scale, forest liming is a technical measure to preserve sensitive forest soils from acidification (see Chap. 2), in addition to forest transformation activities (see Sect. 12.9). Since the beginning of the 1980s, forest sites in several German regions on both public and private forest land have been repeatedly limed, generally financed, or subsidized by public sources.

The results of the NFSIs reveal the success of these measures displaying an increase of soil pH values, not only in the organic layer but also in the mineral topsoil (see Chap. 4; Fig. 12.2). Only on limed forest sites, this is linked to a raise of base saturation in the upper layers of the mineral soil (Fig. 12.2) due to a surplus of calcium carbonate. This underlines the benefits of forest liming, in particular on soils with a high sensitivity of acidification

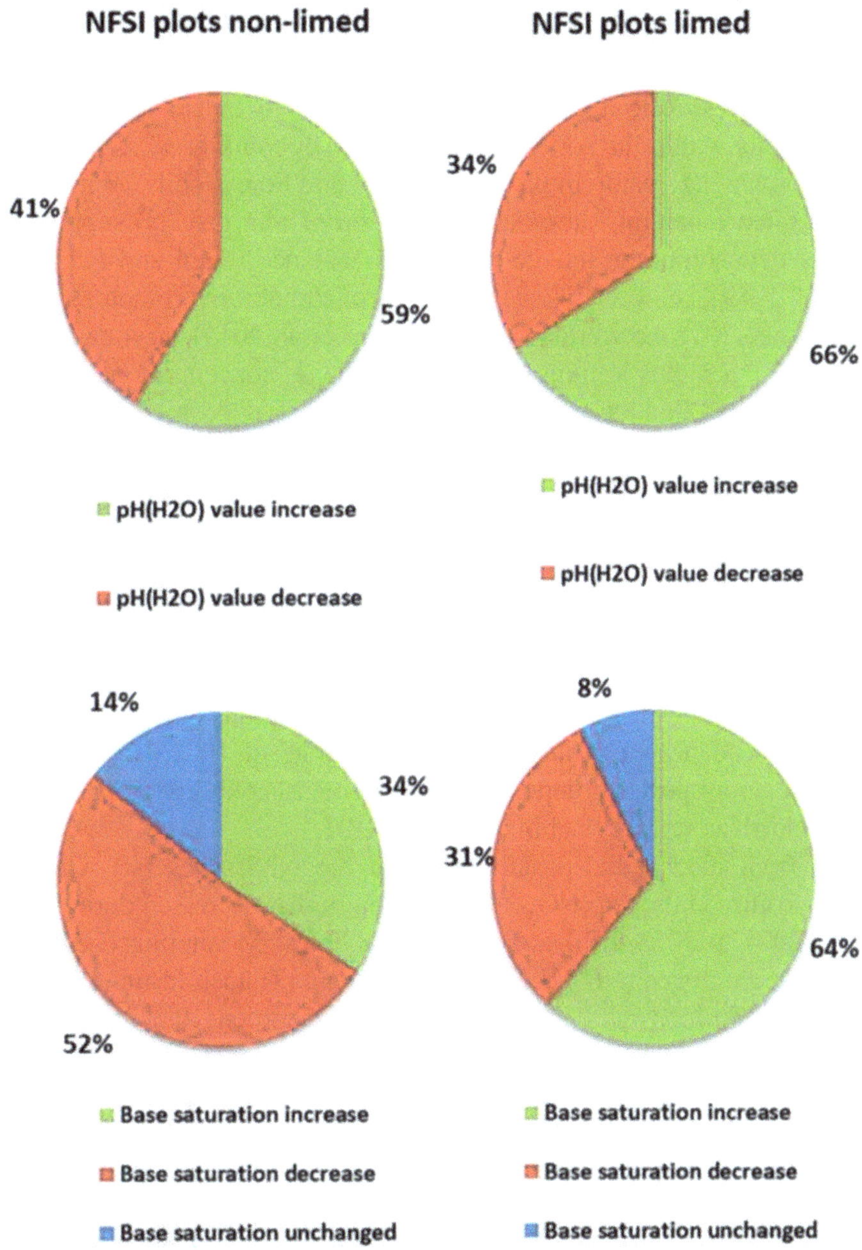

Fig. 12.2 Changes of soil pH (in H_2O, above) and base saturation (below) in NFSI plots in the top mineral soil (10–30 cm depth) from NFSI I to NFSI II, non-limed (left) and limed (right)

We can conclude that clean air policies have been successful in terms of reducing the acidifying sulfur input in forest ecosystems, but efforts have to be increased to reduce nitrogen emissions, to lower their contribution to soil acidification and eutrophication (see Sect. 12.3). Forest liming of soils with considerable acidification is furthermore recommended to balance negative impacts on soil functioning, the vitality, and growth of forests. Both clean air policies and forest liming contributed considerably to the improvement of forest conditions and forest soils in times of *forest decline*, as well as to the prevention of further decline dynamics in Germany.

12.3 Nitrogen Eutrophication Remains Challenging

Eutrophication of terrestrial ecosystems occurs due to the enrichment of priory limiting nutrients within the soil or a water body (Bobbink et al. 2010) and is still considered of having major impact on forests and forest soils. While in forest ecosystems often a natural, site-related limitation of nitrogen (N) supply prevails, enhanced N deposition can induce nutrition imbalance (Oren and Schulze 1989; Schulze et al. 1989), i.e., a relative deficiency of other nutrients compared to nitrogen (Vitousek et al. 1997; Aber et al. 1998; Waldner et al. 2015), or it can improve N availability for trees, thus leading to increased foliar N concentrations (Tietema and Beier 1995). The inclination of the balance is plaid by deposition of both reduced and oxidized N compounds (NO_x, NH_x), with the maximum uptake capacity for nitrogen reached at saturation (Ågren and Bosatta 1988; Aber et al. 1989; De Vries and Schulte-Uebbing 2019). Due to this, Eichhorn (1995) and Cole (1992) define nitrogen saturation as a status at which nitrogen input from deposition and mineralization exceed the retention capacity, and high nitrate leaching occurs (>5 kg ha^{-1} a^{-1}, Block et al. 2000; >10 kg ha^{-1} a^{-1}; Vanguelova et al. 2010). This can also affect species composition of forest ground vegetation including the loss of rare species adapted to nitrogen-poor site conditions (Bobbink et al. 1998). Already an exceedance of a nitrogen deposition rate of 10–15 kg ha^{-1} a^{-1} can change forest floor species composition, increase susceptibility to pathogens, and vary the mycorrhiza regime (Bobbink et al. 2010).

Varying from modelled deposition rates at the NFSI sites (see Chap. 1 and Sect. 12.2), oxidized nitrogen (NO_x-N) deposition further decreased during the period from 2002–2004 to 2012–2014 at the Intensive Forest Monitoring (Level II) sites, whereas reduced nitrogen (NH_4-N) remained constant (Thünen Institute 2018). Mean total nitrogen deposition rates in Germany ranged between 10.6 kg ha^{-1} a^{-1} and 40.7 kg ha^{-1} a^{-1} (median 19.7 kg ha^{-1} a^{-1}) in the period from the first (1990) to the second NFSI (2007) (see Chap. 5). Compared to other European countries, forests in Germany were therefore among the more severely polluted in Europe (Michel and Seidling 2016) with regions exceeding 50 kg°ha^{-1}°year^{-1} (Meesenburg et al. 2005, 2016).

At NFSI sites, extreme exceedances (>10 kg ha^{-1} a^{-1}) of critical loads for eutrophying nitrogen (CL_{nut}(N); see Chap. 2) were observed on 85% of the sites in 1990 with a reduction to less than 50% in 2006 and less than 20% in 2015 (Table 12.1).

Table 12.1 Exceedance risk of eutrophying N (CL_{nut}(N), *Ex.* exceedance, *Pot. Ex.* potential exceedance) on NFSI plots in 1990, 2006, and 2015

Classes (kg ha^{-1} a^{-1})	1990	2006	2015
No exceedance	0.1	0.4	1.1
Pot. Ex. ≤ 10	2.9	11.9	19.4
Pot. Ex. > 10	3.4	2.3	1.0
Ex. ≤ 10	8.7	36.6	59.1
Ex. > 10	84.9	48.8	19.4

Potential exceedance is calculated from the conservative approach with N_{crit} = 0.2–0.4 mg l^{-1} N

Nevertheless, sites with no exceedance are still rare. CL_{nut}(N) exceedances at the time of the NFSI II (2006) were found to increase the occurrence of nitrophilic plant species in montane Norway spruce forests, particularly competitive on eutrophic sites (Ziche et al. 2016).

Within the period nitrogen stocks in forest soils increased in the top mineral soil layer (significantly in 0–5 cm soil depth, slightly in 5–10 cm soil depth). The decrease of nitrogen stock in the layers below 30 cm may be caused by nitrate leaching or gaseous nitrogen losses. Forest soil liming, in particular when repeated, tends to facilitate the increase of nitrogen stock in the mineral soil (Melvin et al. 2013) and may result in nitrate leaching (Durka and Schulze 1992).

Overall, elevated nitrate leaching (>5 kg ha^{-1} a^{-1}) is not expected at the vast majority of NFSI sites (see Chap. 5). However, it is possible on several N-saturated sites with continuously high N deposition or soils with low buffer capacity (UBA 2013). Excess nitrogen nutrition can be observed in Scots pine and oak and to a lesser extent in Norway spruce and European beech. Foliar nitrogen concentration increased in spruce and beech between the NFSI I (1990) and NFSI II (2006) with soil liming showing no distinct effect. With a balanced supply of nutrients being important for plant growth and performance, a high nitrogen status can be linked to nutrition deficiency, in particular of phosphorus (P) in all major tree species and potassium (K) in spruce and beech on one quarter of the sampled sites (see Chap. 8; Braun et al. 2010).

Nitrogen deposition led to a distinct nitrogen eutrophication of forest ecosystems in parts of Germany. Heterogeneous spatial patterns of forest ecosystems with and without nitrogen eutrophication may also be due to varying, distinct nitrogen losses in the past, linked to the practice of litter raking inducing enormous nitrogen exports (up to 1.7 t N ha^{-1} a^{-1}, Kreutzer (1972)), as well as clear-cuts, and stand disturbances stimulating the decomposition of soil organic matter. Nitrogen oversupply (SRU 2015) has several negative effects on forest biodiversity, partly on tree nutrition, and under condition of N saturation, it induces undesired nitrogen outputs, e.g., nitrate leaching into the groundwater and emission of nitrous oxide from the soil into the atmosphere. Due to this, the reduction of nitrogen emissions, in particular of ammonia, should have a particularly high priority in clean air policies, i.e., compliance with emission standards according to the multicomponent protocol of the UN/ECE convention on Long-Range Transboundary Air Pollution (CLRTAP) and the EU National Emission Ceilings Directive (EU NEC Directive).

12.4 Nutrient Sustainability Limits Biomass Harvest Options

Nitrogen (N), phosphorus (P), magnesium (Mg), potassium (K), and calcium (Ca) are macronutrients crucial for tree growth, and their deficiency can limit terrestrial ecosystems functioning (Ellenberg et al. 1986). Harvest losses of nutrients,

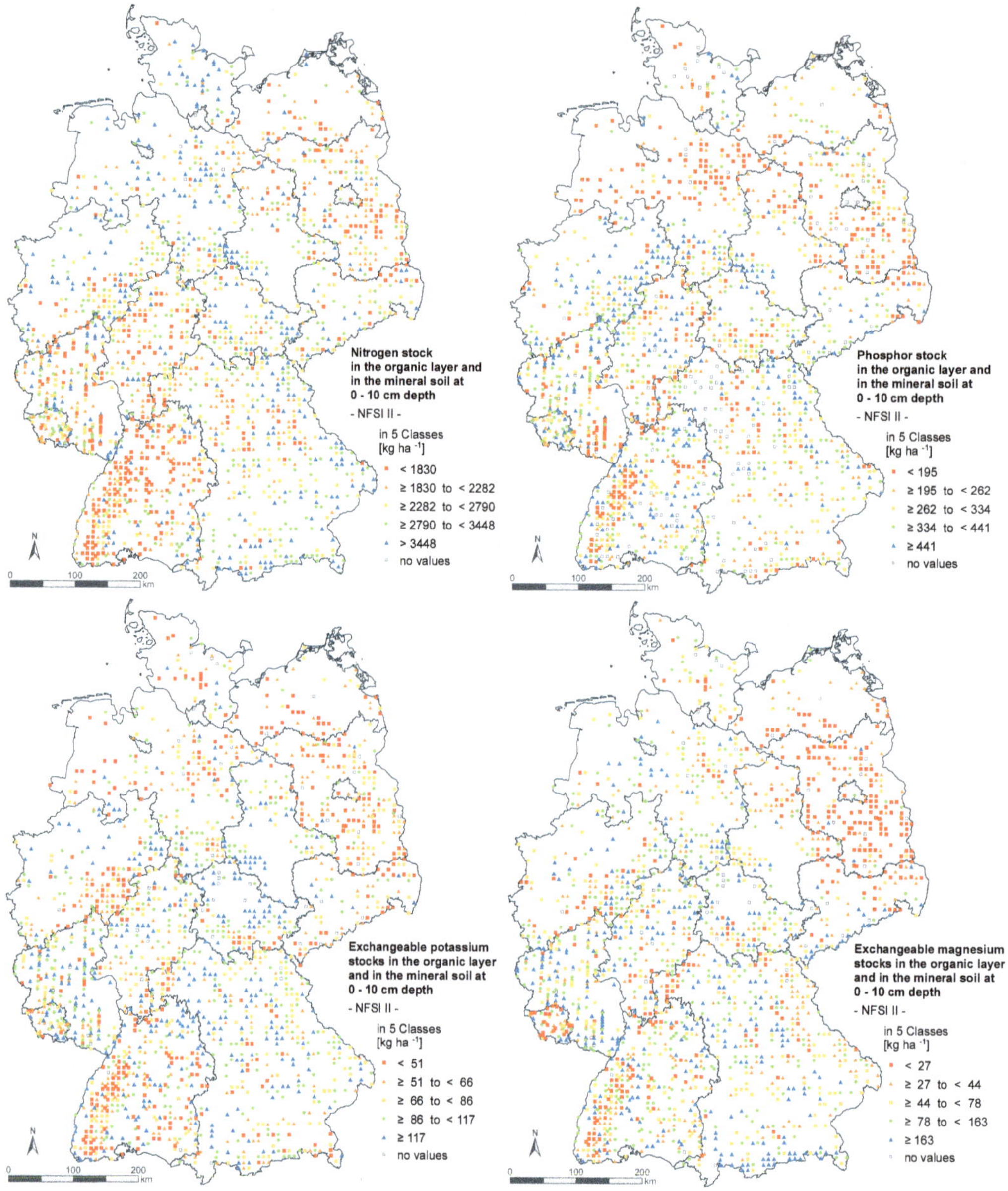

Fig. 12.3 Total nitrogen (N, above left) and phosphorus (P, above right) stocks, as well as exchangeable potassium (K, below left) and magnesium (Mg, below right) in the organic layer and mineral top soil (down to 10 cm depth)

not balanced by atmospheric deposition, can only be resupplied through weathering of soil minerals. However, liming of an overall of 21% of the NFSI II sites, with 51% being acid-sensitive, changed the base saturation and in particular Ca availability considerably (see Chaps. 2 and 4).

Below average N, P, Mg, and K stocks are visible in the organic and the mineral soil layer down to 10 cm depth in large parts of the northeastern lowlands (Fig. 12.3). However, late Pleistocene sites in eastern Schleswig-Holstein and northern Mecklenburg-West Pomerania are an exception with higher values. Above average nutrient stock is found in most areas of central and southern Germany, with low

Table 12.2 Percentage (%) of NFSI II plots with nutrient deficiency (incl. latent range)/luxury nutrition (surplus range) based on foliar threshold values, according to Göttlein (2015); all deficiency percentage values >15% in bold

	N	P	K	Ca	Mg	S
Norway spruce	9/**29**	**20**/12	**22**/9	2/**43**	6/**24**	**43**/0
Scots pine	5/**54**	**19**/1	13/1	6/**20**	**25**/4	13/0
European beech	2/**26**	**61**/1	**29**/7	**17/26**	**27/36**	**46**/1
Oak species	1/**55**	**38**/7	10/8	**16/26**	**16**/5	0/0

values in the Black Forest, several sites in the Rhine-Main area, the Palatinate Forest (low N, P), eastern North Rhine-Westphalia (low K), and the Saarland region (low Mg). A high amount of nutrients were transferred from the humus layer to the mineral soil between the periods of both NFSIs due to a shift of organic matter (Grüneberg et al. 2014; see Chap. 5), thus facilitating nutrient uptake of trees.

Foliar nutrient concentration of the major tree species is low in P (all species) and S (European beech, Norway spruce) at a larger number of sites. Moreover, K deficiency is particularly visible at several spruce and beech sites. Overall, forest foliar nutrition is following the sequence European beech > Norway spruce > Oak spec. > Scots pine (Table 12.2). In contrast, nitrogen nutrition is mostly luxurious and often induces P deficiency relative to N concentration (high N/P ratio; see Chap. 8; Braun et al. 2010; Sardans et al. 2016). Many forest sites have a limited natural supply of Ca, Mg, and K, which is only supplied through mineral weathering. Ecosystem balances of these important nutrients and base cations are often negative partially due to their export by drainage water in relation to a surge of accompanying mobile anions like nitrate and sulfate in the soil solution (see Chap. 3; Ulrich 1986b; Likens et al. 1996).

Due to this sensitive balance in nutrient, harvest intensity, in particular extended harvest of biomass, must be carefully adjusted to the specific site conditions and the potential for replacing exported nutrients (Block and Meiwes 2013). Whole-tree harvesting (WTH, i.e., total biomass harvest above ground) which is rarely practiced in Germany exports the nutrient-rich tree compartments like bark, branches, and often needles or leaves from the forest. Thus, WHT should be restricted to sites with a sufficient nutrient provision ensuring the replacement of lost nutrients and sustainable site productivity.

12.5 Forest Soils Absorb Heavy Metals

Both, geogenic concentration of parent material and atmogenic deposition, determine the concentration of heavy metals (As, Pb, Cd, Cr, Cu, Ni, Hg, Zn) in forest soils. Significant deposition effects, particularly a variance in top soil concentration, were observed for lead (Pb), cadmium (Cd), and mercury (Hg), whereas such effects are less evident for arsenic (As) and copper (Cu). Soil concentrations of the other elements

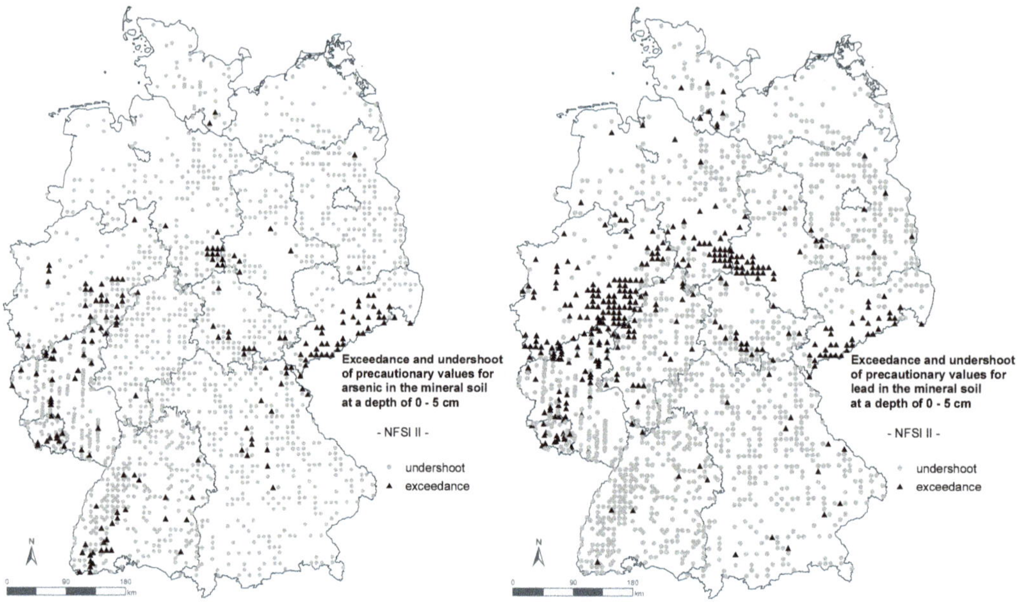

Fig. 12.4 Exceedance of precautionary values for arsenic (As, left) and lead (Pb, right) in the upper mineral soil (0–5 cm depth)

[chromium (Cr), nickel (Ni), zinc (Zn)] are mainly determined by geogenic base concentration (see Chap. 6; Luster et al. 2006). In forest soils, heavy metals are absorbed either by mineral surfaces (following the sequence Cd, Zn < Cu < Ni, Cr < Pb) or by complexation with humic substances in the organic soil compounds (stable complexes, Pb, Cr, Ni, and Cu; weak complexes, Cd and Zn, König et al. 1986).

The heavy metal concentration (Pb, Cd, Cr, Cu, Zn) in the organic layer decreased significantly from the first (1990) to the second (2006) NFSI ranging from −33% (Pb) to −11% (Zn). This indicates a decreased deposition of these elements between both NFSIs as the decrease was lower (−2%) in humus storage. An indication for a translocation of heavy metals into the mineral soil is missing, except for the relatively mobile Cd. The only element which increased within the soil concentration is Hg, although deposition has decreased in central Europe (Ilyin et al. 2016).

Only arsenic (As) with 11.2% and lead (Pb) with 22.1% exceed the precautionary values according to the Federal Soil Protection Act (BBodSchV) at a national scale. Such critical values are mainly recorded in specific regions like the Ore Mountains (As, Pb), the Harz, and the Rhenish Slate Mountains (Pb), resulting from both mining activity over the past century (Fig. 12.4) (Medyńska-Juraszek and Kabała 2012; Perković et al. 2017) and elevated atmospheric contamination. For Pb, past atmospheric inputs of leaded gasoline still play an important role as shown by isotope analyses within the French forest monitoring network (Hernandez et al. 2003).

Heavy metals are mainly absorbed in the organic compounds of the humus layer and in particular in the mineral soil (Luster et al. 2006). Therefore, a critical translocation within the ecosystem and a migration of these toxic elements into the groundwater are prevented. All silvicultural operations increasing and stabilizing the soil organic compound are thus recommended. These include low-disturbance stand management without clear-cuts and mechanical soil treatments. Liming can stabilize the soil organic matter (see Sect. 12.7) and prevent soil acidification (see Sect. 12.2) as an additional measure to keep and develop the absorptive function of forest soils for heavy metals. Nevertheless, all emission reduction options should be used in order to further reduce or even stop future heavy metal emissions. In doing so, special attention should be paid to mercury (Hg) due to its increasing concentration in forest soils.

12.6 Organic Pollutants (POPs) Persist Long Term in Forest Soils

Persistent organic pollutants (POPs) can have harmful effects on humans and the environment paired with a low decomposition rate. These substances can accumulate in various environmental compartments, particularly in soils and sediments. POPs include polycyclic aromatic hydrocarbons (PAHs), polychlorinated biphenyls (PCBs), dichlorodiphenyltrichloroethane (DDT), and hexachlorobenzene (HCB) emitted for several decades and deposited throughout Germany (see Chap. 8).

In general, low concentrations are visible for the investigated POPs in German forest soils. For PAHs and PCBs, lower concentrations than the precautionary values of the German soil protection legislation (BBodSchV) were found. Densely industrialized and urbanized regions show comparatively enhanced PAHs and PCBs values, e.g., in vicinity of brown-coal strip-mining areas (Aichner et al. 2015; Pandelova et al. 2018). The long persistence of POPs in forest soils can be seen for DDx applied as pesticide in the 1980s and still detected in the organic layer, particularly in the Eastern Federal States. The deposition and concentration of POPs are not significantly linked to forest type and environmental factors but on the presence and distance of mostly local or regional emission sources (see Chap. 7; Aichner et al. 2013).

It is difficult to rule out deposition from various sources completely. However, significant efforts have to be taken in order to reduce any additional deposition of POPs in forest ecosystems due to their decade long persistence with the ecosystems. The application of pesticides with persistent organic compounds in forests should be strictly minimized and restricted to cases where other options are not available. Benefits of POP-pesticide application have to be carefully balanced against the threats. Since the concentration of most POPs is linked to the soil organic concentration (SOC) of the organic layer and the top mineral soil, high-disturbance stand management and soil treatments leading to decomposition of soil organic matter should be handled with care.

12.7 Carbon Sequestration in Forest Soil Supports Climate Protection

Forest and forest soils in Germany sequester 52–58 Tg (= million t) CO_2-equivalent per year (reference period 1990 to 2012, year 2015) and thus play an important role in climate protection (Wellbrock et al. 2017; UBA 2017). Including the benefit of wood products from forest utilization, an additional emission amount of 36 Tg CO_2-equivalent is annually avoided within Germany due to the substitution of fossil fuel through firewood (energetic substitution), approximately 30 Tg CO_2 equivalent by the substitution of energy-intensive materials by wood products and material (material substitution) and 3 Tg CO_2 equivalent through increased wood products storage (WBAE/WBW 2017). Thus, forest and sustainable forest utilization amounts to far more than 100 Tg CO_2-equivalet of actual yearly greenhouse gas mitigation achievements in Germany. However, in contrast to the forest carbon sink status of forests in Germany, large-scale deforestation and forest degradation in other parts of the world result in a global net emission of 4900 Tg CO_2-equivalent per year (year 2010, Tubiello et al. 2015).

The organic and the mineral layer (up to 90 cm depth) stored 1023 Tg C (year 2012) and thus represent the second largest carbon pool in forests after forest biomass (1149 Tg C). In total, German forest stands and soils contained approx. 2384 Tg C after 1951 Tg C in 1990 (Fig. 12.5), (Wellbrock et al. 2017).

In mineral forest soils (up to 30 cm depth), the annual increase of carbon storage amounts to 0.4 t ha^{-1} a^{-1} between 1990 and 2006. Concurrently the organic layer lost 0.02 t C ha^{-1} a^{-1} (see Chap. 5; Grüneberg et al. 2014). This relates to a yearly carbon sink rate of 15 Tg CO_2-equivalent for the top soils down to 30 cm depth throughout Germany.

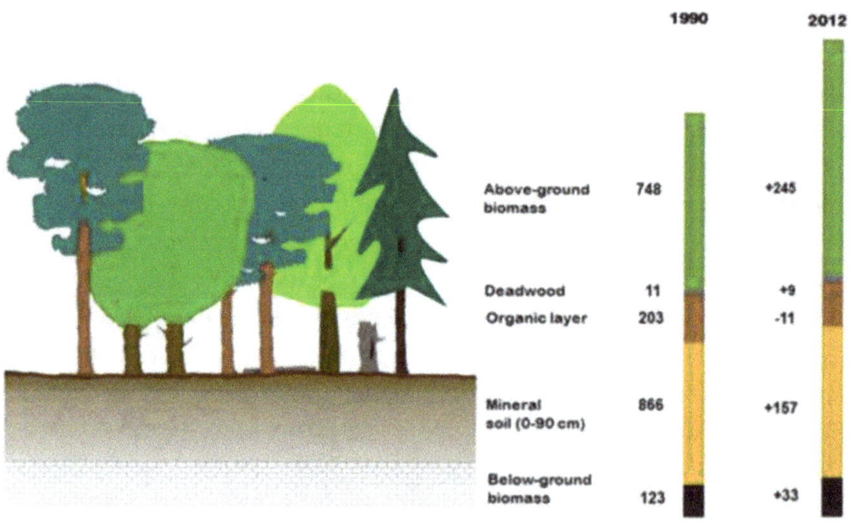

Fig. 12.5 Carbon storage [in Tg C] in forests and forest soils in Germany according to the five carbon pools to be reported in greenhouse gas reporting under LULUCF (Wellbrock et al. 2017)

These results reveal that the sustainable management of forests in Germany over the last decades led to carbon sequestration both in forest and forest soils as well as in wood products. This is complemented with a considerable avoidance of greenhouse gas emissions due to a substitution of fossil fuels and other energy-intensive materials. Thus, the combination of site-adapted, sustainable forest management combined with the construction of long-lived wood materials and products and the option of a final, energetic use (cascade utilization) provides evident benefits for climate protection and should be continued. Large-scale restrictions for forest management or even the ban of timber use would counteract climate protection through decreased energetic and material substitution of wood and wood products (WBAE/WBW 2017), as well as wood imports from larger areas outside Germany (Schulze et al. 2016). Forest transformation with an increase of deciduous tree species and soil liming has no significant effect on the amount of carbon in the organic layer and the mineral soil. However, both management activities contribute to a shift of carbon from the organic layer to the mineral soil, thus possibly stabilizing carbon storage in forest soils. The advantage of an increased share of deciduous trees, however, has to be traded off against the disadvantage of a lower demand of hardwood for wood products and material substitution (Frühwald and Knauf 2013) and the susceptibility of some of these species to climate extremes (see Chap. 11). Different intensities of practiced continuous cover forestry (without clear-cuts) are not significantly affecting carbon sequestration in forest soils (Mund and Schulze 2006; Jandl et al. 2011; Wäldchen et al. 2013; Grüneberg et al. 2013).

Thus, sustainable forestry in Germany produced considerably positive effects on climate protection, and there are overall benefits of continuing the existing practice of forest management. Timber should be used as efficiently as possible, and its use for wood products should always be preferred to an energetic use. Initiatives aiming at the implementation of multiple and cascading uses of timber should be supported through structural measures and incentives (WBAE/WBW 2017). Due to the benefits of wood use, harvest restrictions have to be carefully considered.

12.8 Atmospheric Pollution Interacts with Climate Change Impacts

Impacts of air pollutants and atmospheric deposition on trees can change the response to climate and especially to extreme weather events like heat waves and droughts (Paoletti et al. 2007). Bytnerowicz et al. (2007) mention an increased sensitivity of trees with nitrogen oversupply to late frost, pathogens, and drought (Dziedek et al. 2016). Moreover, a shallower rooting after soil acidification can induce higher drought sensitivity (Ulrich 1986a). Drought is affecting trees and forests by its timing, duration, and intensity (see Chap. 3); a soil water availability of less than 20% can induce severe consequences for forest ecosystem productivity and integrity (Granier et al. 2007), as well as for young trees' survival (Bolte et al. 2016).

The crown condition assessment at NFSI sites provides relevant insights into the risk status of forests to air pollution and climatic impacts (see Chap. 11).

Norway spruce trees, with excessive nitrogen needle concentrations, show a significantly higher defoliation, while their condition is better at limed sites. Norway spruce growing on soils of low water storage capacity and high evapotranspiration potential exhibit more often needle losses than those on other sites. Spruce mortality is often high for several years after warm and dry years or storm events likely linked to increased bark beetle infestations (Bolte et al. 2014). Dominating risks for spruce are intensifying nitrogen eutrophication and climate warming with accompanying biotic threats.

For European beech, similar impact factors on crown condition are evident due to a distinct defoliation peak after the drought in 2003 (Seidling 2007; Eichhorn and Roskams 2013); however, pathogens do not play an important role and overall mortality is low.

Crown condition of oak species is similarly affected by climate warming showing higher defoliation values after the 2003 drought year. Oaks show and have been showing the highest defoliation of all four major tree species in Germany for decades. This is mainly due to pathogen impact of leaf-eating insect communities stimulated by warmer and longer growing seasons (see Chap. 11) which can produce a feedback loop on carbon cycling with enhanced releases of CO_2 and dissolved organic carbon (DOC) from forest soils after decomposition rise of organic matter (Arnold et al. 2016).

These results show that a significant lowering of greenhouse gas emissions is urgently needed to limit ongoing climate change to an extent ($\leq$2 °C aim) to which forests in Germany are able to adapt. This is in particular needed, as past and current air pollutants have predisposed many forests to decline.

12.9 Forest Transformation Affects Forest Soils Positively

Forest transformation from mono-species coniferous forests with Norway spruce and Scots pine to broadleaved and mixed forests is a common practice of forest management in Germany and changed forest structures considerably. Forest cover with spruce and pine has decreased by 329,000 ha, whereas areas with European beech, oak, and other broadleaved species increased by 315,000 ha from 2002 to 2012 (BMEL 2015).

This resulted in positive effects for forest soils: on average broadleaved forests exhibit lower soil acidification and higher base saturation compared to coniferous forests and sequester more organic matter and carbon (SOC) in the mineral soil (see Chaps. 2, 4, and 6). The immobilization of SOC in mineral soil horizons counteracts disturbances of the organic layer and top soil with SOC losses. Moreover, organically bound nutrients are better available for root uptake.

Due to the strongly prevailing economic role of coniferous timber for the wood processing sector (Dieter and Janzen 2015) and the required diversification of risks for climate change impacts (Kolström et al. 2011), mixed forests are preferable with an adequate proportion of coniferous trees (WBAE/WBW 2017) adaptive to future climate and site conditions. Structural measures and incentives should aim at increasing the mixed forest area in Germany including stress-tolerant native species like Silver fir but also non-native species like Douglas and Grand fir. Therefore, ungulate browsing adverse to mixed stand regeneration and development as well as species diversity (Ammer 1996; Schulze et al. 2014) has to be controlled.

12.10 Conclusions and Outlook

Anthropogenic induced site variations due to acid and nitrogen deposition, as well as a changing climate, have modified forest soil traits more rapidly than expected, thus producing forests and forest soils with a changed ecosystem service and risk portfolio.

The results of the repeated NFSIs produced new opportunities to assess the effects of environmental change and management on forests and forest soils. For the first time, status variation of forest soils and corresponding forest ecosystem traits were analyzed in a systematic and representative manner for Germany on a national and regional scale. Both the number of 1900 forest sites and the extent of the various parameters for crown condition, forest stand, vegetation, nutrition, and soil status have provided novel options to link their results with the findings of the Intensive Forest Monitoring (ICP Forests, Level II), in particular for nutrient balances and soil processes, as well as with model approaches as demonstrated in this volume, e.g., for nitrogen cycling and budgeting (see Chap. 5). New possibilities also address the integrative assessment of forest and agricultural sites after the Agricultural Soil Inventory for Germany (Bach et al. 2011) will be finalized. This will enable insights in the sustainable development of soil organic carbon (SOC) in forest and agricultural soils with varying environmental conditions and land uses and could also allow for an economic evaluation of different land-use scenarios.

The NFSI is an important monitoring instrument of the national forest assessment system in Germany which is codified in the Federal Forest Act (FFA, *Bundeswaldgesetz* § 41a: *Walderhebungen*) and of specific meaning for political consultation of the evaluation of forest and soil status changes, in their extent, dynamic, and regional patterns based on time series produced by repeated sampling of the same forest sites with comparable methodologies. Based on the NFSI results, a repetition period of 15–20 years is recommended (Fig. 12.6). Due to this, a new federal ordinance for the third NFSI is currently prepared for the period 2022–2024 in order to picture the dynamic development of soils by monitoring.

The findings show that both, further efforts to reduce air pollution and the consequent continuation of a sustainable and soil conservative forest management, are needed to maintain and develop vital and productive forests with fertile soils. Therefore, the advancement and implementation of politics and structural measures are important. The achievements in reducing the acid burden through clean air

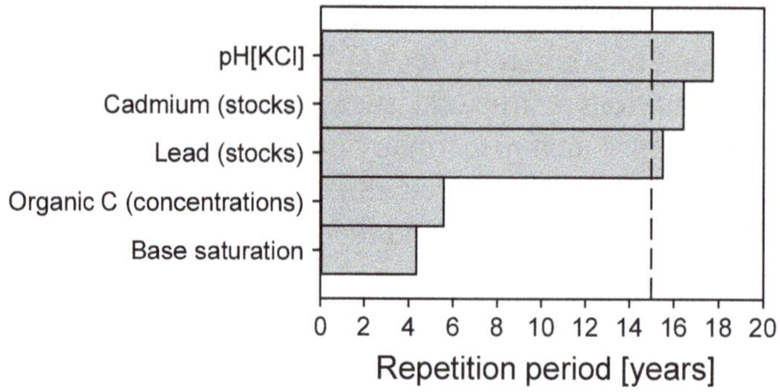

Fig. 12.6 Repetition period for significant changes of different soil status parameters based on analyses of the first and second NFSI. Lead (Pb), cadmium (Cd) storages are derived from the organic layer, whereas pH(KCl), SOC content, and base saturation are related to the upper mineral soil (to 10 cm depth). The dashed line indicates a period of 15 years above which significant changes can be expected for all shown parameters

policies and forest soil liming demonstrate the options of concerted political action and local management. Only the results of the NFSI could provide evidence for these achievements.

The NFSI also demonstrates the successful collaboration of the German Federal and the Federal State authorities in this nation-wide assessment of forest and soil status both on a national and regional scale. Not all research questions can be answered with the NFSI approach of a systematical grid sampling. Thus, several detailed questions like the impacts of forest soil compaction by heavy harvest technique require additional research with alternative approaches. However, the integration of results of additional soil research in the NFSI evaluation provides an advanced benefit.

Finally, the major gain of the NFSI emerges from the attention and implementation of the results in policy, administrative processes, and forest management.

References

Aber JD, Nadelhoffer KJ, Steudler P, Melillo JM (1989) Nitrogen saturation in northern forest ecosystems. Bioscience 39(6):378–386. https://doi.org/10.2307/1311067

Aber JD, McDowell W, Nadelhoffer K, Magill A, Berntson G, Kamakea M, McNulty S, Currie W, Rustad L, Fernandez I (1998) Nitrogen saturation in temperate forest ecosystems—hypotheses revisited. Bioscience 48(11):921–934. https://doi.org/10.2307/1313296

Ågren GI, Bosatta E (1988) Nitrogen saturation of terrestrial ecosystems. Environ Pollut 54(3–4):185–197. https://doi.org/10.1016/0269-7491(88)90111-X

Aichner B, Bussian B, Lehnik-Habrink P, Hein S (2013) Levels and spatial distribution of persistent organic pollutants in the environment: a case study of German forest soils. Environ Sci Technol 47(22):12703–12714. https://doi.org/10.1021/es4019833

Aichner B, Bussian BM, Lehnik-Habrink P, Hein S (2015) Regionalized concentrations and fingerprints of polycyclic aromatic hydrocarbons (PAHs) in German forest soils. Environ Pollut 203:31–39. https://doi.org/10.1016/j.envpol.2015.03.026

Almer B, Dickson W, Ekström C, Hörnström E, Miller U (1974) Effects of acidification on Swedish lakes. Ambio 3(1):30–36

Ammer C (1996) Impact of ungulates on structure and dynamics of natural regeneration of mixed mountain forests in the Bavarian Alps. For Ecol Manag 88(1–2):43–53. https://doi.org/10.1016/s0378-1127(96)03808-x

Arnold AIM, Grüning M, Simon J, Reinhardt AB, Lamersdorf N, Thies C (2016) Forest defoliator pests alter carbon and nitrogen cycles. R Soc Open Sci 3(10). https://doi.org/10.1098/rsos.160361

Bach M, Freibauer A, Siebner C, Flessa H (2011) The German agricultural soil inventory: sampling design for a representative assessment of soil organic carbon stocks. Procedia Environ Sci 7:323–328. https://doi.org/10.1016/j.proenv.2011.07.056

Block J, Meiwes KJ (2013) Erhaltung der Produktivität der Waldböden bei der Holz-und Biomassenutzung. In: Bachmann G, König W, Utermann J (eds) Bodenschutz – Ergänzbares Handbuch der Maßnahmen und Empfehlungen für Schutz, Pflege und Sanierung von Böden, Landschaft und Grundwasser. Erich Schmidt Verlag, Berlin

Block J, Eichhorn J, Gehrmann J, Kölling C, Matzner E, Meiwes KJ, von Wilpert K, Wolff B (2000) Kennwerte zur Charakterisierung des ökochemischen Bodenzustandes und des Gefährdungspotentials durch Bodenversauerung und Stickstoffsättigung an Level II-Waldökosystem-Dauerbeobachtungsflächen. Bundesministerium für Ernährung. Landwirtschaft und Forsten, Bonn

BMEL (2015) The forests in Germany—selected results of the Third National Forest Inventory. BMEL, Berlin

Bobbink R, Hornung M, Roelofs JGM (1998) The effects of air-borne nitrogen pollutants on species diversity in natural and semi-natural European vegetation. J Ecol 86(5):717–738. https://doi.org/10.1046/j.1365-2745.1998.8650717.x

Bobbink R, Hicks K, Galloway J, Spranger T, Alkemade R, Ashmore M, Bustamante M, Cinderby S, Davidson E, Dentener F, Emmett B, Erisman JW, Fenn M, Gilliam F, Nordin A, Pardo L, de Vries W (2010) Global assessment of nitrogen deposition effects on terrestrial plant diversity: a synthesis. Ecol Appl 20(1):30–59. https://doi.org/10.1890/08-1140.1

Bolte A, Hilbrig L, Grundmann BM, Roloff A (2014) Understory dynamics after disturbance accelerate succession from spruce to beech-dominated forest—the Siggaboda case study. Ann For Sci 71(2):139–147. https://doi.org/10.1007/s13595-013-0283-y

Bolte A, Czajkowski T, Cocozza C, Tognetti R, de Miguel M, Pšidová E, Ditmarova L, Dinca L, Delzon S, Cochard H, Ræbild A, de Luis M, Cvjetkovic B, Heiri C, Müller J (2016) Desiccation and mortality dynamics in seedlings of different European beech (*Fagus sylvatica* L.) populations under extreme drought conditions. Front Plant Sci 7:1–12. https://doi.org/10.3389/fpls.2016.00751

BR (1983) Dreizehnte Verordnung zur Durchführung des Bundes-Immissionsschutzgesetz (Verordnung über Großfeuerungsanlagen – 13. BImSchV). Bundesgesetzblatt 26, Bonn

Braun S, Thomas VF, Quiring R, Flückiger W (2010) Does nitrogen deposition increase forest production? The role of phosphorus. Environ Pollut 158(6):2043–2052. https://doi.org/10.1016/j.envpol.2009.11.030

Bytnerowicz A, Omasa K, Paoletti E (2007) Integrated effects of air pollution and climate change on forests: a northern hemisphere perspective. Environ Pollut 147(3):438–445. https://doi.org/10.1016/j.envpol.2006.08.028

Carreira JA, Harrison AF, Sheppard LJ, Woods C (1997) Reduced soil P availability in a Sitka spruce (Picea sitchensis (Bong) Carr) plantation induced by applied acid-mist: significance in forest decline. For Ecol Manag 92(1–3):153–166. https://doi.org/10.1016/s0378-1127(96)03914-x

Cole DW (1992) Nitrogen chemistry, deposition, and cycling in forests. In: Johnson DW, Lindberg SE (eds) Atmospheric deposition and forest cycling. Ecological Studies (Analysis and Synthesis), vol 91. Springer, New York, pp 150–213. https://doi.org/10.1007/978-1-4612-2806-6_6

De Vries W, Schulte-Uebbing L (2019) Impacts of nitrogen deposition on forest ecosystem services and biodiversity. In: Schröter M, Bonn A, Klotz S, Seppelt R, Baessler C (eds) Atlas of ecosystem services: drivers, risks, and societal responses. Springer International Publishing, Cham, pp 183–189. https://doi.org/10.1007/978-3-319-96229-0_29

Dieter M, Janzen N (2015) Deutsches Cluster Forst und Holz im internationalen Wettbewerb. AFZ Wald 70(17):13–15

Durka W, Schulze E (1992) Hydrochemie von Waldquellen des Fichtelgebirges. Umweltwiss Schadst Forsch 4(4):217–226

Dziedek C, Hardtle W, von Oheimb G, Fichtner A (2016) Nitrogen addition enhances drought sensitivity of young deciduous tree species. Front Plant Sci 7. https://doi.org/10.3389/fpls.2016.01100

Eichhorn J (1995) Stickstoffsättigung und ihre Auswirkungen auf das Buchenwaldökosystem der Fallstudie Zierenberg. Berichte des Forschungszentrum Waldökosysteme/A, vol 124. Research Centre for Forest Ecosystems of Göttingen University, Göttingen

Eichhorn J, Roskams P (2013) Assessment of tree condition. In: Ferretti M, Fischer R (eds) Forest monitoring—methods for terrestrial investigations in Europe with an overview of North America and Asia. Developments in Environmental Science, vol 12. Newnes, San Diego, pp 139–167. https://doi.org/10.1016/B978-0-08-098222-9.00008-X

Ellenberg H, Mayer R, Schauermann J (1986) Ökosystemforschung: Ergebnisse des Sollingprojekts 1966–1986. Ulmer, Stuttgart

Frühwald A, Knauf M (2013) Verlust an Wertschöpfung und Klimaschutz droht. Holz-Zentralblatt 12:291–293

Göttlein A (2015) Grenzwertbereiche für die ernährungsdiagnostische Einwertung der Hauptbaumarten Fichte, Kiefer, Eiche, Buche. Allg Forst Jagdztg 186(5/6):110–116

Granier A, Reichstein M, Bréda N, Janssens I, Falge E, Ciais P, Grünwald T, Aubinet M, Berbigier P, Bernhofer C (2007) Evidence for soil water control on carbon and water dynamics in European forests during the extremely dry year: 2003. Agric For Meteorol 143(1–2):123–145. https://doi.org/10.1016/j.agrformet.2006.12.004

Grüneberg E, Schöning I, Hessenmöller D, Schulze ED, Weisser WW (2013) Organic layer and clay content control soil organic carbon stocks in density fractions of differently managed German beech forests. For Ecol Manag 303:1–10. https://doi.org/10.1016/j.foreco.2013.03.014

Grüneberg E, Ziche D, Wellbrock N (2014) Organic carbon stocks and sequestration rates of forest soils in Germany. Glob Chang Biol 20(8):2644–2662. https://doi.org/10.1111/gcb.12558

Hernandez L, Probst A, Probst JL, Ulrich E (2003) Heavy metal distribution in some French forest soils: evidence for atmospheric contamination. Sci Total Environ 312(1–3):195–219. https://doi.org/10.1016/s0048-9697(03)00223-7

Hofmann G, Heinsdorf D, Krauss H (1990) Wirkung atmogener Stickstoffeinträge auf Produktivität und Stabilität von Kiefern-Forstökosystemen. Beitr Forstwirtsch 24(2):59–73

Hüttl RF, Schaaf WW (1997) Magnesium deficiency in forest ecosystems. Nutrients in Ecosystems, vol 1. Springer Science & Business Media, Dordrecht. https://doi.org/10.1007/978-94-011-5402-4

Ilyin I, Rozovskaya O, Travnikov O, Varygina M, Aas W, Pfaffhuber KA (2016) Assessment of heavy metal transboundary pollution, progress in model development and mercury research. EMEP Status Report vol 2/2016. MSC-E & CCC, Moscow, Russia & Kjeller, Norway

Jandl R, Alm J, Vesterdal L, Olsson M, Weiss P, Sjögersten S, Rodeghiero M, Leifeld J, Hagedorn F, Bellamy P, Baritz R (2011) Soil carbon in sensitive European ecosystems: from science to land management—a summary. In: Jandl R, Rodeghiero M, Olsson M (eds) Soil carbon in sensitive European ecosystems. Wiley-Blackwell, Chichester, pp 267–281. https://doi.org/10.1002/9781119970255.ch11

Kolström M, Lindner M, Vilén T, Maroschek M, Seidl R, Lexer MJ, Netherer S, Kremer A, Delzon S, Barbati A, Marchetti M, Corona P (2011) Reviewing the science and implementation of climate change adaptation measures in European forestry. Forests 2(4):961–982. https://doi.org/10.3390/f2040961

König N, Baccini P, Ulrich B (1986) Der Einfluß der natürlichen organischen Substanzen auf die Metallverteilung zwischen Boden und Bodenlösung in einem sauren Waldboden. J Plant Nutr Soil Sci 149(1):68–82. https://doi.org/10.1002/jpln.19861490109

Kreutzer K (1972) Über den Einfluß der Streunutzung auf den Stickstoffhaushalt von Kiefernbeständen (*Pinus silvestris* L.). Forstwiss Centralbl 91(1):263–270. https://doi.org/10.1007/BF02741000

Likens GE, Driscoll CT, Buso DC (1996) Long-term effects of acid rain: response and recovery of a forest ecosystem. Science 272(5259):244–246. https://doi.org/10.1126/science.272.5259.244

Luster J, Zimmermann S, Zwicky CN, Lienemann P, Blaser P (2006) Heavy metals in Swiss forest soils: modification of lithogenic and anthropogenic contents by pedogenetic processes, and implications for ecological risk assessment. In: Frossard E, Blum WEH, Warkentin BP (eds) Function of soils for human societies and the environment. Special Publications, vol 266. Geological Society, London, pp 63–78

Matzner E, Zuber T, Alewell C, Lischeid G, Moritz K (2004) Trends in deposition and canopy leaching of mineral elements as indicated by bulk deposition and throughfall measurements. In: Matzner E (ed) Biogeochemistry of forested catchments in a changing environment. Ecological Studies (Analysis and Synthesis), vol 172. Springer, Berlin, pp 233–250. https://doi.org/10.1007/978-3-662-06073-5_14

Medyńska-Juraszek A, Kabała C (2012) Heavy metal pollution of forest soils affected by the copper industry. J Elem 17(3):441–451. https://doi.org/10.5601/jelem.2012.17.3.07

Meesenburg H, Mohr K, Dämmgen U, Schaaf S, Meiwes KJ, Horváth B (2005) Stickstoff-Einträge und -Bilanzen in den Wäldern des ANSWER-Projektes – Eine Synthese. Landbauforsch Volk 279:95–108

Meesenburg H, Ahrends B, Fleck S, Wagner M, Fortmann H, Scheler B, Klinck U, Dammann I, Eichhorn J, Mindrup M (2016) Long-term changes of ecosystem services at Solling, Germany: recovery from acidification, but increasing nitrogen saturation? Ecol Indic 65:103–112. https://doi.org/10.1016/j.ecolind.2015.12.013

Melvin AM, Lichstein JW, Goodale CL (2013) Forest liming increases forest floor carbon and nitrogen stocks in a mixed hardwood forest. Ecol Appl 23(8):1962–1975. https://doi.org/10.1890/13-0274.1

Michel AK, Seidling W (2016) Forest condition in Europe: 2016. Technical report of ICP Forests. Report under the UNECE Convention on Long-Range Transboundary Air Pollution (CLRTAP). BFW-Dokumentation. BFW Austrian Research Centre for Forests, Vienna

Mund M, Schulze ED (2006) Impacts of forest management on the carbon budget of European beech (*Fagus sylvatica*) forests. Allg Forst Jagdztg 177(3–4):47–63

Oren R, Schulze E-D (1989) Nutritional disharmony and forest decline: a conceptual model. In: Schulze E-D, Lange OL, Oren R (eds) Forest decline and air pollution: a study of spruce (*Picea abies*) on acid soils. Ecological Studies (Analysis and Synthesis), vol 77. Springer, Berlin, pp 425–443. https://doi.org/10.1007/978-3-642-61332-6_20

Pandelova M, Henkelmann B, Bussian BM, Schramm KW (2018) Results of the second national forest soil inventory in Germany—interpretation of level and stock profiles for PCDD/F and PCB in terms of vegetation and humus type. Sci Total Environ 610:1–9. https://doi.org/10.1016/j.scitotenv.2017.07.246

Paoletti E, Bytnerowicz A, Andersen C, Augustaitis A, Ferretti M, Grulke N, Gunthardt-Goerg MS, Innes J, Johnson D, Karnosky D, Luangjame J, Matyssek R, McNulty S, Muller-Starck G, Musselman R, Percy K (2007) Impacts of air pollution and climate change on forest ecosystems—emerging research needs. Sci World J 7:1–8. https://doi.org/10.1100/tsw.2007.52

Perković I, Lazić A, Pernar N, Roje V, Bakšić D (2017) Forest soil pollution with heavy metals (Pb, Zn, Cd, and Cu) in the area of the "French Mines" on the Medvednica Mountain, Republic of Croatia. Southeast Eur For 8(1):31–40. https://doi.org/10.15177/seefor.17-08

Sardans J, Alonso R, Janssens IA, Carnicer J, Vereseglou S, Rillig MC, Fernandez-Martinez M, Sanders TGM, Penuelas J (2016) Foliar and soil concentrations and stoichiometry of nitrogen and phosphorous across European *Pinus sylvestris* forests: relationships with climate, N deposition and tree growth. Funct Ecol 30(5):676–689. https://doi.org/10.1111/1365-2435.12541

Schulze ED, Lange OL, Oren R (1989) Forest decline and air pollution: a study of spruce (*Picea abies*) on acid soils. Ecological Studies (Analysis and Synthesis), vol 77. Springer, Berlin. https://doi.org/10.1007/978-3-642-61332-6

Schulze ED, Bouriaud O, Wäldchen J, Eisenhauer N, Walentowski H, Seele C, Heinze E, Pruschitzki U, Danila G, Marin G, Hessenmöller D, Bouriaud L, Teodosiu M (2014) Ungulate browsing causes species loss in deciduous forests independent of community dynamics and silvicultural management in Central and Southeastern Europe. Ann For Res 57(2):267–288. https://doi.org/10.15287/afr.2014.273

Schulze ED, Frör O, Hessenmöller D (2016) Externe ökologische Folgen von Flächenstilllegungen im Wald. AFZ Wald 15:24–26

Seidling W (2007) Signals of summer drought in crown condition data from the German Level I network. Eur J For Res 126(4):529–544. https://doi.org/10.1007/s10342-007-0174-6

SRU (2015) Towards an integrated approach for nitrogen. Partial translation of the special report "Nitrogen: Strategies for resolving an urgent environmental problem". German Advisory Council on the Environment (SRU), Berlin. https://www.umweltrat.de/SharedDocs/Downloads/EN/02_Special_Reports/2012_2016/2015_01_Nitrogen_Strategies_summary.pdf?__blob=publicationFile&v=3. Accessed 28 Mar 2019

Thünen Institute (2018) Belastung durch Luftverunreinigungen in Deutschland. https://blumwald.thuenen.de/level-ii/auswertungen/deposition/. Accessed 06/03/2018

Tietema A, Beier C (1995) A correlative evaluation of nitrogen cycling in the forest ecosystems of the EC projects NITREX and EXMAN. For Ecol Manag 71(1–2):143–151. https://doi.org/10.1016/0378-1127(94)06091-v

Tubiello FN, Salvatore M, Ferrara AF, House JI, Federici S, Rossi S, Biancalani R, Golec RDC, Jacobs H, Flammini A, Prosperi P, Cardenas-Galindo P, Schmidhuber J, Sanchez MJS, Srivastava N, Smith P (2015) The contribution of agriculture, forestry and other land use activities to global warming, 1990–2012. Glob Chang Biol 21(7):2655–2660. https://doi.org/10.1111/gcb.12865

UBA (2013) Genug getan für Mensch und Umwelt? – Wirkungsforschung unter der Genfer Luftreinhaltekonvention. Umweltbundesamt (UBA), Dessau-Roßlau

UBA (2017) Submission under the United Nations Framework Convention on Climate Change and the Kyoto Protocol 2017—National Inventory Report for the German Greenhouse Gas Inventory 1990–2015. Climate Change. Umweltbundesamt (UBA), Dessau-Roßlau

UBA (2018) National trend tables for the German atmospheric emission reporting 1990–2016. Umweltbundesamt. https://www.umweltbundesamt.de/themen/luft/emissionen-von-luftschadstoffen. Accessed 02/03/2017

Ulrich B (1986a) Die Rolle der Bodenversauerung beim Waldsterben: Langfristige Konsequenzen und forstliche Möglichkeiten. Forstwiss Centralbl 105(1):421–435. https://doi.org/10.1007/BF02741750

Ulrich B (1986b) Natural and anthropogenic components of soil acidification. J Plant Nutr Soil Sci 149(6):702–717. https://doi.org/10.1002/jpln.19861490607

UNECE (1979) Convention on Long-Range Transboundary Air Pollution (CLRTAP). Geneva

Vanguelova EI, Benham S, Pitman R, Moffat AJ, Broadmeadow M, Nisbet T, Durrant D, Barsoum N, Wilkinson M, Bochereau F, Hutchings T, Broadmeadow S, Crow P, Taylor P, Houston TD (2010) Chemical fluxes in time through forest ecosystems in the UK—soil response to pollution recovery. Environ Pollut 158(5):1857–1869. https://doi.org/10.1016/j.envpol.2009.10.044

Vitousek PM, Aber JD, Howarth RW, Likens GE, Matson PA, Schindler DW, Schlesinger WH, Tilman D (1997) Human alteration of the global nitrogen cycle: sources and consequences. Ecol Appl 7(3):737–750. https://doi.org/10.2307/2269431

Wäldchen J, Schulze ED, Schöning I, Schrumpf M, Sierra C (2013) The influence of changes in forest management over the past 200 years on present soil organic carbon stocks. For Ecol Manag 289:243–254. https://doi.org/10.1016/j.foreco.2012.10.014

Waldner P, Thimonier A, Pannatier EG, Etzold S, Schmitt M, Marchetto A, Rautio P, Derome K, Nieminen TM, Nevalainen S, Lindroos AJ, Merila P, Kindermann G, Neumann M, Cools N, de Vos B, Roskams P, Verstraeten A, Hansen K, Karlsson GP, Dietrich HP, Raspe S, Fischer R, Lorenz M, Iost S, Granke O, Sanders TGM, Michel A, Nagel HD, Scheuschner T, Simoncic P, von Wilpert K, Meesenburg H, Fleck S, Benham S, Vanguelova E, Clarke N, Ingerslev M, Vesterdal L, Gundersen P, Stupak I, Jonard M, Potocic N, Minaya M (2015) Exceedance of critical loads and of critical limits impacts tree nutrition across Europe. Ann For Sci 72(7):929–939. https://doi.org/10.1007/s13595-015-0489-2

WBAE/WBW (2017) Climate change mitigation in agriculture and forestry and in the downstream sectors of food and timber use. Berichte über Landwirtschaft – Sonderheft. Scientific Advisory Board on Agricultural Policy, Food and Consumer Health Protection (WBAE) and Scientific Advisory Board on Forest Policy (WBW) at the Federal Ministry of Food and Agriculture (BMEL), Berlin. https://doi.org/10.12767/buel.v1i1.175

Wellbrock N, Grüneberg E, Riedel T, Polley H (2017) Carbon stocks in tree biomass and soils of German forests. Cen Eur For J 63(2–3):105–112. https://doi.org/10.1515/forj-2017-0013

Ziche D, Michler B, Fischer HS, Kompa T, Höhle J, Hilbrig L, Ewald J (2016) Boden als Grundlage biologischer Vielfalt. In: Wellbrock N, Bolte A, Flessa H (eds) Dynamik und räumliche Muster forstlicher Standorte in Deutschland. Ergebnisse der Bodenzustandserhebung im Wald 2006 bis 2008. Thünen Report 43. Johann Heinrich von Thünen Institute, Federal Research Institute for Rural Areas, Forestry and Fisheries, Braunschweig, pp 292–342. https://doi.org/10.3220/REP1473930232000

Zimmermann F, Lux H, Reuter F, Wienhaus O (2002) SO2 pollution and forest decline in the Ore Mountains—historical aspects, scientific analysis, future developments. In: Lomsky B, Materna J, Pfanz H (eds) SO2—pollution and forest decline in the ore mountains. Ministry of Agriculture of the Czech Republic, pp 86–116

Appendix

Species	Frequency	Explained deviance (%)	Optimum (%)	Tolerance (%)
Acer campestre	81	21	100	
Acer platanoides	101	9	100	
Actaea spicata	28	27	100	
Aegopodium podagraria	43	20	100	
Ajuga reptans	113	15	87	31.50
Alliaria petiolata	64	15	80	25.97
Allium ursinum	11	23	100	
Anemone nemorosa	175	7	74	32.81
Angelica sylvestris	32	15	100	
Aposeris foetida	12	33	100	
Arum maculatum	30	22	100	
Asarum europaeum	67	31	100	
Atropa bella-donna	32	9	84	32.40
Brachypodium pinnatum agg.	11	22	100	
Brachypodium sylvaticum	243	24	100	34.13
Bromus ramosus	20	9	100	
Calamagrostis varia	18	28	100	
Calamagrostis villosa	42	6	0	
Calluna vulgaris	152	6	0	
Campanula rotundifolia	20	9	100	
Campanula trachelium	25	28	100	
Cardamine bulbifera	36	8	70	27.07
Cardamine flexuosa	71	8	58	22.72
Carex acutiformis	18	13	83	26.50
Carex alba	15	34	100	
Carex arenaria	17	10	21	8.55

(continued)

N. Wellbrock, A. Bolte (eds.), *Status and Dynamics of Forests in Germany*, Ecological Studies 237, https://doi.org/10.1007/978-3-030-15734-0

Species	Frequency	Explained deviance (%)	Optimum (%)	Tolerance (%)
Carex flacca	46	20	100	
Carex montana	14	23	100	
Carex ornithopoda	11	22	100	
Carex pendula	18	7	100	
Carex sylvatica	274	15	86	33.12
Cephalanthera damasonium	18	24	100	
Ceratocapnos claviculata	31	6	38	18.59
Chaerophyllum hirsutum	20	24	100	
Chaerophyllum temulum	21	14	100	
Chrysosplenium alternifolium	11	12	76	23.18
Circaea lutetiana	199	11	72	27.80
Cirsium arvense	38	11	88	32.27
Cirsium oleraceum	26	16	100	
Clinopodium vulgare	13	30	100	
Convallaria majalis	95	5	100	
Cornus sanguinea	43	31	100	
Corylus avellana	113	7	100	
Crataegus laevigata	26	13	100	
Crataegus monogyna	84	23	100	
Dactylis polygama	72	8	76	31.25
Daphne mezereum	36	34	100	
Deschampsia flexuosa	881	17	0	
Digitalis purpurea	204	6	31	26.54
Epilobium parviflorum	20	7	62	23.60
Epipactis helleborine	15	8	100	
Euonymus europaea	13	17	100	
Eupatorium cannabinum	47	9	100	
Euphorbia amygdaloides	11	23	100	
Euphorbia cyparissias	38	18	100	
Festuca gigantea	101	7	68	27.79
Festuca rubra	25	9	100	
Filipendula ulmaria	13	15	100	
Fragaria vesca	187	23	100	41.02
Fragaria viridis	11	17	100	
Frangula alnus	235	6	0	
Fraxinus excelsior	330	21	100	39.21
Galeopsis pubescens	47	6	61	24.87
Galeopsis speciosa	12	9	73	25.25
Galium album	18	16	100	
Galium odoratum	230	18	86	30.46

(continued)

Species	Frequency	Explained deviance (%)	Optimum (%)	Tolerance (%)
Galium rotundifolium	80	6	100	
Galium saxatile	174	5	12	36.22
Galium sylvaticum	47	9	100	
Gentiana asclepiadea	10	15	100	
Geranium robertianum	207	12	83	33.27
Geum rivale	12	15	82	23.08
Geum urbanum	169	23	95	31.00
Glechoma hederacea	67	17	100	
Hedera helix	148	7	100	
Heracleum sphondylium	15	11	100	
Hieracium lachenalii	10	7	21	11.11
Hordelymus europaeus	83	18	100	34.95
Hypericum hirsutum	15	13	81	25.09
Hypericum pulchrum	45	6	46	21.62
Hypochaeris radicata	19	6	29	16.44
Impatiens noli-tangere	151	5	62	27.84
Juglans regia	16	15	100	
Knautia dipsacifolia	23	28	100	
Lamium galeobdolon	127	15	87	31.81
Lamium maculatum	13	9	100	
Lapsana communis	44	7	65	25.31
Lathyrus linifolius	10	5	53	21.92
Lathyrus vernus	30	26	100	
Ligustrum vulgare	27	27	100	
Lilium martagon	10	19	100	
Lonicera xylosteum	41	21	100	
Lycopus europaeus	23	8	100	
Lysimachia nemorum	31	6	100	
Lysimachia nummularia	35	10	100	
Melica nutans	39	14	100	
Melica uniflora	115	9	71	28.05
Mercurialis perennis	119	34	100	
Milium effusum	293	6	66	30.47
Neottia nidus-avis	12	31	100	
Origanum vulgare	14	20	100	
Paris quadrifolia	34	15	100	
Petasites albus	11	10	100	
Phalaris arundinacea	15	6	69	27.13
Phleum pratense	11	8	100	
Phyteuma spicatum	15	13	100	
Plantago lanceolata	12	12	100	
Plantago major	42	8	80	31.79
Poa chaixii	21	5	66	27.24

(continued)

Species	Frequency	Explained deviance (%)	Optimum (%)	Tolerance (%)
Polygala chamaebuxus	15	27	100	
Polygonatum verticillatum	44	9	100	
Potentilla erecta	32	5	100	
Primula elatior	42	29	100	
Prunella vulgaris	39	8	69	26.21
Prunus spinosa	45	13	100	
Pulmonaria officinalis	18	20	100	
Ranunculus ficaria	25	11	69	22.71
Ranunculus repens	82	7	69	28.45
Rhamnus cathartica	12	9	100	
Rosa arvensis	22	11	100	
Rosa canina	44	9	100	
Rubus caesius	33	10	100	
Rumex sanguineus	33	7	74	30.39
Salvia glutinosa	13	20	100	
Sanicula europaea	44	25	100	
Senecio jacobaea	11	12	28	8.72
Sesleria albicans	12	23	100	
Solanum dulcamara	20	11	100	
Solidago virgaurea	33	8	100	
Sorbus torminalis	18	19	86	21.58
Stachys sylvatica	139	12	85	33.10
Stellaria holostea	127	6	68	30.04
Tilia cordata	30	10	78	28.47
Torilis japonica	22	11	73	24.44
Trifolium pratense	20	11	100	
Trifolium repens	22	6	100	
Ulmus glabra	26	11	100	
Urtica dioica	442	13	69	26.68
Valeriana officinalis agg.	14	6	100	
Veronica chamaedrys	43	9	100	
Veronica montana	19	7	67	25.68
Veronica urticifolia	10	38	100	
Viburnum lantana	18	42	100	
Viburnum opulus	18	16	100	
Vicia cracca	22	19	100	
Vicia sepium	91	15	83	29.18
Viola hirta	17	18	100	
Viola reichenbachiana	261	20	100	42.76

Index

N. Wellbrock, A. Bolte (eds.), *Status and Dynamics of Forests in Germany*, Ecological Studies 237, https://doi.org/10.1007/978-3-030-15734-0

www.ingramcontent.com/pod-product-compliance
Lightning Source LLC
LaVergne TN
LVHW082112200926
848484LV00008B/1944